geoENV VII – Geostatistics for Environmental Applications

Quantitative Geology and Geostatistics

VOLUME 16

For other titles published in this series, go to
http://www.springer.com/series/6466

Peter M. Atkinson · Christopher D. Lloyd
Editors

geoENV VII – Geostatistics for Environmental Applications

Proceedings of the Seventh European Conference on Geostatistics for Environmental Applications

Springer

Editors
Prof. Peter M. Atkinson
School of Geography
University of Southampton
Highfield
Southampton
United Kingdom SO17 1BJ
P.M.Atkinson@soton.ac.uk

Dr. Christopher D. Lloyd
School of Geography, Archaeology
and Palaeoecology
Queen's University
Belfast
United Kingdom BT7 1NN
c.lloyd@qub.ac.uk

Cover figure: Fig. 5 on p. 208 in this book.

ISBN 978-94-007-3221-6 ISBN 978-90-481-2322-3 (eBook)
DOI 10.1007/978-90-481-2322-3
Springer Dordrecht Heidelberg London New York

Cover design: deblik

Printed on acid-free paper

Springer is part of Springer Science+Business Media (www.springer.com)

Foreword

Characterising spatial and temporal variation in environmental properties, generating maps from sparse samples, and quantifying uncertainties in the maps, are key concerns across the environmental sciences. The body of tools known as geostatistics offers a powerful means of addressing these and related questions. This volume presents recent research in methodological developments in geostatistics and in a variety of specific environmental application areas including soil science, climatology, pollution, health, wildlife mapping, fisheries and remote sensing, amongst others.

This book contains selected contributions from geoENV VII, the 7th International Conference on Geostatistics for Environmental Applications, held in Southampton, UK, in September 2008. Like previous conferences in the series, the meeting attracted a diversity of researchers from across Europe and further afield. A total of 82 abstracts were submitted to the conference and from these the organisation committee selected 46 papers for oral presentation and 30 for poster presentation.

The chapters contained in the book represent the state-of-the-art in geostatistics for the environmental sciences. The book includes 35 chapters arranged according to their main focus, whether methodological, or in a particular application. All of the chapters included were accepted after review by members of the scientific committee and each chapter was also subject to checks by the editors.

The editors wish to acknowledge the reviewers and the authors of the chapters that make up this book; it would not have existed without their efforts. The editors would also like to thank the sponsors of the conference, who included Cambridge University Press, Wiley, Taylor and Francis, the Ordnance Survey, the GeoData Institute, University of Southampton, School of Geography, University of Southampton and the Remote Sensing and Photogrammetry Society.

Southampton, May 2009 — *Peter Atkinson*
Belfast — *Christopher Lloyd*

Organising Committee

Peter Atkinson, *University of Southampton, UK (Chairman)*
Peter Gething, *University of Southampton, UK*
Julia Branson, *University of Southampton, UK*
Philippe Renard, *University of Neuchâtel, Switzerland*
Roland Froidevaux, *FSS International, Geneva, Switzerland*
Christopher Lloyd, *Queen's University, Belfast, UK*

Scientific Committee

The editors are grateful to the following persons for their work as referees:
Rachid Ababou, *IMFT, Toulouse, France*
Denis Allard, *INRA, Avignon, France*
Andras Bardossy, *University of Stuttgart, Germany*
Ana Bio, *Instituto Superior Técnico, Portugal*
José Capilla, *Universidad Politécnica de Valencia, Spain*
Eduardo Cassiraga, *Universidad Politécnica de Valencia, Spain*
Olaf Cirpka, *Universtät Tübingen, Germany*
Isobel Clark, *Geostokos Limited, Alloa, UK*
Noel Cressie, *Ohio State University, USA*
Vladimir Cvetkovic, *KTH Royal Institute of Technology, Stockholm, Sweden*
Chantal de Fouquet, *École des Mines de Paris, France*
Hélène Demougeot-Renard, *FSS International, Neuchâtel, Switzerland*
Souheil Ezzedine, *Lawrence Livermore National Laboratory, USA*
Tilmann Gneiting, *University of Washington, USA*
Pierre Goovaerts, *Biomedware, Ann Arbor, USA*
Alberto Guadagnini, *Politecnico de Milano, Italy*
Gilles Guillot, *INRA, Paris, France*
Jaime Gómez-Hernández, *Universidad Politécnica de Valencia, Spain*
Harrie-Jan Hendricks Franssen, *ETH Zürich, Switzerland*
Phaedon Kyriakidis, *University of California Santa Barbara, USA*
Denis Marcotte, *Ecole Polytechnique Montréal, Canada*
Pascal Monestiez, *INRA, Avignon, France*
Margaret Oliver, *University of Reading, UK*
Maria João Pereira, *Instituto Superior Técnico, Portugal*
Monica Riva, *Politecnico de Milano, Italy*
Paul D. Sampson, *University of Washington, USA*
Mohan Srivastava, *FSS Canada, Canada*
Fritz Stauffer, *ETH Zürich, Switzerland*
Xavier Sánchez-Vila, *Universidad Politécnica de Catalunya, Spain*
Nick Tate, *University of Leicester, UK*
Hans Wackernagel, *École des Mines de Paris, France*
Richard Webster, *Rothamsted Experimental Station, UK*

Contents

Contributors

Peter M. Atkinson School of Geography, University of Southampton, Highfield, Southampton, United Kingdom SO17 1BJ, P.M.Atkinson@soton.ac.uk

José Manuel Mirás Avalos Facultad de Ciencias, Universidade da Coruña, UDC, Campus A Zapateira C.P. 15071, A Coruña, Spain, jmirasa@udc.es

Paula Helena Ávila LNEG – S. Mamede Infesta Laboratory, Porto, Portugal
and
GeoBioTec – GeoBiosciences, Technologies and Engineering Research Unit, Aveiro, Portugal

Jean-Nöel Bacro I3M, Université de Montpellier 2, Place Eugène Bataillon, 34 095 Montpellier Cedex 5, France, bacro@math.univ-montp2.fr

András Bárdossy Institut fúr Wasserbau, Pfaffenwaldring 61, 70569 Stuttgart, Germany, Andras.Bardossy@iws.uni-stuttgart.de

Lucy Bastin Knowledge Engineering Group, School of Engineering and Applied Science, Aston University, Birmingham B4 7ET, UK

Edwige Bellier Biostatistique et Processus Spatiaux, INRA, Domaine Saint-Paul, Site Agroparc, 84914 Avignon cedex 9, France
and
DIMAR, CNRS UMR 6540, Centre d'Océanologie de Marseille, Parc Scientifique et Technologique de Luminy, Case 901, 13288 Marseille Cedex 9, France

Riccardo Bersezio Dipartimento Scienze della Terra – Università di Milano, via Mangiagalli 34, 20133 I-Milano, Italy

Jean-François Bourillet IFREMER, Dép. Géosciences Marines, Laboratoire Environnements Sédimentaires, BP70, 29280 Plouzané – France, Jean.Francois.Bourillet@ifremer.fr

Maria Helena Caeiro Centro de Recursos Naturais e Ambiente, Instituto Superior Técnico, Universidade Técnica de Lisboa, Lisbon, Portugal

José Esteban Capilla Romá Instituto de Ingeniería del Agua y Medio Ambiente, Universidad Politécnica de Valencia, Camino de Vera s/n, 46071-Valencia, Spain, jcapilla@upv.es

Pierrette Chagneau CIRAD, UR 37, Campus international de Baillarguet, 34 398 Montpellier Cedex 5, France, pierrette.chagneau@cirad.fr

Mike Christie Institute of Petroleum Engineering, Heriot-Watt University, Edinburgh, UK

Dan Cornford Non-linearity and Complexity Research Group (NCRG), Aston University, Aston Triangle, Birmingham, B4 7ET, UK, d.cornford@aston.ac.uk

Lehel Csató Faculty of Mathematics and Informatics, Universitatea BABES-BOLYAI, Str. Mihail Kogalniceanu, Nr. 1 RO-400084 Cluj-Napoca, Romania, lehel.csato@cs.ubbcluj.ro

Eduardo Ferreira da Silva GeoBioTec – GeoBiosciences, Technologies and Engineering Research Unit, Aveiro, Portugal

Jorge Dafonte Dafonte Universidad de Santiago de Compostela (USC), Escuela Politécnica Superior, 27002, 13088-300, Lugo, Spain, jorge.dafonte@usc.es

Alain Dassargues Katholieke Universiteit Leuven, Hydrogeology & Engineering Geology, Redingenstraat 16, B-3000 Leuven, Belgium, alain.dassargues@geo.kuleuven.ac.be

Diana dell'Arciprete Dipartimento Scienze della Terra – Università di Milano, via Mangiagalli 34, 20133 I-Milano, Italy, diana.dellarciprete@unimi.it

Vasily Demyanov Institute of Petroleum Engineering, Heriot-Watt University, Edinburgh, UK, vasily.demyanov@pet.hw.ac.uk

Ainslie Denham School of Engineering, Edith Cowan University, Perth, Western Australia, a.denham@ecu.edu.au

Peter J. Diggle Department of Medicine, Lancaster University, Lancaster LA1 4YF, UK, p.diggle@lancaster.ac.uk

Rita Durão Instituto Superior Técnico, CERENA, Av. Rovisco Pais, 1049-001 Lisboa, Portugal, rmdurao@ist.utl.pt

Fabrizio Felletti Dipartimento Scienze della Terra – Università di Milano, via Mangiagalli 34, 20133 I-Milano, Italy

Loris Foresti Institute of Geomatics and Analysis of Risk, University of Lausanne, Switzerland, Loris.Foresti@unil.ch

Patricia Sande Fouz Facultad de Ciencias, Universidade da Coruña, UDC, Campus A Zapateira C.P. 15071, A Coruña, Spain

Jianlin Fu Department of Hydraulic and Environmental Engineering, Universidad Politécnica de Valencia, 46071 Valencia, Spain, jianfu@dihma.upv.es

Edith Gabriel IUT STID – LANLG, Université d'Avignon, BP 1207, 84911 Avignon, France, edith.gabriel@univ-avignon.fr

Daniel Giménez Department of Environmental Sciences, Rutgers, The State University of NJ, 14 College Farm Road, New Brunswick, NJ, USA

Joaquim Góis Engineering Faculty of Porto University, Mining Department – Rua Dr. Roberto Frias. 4200-465 Porto, Portugal

J. Jaime Gómez-Hernández Department of Hydraulic and Environmental Engineering, Universidad Politécnica de Valencia, 46071 Valencia, Spain, jaime@dihma.upv.es

Pierre Goovaerts Biomedware, Inc., 516 North State Street, Ann Arbor, MI, USA

Carol A. Gotway Centers for Disease Control and Prevention, Office of Workforce and Career Development, 1600 Clifton Rd, NE, MS E-94, Atlanta, GA 30333, USA, cdg7@cdc.gov

Annelies Govaerts Research unit mining, K.U. Leuven, Kasteelpark Arenberg 40 bus 2448, B-3001 Leuven, Belgium, annelies.govaerts@bwk.kuleuven.be

Célia Regina Grego EMBRAPA-CNPM, Av. Dr. Júlio Soares de Arruda, 803, Parque São Quirino, CEP 13088-300, Campinas, SP, Brazil, crgrego@cnpm.embrapa.br

Christophe Guinet Centre d'Etudes Biologiques de Chizé, CNRS, 79360 Villiers-en-Bois, France

Claus P. Haslauer Department of Hydrology and Geohydrology, Institute of Hydraulic Engineering, University of Stuttgart, Germany, claus.haslauer@iws.uni-stuttgart.de

Ana Horta Centro de Recursos Naturais e Ambiente, Instituto Superior Técnico, Universidade Técnica de Lisboa, Lisbon, Portugal, ahorta@ist.utl.pt

Marijke Huysmans Katholieke Universiteit Leuven, Applied Geology and Mineralogy, Celestijnenlaan 200 E - Bus 2408, 3001 Heverlee, Belgium, marijke.huysmans@ees.kuleuven.be

Ben Ingram Neural Computing Research Group, Aston University, Aston Street, Birmingham B4 7ET, UK
and
Knowledge Engineering Group, School of Engineering and Applied Science, Aston University, Birmingham B4 7ET, UK, B.R.Ingram@aston.ac.uk

Nicolas Jeannee GEOVARIANCES, 49bis Avenue Franklin Roosevelt, BP91, 77212 Avon – France, jeannee@geovariances.com

Chockalingam Jeganathan School of Geography, University of Southampton, Highfield, Southampton SO17 1BJ, UK

Mikhail Kanevski Institute of Geomatics and Analysis of Risk, University of Lausanne, Switzerland

Sergey Kazakov Nuclear Safety Institute Russian Academy of Sciences, B. Tulskaya 52, 113191, Moscow, Russia

Hannes Kazianka Institute of Statistics, University of Klagenfurt, Universitätsstraße 65-67, 9020 Klagenfurt, Austria, hannes.kazianka@uni-klu.ac.at

Ruth Kerry Department of Geography, Brigham Young University, Provo, UT, USA
and
CRSSA, Rutgers, The State University of NJ, 14 College Farm Road, New Brunswick, NJ, USA

Olivier Le Moine IFREMER, Laboratoire Environnement-Ressource des Pertuis Charentais, Avenue de Mus de Loup, 17390 La Tremblade – France, olemoine@ifremer.fr

Oy Leuangthong Centre for Computational Geostatistics (CCG), Department of Civil and Environmental Engineering, University of Alberta, Canada, oy.leuangthong@ualberta.ca

Jing Li Institut für Wasserbau, Pfaffenwaldring 61, 70569 Stuttgart, Germany

Carlos Llopis-Albert Instituto de Ingeniería del Agua y Medio Ambiente, Universidad Politécnica de Valencia, Camino de Vera s/n, 46071-Valencia, Spain, cllopisa@gmail.com

Christopher D. Lloyd School of Geography, Archaeology and Palaeoecology, Queen's University, Belfast, United Kingdom BT7 1NN, c.lloyd@qub.ac.uk

Cedric Magneron ESTIMAGES, 10 Avenue du Québec, 91140 Villebon-sur-Yvette – France, cedric.magneron@estimages.com

Jennifer McKinley School of Geography, Archaeology and Palaeoecology, Queen's University, Belfast, BT7 1NN, UK, j.mckinley@qub.ac.uk

Maurici Monego Faculty of Engineering of University of Porto, Rua Dr. Roberto Frias, 4200-465 Porto, Portugal, mdmonego@fe.up.pt

Pascal Monestiez Biostatistique et Processus Spatiaux, INRA, Domaine Saint-Paul, Site Agroparc, 84914 Avignon cedex 9, France

Frédéric Mortier CIRAD, UR 39, Campus international de Baillarguet, 34 398 Montpellier Cedex 5, France, frederic.mortier@cirad.fr

Ute Mueller School of Engineering, Edith Cowan University, Perth, Western Australia, u.mueller@ecu.edu.au

Mário Neves Faculty of Engineering of University of Porto, Rua Dr. Roberto Frias, 4200-465 Porto, Portugal, mjneves@fe.up.pt

Ruben Nunes Centro de Recursos Naturais e Ambiente, Instituto Superior Técnico, Universidade Técnica de Lisboa, Lisbon, Portugal

Julián M. Ortiz Department of Mining Engineering, University of Chile, Av. Tupper 2069, Santiago, 837-0451, Chile, jortiz@ing.uchile.cl

Peter Oudemans Department of Plant Biology & Pathology, Rutgers, The State University of NJ, 59 Dudley Road, New Brunswick, NJ, USA

Oscar Peredo Department of Mining Engineering, University of Chile, Av. Tupper 2069, Santiago, 837-0451, Chile, operedo@dcc.uchile.cl

Henrique Garcia Pereira CERENA – Natural Resources and Environment Center of IST, Lisboa, Portugal, henrique.pereira@ist.utl.pt

Maria João Pereira CERENA, Instituto Superior Técnico, Av. Rovisco Pais, 1. 1049-001 Lisboa, Portugal

Nicolas Picard CIRAD, UR 37, Campus international de Baillarguet, 34 398 Montpellier Cedex 5, France, nicolas.picard@cirad.fr

Luiza Honora Pierre Instituto Agronômico (IAC), Av. Barão de Itapura, 1481 CP28, CEP 13020-902, Campinas, SP, Brazil

Jürgen Pilz Institute of Statistics, University of Klagenfurt, Universitätsstraße 65-67, 9020 Klagenfurt, Austria, juergen.pilz@uni-klu.ac.at

Alexei Pozdnoukhov Institute of Geomatics and Analysis of Risk, University of Lausanne, Switzerland, alexei.pozdnoukhov@unil.ch

Patrícia Ramos Faculty of Engineering of University of Porto, Rua Dr. Roberto Frias, 4200-465 Porto, Portugal
and
Institute of Accountancy and Administration of Porto, Department of Mathematics, R. Jaime Lopes Amorim, 4465-004 S. M. Infesta, Portugal, patricia@fe.up.pt

Ana Russo CERENA, Instituto Superior Técnico, Av. Rovisco Pais, 1. 1049-001 Lisboa, Portugal

Ana Rita Salgueiro CERENA – Natural Resources and Environment Center of IST, Lisboa, Portugal, rita.salgueiro@ist.utl.pt

Elena Savelyeva Nuclear Safety Institute Russian Academy of Sciences, B. Tulskaya 52, 113191, Moscow, Russia, esav@ibrae.ac.ru

Glécio Machado Siqueira Universidad de Santiago de Compostela (USC), Escuela Politécnica Superior, 27002, 13088-300, Lugo, Spain, glecio.machado@rai.usc.es

Amílcar Soares Instituto Superior Técnico, CERENA, Av. Rovisco Pais, 1049-001 Lisboa, Portugal
and
Centro de Recursos Naturais e Ambiente, Instituto Superior Técnico, Universidade Técnica de Lisboa, Lisbon, Portugal, asoares@ist.utl.pt

Filip Tack Department of Applied Analytical and Physical Chemistry, Ghent University, Faculty of Bioscience Engineering, Coupure 653, 9000 Gent, Belgium

Meklit Tariku Department of Soil Management, Faculty of Bioscience Engineering, Ghent University, Coupure 653, 9000 Gent, Belgium

Ricardo M. Trigo Centro de Geofísica da Universidade de Lisboa, Ed. C8, Campo Grande, 1749-016 Lisboa, Portugal

Devis Tuia Institute of Geomatics and Analysis of Risk, University of Lausanne, Switzerland

Sergey Utkin Nuclear Safety Institute Russian Academy of Sciences, B. Tulskaya 52, 113191, Moscow, Russia

Marc Van Meirvenne Department of Soil Management, Faculty of Bioscience Engineering, Ghent University, Coupure 653, 9000 Gent, Belgium, marc.vanmeirvenne@ugent.be

André Vervoort Research unit mining, K.U. Leuven, Kasteelpark Arenberg 40 bus 2448, B-3001 Leuven, Belgium, andre.vervoort@bwk.kuleuven.be

Sidney Rosa Vieira Instituto Agronômico (IAC), Av. Barão de Itapura, 1481 CP28, CEP 13020-902, Campinas, SP, Brazil, sidney@iac.sp.gov.br

Eva Vidal Vázquez Facultad de Ciencias, Universidade da Coruña, UDC, Campus A Zapateira C.P. 15071, A Coruña, Spain

Matthew Williams Knowledge Engineering Group, School of Engineering and Applied Science, Aston University, Birmingham B4 7ET, UK, williamw@aston.ac.uk

Linda J. Young Department of Statistics, 404 McCarty Hall C, P.O. Box 110339, University of Florida, Gainesville, FL 32611-0339, USA, LJYoung@ufl.edu

Geostatistical Modelling of Wildlife Populations: A Non-stationary Hierarchical Model for Count Data

Edwige Bellier, Pascal Monestiez, and Christophe Guinet

Abstract We propose a hierarchical model coupled to geostatistics to deal with a non-gaussian data distribution and take explicitly into account complex spatial structures (i.e. trends, patchiness and random fluctuations). A common characteristic of animal count data is a distribution that is both zero-inflated and heavy tailed. In such cases, empirical variograms are no more robust and most structural analyses result in poor and noisy estimated spatial variogram structures. Thus kriged maps feature a broad variance of prediction. Moreover, due to the heterogeneity of wildlife population habitats, a nonstationary model is often required. To avoid these difficulties, we propose a hierarchical model that assumes that the count data follow a Poisson distribution given a theoretical sighting density which is a latent variable to be estimate. This density is modelled as the product of a positive long range trend by a positive stationary random field, characterized by a unit mean and a variogram function. A first estimate of the drift is used to obtain an estimate of the variogram of residuals including a correction term for variance coming from the Poisson distribution and weights due to the non-constant spatial mean. Then a kriging procedure similar to a modified universal kriging is implemented to directly map the latent density from raw count data. An application on fin whale data illustrates the effectiveness of the method in mapping animal density in a context that is presumably non-stationary.

E. Bellier and P. Monestiez
Biostatistique et Processus Spatiaux, INRA, Domaine Saint-Paul, Site Agroparc, 84914 Avignon cedex 9, France

E. Bellier (✉)
Norwegian Institute for Nature Research - NINA, NO-7485 Trondheim, NORWAY
e-mail: edwige.bellier@nina.no

C. Guinet
Centre d'Etudes Biologiques de Chizé, CNRS, 79360 Villiers-en-Bois, France

P.M. Atkinson and C.D. Lloyd (eds.), *geoENV VII – Geostatistics for Environmental Applications*, Quantitative Geology and Geostatistics 16,
DOI 10.1007/978-90-481-2322-3_1,

1 Introduction

Current wildlife research relies heavily on population monitoring, sometimes performed over large areas (Pollock et al., 2002). Counts provided by field surveys can be used to estimate population sizes (Kingsley and Reeves, 1998; Grigg et al., 1999) or to characterize spatial structures in populations (Isaak and Russel, 2006). The latter has received much recent interest because animals respond to spatial heterogeneity at different spatial scales (Kotliar and Wiens, 1990; Levin, 1992). Therefore, ecological data often include several spatial patterns, which can be regarded as trends at broad scales, patchiness at intermediate and local scale, and random fluctuations or noise at fine scales (Fortin and Dale, 2005). Furthermore, an additional common characteristic of ecological count data is that they are positively skewed and contain much more zeros than would be expected in classical data distribution (Clarke and Green, 1998; Flechter et al., 2005). The form of the distribution is usually due to the patchy nature of the environment and/or the inherent heterogeneity of the species distribution and to sampling coupled to observations processes (Martin et al., 2005). However, standard spatial statistical tools cannot easily deal with count data. When the data are non-Gaussian, hierarchical modelling may be a useful alternative for modelling the spatial distribution of count data (Latimer et al., 2006; Thogmartin et al., 2004). Indeed, ecological approaches and sampling situations should naturally lead to a hierarchical construction (Royle et al., 2005). Although most recent publications solve hierarchical models within a Bayesian framework, hierarchical modelling is not necessarily restricted to Bayesian statistics (Ver Hoef and Frost, 2003; Thogmartin et al., 2004; Cunningham and Lindenmayer, 2005). In a frequentist context, Monestiez et al. (2006) proposed a corrected variogram estimator that takes into account the variability added by the Poisson observation process in order to produce maps of relative abundance.

This paper presents a generalization of Poisson kriging introduced in Monestiez et al. (2006) based on a spatial hierarchical model. The model we propose has two levels: the first level deals with the variability resulting from the heterogeneity of the observation effort and ecological process (i.e. the variability resulting from the sighting process and ecological process themselves), which can naturally be modeled by a Poisson distribution (Monestiez et al., 2006). In the second level we take account of the non-stationarity of the species distribution (i.e. in most situations, populations show a trend in their spatial distribution [Fortin and Agrawal 2005]) by decomposing the spatially non-constant mean, by multiplication of a spatial trend by a stationary field.

Our method can be help to characterize spatial distribution and to address wildlife population spatial distributions through mapping which could be of great interest for management or conservation purposes. Our approach typically applies to animal count data and especially to field transect surveys, a popular method to count animals – including marine mammals (e.g. dugong (Pollock et al., 2006); small cetaceans (Hammond et al., 2002); manatees (Wright et al., 2002)), seabirds (Tasker et al., 1984; Briggs et al., 1985) and terrestrial mammals (e.g. kangaroos

[Caughley and Grigg 1981], impala [Peel and Bothma 1995]) in which individuals or groups of individuals (i.e. "sightings") are recorded at discrete locations.

We provide a case study, with an application based on the spatial distribution of fin whales in a context that is presumably non-stationary.

2 Model

2.1 Hierarchical Model for Animals Sightings

We define a spatial hierarchical model with two levels. The first one models the number of sightings Z into an 1 km-long strip by a Poisson distribution whose parameter Y is a non stationary random field. The second level models Y as the product of a regional drift m and a latent variable X.

For all sites s, we model the number of observed sightings Z knowing Y the latent variable which represents the theoretical sighting density, by independent Poisson random variables:

$$\begin{cases} Z_s | Y_s \sim \mathcal{P}(Y_s) \\ Y_s \quad = m_s X_s \end{cases} \tag{1}$$

where Y_s is the product of a deterministic drift m_s by a positive stationary random field X with unit mean, variance σ_X^2, and covariance function $C_X(s - s') = \mathsf{Cov}[X_s, X_{s'}]$, noted $C_{ss'}$ to simplify notation.

The covariance function $C_X(s - s')$ may be replaced by the variogram function $\gamma_X(s - s') = \frac{1}{2}\,\mathsf{E}\big[(X_s - X_{s'})^2\big]$.

There is no distributional hypothesis on X but the inequality $X \geq 0$.

2.2 Expectation and Variance of Z_s

From Equation (1), it follows directly that:

$$\begin{aligned} \mathsf{E}[Z_s | X_s] &= Y_s = m_s\, X_s \\ \mathsf{Var}[Z_s | X_s] &= Y_s = m_s\, X_s \\ \mathsf{E}\big[(Z_s)^2 \big| X_s\big] &= Y_s + Y_s^2 = m_s\, X_s + m_s^2\, X_s^2 \end{aligned} \tag{2}$$

and when deconditioning:

$$\begin{aligned} \mathsf{E}[Z_s] &= m_s \\ \mathsf{Var}[Z_s] &= m_s^2\, \sigma_X^2 + m_s \end{aligned} \tag{3}$$

For the covariance expression, the conditional independence of observations at different sites leads to:

$$\begin{aligned} \mathsf{E}\Big[Z_s\, Z_{s'} | X\Big] &= \mathsf{Cov}\big[Z_s, Z_{s'} | X\big] + \mathsf{E}[Z_s | X_s]\, \mathsf{E}[Z_{s'} | X_{s'}] \\ &= \delta_{ss'}\, m_s X_s + m_s\, m_{s'}\, X_s X_{s'} \end{aligned} \tag{4}$$

where $\delta_{ss'}$ is the Kronecker delta which is 1 if $s = s'$ and 0 otherwise.

2.3 Variogram Expressions

In order to characterize the relationship between the variograms of Z and X, we develop the expressions of the two first moments of $(Z_s - Z_{s'})$.

$$\begin{aligned} \mathsf{E}\big[Z_s - Z_{s'} | X\big] &= \mathsf{E}[Z_s | X_s] - \mathsf{E}[Z_{s'} | X_{s'}] = m_s\, X_s - m_{s'}\, X_{s'} \\ \mathsf{E}\big[Z_s - Z_{s'}\big] &= \mathsf{E}\big[X\big]\, \big(m_s - m_{s'}\big) = m_s - m_{s'} \end{aligned} \tag{5}$$

The second order moment can be derived from Equations (2) and (4).

$$\begin{aligned} \mathsf{E}\Big[(Z_s - Z_{s'})^2\, | X\Big] &= \mathsf{E}\Big[(Z_s)^2 | X_s\Big] + \mathsf{E}\Big[(Z_{s'})^2 | X_{s'}\Big] - 2\,\mathsf{E}\Big[Z_s\, Z_{s'} | X\Big] \\ &= \big(Y_s + Y_{s'} - 2\delta_{ss'}\, Y_s\big) + \big(Y_s - Y_{s'}\big)^2 \\ \mathsf{E}\Big[(Z_s - Z_{s'})^2\Big] &= \Big(m_s + m_{s'} - 2\,\delta_{ss'}\, m_s)\Big) + \mathsf{E}\Big[\Big(m_s\, X_s - m_{s'}\, X_{s'}\Big)^2\Big] \end{aligned}$$

When m_s is assumed to be known and different everywhere (i.e. $m_s = m_{s'}$), we have to develop the two first moments of $(\frac{Z_s}{m_s} - \frac{Z_{s'}}{m_{s'}})$:

$$\begin{aligned} \mathsf{E}\left[\frac{Z_s}{m_s} - \frac{Z_{s'}}{m_{s'}} \middle| X\right] &= \frac{1}{m_s}\, \mathsf{E}[Z_s | X_s] - \frac{1}{m_{s'}}\, \mathsf{E}[Z_{s'} | X_{s'}] = X_s - X_{s'} \\ \mathsf{E}\left[\frac{Z_s}{m_s} - \frac{Z_{s'}}{m_{s'}}\right] &= 0 \end{aligned} \tag{6}$$

The expression of the non-conditional order-2 moment is derived from Equations (2) and (4).

$$\begin{aligned} \mathsf{E}\left[\left(\frac{Z_s}{m_s} - \frac{Z_{s'}}{m_{s'}}\right)^2 \middle| X\right] &= \frac{1}{m_s^2}\, \mathsf{E}\Big[(Z_s)^2 | X_s\Big] + \frac{1}{m_{s'}^2}\, \mathsf{E}\Big[(Z_{s'})^2 | X_{s'}\Big] - \frac{2\,\mathsf{E}\big[Z_s\, Z_{s'} | X\big]}{m_s\, m_{s'}} \\ &= \frac{X_s}{m_s} + \frac{X_{s'}}{m_{s'}} - 2\delta_{ss'}\, \frac{X_s}{m_s} + (X_s - X_{s'})^2 \end{aligned}$$

$$\frac{1}{2}\,\mathsf{E}\left[\left(\frac{Z_s}{m_s}-\frac{Z_{s'}}{m_{s'}}\right)^2\right]=\frac{1}{2}\left(\frac{m_s+m_{s'}}{m_s\,m_{s'}}\right)-\delta_{ss'}\,\frac{1}{m_s}+\gamma_X(s-s') \tag{7}$$

Let $\gamma_{Z/m}(s-s')$ denote the non-stationary theoretical variogram corresponding to the random field (Z_s/m_s), we get for $s \neq s'$ the relationship:

$$\gamma_X(s-s') = \gamma_{Z/m}(s-s')-\frac{1}{2}\left(\frac{m_s+m_{s'}}{m_s\,m_{s'}}\right) \tag{8}$$

We can check for $s = s'$ that Equation (7) reduces to $\gamma_X(0) = \gamma_Z(0) = 0$

For $s \neq s'$, we also have:

$$\begin{aligned}\mathsf{Var}\left[\frac{Z_s}{m_s}-\frac{Z_{s'}}{m_{s'}}\middle|X\right] &= \mathsf{E}\left[\left(\frac{Z_s}{m_s}-\frac{Z_{s'}}{m_{s'}}\right)^2\middle|X\right]-\mathsf{E}^2\left[\frac{Z_s}{m_s}-\frac{Z_{s'}}{m_{s'}}\middle|X\right]\\ &= \frac{X_s}{m_s}+\frac{X_{s'}}{m_{s'}}+(X_s-X_{s'})^2-(X_s-X_{s'})^2\\ &= \frac{X_s}{m_s}+\frac{X_{s'}}{m_{s'}}\end{aligned}$$

$$\mathsf{E}\left[\mathsf{Var}\left[\frac{Z_s}{m_s}-\frac{Z_{s'}}{m_{s'}}\middle|X\right]\right]=\mathsf{E}\left[\frac{X_s}{m_s}+\frac{X_{s'}}{m_{s'}}\right]=\left(\frac{m_s+m_{s'}}{m_s\,m_{s'}}\right) \tag{9}$$

2.4 Estimation of $\gamma_X(h)$

Let $Z_\alpha, \alpha = 1,\ldots,n$ be the n measurements of $Z(s_\alpha)$. Because of the non-constant mean $m(s)$, it is not meaningful to directly compute experimental variogram on Z_α's, even on the corrected values Z_α/m_α. So we propose a modified experimental variogram for X:

$$\gamma_X^*(h) = \frac{1}{2\,N(h)}\sum_{\alpha,\beta}\left(\frac{m_\alpha\,m_\beta}{m_\alpha+m_\beta}\left(\frac{Z_\alpha}{m_\alpha}-\frac{Z_\beta}{m_\beta}\right)^2-1\right)\mathbb{I}_{d_{\alpha\beta}\sim h} \tag{10}$$

where $\mathbb{I}_{d_{\alpha\beta}\sim h}$ is the indicator function of pairs (s_α, s_β) whose distance is close to h, where $N(h) = \sum_{\alpha,\beta}\frac{m_\alpha\,m_\beta}{m_\alpha+m_\beta}\,\mathbb{I}_{d_{\alpha\beta}\sim h}$ is a normalizing constant. The weight system directly derives from Equation (9) and the minus-one bias-correction term from Equation (8).

Such estimates can show great sensitivity to rare positive data that neighbour areas with extremely low local mean. It may be necessary to increase the robustness of such estimate by limiting minimum values of m_s (positive and not too close to zero).

A simpler estimates of γ_X can be proposed on subareas where the mean m_s can be assumed constant or when the empirical variogram estimate $\gamma_Z^*(h)$ is restricted to pairs of sampled sites with the same mean m_s:

$$\gamma_X^*(h) = \frac{1}{m^2}\left[\gamma_Z^*(h) - m\right] \tag{11}$$

where m is the locally constant value of m_s.

2.5 Mapping Y by Multiplicative Poisson Kriging

The spatial interpolation of Y is implemented through a Poisson Kriging (PK) at any site $s_o \in \mathcal{D}$. This kriging is a linear predictor of Y_o combining the observed data Z_α weighted by the drift terms $m(s_\alpha)$ and $m(s_o)$ respectively noted m_α and m_o.

$$Y_o^* = \sum_{\alpha=1}^{n} \lambda_\alpha \frac{m_o Z_\alpha}{m_\alpha} \tag{12}$$

The unbiasedness of Y_o^* leads to the usual condition on λ_α's.

$$\sum_{\alpha=1}^{n} \lambda_\alpha = 1 \tag{13}$$

The expression of the Mean Square Error of Prediction (MSEP) can also be derived from the kriging estimate expression.

$$\mathsf{E}\left[(Y_o^* - Y_o)^2\right] = m_o^2 \left(\sigma_X^2 + \sum_{\alpha=1}^{n} \frac{\lambda_\alpha^2}{m_\alpha} + \sum_{\alpha=1}^{n}\sum_{\beta=1}^{n} \lambda_\alpha \lambda_\beta C_{\alpha\beta} - 2\sum_{\alpha=1}^{n} \lambda_\alpha C_{\alpha o}\right) \tag{14}$$

By minimizing this expression (14) on λ_i's subject to the unbiasedness constraint, we obtain the following kriging system of $(n+1)$ equations where μ is the Lagrange multiplier.

$$\begin{cases} \sum_{\beta=1}^{n} \lambda_\beta C_{\alpha\beta} + \dfrac{\lambda_\alpha}{m_\alpha} + \mu = C_{\alpha o} \quad \text{for} \quad \alpha = 1, \ldots, n \\ \sum_{\alpha=1}^{n} \lambda_\alpha = 1 \end{cases} \tag{15}$$

The kriging system expressed with covariance is preferably used for computation when both variogram and covariance exist. The kriging system may be expressed from the variogram using the usual relation $C_{ss'} = \sigma_X^2 - \gamma_X(s - s')$.

The expression of the prediction variance resulting from this kriging system reduces after calculation to:

$$\mathrm{Var}(Y_o^* - Y_o) = m_o^2\Big(\sigma_X^2 - \sum_{\alpha=1}^{n} \lambda_\alpha C_{\alpha o} - \mu\Big) \tag{16}$$

It can be easily shown that the kriging of X_o defined as $X_o^* = \sum_{\alpha=1}^{n} \lambda_\alpha \frac{Z_\alpha}{m_\alpha}$ gives the same solutions in λ's and μ, so krigings of Y_o^* or X_o^* becomes equivalent using the relationship $Y_o^* = m_o\, X_o^*$.

3 Fin Whale Abundance in Pelagos Sanctuary

In the Mediterranean Sea, the fin whale (*Balenoptera physalus*) is the largest marine predator commonly observed. Several hundred to several thousand individuals were estimated to be present in the western Mediterranean Sea during summer (Forcada et al., 1996).

The sighting database used in this study as an illustrative example merges data from different sources, and is fully described in Monestiez et al. (2006). The fin whale surveys mainly focused on the northwestern Mediterranean Sea. Count data were aggregated on cells of 0.1° of longitude by 0.1° of latitude (approximately 90 km^2) in a regular grid. For each year from 1993 to 2001, July and August data were assembled and we computed in each cell the total number of fin whale sightings and the corresponding total searching effort defined as the time (in hours) spent searching inside the cell. So the number of sightings per unit effort or, with some assumptions, the relative abundance can be computed.

In this study, we focused particulary on the Pelagos sanctuary (International Cetacean Sanctuary of the Mediterranean), which was established on November 25th, 1999 by the governments of Italy, France and Monaco. The sanctuary limits are shown in Fig. 1, with the map of searching efforts.

The objective is to map the spatial distribution of fin whales inside the Pelagos sanctuary during the summer of 2001. Due to limitation of the available data subset, we have to assume values for some parameters: mean boat speed is fixed to six nautical knots (11.1 km/h), effective distance of detection to 750 m and mean school size to 1.6 in order to transform hours of searching in surveyed areas and to compute relative abundance estimates from raw sightings of whale schools. For sampled cells, the searching effort was not always exactly the same, so we had to introduce this effort as a factor to the multiplicative drift m_α in order to normalise sighting counts for unit effort. Except this change on m terms, previous estimate expressions and the kriging system remains globally the same.

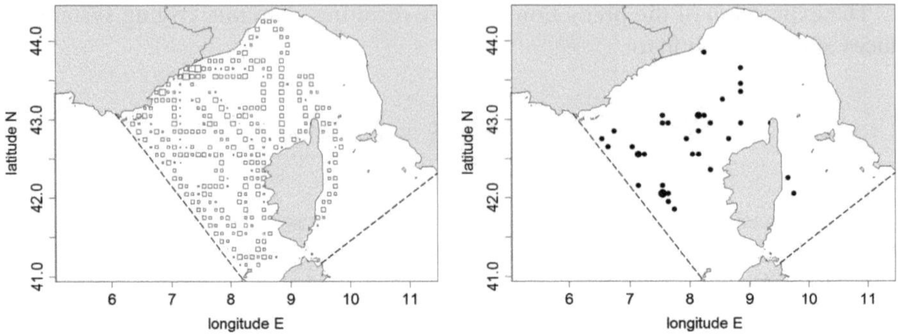

Fig. 1 Map of searching efforts for year 2001 (*left*, the *largest square symbols* are for 3 h of efforts in a pixel of 0.1° by 0.1°) and map of fin whale sightings in 2001 (*right*, number of schools ranging for 1 to 3) that will be used in the multiplicative kriging of the relative abundance. *Dashed lines* give Pelagos sanctuary limits

4 Results

We mapped the spatial distribution of whales by using multiplicative kriging in order to take into account the spatial trend of the fin whales distribution in the northwestern Mediterranean sea.

We first estimate the spatial drift by extracting a smooth long-range spatial component by filter kriging (Wackernagel, 2003) from the 1993–2000 pooled data (excluding the 2001 ones). The resulting map is displayed on Fig. 2 and seems representative of the permanence of the fin whale spatial distribution over years. This long-range component reveals the non-stationarity in fin-whale spatial distribution and could be considered as the potential habitat of fin whales in the northwestern Mediterranean Sea. It is modelled as a deterministic drift. Then the experimental variogram is fitted by a spherical model (Fig. 2) and multiplicative Poisson kriging is applied to fin whale count data.

The two maps of kriged expectations of whale sightings obtained from multiplicative kriging (i.e. taking account for non-stationarity) and from Poisson kriging show some difference (Fig. 3), especially in the western area outside of the Pelagos sanctuary and on the eastern part of the sanctuary which was not surveyed. This observed difference seems be due to the considerations of the deterministic drift in the multiplicative model, since this pattern shows some similarities with the map of the potential habitat. In other respect, the two methods differ in extrapolating context due to the deterministic drift but gives quite close result where the sighting effort is dense enough.

Maps of standard error differ a lot more. It is clear for multiplicative kriging that the drift had a real influence, leading to smaller errors in region of lower whale density and potentially very large errors when extrapolating on high density areas (western region outside Pelagos). The standard error map of Poisson ordinary kriging reflects more conventionally the spatial distribution of sighting efforts with poor

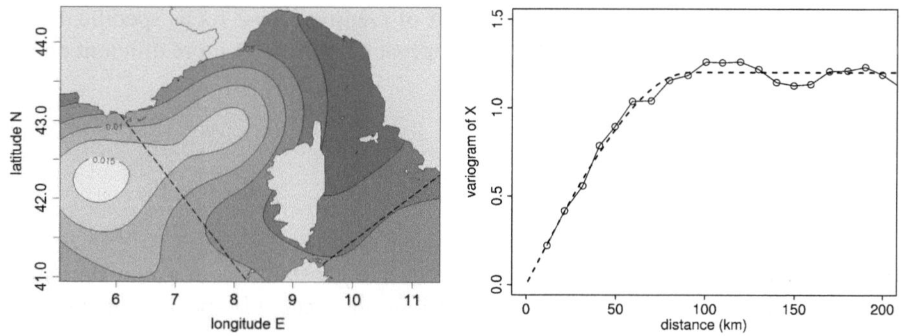

Fig. 2 Map of the drift term (*left*, number of whale schools per square kilometer) and the modified experimental variogram calculated from Equation (10)

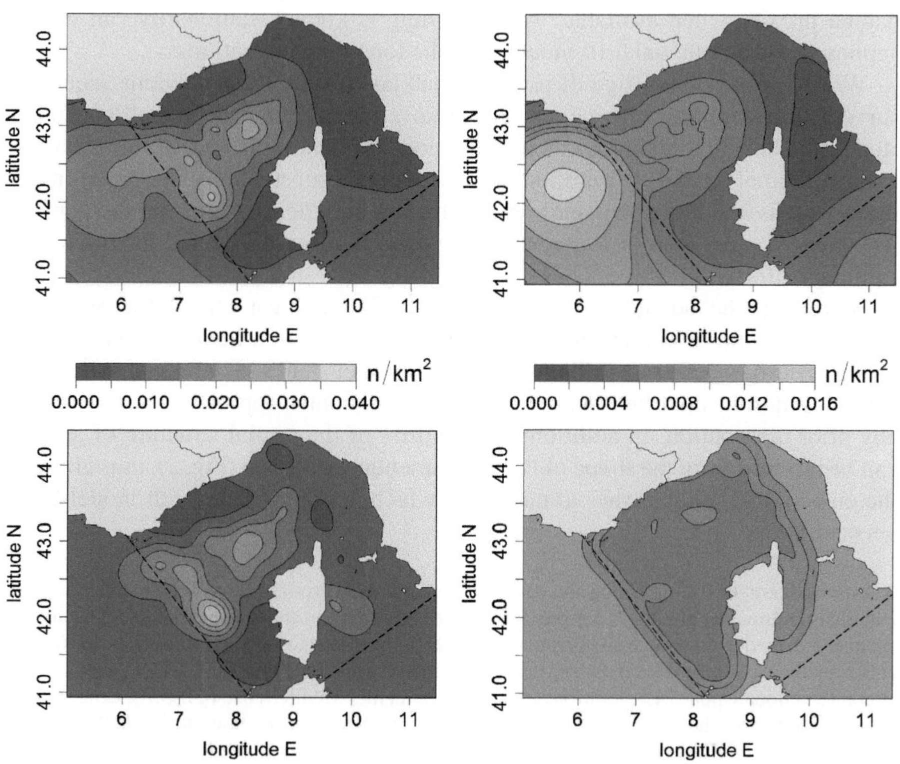

Fig. 3 Maps of kriged expectation of whale school sighting (*left column*, mean number of school per square km) and associated maps of standard error (*right column*, same unit) obtained from multiplicative kriging (*top row*) and from Poisson ordinary kriging (*bottom row*). Map legends are specific to the variables (expectation or standard error) but are identical for both methods

performance on the eastern part of the Golf of Genova. If we focus specifically on the Pelagos sanctuary, the multiplicative approach seems a lot more efficient due to drift information.

5 Discussion

In this study we showed that it is possible to use geostatistics in a non-stationary context of count data and zero inflated distributions since it is specified in a hierarchical spatial model.

It seems also essential to take into account the non-stationarity in the proposed multiplicative kriging because it is a reality for many animal spatial distribution. This non-stationarity can be generally modeled from previous surveys or from habitat proxies when available. When nothing is known, stationarity can be first supposed and a potential drift modeled as the long range variations.

When a good knowledge of potential habitats results from previous sequential surveys, the sampling scheme can be improved using the drift modelling. In this study, we show that taking account of the non-stationarity had a real impact on the map of animal spatial distribution since it reduces substantially the error on low density areas and larger standard error values in high density area; on the contrary the standard error map of Poisson kriging reflects more the spatial distribution of sightings efforts.

Moreover, the advantage of developing a hierarchical model for modelling species distribution in a frequentist context rather than in a Bayesian one is that it avoids specifying the Y distribution unlike Diggle et al. (1998) who had to hypothesize a log-normal distribution for Y; indeed, a frequentist approach does not require any prior distribution. In addition, a diagnostic of the spatial structure of animals can be inferred from the shape of the experimental variogram (Fig. 2), thus allowing the choice of a suitable variogram model which is not the case with model-based geostatistics.

Acknowledgements Concerning the case study, the authors would like to acknowledge Laurent Dubroca (Dubroca et al., 2004) for providing access to a part of the original dataset. The authors are also indebted to all people and organisations who contributed to the sighting data collection and collation: the Centre de Recherche sur les Mammifres Marins (CRMM), the CETUS, the Commission Internationale pour l'Exploration Scientifique de la mer Méditerranée (CIESM), Conservation Information Recherche sur les Cétacés (CIRC), Delphinia Sea Conservation, the Ecole Pratique des Hautes Etudes (EPHE) particularly P.C. Beaubrun, L. David and N. Di-Mglio, the Groupe d'Etude des Cétacés de Méditerranée (GECEM), the Groupe de Recherche sur les Cétacés (GREC), the Institut Franais de Recherche pour l'Exploitation de la Mer (IFREMER), the Muse Océanographique de Monaco, the Réserve Internationale en Mer Méditerranée Occidentale (RIMMO), the Supreme Allied Commander Atlantic Undersea Research Centre (SACLANT), the Société de Navigation Corse Méditerranée (SNCM), the Swiss Cetacean Society (SCS) and the WWF-France.

References

Briggs K, Tyler W, Lewis D (1985) Comparison of ships and aerial surveys of birds at sea. J Wildl Manag 49:405–411

Caughley G, Grigg G (1981) Surveys of the distribution and density of kangaroos in the pastoral zone in south australian and their bearing on the feasibility of aerial survey in large remote areas. Aust Wildl Res 8:1–11

Clarke K, Green R (1998) Statistical design and analysis for a "biological effects" study. Mar Ecol Prog Ser 46:213–226

Cunningham R, Lindenmayer D (2005) Modelling count data of rare species: some statistical issues. Ecology 86(5):1135–1142

Diggle J-P, Tawn J, Moyeed R (1998) Model based geostatistics. Appl Stat 47:299–350

Dubroca L, André J-M, Beaubrun P, Bonnin E, David L, Durbec J-P, Monestiez P, Guinet C (2004) Summer fin whales (Balaenoptera physalus) distribution in relation to oceanographic conditions: implications for conservation. CIESM Workshop Monographs n°25. Monaco pp 77–84. (Proc. Venice, 28–31 January 2004)

Flechter D, Mackenzie D, Villouta E (2005) Modelling skewed data with many zeros: a simple approach combining ordinary and logistic regression. Environ Ecol Stat 12:45–54

Forcada J, Aguilar A, Hammon P, Pastor X, Aguilar R (1996) Distribution and abundance of fin whales (balaenoptera physalus) in the western mediterranean sea during the summer. J Zool 238:23–34. Part 1

Fortin M, Agrawal A (2005) Landscape ecology come of age. Ecology 86:1965–1966

Fortin M-J, Dale MRT (2005) Spatial analysis: a guide for ecologists. Cambridge: Cambridge University Press

Grigg G, Beard L, Alexander P, Pople A, Cairns S (1999) Survey kangaroos in south australia 1978–1998 a brief report focusing on methodology. Aust Zool 31:292–300

Hammond P, Berggren P, Benke H, Borchers D, Collet A, Heide-Jorgensen M, Heimlich S, Hiby A, Leopold M, Oien N (2002) Abundance of harbour porpoise and other cetaceans in the north sea and adjacent waters. J Appl Ecol 39:361–376

Isaak D, Russel R (2006) Network-scale spatial and temporal variation in chinook salmon (oncorhynchus tshawytscha) redd distributions: patterns inferred from spatially continuous replicate surveys. Can J Fisheries and Aquat Sci 63:285–296

Kingsley M, Reeves R (1998) Aerial surveys of cetaceans in the gulf of st lawrence in 1995 and 1996. Can J Zool 76:1529–1550

Kotliar N, Wiens J (1990) Multiples scales of patchiness and patch structure: a hierarchical framework for the study of heterogeneity. Oikos 59:253–260

Latimer M, Wu S, Gelfand A, Silander J (2006) Building stistical models to analyse species distributions. Ecol Appl 16(1):33–50

Levin S (1992) The problem of pattern in ecology. Ecology 73(6):1943–1967

Martin TG, Brendan A, Wintle J, Rhodes R, Kuhnert PM, Field S, Low-Choy S, Tyre A, Possingham HP (2005) Zero tolerance ecology: improving ecological inference by modelling the source of zero observations. Ecol Lett 8:1236–1254

Monestiez P, Dubroca L, Bonnin E, Durbec J-P, Guinet C (2006) Geostatistical modelling of spatial distribution of *Balaenoptera phylasus* in the northwertern mediterranean sea from sparse count data and heterogeneous observation efforts. Ecol Model 193:615–628

Peel M, Bothma J (1995) Comparison of the accuracy of four methods commonly used to count impala. S Afr J Wildl Res 25:41–43

Pollock K, Marsh H, Lawler I, Alldredge M (2006) Estimating animal abundance in heterogeneous environments : an application to aerial surveys for dugongs. J Wildl Manag 70(1):255–262

Pollock K, Nichols J, Simons T, Farnsworth G, Bailey L, Sauer J (2002) Large scale wildlife monitoring studies: statistical methods for design and analysis. Environmetrics 13:105–119

Royle J, Nichols J, Kry M (2005) Modelling occurence and abundance of species when detection is imperfect. Oikos 110:353–359

Tasker M, Jones P, Dixon T, Blake B (1984) Counting seabird at sea from ships: a review of methods employed and a suggestion for standardized approach. The Auk 101:567–577

Thogmartin W, Sauer J, Knutson M (2004) A hierarchical spatial model of avian abundance with the application to cerulean warblers. Ecol Appl 14(6):1766–1779

Ver Hoef J, Frost K (2003) A bayesian hierarchical model for monitoring harbor seals changes in prince william sound, alaska. Environ Ecol Stat 10:201–219

Wackernagel H (2003) Multivariate geostatistics 3rd edn. Springer, Berlin

Wright I, Reynolds J, Ackerman B, Ward L, Weigle B (2002) Trends in manatee (trichechus manatus latirostris) counts and habitat use in tampa bay 1987–1994: implication for conservation. Mar Mammals Sci 18:259–274

Incorporating Survey Data to Improve Space–Time Geostatistical Analysis of King Prawn Catch Rate

Ainslie Denham and Ute Mueller

Abstract Commercial fishing logbook data from the Shark Bay managed prawn fishery in Western Australia provide king prawn catch rate data densely informed and irregularly spaced in both the spatial and temporal domains. Space–time geostatistical analysis for the data from the 2001 to 2004 fishing seasons has shown that short term catch rate prediction is possible with the use of the product-sum covariance model and the subsequent kriging estimation process. However the operation of closure lines within the fishery makes it difficult to capture the high catch rate behaviour in areas as they first open to trawling. One of these regions is the Extended Nursery Area which usually opens in the first week of May. Analysis of the survey trawls from seasons 2001 to 2003 in this region in March and April shows there is a moderate positive correlation between the actual catch rate and the survey catch rate. By using the survey catch rate data as additional data in space–time geostatistical estimation of the catch rates for May 2004, the space–time behaviour of the king prawn catch rate data is more successfully captured.

1 Introduction

We consider king prawn logbook catch rate data from the Shark Bay Prawn Managed Fishery in Western Australia and incorporate catch rate data from survey trawls in the preceding months to more accurately reproduce the space–time behaviour of the prawn catch rate in the fishing region. The king prawn catch rate data are densely informed in both the spatial and temporal domains and involve varying locations at successive time instants. Space–time geostatistical analysis for king prawn catch rate data from the 2001 to 2004 fishing seasons, incorporating traditional time series modelling of annual king prawn catch rate trends, has shown that short term catch rate prediction is possible with the use of the product-sum covariance model and subsequent kriging. However, time-limited closure lines operate

A. Denham (✉) and U. Mueller
School of Engineering, Edith Cowan University, Perth, Western Australia
e-mail: a.denham@ecu.edu.au; u.mueller@ecu.edu.au

P.M. Atkinson and C.D. Lloyd (eds.), *geoENV VII – Geostatistics for Environmental Applications*, Quantitative Geology and Geostatistics 16,
DOI 10.1007/978-90-481-2322-3_2,

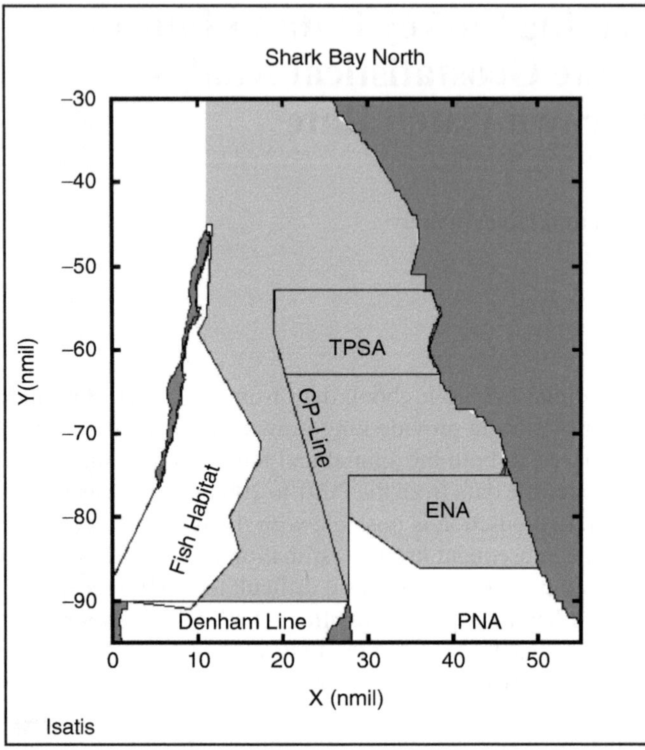

Fig. 1 Shark Bay North fishing region (*light grey*) and permanent and temporary closure lines for the fishery

in the fishery and the timing of the closures is dependent on the lunar phase and survey results. It is therefore difficult to capture successfully the high king prawn catch rate behaviour in areas as they first open to trawling.

Of particular importance is the opening of the extended nursery area (ENA) (Fig. 1) at the start of the last quarter in May producing high catch rates in the newly opened region. Using the catch rate logbook and survey data from the 2001 to 2003 seasons along with the logbook and survey catch rate data from the lunar months of March and April 2004, we investigate to what extent their use improves the reproduction of the catch rate data for the (lunar) month of May of season 2004. The ENA is surveyed in March and April of each season and analysis of data from seasons 2001 to 2004 shows that there is a moderate positive correlation between the actual catch rate and the survey catch rate from preceding months.

Space–time geostatistical estimation of king prawn catch rate for May 2004 is performed using the survey catch rate data as additional information. Multiplicative trend models are employed involving a polynomial trend model and (lunar) weekly seasonal indices obtained from classical decomposition. Spatio-temporal semivariograms of the combined detrended and deseasonalised data for 2001 to 2003 are computed and modelled using product-sum covariance models (De Iaco et al., 2001;

De Cesare et al., 2002). Cross-validation (Mueller et al., 2008) has shown that these semivariogram models capture the properties of the sample data and supports their use for estimation and smoothing of the king prawn catch rate data. We compare the estimates with those previously obtained using no survey data and with the actual catch data for 2004 and show that the space–time behaviour of the king prawn catch rate data is captured accurately with the use of the additional survey catch rate data in an area which has just opened to trawling.

2 Data Description

The data in this study are king prawn catch rate logbook and survey data from 2001 to 2004 from the Shark Bay North fishing region of the Shark Bay prawn fishery. For our analysis the catch locations were converted to nautical miles and a local coordinate system with origin at 24° southern latitude and 113° eastern longitude. Records without coordinates were eliminated from the data sets and the remaining records were aggregated to a single centroidal location for each vessel per night. This resulted in 90% of the data being used. The survey data consist of 17 locations across the study region sampled around the third moon phase in the months of March and April of each season. Spatial maps of the fishing locations for seasons 2001 to 2003, including the permanent closure lines for the fishery are shown in Fig. 2 along with the survey locations.

The means and medians of the daily king prawn catch rate are similar in 2001 and 2003 with 2002 showing a slightly larger mean and median (Table 1). The variance of the 2001 data set is considerably smaller than that of 2002 and 2003. The 2001 season also has a smaller range than the 2002 and 2003 seasons. The catch rate data for all three seasons have a moderate positive skew. The catch rate data were averaged over each lunar week to obtain a time series for each season to be used to model the temporal trend. These annual time series show similar means, medians and positive skewness to the corresponding daily data sets they were computed from (Table 1). Their variances, as expected by the averaging process, are smaller. Similarly, the minima/maxima of the averaged weekly data are larger/smaller than the corresponding daily data. Of the fishing weeks evident in each of the fishing seasons, there are a number of weeks for which there are no fishing data because of a closure period around the full moon of each month and also, in some weeks the fleet concentrates on the Denham Sound region.

3 Temporal Trend Modelling

Previous analysis (Harman, 2001; Mueller et al., 2008) has demonstrated that multiplicative classical decomposition models are appropriate for modeling the temporal trend in the king prawn catch rate data using the 4 week lunar cycle as the seasonal

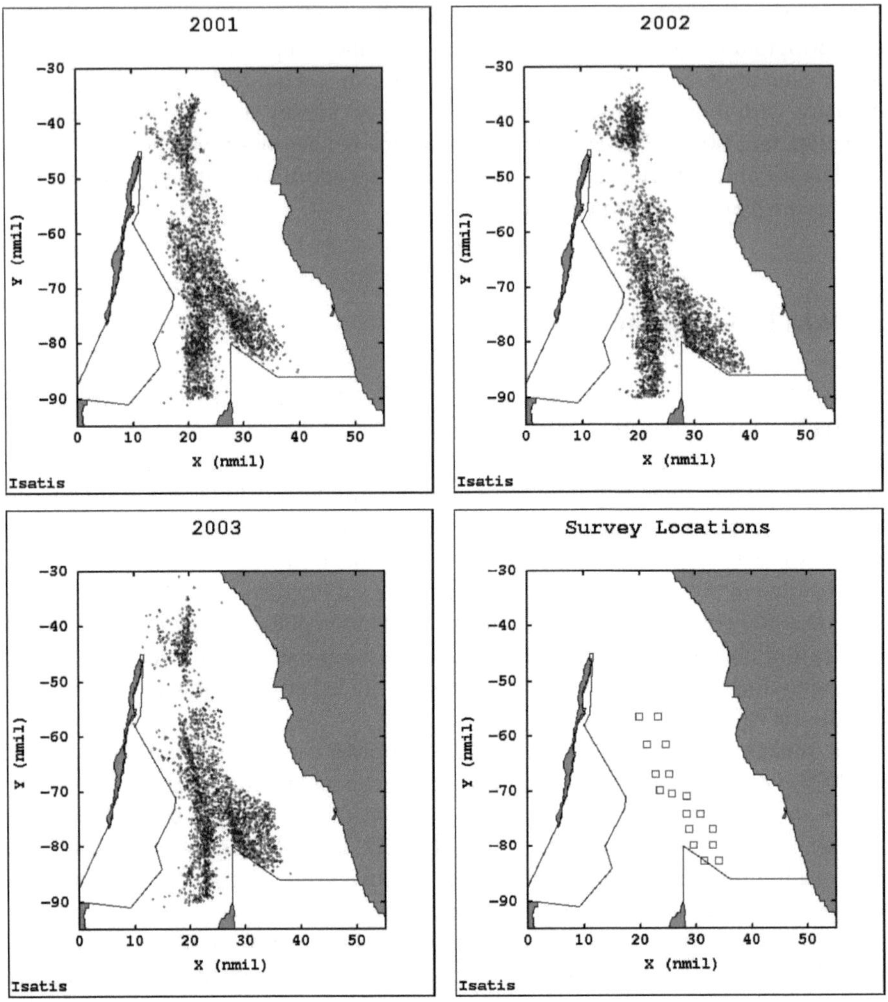

Fig. 2 Fishing locations in Shark Bay North for seasons 2001 to 2003, and survey locations

factors. Classical decomposition was performed on the weekly averaged king prawn catch rate data for seasons 2001 to 2003. A four point centred moving average was used to remove the annual effects of the 4 week lunar cycle and to identify underlying trends in the data (Fig. 3). As a function of the number of weeks the catch rate trend first increases until a maximum is reached, and then is a decreasing function of time. This pattern was also evident in the research on previous king prawn catch rate data in Harman (2001). This trend is modelled later by fitting a polynomial to the deseasonalised data, for which we must first calculate the seasonal factors.

The weekly average data were divided by the centred moving average to obtain the seasonal index component which was used to determine the seasonal factors for

Table 1 Summary statistics of daily and average weekly King Prawn catch rate for seasons 2001 to 2003

	Daily data			Average weekly data		
Season	2001	2002	2003	2001	2002	2003
Mean	28.03	34.25	29.43	24.69	28.24	23.48
Median	23.68	26.65	22.62	22.87	24.25	19.04
Variance	304.59	661.30	581.50	119.88	231.24	246.99
Skewness	1.78	3.13	2.92	1.28	1.65	2.75
Minimum	1.07	0.92	1.08	11.50	12.00	9.68
Maximum	146.07	440.32	266.63	56.99	72.13	83.86
Count	3,346	3,276	2,892	30(31)	29(31)	27(31)

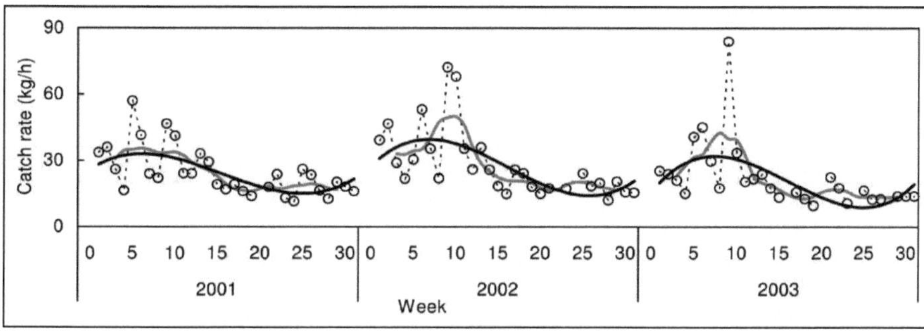

Fig. 3 Average weekly king prawn catch rate for seasons (*white circles*), centred moving average (*solid grey line*) and fitted deseasonalised trend (*solid black line*) for seasons 2001–2003

each of the four lunar phases. The Classical Decomposition seasonal factors for the king prawn catch rate for each season (Fig. 4) are similar between years. For all years the factor for the last quarter moon week is largest whilst the lowest annual factors are for the full moon period when the fishery is closed for 3 to 7 days due to the expected low catch rate (Sporer et al., 2007). The last quarter moon and new moon week factors are greater than one for all seasons whilst the full moon and first quarter moon week factors are below one for all seasons. Due to the similarity of factors across the three seasons we also compute average factors obtained by averaging across the three seasons (Fig. 4) for use in an average classical decomposition model for all three seasons.

The deseasonalised data for seasons 2001 to 2003 were calculated by dividing the catch rate data by the seasonal factors obtained for the individual seasons, and then modeled using polynomial trend lines (Fig. 3). The equations and accuracy measures are shown in Table 2. For all years a cubic function was appropriate for modelling the trend. The model for 2001 has the largest R^2 value of 0.717 indicating a large correlation with the deseasonalised data. It also shows the smallest mean error, mean percentage error and mean absolute deviation across the three seasons. The models for seasons 2002 and 2003 have slightly smaller R^2 values, still showing moderate correlation with the deseasonalised data. As the shapes and equations of

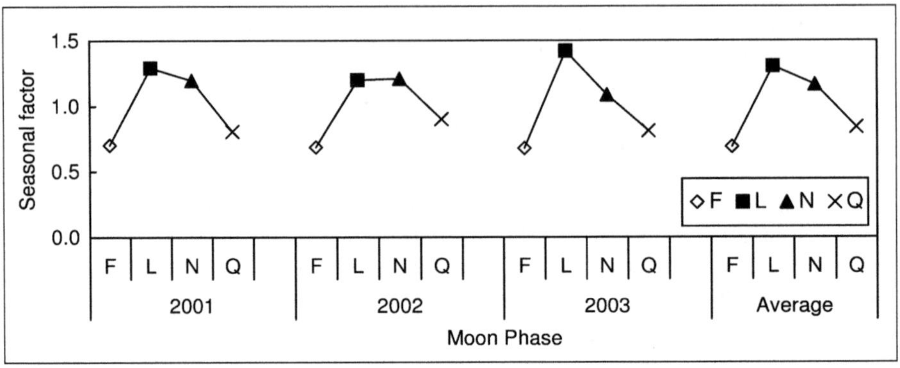

Fig. 4 Seasonal factors for seasons 2001 to 2003 based on moon phase

Table 2 Equations of fitted third order polynomial deseasonalised trend models, 2001–2003 and average model and accuracy measures

	Equation	Mean error	Mean % error	Mean abs. deviation	R^2
2001	$0.005t^3 - 0.251t^2 + 2.476t + 26.071$	0.002	−0.508	3.214	0.717
2002	$0.007t^3 - 0.365t^2 + 3.923t + 27.437$	−0.002	−1.114	5.530	0.612
2003	$0.009t^3 - 0.425t^2 + 4.879t + 15.687$	0.071	−2.598	5.200	0.566
Average	$0.007t^3 - 0.347t^2 + 3.760t + 23.065$	–	–	–	–

the trends were similar for the three seasons, an average model was computed by averaging the polynomial coefficients across the three seasons.

Multiplicative classical decomposition models for seasons 2001 to 2003 were obtained by multiplying the deseasonalised trend by the relevant (lunar) seasonal factor. Individual classical decomposition models were calculated for each season along with an average model using the average polynomial trend and average seasonal factors (Fig. 5). These models replicate the catch rate time series well. The noticeable differences exist for peaks evident in the data in weeks 5 and 9, which correspond to the opening times of two closure lines near the nursery areas. The Carnarvon-Peron line opened in week 5 in season 2001 and 2003 and in week 6 in season 2002 whilst the ENA opened in week 9 for all three seasons.

Accuracy measures for the classical decomposition models (Table 3) show the errors of the average models are greater in magnitude than their corresponding individual model, with the exception of the mean percentage error for season 2001. The large mean percentage error of the 2003 average model is due to the contribution of the first 3 weeks where the model is significantly higher than actual values. The R^2 values of the average models are only slightly smaller than their corresponding

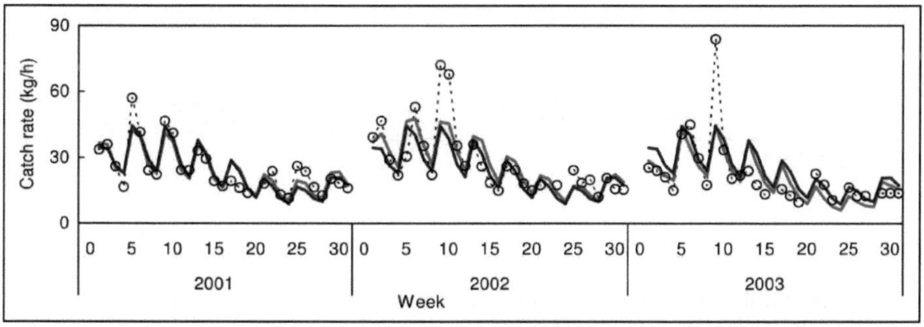

Fig. 5 Individual season classical decomposition model (*solid grey line*), average classical decomposition model (*solid black line*) and weekly king prawn catch rate (*circles*) for seasons 2001 to 2003

Table 3 Accuracy of classical decomposition models (individual and average), 2001–2003

	2001		2002		2003	
	Individual	Average	Individual	Average	Individual	Average
Mean error	−0.065	−0.281	−0.256	−3.480	−0.353	3.147
Mean % error	2.969	0.663	4.671	−7.177	5.704	25.790
Mean abs deviation	3.474	3.797	5.850	6.418	5.976	6.969
R-squared	0.813	0.794	0.696	0.675	0.610	0.567

individual model. Therefore, the average model was chosen to remove the temporal trend from the king prawn catch rate data to obtain the adjusted king prawn catch rate data.

4 Variography

Space–time semivariograms were computed for the adjusted king prawn catch rate data for the individual seasons 2001 to 2003. Although there was slight evidence of anisotropy in the spatial direction, it was regarded as an artifact of the shape of the fishing region and so disregarded in the modelling. For all three seasons the structure in both the temporal and spatial directions is similar and so a model was computed for the combined 2001 to 2003 seasons. The marginal spatial and temporal experimental semivariograms along with their fitted models are shown in Fig. 6. The spatiotemporal experimental semivariogram and its fitted semivariogram model are shown in Fig. 7. A generalized product-sum model was used (De Iaco et al., 2001) and the semivariogram model parameters are shown in Table 4. The marginal spatial semivariograms consist of a nugget effect and a long range spherical structure. The marginal temporal semivariogram consists of a nugget effect, a short range spherical structure and a long range spherical structure. The global sill of the spatiotemporal semivariogram is fitted to the data variance.

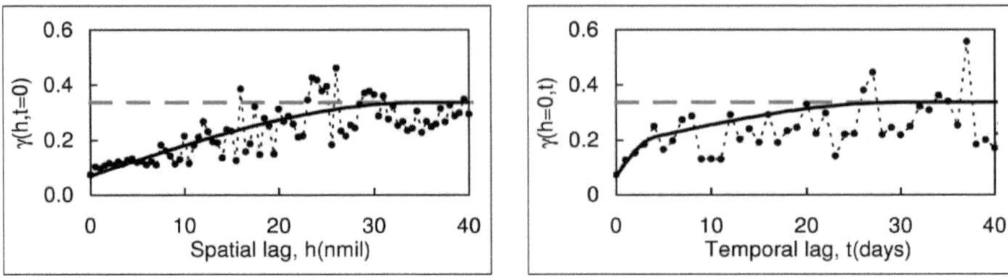

Fig. 6 Experimental marginal spatial semivariogram (*left, black circles*) and marginal temporal semivariogram (*right, black circles*) with fitted models (*solid black line*) and data variance (*grey dashed line*) for adjusted king prawn catch rate data of combined seasons 2001 to 2003

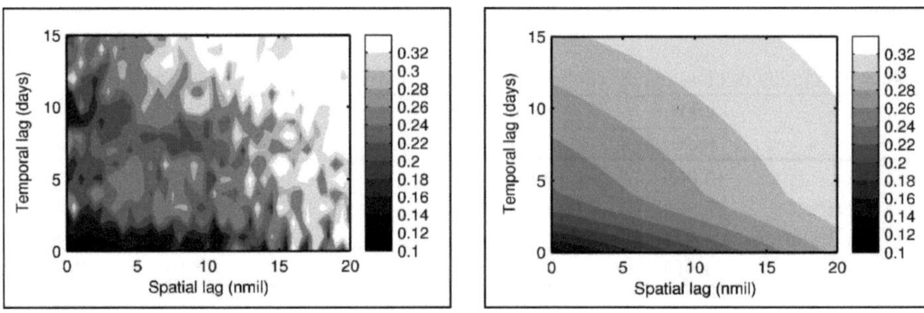

Fig. 7 Space–time semivariogram for seasons 2001 to 2003, experimental (*left*) and fitted model (*right*)

Table 4 Semivariogram model parameters for adjusted king prawn catch rate, seasons 2001–2003

			First spherical structure		Second spherical structure	
Season	Semivariogram	Nugget	Range	Sill	Range	Sill
2001	Spatial	0.05	35.0	0.15	–	–
	Temporal	0.05	1.5	0.04	30.0	0.11
2002	Spatial	0.07	35.0	0.28	–	–
	Temporal	0.07	1.5	0.04	30.0	0.24
2003	Spatial	0.07	35.0	0.29	–	–
	Temporal	0.07	1.5	0.04	30.0	0.25
2001–2003	Spatial	0.07	35.0	0.26	–	–
	Temporal	0.07	4.0	0.11	30.0	0.15

5 Opening of Extended Nursery Area

There are a number of closure lines implemented in the fishery. The ENA closure is one such closure line that opens just before the last quarter moon phase in May in all three seasons. This corresponds to the peak seen in May (Week 9) in previous

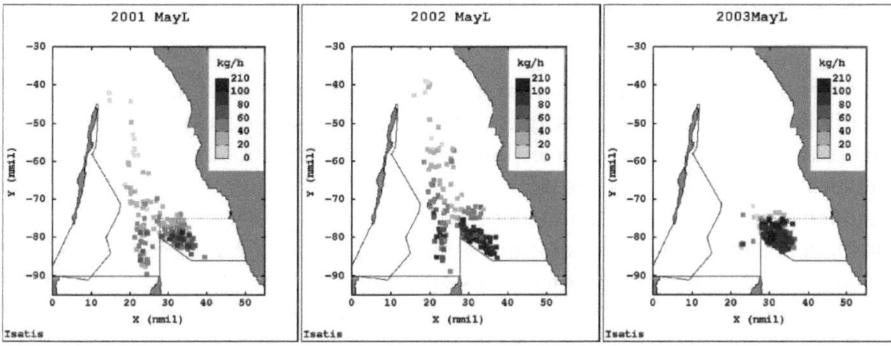

Fig. 8 Logbook catch rates for week of last quarter moon phase in seasons 2001–2003, ENA shown by dotted line

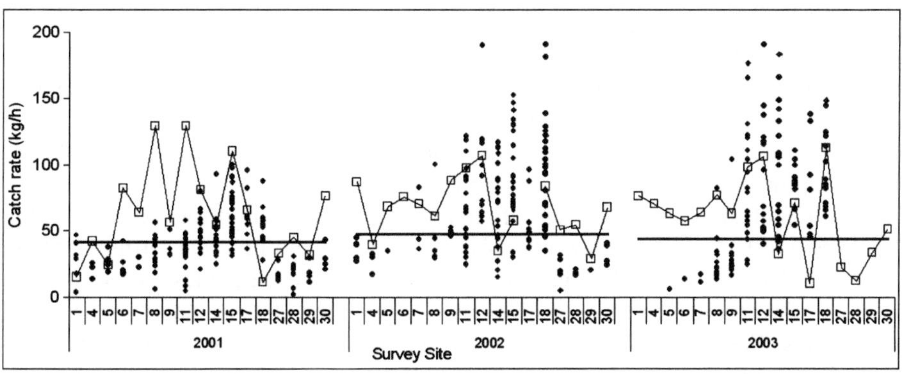

Fig. 9 Average survey catch rates (*white squares*), logbook catch rates at survey locations (*black circles*) and classical decomposition model fit (*solid line*) for week of last quarter moon phase in season 2001–2003

plots. Spatial maps of the catch rate data for this week in seasons 2001 and 2003 are shown in Fig. 8. The catch rates in the ENA are relatively high compared to those further away from the ENA and they are significantly higher than the estimate of the classical decomposition model (Fig. 9). The March and April survey data for seasons 2001 to 2003 showed similarities with the actual catch rates within the ENA in the first week it is open. It was decided that the average of the 2 months was the most reasonable indicator of the catch rate values in the ENA (Fig. 9) and the use of the average survey data in the estimation process would help to reproduce the high catch rate behaviour in the ENA.

6 Estimation

Short term catch rate prediction is possible with the use of the product-sum covariance model and the subsequent kriging estimation process. We predict the king prawn catch rate data for the (lunar) month of May 2004 by space–time geostatistical estimation using the March and April logbook catch rate data and the spatiotemporal semivariogram model obtained from the 2001 to 2003 fishing seasons. Grid catch rate estimates for the fishing region and jacknife estimates for the actual logbook catch rate data locations are shown in Fig. 10 for the week of the last quarter moon phase of May as the ENA is opened to trawling. It is evident that this method does not adequately capture the relatively high catch rates in the ENA.

As the average of the March and April survey data give a good indication of the catch rate levels seen in the ENA, we re-estimated the catch rates in May 2004 including the average survey data in the kriging process as additional data along with the March and April logbook catch rate data. The survey data were detrended and deseasonalised using the trend and seasonal index for the last quarter moon phase of May, but were allocated a date from the preceding week to enable its use in the estimation process which was directly affected by the short temporal range of the semivariogram model. Estimates over the fishing region and jacknife estimates for the actual logbook catch rate data locations in Fig. 11 demonstrate the ability to better capture the high catch rates in the ENA.

While inclusion of the survey data improved the estimation for the last quarter, this was not the case for the weeks of the new moon and first quarter moon phase of May 2004. The relatively high catch rates are much fewer in these weeks and the estimates involving the survey data are much higher than those evident in the actual catch rates (Fig. 12). The estimates involving no survey data are more representative of the actual catch rates. Accuracy measures for the jacknife estimates (Table 5) support the use of the survey data to estimate for the last quarter moon week. Estimation using the survey data decreases the magnitude of the errors for the week of

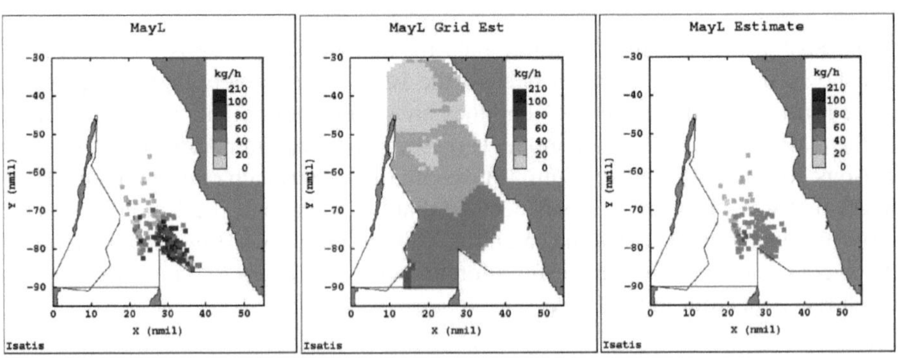

Fig. 10 Logbook catch rates (*left*), grid estimates (*centre*) and jacknife estimates (*right*) of the king prawn catch rate for the week of last quarter moon phase of May 2004

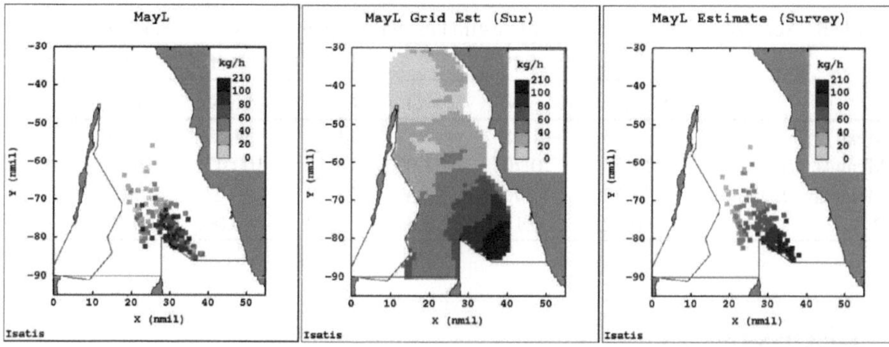

Fig. 11 Logbook catch rates (*left*), grid estimates (*centre*) and jacknife estimates (*right*) of the king prawn catch rate using average survey data for the week of last quarter moon phase of May 2004

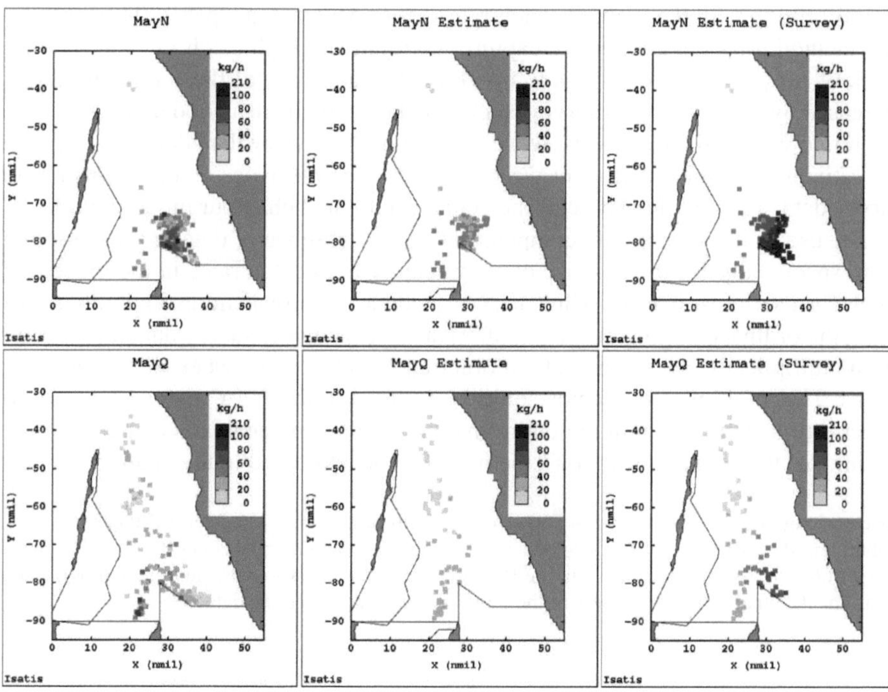

Fig. 12 Logbook catch rates (*left*), grid estimates (*centre*) and jacknife estimates (*right*) of the king prawn catch rate using average survey data for the week of new moon (*top*) and first quarter moon phase (*bottom*) of May 2004

the last quarter moon phase but increases the magnitude of the errors for the weeks of the new moon and first quarter moon phase. The R^2 value of the estimates using survey data increases for the last quarter moon phase week but decreases for the weeks of the new moon and first quarter moon phase.

Table 5 Accuracy measures for jacknife estimates, with and without survey data, May 2004

	Jacknife estimates			Jacknife estimates (average survey)		
	MayL	MayN	MayQ	MayL	MayN	MayQ
Mean error	−21.65	−4.48	−11.13	11.22	34.27	4.61
Mean % error	−17.77	0.70	−23.00	15.45	100.41	36.68
Mean abs deviation	27.72	10.90	12.60	23.47	36.23	18.00
R^2	0.22	0.13	0.66	0.30	0.002	0.09

7 Discussion

We have shown that it is possible to predict the king prawn catch rates for May 2004 using a spatiotemporal geostatistical model obtained using the data of seasons 2001 to 2003, along with the logbook catch rate data of March and April 2004. However, this method does not adequately capture the relatively high catch rates in the first week of May as the ENA opens to trawling. The accuracy of the estimates can be increased by using the April survey catch rate data, which are a good indicator of the catch rate values in the ENA. However, including the survey data does not improve the estimates for the subsequent weeks in May. The estimates computed using no survey data are more indicative of the actual catch rate behaviour in these weeks.

The use of the survey data compensates for the absence of data in the ENA and more specifically for the absence of significantly high catch rates in the preceding week for use in estimation. Thus, they provide a more realistic estimate than just using the ordinary kriging mean. An alternative to using the survey data might be to use a multiplicative factor in the temporal trend model for the week where the ENA opens. This multiplicative factor could be isolated to the ENA region so as not to affect estimates throughout the entire fishing region. The peak associated with the opening of the Carnarvon-Peron Line may also be addressed in this manner.

Acknowledgements The authors acknowledge helpful discussions with Mervi Kangas and Nick Caputi and the assistance of Errol Sporer in the logbook program and Joshua Brown and Gareth Parry from the WA Fisheries and Marine Research Laboratories who extracted the logbook catch data. Thanks go also to the skippers of the trawl fleet who collected the data.

References

De Cesare L, Myers DE, Posa D (2002) FORTRAN programs for space-time modeling. Comput Geosci 28:205–212

De Iaco S, Myers DE, Posa D (2001) Space-time analysis using a general product-sum model. Stat Probab Lett 52:21–28

Harman TS (2001) The effect of the moon phase on the daily catch rate of king, tiger and endeavour prawns in the Shark Bay and Exmouth Gulf fisheries. Honours Thesis, Edith Cowan University

Mueller U, Kangas M, Dickson J, Denham A, Caputi N, Bloom L, Sporer E (2008) Spatial and temporal distribution of western king prawns (Penaeus latisulcatus), brown tiger prawns (Penaeus esculentus), and saucer scallops (Amusium balloti) in Shark Bay for fisheries management. Project No. 2005/038, Fisheries Research and Development Corporation, Department of Fisheries, Government of Western Australia and Edith Cowan University

Sporer E, Kangas M, Brown S (2007) Shark Bay Prawn Managed Fishery status report. In: Fletcher WJ and Santoro K (eds) State of the Fisheries Report 2006/07. Department of Fisheries, Western Australia, pp 94–99

Multivariate Interpolation of Monthly Precipitation Amount in the United Kingdom

Christopher D. Lloyd

Abstract Many different interpolation procedures have been used to generate maps of precipitation amount from point data. Several case analyses have shown that making use of covariates such as elevation may increase the accuracy of predictions. Kriging-based approaches, as often employed for mapping precipitation amount, are usually based on a global variogram model and the assumption is made that spatial variation is the same at all locations. This chapter assesses the impact on prediction accuracy of using (i) local variogram models as against a global variogram model and (ii) multivariate approaches as against univariate approaches. Various kriging-based interpolation procedures are applied along with inverse distance weighting and regularised splines with tension. The results suggest that multivariate approaches such as kriging with an external drift may provide more accurate predictions than standard univariate approaches such as ordinary kriging. In addition, kriging based on local variogram models, rather than a global variogram model, is shown to provide smaller prediction errors.

1 Introduction

Maps of precipitation amount have been generated from sparse samples using a variety of univariate and multivariate interpolation procedures. The relationship between altitude and precipitation amount, which has been observed in many case studies, has been exploited and digital elevation models (DEMs) have been used to inform the prediction procedure. In such cases, increases in prediction accuracy over methods that do not make use of secondary data such as elevation have been observed (Goovaerts, 2000; Lloyd, 2005). Methods like simple kriging with locally varying means (SKlm), kriging with an external drift (KED) and cokriging (CK)

C.D. Lloyd (✉)
School of Geography, Archaeology and Palaeoecology, Queen's University,
Belfast, United Kingdom BT7 1NN
e-mail: c.lloyd@qub.ac.uk

P.M. Atkinson and C.D. Lloyd (eds.), *geoENV VII – Geostatistics for Environmental Applications*, Quantitative Geology and Geostatistics 16,
DOI 10.1007/978-90-481-2322-3_3,

allow for the integration of elevation data into the mapping of precipitation amount from sparse samples. Comparison of results using univariate and multivariate methods is one objective of this chapter.

With kriging-based approaches, the variogram is usually estimated from all of the available data and it is assumed that the spatial structure of the variable (here, precipitation amount) is the same for all areas. Where this is not the case some procedure is required for estimating the nonstationary variogram. This study assesses the use of locally estimated and modelled variograms for mapping monthly precipitation in the United Kingdom. Ordinary kriging (OK), and SKlm are utilised with kriging weights obtained using both global and local variogram models (the latter fitted automatically by maximum likelihood) while KED is employed using local variogram models only. Variations in spatial structure, as observed using the local variogram models, are explored and discussed and some problems with automated model fitting highlighted. Univariate and multivariate regularised splines with tension (RST) are also used for comparative purposes, as there is no need to estimate the variogram in those cases. The prediction accuracy of the methods is compared through cross-validation and the spatial distribution of prediction errors is assessed.

This paper builds on research presented by Lloyd (2002, 2005, 2009) and expands on the range of methods applied and the exploration of spatial variation in prediction errors. The basis of this chapter is a comparison of selected univariate and multivariate interpolation procedures including inverse distance weighting (IDW), OK, CK, SKlm, KED, global regression (GR), moving window regression (MWR), geographically weighted regression (GWR) (with elevation as the independent variable and precipitation amount as the dependent variable) and RST. In addition, locally estimated and modelled variograms are used for OK and SKlm and the results compared with predictions made using the global variogram model. Data for July 2006 provide the basis of the analysis. Goovaerts (2000) used a variety of geostatistical methods to make predictions of precipitation amount. RST has been used before to map precipitation amount with elevation as a covariate (Hofierka et al., 2002). Brunsdon et al. (2001) applied GWR for the exploration of spatial variations in the relationship between altitude and precipitation amount. Lloyd (2005) provides a review of these and other related applications.

Some pros and cons of the alternative procedures are identified following a variety of criteria. Finally, some ideas for future research, including the extension of the methods to prediction of precipitation amounts for shorter time periods (e.g., a day), are outlined briefly.

2 Study Area and Data

The analysis is based on precipitation amount measurements, for July 2006, made across the UK and The Isle of Man under the auspices of the UK Meteorological Office. The data were obtained from the British Atmospheric Data Centre (BADC) web site (http://www.badc.rl.ac.uk). The data are referenced using the Ordnance

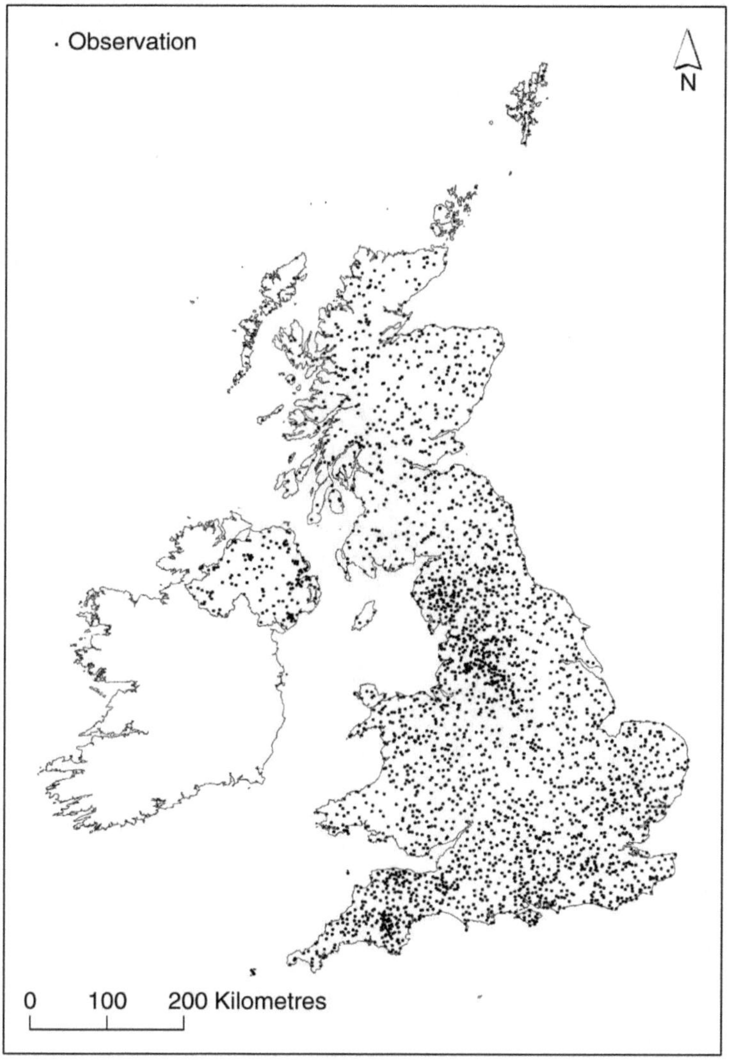

Fig. 1 Locations of precipitation measurements for July 2006

Survey British grid system. The data for Northern Ireland are referenced using the Irish Grid co-ordinate system and these were converted to British Grid co-ordinates using ArcGIS^{TM}. The locations of all of the measurements are shown in Fig. 1, while summary statistics are given in Table 1. A map of July 2006 precipitation amounts generated with IDW is given in Fig. 2 (note the 'clumping' of values around observation locations, characteristic of IDW). For July 2006, precipitation amounts are most consistently large in the west of Scotland, part of the north west of England, part of the south west of England, in addition to a few other small areas including the south east of Northern Ireland.

Table 1 July precipitation amount summary statistics

Number	Minimum (mm)	Maximum (mm)	Mean (mm)	Standard deviation (mm)
2,924	2.000	224.000	45.238	29.329

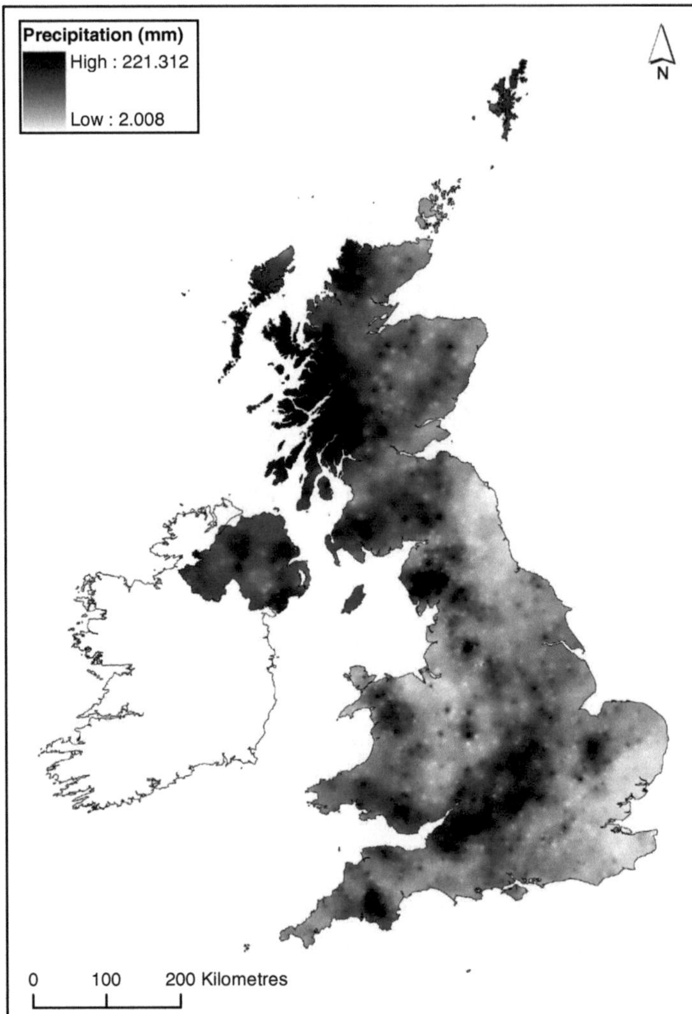

Fig. 2 Precipitation amounts for July 2006: IDW with 16 nearest neighbours

3 Methods

The analysis makes use of a variety of widely-used interpolation procedures. These include IDW, RST, OK, CK, SKlm and KED. Two variants of RST were used – standard univariate RST (referred to as RST2D) and three-dimensional RST which accounts for elevation in the precipitation prediction process (termed RST3D). In addition, GR, MWR and GWR were used with elevation as the independent variable and precipitation amount as the dependent variable. IDW is well known and an introductory account is provided by Burrough and McDonnell (1998). A summary of some key variants of thin plate splines, including RST2D, is provided by Lloyd (2006) (where RST is referred to as the completely regularised spline). Hofierka et al. (2002) describe RST3D and present an application. Introductions to OK, CK, SKlm and KED are provided by Goovaerts (1997), Wackernagel (2003) and Lloyd (2006). MWR is simply regression conducted using the n paired observations in a moving window. GWR is described by Fotheringham et al. (2002) and in the present analysis an adaptive bi-square kernel was used such that the size of the kernel varies according to the density of observations locally.

IDW and MWR were conducted with purpose-written Fortran 77 code, GWR was conducted using the GWR software detailed by Fotheringham et al. (2002). Gstat (Pebesma et al., 1998), was used for global variogram estimation (for CK, autovariograms and cross-variograms) and for fitting models using weighted least squares (WLS). The GSLIB (Deutsch and Journel, 1998) routine kt3d was used for OK, SKlm and KED. OK and SKlm were implemented using global variograms and variograms estimated locally and modelled automatically using the MLREML maximum likelihood routine written by Pardo-Igúzquiza (1997), while KED was conducted using the latter approach only since the KED prediction neighbourhood should ideally correspond to the neighbourhood over which the trend-free variogram is estimated (Hengl, 2007). A Fortran 77 program was written to visit each observation in the precipitation dataset and extract the n nearest neighbours to each observation location, after which the MLREML routine was called to estimate the variogram and fit a model. The simplex method for function minimisation (Pardo-Igúzquiza, 1997) was applied to fit variogram models in this analysis. A nugget effect and a spherical model were fitted using this procedure. A modified version of the GSLIB routine kt3d was used for LVOK, LVSKlm and LVKED (where LV indicates local variogram). The GRASS GIS RST routines s.surf.rst and s.vol.rst were used for RST2D and RST3D respectively. Routines with the same functionality (but for GRASS site files rather than vector point files as used in this analysis) are described by Neteler and Mitasova (2004). With RST, selection of a tension parameter, a smoothing parameter (for a smoothing parameter of zero, RST is an exact predictor) and a z factor is necessary, as described by Hofierka et al. (2002).

4 Analysis

The present section focuses first on the analysis of local spatial variation in (i) the relationship between elevation and precipitation amount and (ii) the spatial structure of precipitation amount. Given that the relationship between altitude and precipitation is utilised for prediction purposes, a DEM for the UK is given in Fig. 3 (GTOPO 30 DEM: http://edc.usgs.gov/products/ elevation/gtopo30/gtopo30.html). Figure 4

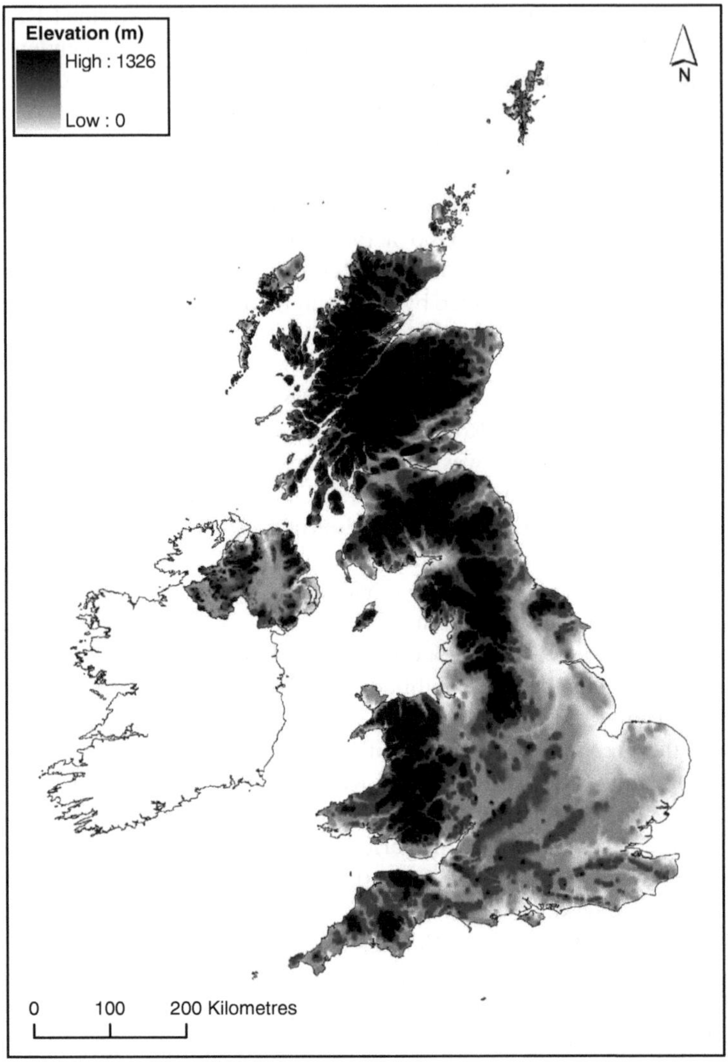

Fig. 3 DEM for the UK. Data available from U.S. Geological Survey/EROS, Sioux Falls, SD

gives the GWR coefficient of determination (using the bi-square kernel with 150 nearest neighbours). There are large r^2 values in the west of Scotland, the east of Northern Ireland, south west England and parts of the English midlands. With the exception of the latter, these are areas with high elevations. Figure 5 gives the ranges of variogram models, with a spherical structured component, fitted to variograms estimated from 128 nearest observations. Figure 5 shows that the local range varies markedly across the UK. As an example, in Scotland, there are three distinct regions – with predominantly medium range variation in the north, short range variation in

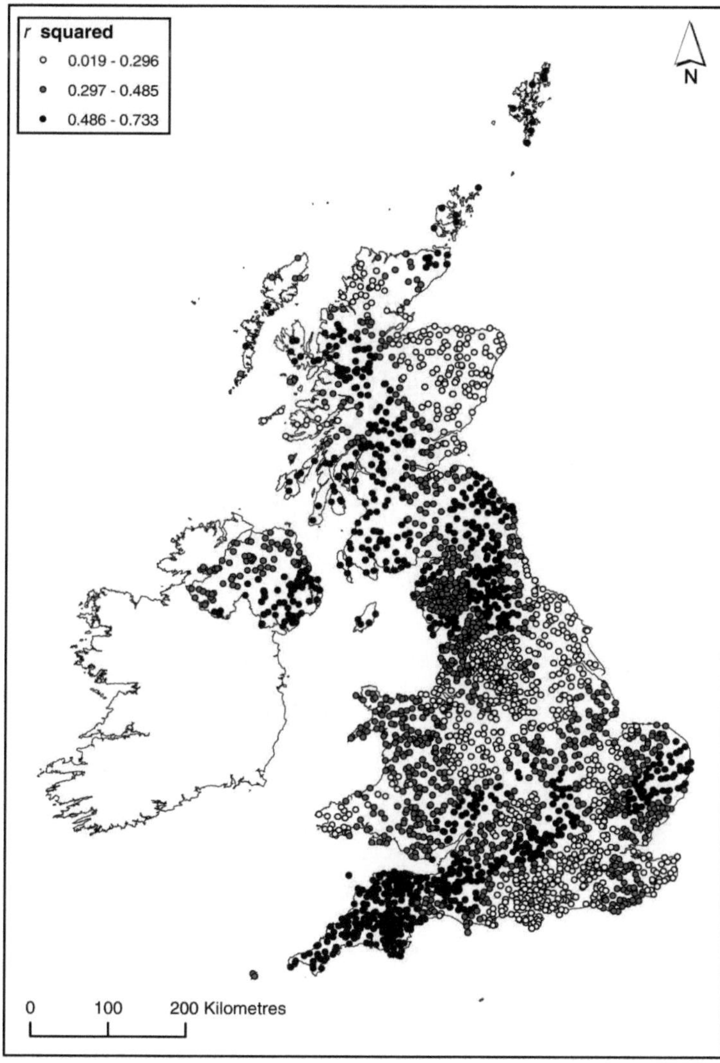

Fig. 4 GWR coefficient of determination

the east and south and long range variation in the west of the region. The models from which the ranges in Fig. 5 are derived were used for for LVOK. For LVSKlm and LVKED, local variogram models, comprising a nugget effect and a spherical component, were derived using ML with GLS regression being used to model the elevation–precipitation relationship.

The remainder of the section focuses on the assessment of spatial interpolation procedures using cross-validation. With IDW an exponent of two was used and the number of nearest neighbours increased from eight to 16, 32, 64 and 128.

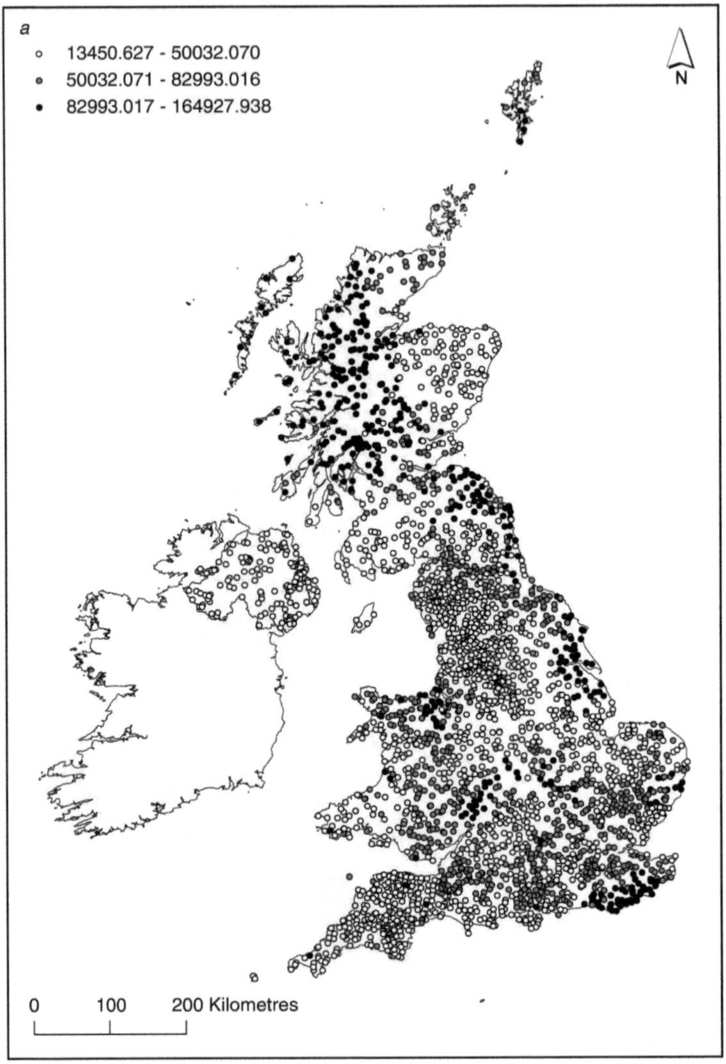

Fig. 5 Ranges of models fitted to variograms estimated from 128 nearest observations (spherical model)

The cross-validation RMSE was smallest for eight nearest neighbours and the corresponding errors are summarised in Table 2. For the kriging variants, a similar procedure was followed. The variogram model for OK was fitted using the WLS functionality of Gstat. In the case of CK, the precipitation autovariogram model was the same as for OK and, following the linear model of coregionalisation (Goovaerts, 1997), the same number and type of structures and range parameters were fitted to the elevation autovariogram and the cross-variogram. For SKlm, only

Table 2 July precipitation amount cross-validation prediction summary statistics

Method	NNN	Maximum negative (mm)	Maximum positive (mm)	Mean (mm)	RMSE (mm)
IDW	8	−86.665	116.976	0.640	13.700
GR	All	−161.286	61.620	−0.004	27.519
MWR	8	−93.528	80.869	0.224	14.626
GWR	150	−112.217	52.932	−0.109	17.158
RST2D	32[a]	−83.872	92.871	0.239	13.139
RST3D	512[a]	−90.782	80.395	0.077	12.967
OK	16	−86.216	96.314	0.161	13.207
LVOK	128	−87.274	86.433	0.153	13.002
CK	8	−87.649	89.716	0.128	13.228
SKlm	128	−86.771	90.278	0.042	13.148
LVSKlm	128	−85.298	89.306	0.112	12.971
LVKED	128	−79.923	87.571	0.074	12.524

[a] For RST, NNN has a different definition – it is the minimum number of points used for prediction

64 and 128 nearest neighbours were utilised for prediction. For SKlm, the variogram was estimated from residuals of an OLS regression of elevation and precipitation. The resulting model coefficients were used to derive new β coefficients using GLS in Gstat (a similar process is described by Hengl (2007)). The SKlm local means were the GLS regression predictions. Lloyd (2009) used several methods for deriving the local mean, but only one is presented in this study (in addition to that given the local variogram procedure, outlined above). With RST2D and RST3D a variety of different tension and smoothing parameters were applied and assessed along with varying the z factor for RST3D. For RST2D, a tension parameter of 70 and a smoothing parameter value of 0.7 provided the smallest cross-validation RMSE. For RST3D, a tension parameter of 35 and a smoothing parameter value of 0.1 with a z multiplier of 20 provided the smallest cross-validation RMSE. Table 2 summarises the cross-validation errors for each approach with the figures only given for the combination of the number of nearest neighbours and model parameters that resulted in the smallest RMSE. Selection of methods was systematic, but clearly other combinations of parameters may result in more accurate cross-validation predictions.

In terms of RMSE values, the most accurate cross-validation predictions are for LVKED, with RST3D and LVSKlm next in line. The mean error closest to zero is for GR, with SKlm having the second smallest (absolute) value. The smallest maximum absolute error is for LVOK. It is notable that the RMSE for CK is larger than that for OK. This is probably due, at least in part, to the weak global relationship between elevation and precipitation amount.

The absolute differences between OK 16 and LVKED 128 cross-validation errors were computed and interpolated using IDW based on four nearest neighbours (this was done purely for visualisation purposes). The resulting map is shown in Fig. 6. Obviously, positive values in Fig. 6 indicate instances where LVKED predictions were more accurate than OK predictions, while negative values show the

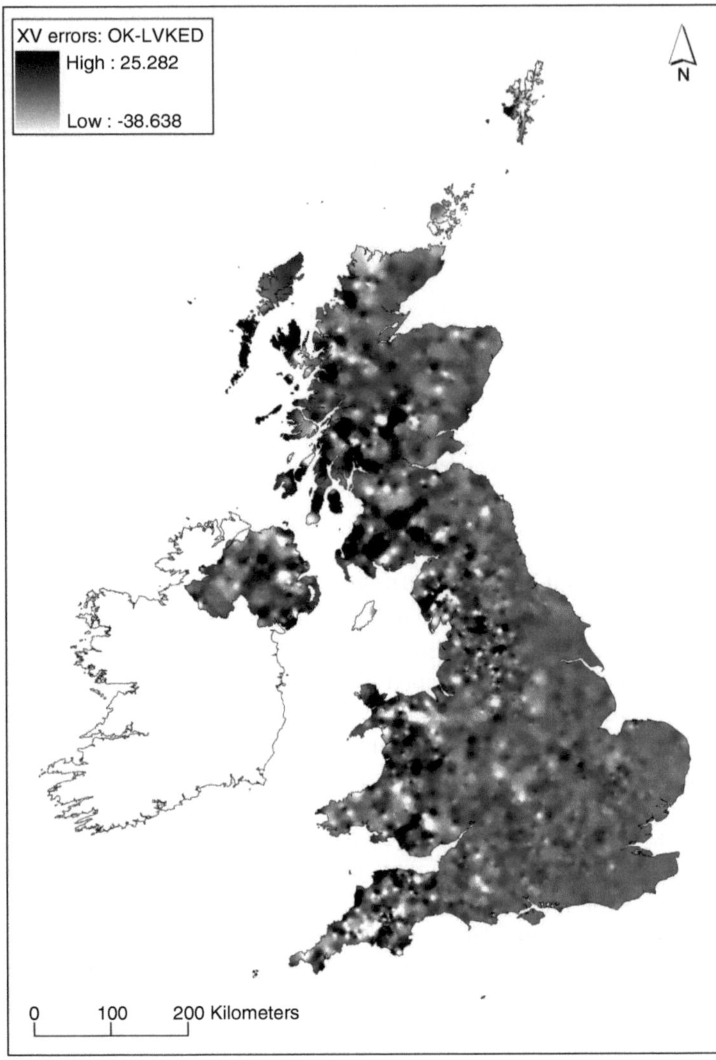

Fig. 6 OK 16 XV absolute errors – LVKED 128 absolute errors (generated using IDW 4)

reverse. In most areas there are only small differences between the predicted values. The most obvious differences are in the west of Britain and the east of Northern Ireland. Variations in elevations and the effect of this on precipitation amounts (i.e., the relationship between altitude and precipitation amount) in some areas is one explanation for these differences. The range or spatial frequency of variation in precipitation amounts varies at a local scale more markedly in some places than others (see Fig. 5) and this will impact on the degree to which local variograms may offer benefits in providing more appropriate weights for kriging.

5 Discussion

Adjustment of the parameters for each technique would undoubtedly result in a different ranking of methods in terms of the RMSE, mean and maximum absolute error (see Davis [1987] for a discussion about the use of cross-validation for assessing prediction accuracy). However, the results do at least indicate approaches that seem likely to provide predictions with a similar level of accuracy. The results demonstrate potential benefits in making use of elevation data as a part of the precipitation interpolation procedure. There are also suggestions that kriging based on locally estimated and modelled variograms may potentially provide accurate predictions than kriging based on global variogram models. However, automated modelling of variograms is computationally demanding. In this study, only a nugget effect and a single spherical component were fitted to the local variograms. In cases where the spatial variation is complex, two or more structured components may be desirable. In some areas a model other than the spherical model may have been preferable. RST requires interaction in terms of selection of tension and smoothing parameters, but this is more straightforward than the variogram estimation and modelling procedure required for kriging. In addition, use of a secondary variable is possible using RST3D and cross-validation errors are comparable with those obtained using kriging.

With the most obvious exception of GR, most of the approaches provide broadly comparable results in many respects and criteria other than simply cross-validation RMSEs are important considerations. If there is a desire to avoid common problems like clumping around observation locations when using IDW, then prediction accuracy (judged by cross-validation or otherwise) may not be the greatest concern. The most appropriate prediction approach is likely to vary between seasons as demonstrated by Lloyd (2005, 2009). This is due, in part, to seasonal differences in precipitation intensity. There is little evidence of spatial structure (i.e., positive spatial autocorrelation) in the cross-validation errors. However, some general trends are observable. For example, errors in the low-lying south and east of Britain tend are consistently smaller than those in the generally more highly-elevated areas in the north and west of Britain (where precipitation amounts tend to be greater). There is an element of circularity in the assessment of SKlm using cross-validation in that the local means are derived using all available data (i.e., the observation at the prediction location is used also). The local variograms for LVOK, LVSKlm and LVKED are also derived using all available observations. Testing suggested that these factors had little impact on results.

6 Conclusions

This chapter indicates that more complicated multivariate approaches may be likely to offer benefits over simpler univariate approaches for mapping monthly precipitation amount. In terms of kriging-based approaches, there is a suggestion that local variogram estimation and modelling may offer benefits. The results show that, judging by cross-validation errors, RST provides results comparable to kriging. Expanding the analysis to include other months or to analyse data for shorter time periods (e.g., weeks or days, where the altitude–elevation relationship is likely to less strong than for months) would be sensible foci for future research. An alternative means of comparing approaches to spatial interpolation is to divide the data set into two, and predict at one set of locations using the second set of data (this is termed jackknifing). This approach is being used along with cross-validation in ongoing research.

Acknowledgements The British Atmospheric Data Centre is thanked for allowing access to the United Kingdom Meteorological Office (UKMO) Land Surface Observation Stations Data. The UKMO is acknowledged as the originator of these data. Eulogio Pardo-Igúzquiza is thanked for providing a copy of his MLREML Fortran code.

References

Brunsdon C, McClatchey J, Unwin DJ (2001) Spatial variations in the average rainfall-altitude relationship in Great Britain: an approach using geographically weighted regression. Int J Climatol 21:455–466

Burrough PA, McDonnell RA (1998) Principles of geographical information systems. Oxford University Press, Oxford

Davis BM (1987) Uses and abuses of cross-validation in geostatistics. Math Geol 29:241–248

Deutsch CV, Journel AG (1998) GSLIB: Geostatistical Software Library and User's Guide 2nd edn. Oxford University Press, New York

Fotheringham AS, Brunsdon C, Charlton M (2002) Geographically weighted regression: the analysis of spatially varying relationships. Wiley, Chichester

Goovaerts P (1997) Geostatistics for natural resources evaluation. Oxford University Press, New York

Goovaerts P (2000) Geostatistical approaches for incorporating elevation into the spatial interpolation of rainfall. J Hydrol 228:113–129

Hengl T (2007) A practical guide to geostatistical mapping of environmental variables. Office for Official Publications of the European Communities, Luxembourg. http://eusoils.jrc.it/ESDB_Archive/eusoils_docs/other/EUR22904en.pdf; last accessed 24/11/2008

Hofierka J, Parajka J, Mitasova M, Mitas L (2002) Multivariate interpolation of precipitation using regularized spline with tension. Trans GIS 6:135–150

Lloyd CD (2002) Increasing the accuracy of predictions of monthly precipitation in Great Britain using kriging with an external drift. In: Foody, GM Atkinson PM (eds) Uncertainty in remote sensing and GIS Wiley, Chichester, pp 243–267

Lloyd CD (2005) Assessing the effect of integrating elevation data into the estimation of monthly precipitation in Great Britain. J Hydrol 308:128–150

Lloyd CD (2006) Local models for spatial analysis. CRC Press, Boca Raton, FL

Lloyd CD (2009) Nonstationary models for exploring and mapping monthly precipitation in the United Kingdom. Int J Climatol, in press

Neteler M, Mitasova H (2004) Open source GIS: a GRASS approach, 2nd edn. Springer New York

Pardo-Igúzquiza E (1997) MLREML: a computer program for the inference of spatial covariance parameters by maximum likelihood and restricted maximum likelihood. Comput Geosci 23:153–162

Pebesma EJ, Wesseling CG (1998) Gstat, a program for geostatistical modelling, prediction and simulation. Comput Geosci 24:17–31

Wackernagel H (2003) Multivariate geostatistics: an introduction with applications, 3rd edn. Springer Berlin

Extreme Precipitation Modelling Using Geostatistics and Machine Learning Algorithms

Loris Foresti, Alexei Pozdnoukhov, Devis Tuia, and Mikhail Kanevski

Abstract The paper presents an approach for mapping of precipitation data. The main goal is to perform spatial predictions and simulations of precipitation fields using geostatistical methods (ordinary kriging, kriging with external drift) as well as machine learning algorithms (neural networks). More practically, the objective is to reproduce simultaneously both the spatial patterns and the extreme values. This objective is best reached by models integrating geostatistics and machine learning algorithms. To demonstrate how such models work, two case studies have been considered: first, a 2-day accumulation of heavy precipitation and second, a 6-day accumulation of extreme orographic precipitation. The first example is used to compare the performance of two optimization algorithms (conjugate gradients and Levenberg-Marquardt) of a neural network for the reproduction of extreme values. Hybrid models, which combine geostatistical and machine learning algorithms, are also treated in this context. The second dataset is used to analyze the contribution of radar Doppler imagery when used as external drift or as input in the models (kriging with external drift and neural networks). Model assessment is carried out by comparing independent validation errors as well as analyzing data patterns.

1 Introduction

Spatial interpolation of precipitation is one of the most challenging fields of research for geostatisticians, meteorologists, climatologists and natural hazards practitioners. Usually, the prediction of precipitation is performed with physical models, mainly because of the high spatial variability and nonlinearity of the problem. The limits of physical models (mainly computational) are encountered when prediction is performed at scales that are smaller than 2–3 km. Geostatistics offers different methods to deal with this problem and provides interesting results by several authors

L. Foresti (✉), A. Pozdnoukhov, D. Tuia, and M. Kanevski
Institute of Geomatics and Analysis of Risk, University of Lausanne, Switzerland
e-mail: Loris.Foresti@unil.ch; Alexei.Pozdnoukhov@nuim.ie; Devis.Tuia@unil.ch; Mikhail.Kanevski@unil.ch

P.M. Atkinson and C.D. Lloyd (eds.), *geoENV VII – Geostatistics for Environmental Applications*, Quantitative Geology and Geostatistics 16,
DOI 10.1007/978-90-481-2322-3_4,

(Attore et al., 2007; Daly et al., 1997; Dobesch et al., 2007; Dubois et al., 2003; Goovaerts, 2000; Jeffrey et al., 2001; Lloyd, 2007). Nevertheless, the high spatial variability and noise in the precipitation data have led to an increasing use of machine learning (ML) methods (Antonic et al., 2001; Bryan and Adams, 2002; Demyanov et al., 1998; Parkin and Kanevski, 2008). The motivation of applying these methods is also given by an increasing volume and availability of data. The objective of the paper is to show how machine learning algorithms can be used to perform spatial prediction of precipitation and how geostatistics can be applied in a complementary way with them.

Section 2 is an introduction to the theory of Multi Layer Perceptron, the neural network used in this study. A short overview of optimization methods is also necessary to understand the first case study. A complete introduction to neural networks and related optimization methods can be found in Bishop (1995) and Haykin (1998). Theory of geostatistics such as ordinary kriging, kriging with external drift (KED) and sequential Gaussian simulations (SGS) can be found in Deutsch and Journel (1997), Hengl et al. (2003), Isaaks and Srivastava (1989), Kanevski (2004), and Wackernagel (2003), Section 3 presents briefly the case studies and the applied methodology. Section 4 illustrates the results. Conclusions are presented in Section 5.

2 Multi Layer Perceptron

Multi Layer Perceptron (MLP) is a neural network belonging to the family of machine learning algorithms and it has been applied in a variety of fields such as computer science, speech recognition, finance, environmental sciences, remote sensing, data mining, etc. Learning about statistical relationships between variables, MLP aims at the prediction of continuous or discrete data, respectively for non-linear regression and classification tasks. In this paper, only regression is considered.

MLP architecture is composed of an input, one or more hidden and output layers. Predictors for regression are contained in the input layer and the target variable in the output layer. The hidden layer connects the input layer to the output layer by means of weights. Moreover, each neuron of the hidden layer has an activation function that is responsible for the non-linearity of the network. The latter transforms a weighted linear summation of the inputs (the predictors) with a nonlinear function (often a logistic or hyperbolic transformation). The aim of using MLP is to find an optimal configuration of weights that can reproduce the functional relationship between the inputs and the output variables. MLP typical architecture is shown in Fig. 1.

The output part of MLP is defined as follows:

$$y_i^{out} = g^{out}\left[\sum_j w_{ij}^{(2)} \cdot g\left(\sum_k w_{jk}^{(1)} \cdot x_k^{(0)}\right)\right] \tag{1}$$

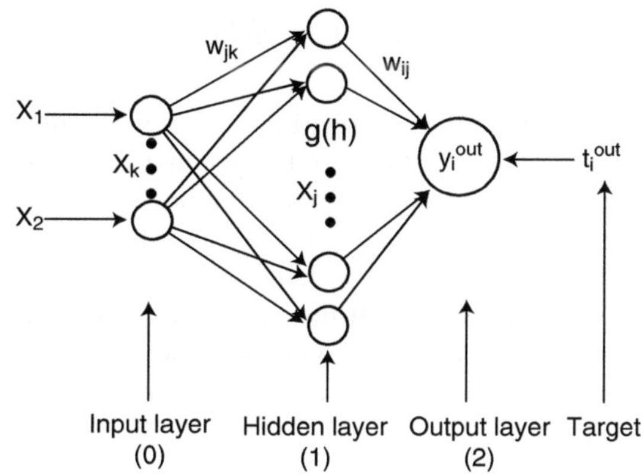

Fig. 1 MLP is composed of an input layer, one or more hidden layers and an output layer that are connected by weights

where w_{ij} is the weight between the *jth* neuron and the *ith* neuron, x_k is the *kth* input neuron and the activation function g applies a non-linear transformation of the weighted linear summation ξ:

$$g(\xi) = \frac{1}{1 + e^{-\xi}} \qquad \text{or} \qquad g(\xi) = \frac{e^{\xi} - e^{-\xi}}{e^{\xi} + e^{-\xi}} \tag{2}$$

The weighted sum of inputs is therefore transformed in the hidden layer via the activation function (brackets in Equation (1)). A weighted linear summation of the hidden neurons is also carried out in the output layer (square brackets in Equation (1)). g^{out} can be either a linear or a non-linear function.

The main time consuming phase of MLP is the training procedure where the optimal values of weights in the network are sought. First, the weights are initialized randomly (or by using the annealing algorithm as it is proposed in Kanevski [2004]) between some specified limits and the inputs have to be scaled between some specified boundaries depending on the choice of the activation function (usually 0–1). Then, the feed-forward part is computed by introducing the input vector **x** in the neural network. The dimension of the input vector is equal to the number *k* of predictors kept for the analysis. The procedure is then iterated for every training sample. Then, a measure of dissimilarity between the prediction and the target values is calculated and minimized during training. For the regression problems, the mean squared error cost function is used:

$$E(w) = \frac{1}{2}\frac{1}{m}\frac{1}{n}\sum_{i=1}^{m}\sum_{s=1}^{n}\left(t_{is}^{out} - y_{is}^{out}\right)^2 \tag{3}$$

where t_{is}^{out} is the *i*th target value for training sample *s*, y_{is}^{out} is *i*th output for training sample *s*. In this paper only one-output MLP is considered (*i* is equal to 1). It is evident that the minimization of the cost function is the aim of the training phase of MLP. Finally, let us recall that MLP is a universal approximator (Bishop, 1995; Haykin, 1998).

2.1 *Optimization Algorithms*

The cost function of Equation (3) has to be minimized by optimizing the values of weights. The cost function is not convex, has many local minima and therefore the solution is not unique. To solve this problem several optimization algorithms can be applied in order to approach the true minimum of the cost function. The back-propagation of the error defines how weights are changed after each iteration. An in-depth description of the optimization algorithms can be found in LeCun et al. (1998). Here, only the conjugate gradient and the Levenberg-Marquardt optimization are briefly described.

2.1.1 Conjugate Gradient Algorithm

Conjugate gradient (CG) is a first order optimization algorithm that computes conjugate directions by a combination of consecutive directions and exploits the line search method as it is explained in LeCun et al. (1998). At the first iteration the gradient of the cost function is computed. A first minimum of the cost function can be found in the direction which is contrary to the gradient. Several numerical evaluations of the cost function along this direction allow identification of this minimum (line search). At the second iteration, the new direction is computed by a combination between the previous and the current direction:

$$\mathbf{d}_{t+1} = -\nabla E(w_{t+1}) + \beta_{t+1}\mathbf{d}_t \quad (4)$$

where $\nabla E(w_{t+1})$ is the gradient of the cost function evaluated at iteration *t+1*, β defines the type of combination between two consecutive directions LeCun et al. (1998) and **d** is the direction. Line search is now applied along this conjugate direction and the minimum along it yields the new weight vector.

CG is widely used to train MLP because of its ability to find solutions with weights that are centered around the origin of the cost function. In fact, the algorithm does not find solutions with large and biased weights that can lead to overfitting of training data.

2.1.2 Levenberg-Marquardt Algorithm

Levenberg-Marquardt (LM) is a second order optimization algorithm that is used in the attempt to speed up the training phase by taking into account the information about the curvature of the error function (second derivatives). A local quadratic approximation of the error function is carried out in order to find its minimum in only one iteration. The method is a combination of the steepest descent (simple gradient descent with a momentum term) and the local quadratic optimization. The update rule for the weights is

$$w_{t+1} = w_t - (\mathbf{H} + \lambda \cdot diag(\mathbf{H}))^{-1}\nabla E(w_t) \quad (5)$$

where **H** is an approximation of the Hessian matrix. The parameter λ defines a tradeoff between the steepest descent and the quadratic approximation. In other words, the optimization is similar to the steepest descent in the regions where the error increases and similar to the quadratic optimization nearby the minimum. A complete description of this algorithm can be found in Roweis (2000).

It is often noted that after the training LM algorithm tends to find lower minimums of the cost function than the CG algorithm even with the same number of hidden neurons. Hence, LM can fit better to training data than CG and exploits more the flexibility of MLP by finding solutions with larger and more biased weights. The effect is an increasing risk of overfitting training data and the resulting loss of patterns.

3 Case Studies and Methodology

3.1 Case Study: Precipitation of 2nd and 3rd October 2006

This case study was an extreme event provoked by a cold front that affected Switzerland during 2 days. It generated a very anisotropic narrow line of heavy precipitation on the southern part of the Alps because of particular orographic conditions. The phenomenon was monitored by 413 rain gauges whose summary statistics are presented in Table 1. For modeling purposes and for checking generalization abilities on new data, they were divided into 360 used for training and 53 for validation.

This particular situation under study is used to compare the performance of two optimization algorithms (CG and LM). Training of the neural network is carried out by applying an early stopping criterion: the training procedure ends when training error equalizes the quantity of noise present in the data. This procedure of model selection avoids the need to take a testing subset from training data. Noise was estimated by modelling the raw variogram and it is simply the square root of the nugget. A two-inputs MLP (X,Y coordinates), with 15 hidden neurons is trained using the cited algorithms. With less than 15 hidden neurons, both LM and CG were not able to reduce the training error below the noise level. Independent validation errors, patterns and the ability of reproducing extreme values are compared and the peculiarities of the two optimization methods are pointed out.

Finally, hybrid models (combining geostatistics and machine learning) are considered. Neural Networks are known to be very useful for modelling of patterns in data. However, they risk to overfit the data when trying to reproduce extreme values. Geostatistical techniques can partially solve this problem by means of predictions and simulations. Therefore, MLP can be applied in order to solve problems of spatial non-stationarity by modelling trends. The obtained stationary residuals can then be interpolated/simulated using the family of kriging methods.

Table 1 Summary statistics of 2nd and 3rd of October 2006, mm of water

Mean	Variance	Skewness	Kurtosis	Min	Max
22.3	577.8	4.22	22.61	0	218.4

3.2 Case Study: Precipitation from the 18th to the 23rd of August 2005

The second case study considers an important orographic precipitation event on the northern side of the Alps. This situation was recorded by 439 rain gauges (302 were used for training and 137 for validation) and by three atmospheric Doppler radars, giving an image at 1 km resolution (see Fig. 2). A Radar image was provided and pre-processed by the Federal Office of Meteorology and Climatology (MeteoSwiss). Details on pre-processing of radar image are explained in Germann et al. (2006) and Joss et al. (1997). In the context of this study radar data is used as auxiliary information. Table 2 shows summary statistics. It can be noticed that there are some differences between rain gauge and radar statistics mainly because of their spatial resolutions. Radar is more likely to be able to detect local patterns of precipitation. On the contrary, the rain gauge network topology is not dense enough to measure such phenomena and it misses spatial extremes.

Radar is an important source of information for improving spatial prediction of precipitation and it can therefore be integrated as an external drift or an input in the neural models. This study is an attempt to compare two different model approaches: geostatistics by using kriging with external drift and machine learning by means of artificial neural networks (ANN) with an external drift as in Parkin and Kanevski (2008) (ANNEX, inputs are X,Y and radar image). The main goal is to study the effect of the drift in different models. Mapping of the differences between the drift and the prediction, which highlights radar and rain gauges biases, is a interesting way to do the analysis.

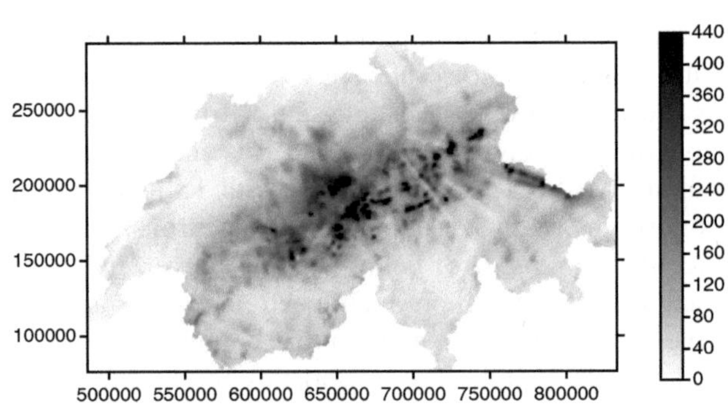

Fig. 2 One kilometer grid radar image of 6 days of precipitation in Switzerland (mm of water)

Table 2 Summary statistics from 18th to 23rd of August 2005, mm of water

	Mean	Variance	Skewness	Kurtosis	Min	Max
Rain gauge	98.99	3,931	1.08	0.57	14.3	324.3
Radar	107.73	4,581	1.36	1.81	18.5	524.5

4 Results and Discussions

For the applications presented, model assessment is carried out using a randomly selected validation set. Model selection is performed by cross-validation techniques or by using an early stopping criterion, which helped to avoid a typical split into three datasets (training, testing and validation), which would have led to a loss of predictive power because of the lack of data.

4.1 Precipitation of 2nd and 3rd October 2006

As stated in the previous section, the first case study aims to compare different optimization algorithms for modeling an extreme precipitation event using MLP. The analysis was carried out with a two-inputs MLP only taking into account geographic coordinates (X,Y). Using the LM algorithm, the MLP can easily model extreme values with the same number of hidden neurons as CG without overfitting data as shown in Fig. 3 (MLP predictions cover whole of Switzerland but for the visualization purposes only a selected interesting area is shown).

Table 3 shows RMSE calculated over the validation dataset. LM training error is equal to noise level in the data from the early stopping criterion. Ordinary kriging provides more accurate results than MLP-CG but less accurate results than MLP-LM. The large validation errors of the neural net trained with CG are simply the consequence of undertraining. In fact, according to Figs. 3 and 4, the CG algorithm is more adapted to model global trends and it leaves some spatial structure. On the other hand, the LM algorithm is able to model the spatial structure without overfitting data (dark tones in Fig. 3). In fact, the map of LM presents higher values than the one with CG in the zone of heavy precipitation causing a decrease of the validation error. However, if the number of hidden units is strongly increased, for example 40 hidden neurons organized in two hidden layers, CG can find more or

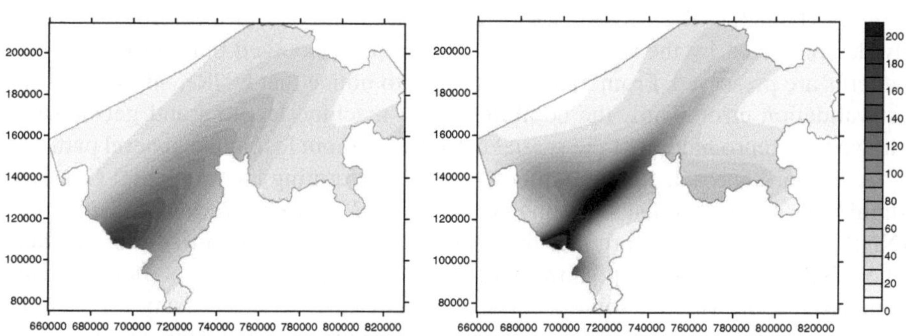

Fig. 3 2-15-1 MLP output trained with CG (on the *left*) and 2-15-1 MLP output trained with LM (on the *right*), mm of water

Table 3 Training and validation RMSE of the different methods

Method	OK	MLP-CG	MLP-LM	NNRK
Training RMSE	0	10.54	6.10	0
Validation RMSE	8.12	11.76	7.23	5.62

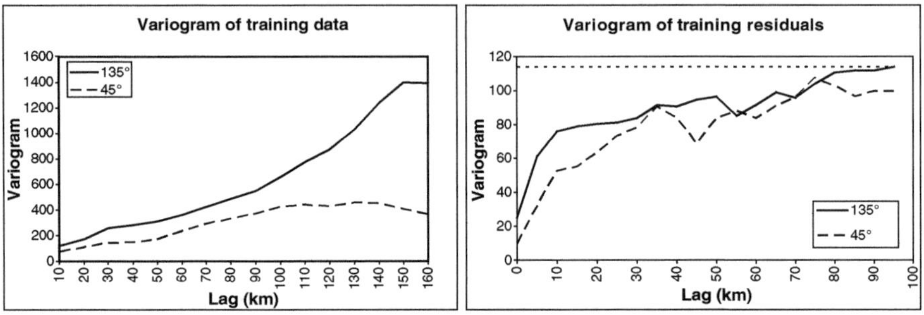

Fig. 4 Variogram for training data (on the *left*) and for training residuals (on the *right*) of MLP-CG

less the same solution as the LM algorithm but it needs more computational time. For the task at hand, LM is preferable because it needs less complex models (less hidden units) and less computational time compared to CG.

The MLP solutions could be improved by the application of a hybrid model, taking advantage of the best characteristics of both the geostatistical methods and the machine learning algorithms (Kanevski, 2004). MLP is well adapted to handle the problem of modeling non-linear trends. A hybrid model can be applied: in a first phase, a non-linear trend is modeled using an MLP architecture which underfits the data (like the one trained with CG). The variogram of the training residuals calculated from the trend shows a short scale range and a stabilization at the a-priori variance of the residuals. Raw and detrended variograms are shown in Fig. 4.

These conditions are optimal for applying kriging to interpolate training residuals. Finally, the map of the residuals is added to the map of the trend. This methodology was called by Kanevski et al. (1996). *Neural Network Residual Kriging* (NNRK). The importance of using this approach in meteorology is easily demonstrated by the fact that extremes are not smoothed out and the original patterns are preserved. From Table 3 it is easy to notice that NNRK shows the lowest validation error. Thus, the combination of machine learning and geostatistics allowed the reproduction of the spatial extremes without losing the general patterns.

A similar approach can be implemented by performing the sequential Gaussian simulation of the residuals leading to the *Neural Network Residual Simulations* (NNRS). The aim is to reproduce the histogram and the variogram of training data. The NNRS simulation map shows more spatial variability and it is more realistic compared to the smooth solution given by NNRK. Computing many realizations permits calculation of probabilities of exceeding some predefined thresholds of precipitation, which is important, for example, for risk assessments. The NNRK mean value and one realization of NNRS are shown in Fig. 5.

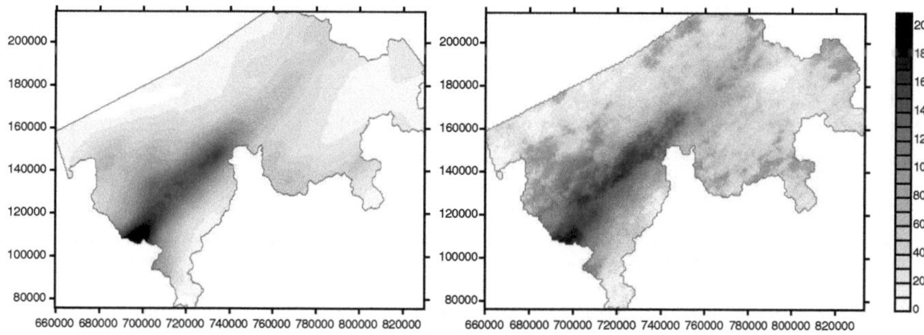

Fig. 5 NNRK mean value (on the *left*) and one realization of NNRS (on the *right*)

4.2 Precipitation of 18th to 23rd of August 2005

Let us compare artificial neural network with an external drift and kriging with an external drift, where the drift is given by the radar. The aim was not to validate the two models and to choose the best one but to analyze the contribution of the radar image to the prediction. The latter depends on the choice of the interpolation algorithm. Since a radar image is more likely to contain uncertainties, the rain gauge measurements are used as the target function. KED parameters were optimized using onefold cross-validation. For MLP it was more difficult to apply the methodology shown in Section 4.1 because it was harder to estimate the noise in data. Therefore, a different early stopping criterion was chosen. For the task, a testing subset was selected from the training base. At the end of the training process, the network weight configuration which minimized the testing error was used for predictions. Validation RMSE values for both methods are not very different, i.e. 21.61 mm for KED and 25.28 mm for MLP.

Outputs of these models are shown in Fig. 6. However, if the drift is substracted from the final prediction, several discrepancies can be seen as shown in Fig. 7. The resulting map is called the *map of the differences*. Differences close to 0 mean that radar contributed the most information to the prediction. Positive or negative differences highlight radar biases or rain gauge measurement errors that, unfortunately, exist. The variance of the differences is equal to 432.2 for KED and 242.4 for MLP. The higher variance of KED differences is explained by the fact that the predicted function must pass through the training points creating hot spots from the drift. On the contrary, MLP yields smooth solutions reducing the variance of differences.

Another analysis deals with the comparison of summary statistics between model outputs and training data. Table 4 shows summary statistics of MLP and KED predictions. A comparison between Table 4 and Table 2 illustrates that MLP results are more smooth than KED predictions (smaller skewness and maximum value mainly). Overall, MLP statistics are closer to those for the rain gauges and KED statistics are closer to those for the radar data. However, if data preprocessing is applied to raw data, for instance a logarithmic transformation, we can significantly improve the results of MLP (see Table 4).

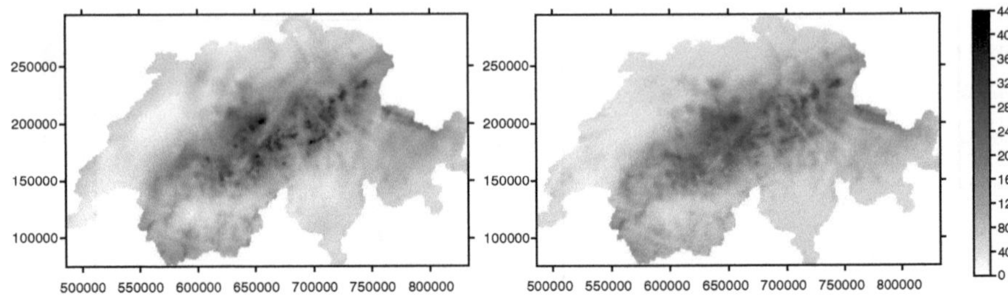

Fig. 6 KED (on the *left*) and MLP (on the *right*) prediction maps; MLP smooths out radar extremes

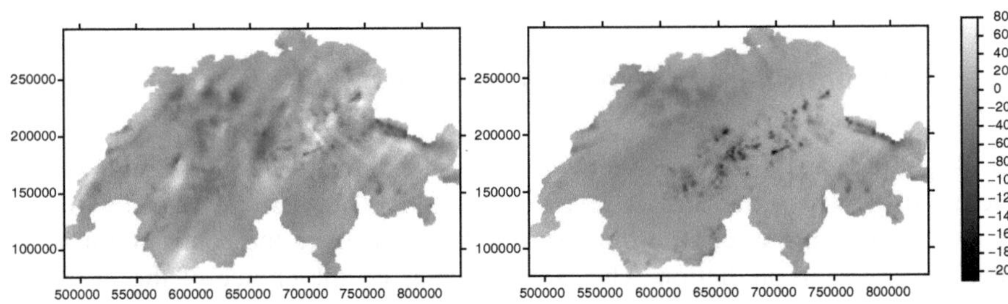

Fig. 7 KED differences (on the *left*) and MLP differences (on the *right*) from the drift; the models overestimate radar measurements in white regions and underestimate them in black regions

Table 4 Summary statistics of KED and MLP outputs, mm

	Mean	Variance	Skewness	Min	Max
KED	104.4	4,171	1.10	13.0	462.0
MLP	103.6	3,731	1.06	25.5	415.6
MLP-LOG	102.2	4,028	1.27	18.4	483.9

Finally, MLP finds the smooth functional relationship between the target (rain gauges) and its inputs (X, Y, radar). KED is an exact interpolator which passes through training samples and which is close to the drift far from the measurements (where kriging variance is high). If rain gauges or radar data contained outliers and measurement errors, MLP would be preferable because of its robustness. The analysis is helpful for guiding meteorologists in the choice of the interpolation algorithm.

5 Conclusions

Machine learning methods and geostatistics have been successfully applied for modelling precipitation fields. The choice of the interpolation algorithm depends on the goal of the study: modeling the patterns or the extreme values. Modeling extremes with a two-inputs neural network, Levenberg-Marquardt algorithm is preferable to conjugate gradient which performs better for modeling the patterns. Hybrid models gave also practical solutions to the reproduction of extremes without the risk of losing patterns. Regarding the introduction of drifts in the models, it was shown that the quality of data, the summary statistics and the analysis of the drift contribution are important for making a choice of model. Future perspectives of the research will consider the modeling of multiscale meteorological phenomena that present global trends and short scale variability due to the influence of topography such as orographic precipitation and temperature inversions. The question of data pre-processing and of the choice of the relevant predictors (geo-features computed from a DEM) is also a future direction for research.

Acknowledgement The study was partially supported by the Swiss National Science Foundation projects *GeoKernels* (project N° 200021-113944) and *Clusterville* (project N° 100012-113506).

References

Antonic O, Krizan J, Marki A, Bukovec D (2001) Spatio-temporal interpolation of climatic variables over large region of complex terrain using neural networks. Ecol Model, 138:255–263

Attore F, Alfo M, Sanctis M, Francesconi F, Bruno F (2007) Comparison of interpolation methods for mapping climatic and bioclimatic variables at regional scale. Int J Climatol 27:1825–1843

Bishop CM (1995) Neural networks for pattern recognition. Oxford University Press, Oxford

Bryan BA, Adams JM (2002) Three-dimensional neurointerpolation of annual mean precipitation and temperature surfaces for China. Geogr Anal 34(2):94–111

Daly C, Taylor GH, Gibson WP (1997) The PRISM approach to mapping precipitation and temperature. 10*th* Conference on applied climatology, Reno, NV, Am Meteor Soc 10–12

Demyanov V, Kanevski M, Chernov S, Savelieva E, Timonin V (1998) Neural network residual kriging application for climatic data. J Geogr Inf Decis Anal 2(2):234–252

Deutsch CV, Journel AG (1997) GSLIB: geostatistical software library and user's guide. Oxford University Press, New York

Dobesch H, Dumolard P, Dyras (eds) (2007) Spatial Interpolation for climate data: the use of GIS in climatology and meteorology. Geographical Information Systems series (ISTE)

Dubois G, Malczewski J, De Cort M (eds) (2003) Mapping radioactivity in the environment. Spatial Interpolation Comparison 97, European Commission, JRC. ISPRA

Germann U, Galli G, Boscacci M, Bolliger M (2006) Radar precipitation measurement in a mountainous region. Q. J. R. Meteorol Soc 132(618):1669–1692

Goovaerts P (2000) Geostatistical approaches for incorporating elevation into the spatial interpolation of rainfall. J Hydrol 228:113–129

Haykin S (1998) Neural networks: a comprehensive foundation. Prentice Hall, Upper Saddle River, NJ

Hengl T, Geuvelink GBM, Stein A (2003) Comparison of kriging with external drift and regression-kriging. Technical note, ITC

Isaaks EH, Srivastava RM (1989) An introduction to applied geostatistics. Oxford University Press, New York

Jeffrey SJ, Carter JO, Moodie KB, Beswick AR (2001) Using spatial interpolation to construct a comprehensive archive of Australian climate data. Environ Model Soft 16:309–330

Joss J, Schädler B, Galli G, Cavalli R, Boscacci M, Held E, Della Bruna G, Kappenberger G, Nespor V, Spiess R (1997) Operational use of radar for precipitation measurements in Switzerland. VDF Hochschulverlag AG, ETH Zürich

Kanevski M (2004) Analysis and modelling of spatial environmental data. EPFL Press, Lausanne

Kanevski M, Arutyunyan R, Bolshov L, Demyanov V, Maignan M (1996) Artificial neural networks and spatial estimations of Chernobyl fallout. Geoinformatics: 7(1–2):5–11
LeCun Y, Bottou L, Orr GB, Müller KR (1998) Efficient BackProp. In: Orr G, Müller K (eds) Neural networks: tricks of the trade. Springer-Verlag, Berlin, Heidelberg
Lloyd CD (2007) Local models for spatial analysis. CRC Press, Boca Raton, FL
Parkin R, Kanevski M (2008) ANNEX model: artificial neural networks with external drift environmental data mapping. In: Pilz J (ed) Interfacing geostatistics and GIS. Springer-Verlag, Berlin, Heidelberg
Roweis S (2000) Levenberg-Marquardt optimization. Lecture notes
Wackernagel H (2003) Multivariate geostatistics. Springer, Berlin

On Geostatistical Analysis of Rainfall Using Data from Boundary Sites

José Manuel Mirás Avalos, Patricia Sande Fouz, and Eva Vidal Vázquez

Abstract This study examines the effect of considering data from rain gauges nearby the boundaries of Galicia (NW Spain) in order to minimize the border effect. Two datasets were considered: the first one comprised 232 climatic stations within Galicia and the second one consisted of 322 rain gauges including the former 232 from Galicia and adding 90 stations from boundary provinces (42 from Asturias, 31 from León and 17 from Zamora). Total monthly rainfall data from 2006 was analyzed and descriptive statistics demonstrated slight differences between both datasets. Theoretical structures were described for all the studied monthly datasets. Spatial dependence analysis showed that the best-fitting semivariogram model structure was the same for both datasets in most of the cases, even though the model parameters showed great differences. Similarly, cross-validation parameter values were clearly distinct among datasets; mostly, the ones corresponding to the 322 stations dataset were closer to the ideal values. Ordinary kriging was performed for both datasets and resulting variance maps showed improvements when the information from boundary regions was taken into account. These improvements can reach up to 25% of the maximum variance value and they were observed in wet months such as January whereas, in dry months such as July, no improvement was observed. Minimum error values were usually lower when extra information was used in the interpolations. In conclusion, a better mapping of the rainfall within a region can be achieved using data from boundary areas, reducing the variance of the estimates.

1 Introduction

Rainfall is a spatio-temporal intermittent phenomenon displaying large spatial and temporal variability whereas rain gauge networks only collect point estimates. In addition, the characterization of rainfall spatial variability is of great interest to

J.M.M. Avalos (✉), P.S. Fouz, and E.V. Vázquez
Facultad de Ciencias, Universidade da Coruña, UDC, Campus A Zapateira C.P. 15071, A Coruña, Spain
e-mail: jmirasa@udc.es; psande@udc.es; evidal@udc.es

P.M. Atkinson and C.D. Lloyd (eds.), *geoENV VII – Geostatistics for Environmental Applications*, Quantitative Geology and Geostatistics 16,
DOI 10.1007/978-90-481-2322-3_5,

water resources planners, regulators, and decision makers (Ali et al., 2000). It is thus necessary to estimate point rainfall at unrecorded locations from values at surrounding sites (Goovaerts, 2000).

This interpolation problem has been largely studied and several authors proved the convenience of performing geostatistical analyses in order to map rainfall at different geographical locations (Abtew et al., 1993; Goovaerts, 2000; Militino et al., 2001; Watkins et al., 2005; Mirás Avalos et al., 2007). However, the use of a complex technique is no warranty of a better performance in a given region (Gómez Hernández et al., 2001).

Enhancing these approaches using secondary information, i.e. altitude from digital elevation models, was assessed by a number of authors (Goovaerts, 2000; Gómez Hernández et al., 2001; Mirás Avalos and Paz González, 2008). An important fact to consider in this kind of multivariate interpolation is the correlation between rainfall and secondary variables such as altitude since the introduction of a secondary attribute in estimation seems worthy only for correlations above 0.4 (Asli and Marcotte, 1995). Furthermore, the use of rainfall data registered at sites neighbouring the studied region in order to improve the estimations has not been properly studied.

Following these lines of research, this work focuses on the study of rainfall variability in Galicia (NW Spain). The main objective of this exercise was to compare the estimations obtained from rainfall datasets within this region and those obtained from an increased dataset with registers from gauges located out of the studied area but at neighbouring sites.

2 Material and Methods

The data sets used in this exercise corresponded to total monthly rainfall (in mm) during the period from January till December 2006 and are referred (i) to 232 rain gauges which are irregularly distributed in Galicia and (ii) to 322 climatic stations including the former 232 from Galicia and adding 90 gauges from neighbouring provinces (42 from Asturias, 31 from León and 17 from Zamora). Figure 1 shows the geographical location of the studied region and the neighbouring provinces within Spain.

The data sets were characterized statistically; this description included the calculation of mean, median, mode, minimum, maximum, coefficient of variation, standard deviation, skewness and kurtosis. Data number varied from 159 to 190 measurements for the first dataset and from 243 to 275 registered values for the second one, depending on the month.

Once the data sets were statistically characterized, the absence of outliers was verified. Then, stationarity was analyzed and, when observed, any drift was filtered. Moreover, an analysis of correlation between rainfall and altitude was carried out in order to incorporate this information as an ancillary variable in the interpolations.

In order to conduct spatial interpolation using geostatistical techniques, a comprehensive analysis of the spatial structure of the data sets was performed using

Fig. 1 Location of the studied region and neighbouring sites on Spain

GSTAT software (Pebesma, 2000) which is integrated in a GIS called PCRaster (Van Deursen and Wesseling, 1992).

The basis for interpolations by both geostatistical and deterministic procedures was a digital elevation model of Galicia (Thonon and Paz González, 2004) which consists of regular cells of 500 by 500 m covering an area of 29, 750 km^2.

Spatial variability was primarily evaluated through semivariogram estimation, graphing, model fitting and comparison for each variable (Burgess and Webster, 1980). Classic criteria for calculating semivariograms were taken into account (Samper Calvete and Carrera Ramírez, 1996; Goovaerts, 1997). The dependence relation has been computed according to Cambardella et al. (1994). The cross-validation technique (Chilés and Delfiner, 1999) was used to check the model performance. Non-dimensional Mean Square Error (NMSE) parameter was the main criterion for deciding which fitted model was the best one for each monthly data set. Other parameters, such as determination coefficient (r^2) and Mean Square Error (MSE) were also taken into account.

Inverse distance weighting (IDW) method was used as a reference for mapping monthly rainfall data: rainfall is estimated as a linear combination of several surrounding observations, with the weights being inversely proportional to the square distance between observations and the point to be estimated (Burrough and McDonnell, 1998). IDW was used in order to get a rapid mapping of rainfall in this region since previous studies (Mirás Avalos et al., 2007; Mirás Avalos and

Paz González, 2008) showed that spatial structures cannot always be described for monthly rainfall in Galicia. In fact, these authors reported a spatial dependence in 69% of their datasets which corresponded to several years. However, this technique has the important lack of not providing a measure of estimation errors, thus, kriging interpolations are preferred.

Interpolation by ordinary kriging (OK) was the geostatistical method applied to these rainfall data. OK is by far the most common type of kriging in practice (Webster and Oliver, 2001). Kriging interpolation methods provide each cell with a local, optimal prediction and an estimation of the error that depends on the variogram and on the spatial configuration of the data (Burrough and McDonnell, 1998). Kriging is a generalized technique that allows one to account for the spatial dependence between observations, as revealed by the semivariogram, into spatial predictions (Goovaerts, 2000). The OK weights are determined such as to minimize the estimation variance (Goovaerts, 2000). Comprehensive theoretical review and mathematical formulation of kriging are beyond the scope of this work and can be found in Goovaerts (1997) and Chilés and Delfiner (1999). In case that data series showed any trend, an interpolation by simple kriging instead of OK was carried out.

To test the goodness of fit of the estimations, the mean squared-error (MSEP) and root mean squared-error (RMSE, the square root of MSEP) were calculated according to Stacey et al. (2006):

$$MSEP = \frac{1}{n}\sum_{i=1}^{n}\left[z\left(x_i\right) - \hat{z}\left(x_i\right)\right]^2$$

Where n is the number of data, $z(x_i)$ is the measured value and $\hat{z}(x_i)$ is the estimated value. The smaller the values of these statistics, the closer the estimation is to the measurement.

3 Results and Discussion

Mean monthly rainfall during the study period varied from 16.71 mm in July to 326.54 mm in October for the Galicia dataset (Table 1). These values were different from those observed on the dataset containing information from boundary sites, July being the driest month with an average of 24 mm and October the wettest with 292.8 mm (Table 2).

Monthly rainfall coefficients of variation, which ranged from 0.34 to 0.8, showed the spatial heterogeneity of the precipitation in both datasets; this variability was higher in the dry season, from June to August (Tables 1 and 2).

Skewness and kurtosis coefficients showed values which were close to those of a standard Gaussian distribution. From this test, it was assumed that monthly rainfall data followed a Gaussian distribution (Tables 1 and 2). However, mean values were usually higher than the median values which indicates a slight deviation from the standard Gaussian distribution. However, no data transformation was performed before the structural analysis.

Table 1 Statistical summary of the 2006 monthly rainfall for the Galicia dataset (SD = standard deviation; CV = coefficient of variation; Min. = minimum; Max. = maximum; Skew. = skewness coefficient, Kurt. = kurtosis coefficient)

Month	N	Mean (mm)	Median (mm)	SD (mm)	CV	Min. (mm)	Max. (mm)	Mode (mm)	Skew.	Kurt.
January	173	56.54	56.30	25.42	0.45	10.4	160.0	83.2	0.71	0.84
February	176	139.49	135.85	53.00	0.38	19.3	281.9	216.0	0.36	−0.10
March	170	223.01	198.95	108.63	0.49	55.4	485.2	216.5	0.54	−0.62
April	174	82.48	79.05	27.68	0.34	19.5	240.3	70.0	1.59	6.30
May	168	31.29	28.10	16.93	0.54	1.8	105.7	17.8	1.10	2.02
June	177	28.73	25.90	19.15	0.67	1.3	94.4	9.0	1.11	1.10
July	165	16.71	12.90	13.35	0.80	1.1	61.0	8.0	1.40	1.62
August	174	44.26	40.70	23.64	0.53	2.0	124.4	24.0	0.92	0.82
September	190	92.46	83.80	40.94	0.44	0.0	225.9	81.0	0.44	−0.18
October	159	326.54	323.00	111.24	0.34	87.0	601.7	320.8	0.07	−0.48
November	174	254.22	233.00	108.76	0.43	11.8	569.8	138.6	0.44	−0.41
December	182	193.48	188.85	87.31	0.45	24.0	432.0	134.0	0.46	−0.19

Table 2 Statistical summary of the 2006 monthly rainfall for the dataset containing information from boundary sites (SD = standard deviation; CV = coefficient of variation; Min. = minimum; Max. = maximum; Skew. = skewness coefficient, Kurt. = kurtosis coefficient)

Month	N	Mean (mm)	Median (mm)	SD (mm)	CV	Min. (mm)	Max. (mm)	Mode (mm)	Skew.	Kurt.
January	257	52.99	51.00	26.32	0.50	3.0	160.0	83.2	0.68	0.55
February	259	136.96	134.80	59.54	0.43	19.3	288.3	65.0	0.33	−0.49
March	255	189.78	155.70	109.65	0.58	18.0	569.4	216.5	0.91	0.16
April	261	75.55	73.00	27.61	0.36	0.4	240.3	83.0	1.41	5.38
May	256	33.40	31.30	17.38	0.52	0.4	105.7	36.1	0.77	0.92
June	265	34.02	31.20	20.80	0.61	0.1	102.4	36.8	0.84	0.54
July	253	24.00	17.20	19.15	0.80	0.0	105.3	14.0	1.12	0.98
August	259	39.67	34.00	22.68	0.57	0.0	137.5	24.0	1.31	2.31
September	275	85.60	77.40	39.34	0.46	0.0	225.9	81.0	0.71	0.23
October	243	292.80	284.00	118.86	0.41	34.9	601.7	320.8	0.34	−0.60
November	260	222.59	201.20	110.05	0.49	11.8	569.8	258.0	0.71	−0.19
December	270	167.22	152.50	90.60	0.54	10.4	546.2	134.0	0.82	0.82

Correlation between rainfall and elevation, as assessed by linear regression analysis, ranged from 0.04 to 0.47, weak for most of the months (data not shown). According to other authors (Asli and Marcotte, 1995; Goovaerts, 2000), the benefit of multivariate techniques can become marginal if correlation between rainfall and elevation (or other environmental descriptors) is too small, as in this study. Thus, interpolations by kriging with external drift or by cokriging were withdrawn.

Inverse distance weighting results proved that this method was suitable for a rapid estimation of rainfall at the studied level. Output maps showed, in general,

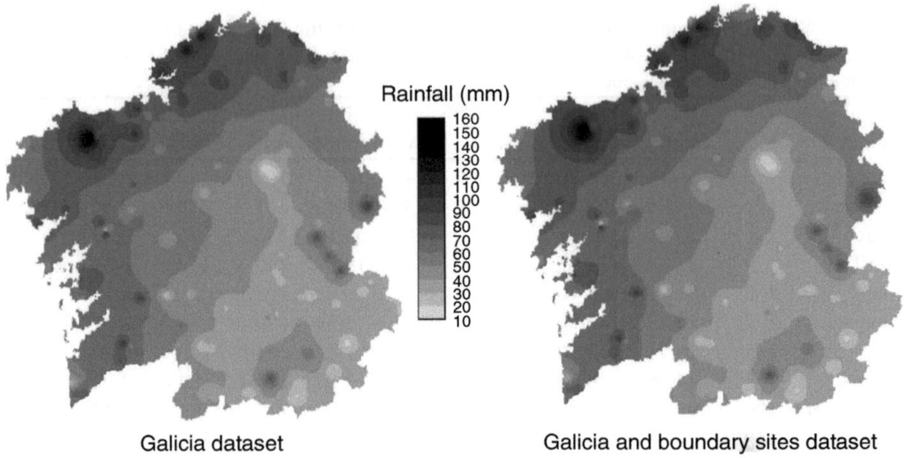

Fig. 2 Resulting maps from inverse distance weighting interpolation for January 2006

a discontinuous appearance. In general, resulting maps were almost identical for both datasets. Generated maps corresponding to January 2006 are shown in Fig. 2, exemplifying this situation.

Different theoretical functions (spherical, exponential and Gaussian) with variable nugget effects depending on the month were fitted to experimental semivariograms.

From an analysis of the semivariograms and their fitted parameters, regarding both datasets, we observe that monthly rainfall showed a variable nugget effect ranging from 0% to 87.3% of the sill value, which is a dependence ratio (Cambardella et al., 1994), and a rather short range of spatial dependence varying from 10.68 to 60.61 km (Tables 3 and 4). Generally, range values increased when boundary information was taken into account; however, the opposite fact was observed in January, June and August (Tables 3 and 4).

Specifically, nugget effects for the Galicia dataset ranged from 0% to 70.6% of the sill value and ranges varied from 11.8 to 55.9 km (Table 3). According to the criteria outlined by Cambardella et al. (1994), 8 months showed strong spatial correlation and the rest presented moderate correlation.

In the case of the Galicia and boundary sites dataset, nugget effects ranged from 0% to 87.3% of the sill value, showing a higher variability than those fitted to data only from Galicia. Nevertheless, range values for this dataset varied from 10.68 to 60.61 km, larger than those observed for the Galicia dataset (Table 4). Regarding the values of the dependence relation (DR), 7 months presented strong spatial correlation, 4 months showed moderate spatial correlation and 1 month showed weak correlation.

In addition, theoretical structures were the same for both datasets in most of the occasions. The exceptions were September, November and December; in the case of September, an exponential model was fitted when the Galicia dataset was

Table 3 Theoretical model parameters fitted to experimental variograms from the Galicia dataset (C_0 = Nugget effect; DR = dependence relation, MSE = mean square error, NMSE = non-dimensional mean square error)

Month	Trend	Model	C_0	Sill	Range (km)	DR	Cross-validation r^2	MSE	NMSE
January	Quadratic	Spherical	110.6	248.1	55.85	30.8	0.29	0.009	1.21
February	Linear	Exponential	200.0	2,200.0	20.00	8.3	0.23	−0.001	1.22
March	Linear	Exponential	0.0	8,919.7	11.82	0.0	0.69	−0.003	1.12
April	–	Exponential	600.0	250.0	20.00	70.6	0.36	0.000	0.98
May	–	Exponential	50.0	240.0	20.00	17.2	0.57	−0.017	1.12
June	–	Spherical	172.7	165.9	48.56	51.0	0.52	−0.001	1.04
July	–	Exponential	50.0	90.0	20.00	35.7	0.70	0.014	1.01
August	–	Exponential	100.0	600.0	20.00	14.3	0.57	0.006	1.10
September	–	Exponential	300.0	900.0	20.00	25.0	0.70	0.009	1.15
October	–	Exponential	2,000.0	8,000.0	30.00	20.0	0.75	0.002	1.10
November	–	Spherical	1,098.3	7,468.6	47.82	12.8	0.80	0.047	1.16
December	–	Spherical	423.0	6,521.4	32.57	6.1	0.66	0.014	1.31

Table 4 Theoretical model parameters fitted to experimental variograms from the Galicia and boundary sites dataset (C_0 = Nugget effect; DR = dependence relation, MSE = mean square error, NMSE = non-dimensional mean square error)

Month	Trend	Model	C_0	Sill	Range (km)	DR	Cross-validation r^2	MSE	NMSE
January	Quadratic	Spherical	160.3	190.1	51.65	45.8	−0.02	0.014	1.10
February	Linear	Exponential	200.0	2,200.0	20.00	8.3	0.31	−0.004	1.16
March	–	Exponential	1,000.0	11,000.0	40.00	8.3	0.89	−0.007	0.90
April	–	Exponential	502.8	295.4	46.36	63.0	0.53	0.000	0.95
May	–	Exponential	459.2	255.1	18.45	64.3	0.63	−0.007	1.03
June	–	Exponential	100.0	300.0	30.00	25.0	0.59	−0.004	1.19
July	Linear	Exponential	969.8	141.6	38.43	87.3	0.41	−0.003	0.99
August	–	Exponential	0.0	514.9	10.68	0.0	0.76	0.015	0.83
September	–	Gaussian	415.0	650.0	20.00	39.0	0.71	0.001	1.10
October	–	Exponential	2,761.2	12,809.1	60.61	17.7	0.78	0.000	1.01
November	–	Exponential	1,000.0	11,000.0	50.00	8.3	0.85	−0.001	1.08
December	–	Exponential	1,000.0	8,000.0	60.00	11.1	0.77	0.002	1.18

taken into account but a Gaussian model was fitted when the boundary sites were accounted for. The variograms for November and December showed a spherical behaviour when the Galicia dataset was analysed and an exponential structure when neighbouring information was used (Tables 3 and 4). Usually, a theoretical model representing a more spatially continuous behaviour was fitted to the data set with information from boundary sites.

Selected cross-validation parameters (r^2, MSE and NMSE) of the fitted semi-variograms are shown in Tables 3 and 4 as well. In general, values for the NMSE parameter were close to the considered ideal value of 1 for both datasets.

In addition, an increase in the accuracy of the estimations was observed when boundary information was used, according to the magnitudes of the cross-validation parameters (Tables 3 and 4), the exception was January.

Maps obtained by ordinary kriging were too smooth for reproducing measured maximum and minimum data; moreover, error maps tended to show a high and uniform uncertainty pattern. Kriging errors were high in most of the study area except in those areas located nearby the rain gauges. An example for November 2006 is depicted on Fig. 3. No differences were observed between maps obtained by OK and those resulting from simple kriging for both datasets which presented any trend (data not shown).

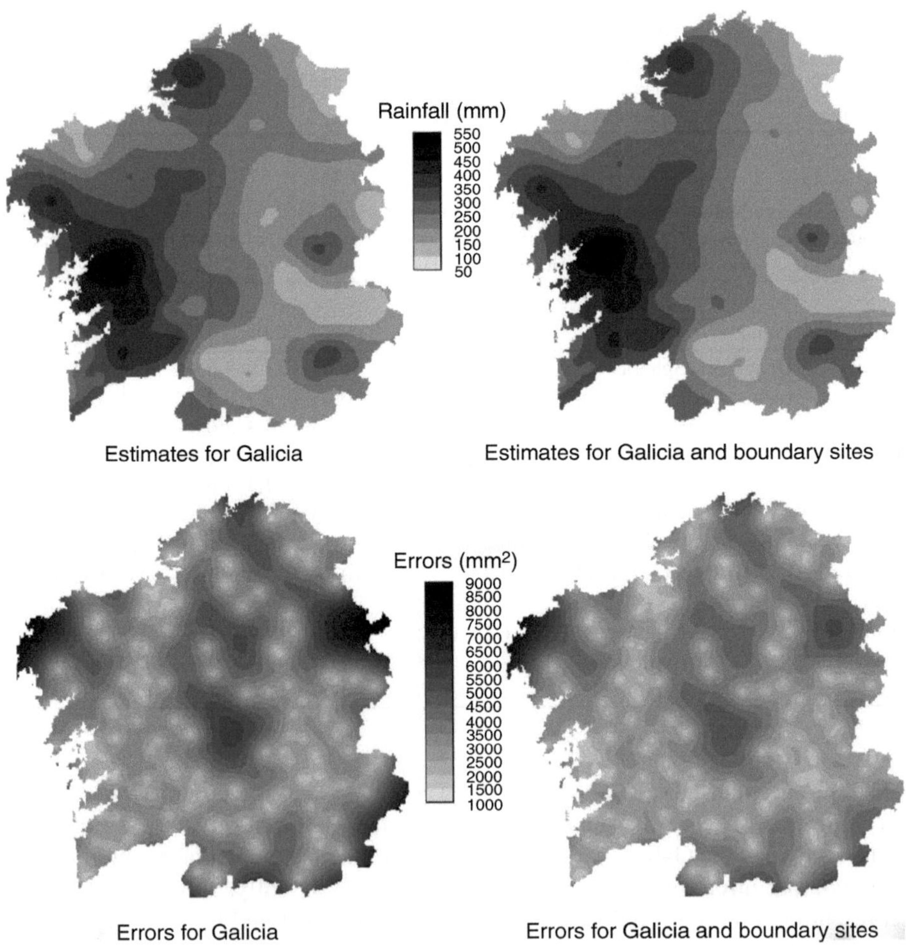

Fig. 3 Example of estimate and kriging error maps generated by ordinary kriging for November 2006

Table 5 Goodness of fit results for the interpolations performed (MSEP = mean squared-error of prediction, RMSE = root mean squared-error)

	IDW Galicia dataset	OK Galicia dataset	IDW Galicia + boundary	OK Galicia + boundary
MSEP	4.72	4.51	11.87	4.53
RMSE	2.17	2.12	3.45	2.13

Differences in performance for both datasets were minimal regarding map estimations but an increase in confidence was observed in those areas located at the borders, as demonstrated by the kriging error maps (Fig. 3). Moreover, the range of rainfall values was lower in the case that boundary information had been taken into account than when the Galicia dataset had been solely considered.

In order to quantify these differences in performances, MSEP and RMSE were calculated for both, inverse distance weighting (IDW) and ordinary kriging (OK) approaches (Table 5).

Results from this analysis showed values of RMSE of 2.17 and 2.12 for inverse distance weighting and OK, respectively when only the Galicia dataset was taken into account. In the case of the increased dataset, those values were 3.45 for inverse distance weighting and 2.13 for OK. Therefore, no important differences were found between the two OK approaches but a slight difference was observed in the case of inverse distance weighting likely due to the mathematical theory behind this method (Table 5).

MSEP values were similar between the estimation approaches for the Galicia dataset; however, a big difference in the values for these statistics was found in the case of the increased dataset (Table 5). Comparing the OK for both datasets, MSEP values were practically identical.

4 Conclusions

Structural analysis of the studied datasets showed slight differences in their parameters when they were fitted to registers from Galicia or to those increased with information from boundary sites. Usually, models fitted to semivariograms from the increased dataset represented a more spatially continuous behaviour of the phenomenon.

Inverse distance weighting maps showed a high degree of similarity for both datasets. Ordinary kriging outputs were slightly improved when neighbouring sites were taken into account as proved by the estimation error maps. However, the quantitative performance of this technique was similar for both datasets.

A better mapping of monthly rainfall in Galicia may be achieved by using data registered at boundary locations, reducing the variance of the estimations and the border effect.

Acknowledgements The authors thank the Centro de Investigaciones Forestales de Lourizán (Spain), the Centro Meteorológico Territorial de Galicia of the Spanish Ministry of Environment and Energy and Meteogalicia of the Galician Government for providing them with data for this study. Two anonymous reviewers are kindly thanked for their suggestions.

References

Abtew W, Obeysekera J, Shih G (1993) Spatial analysis for monthly rainfall in South Florida. J Am Water Resour Assoc 29(2):179–188

Ali A, Abtew W, Van Horn S, Khanal N (2000) Temporal and spatial characterization of rainfall over Central and South Florida. J Am Water Resour Assoc 36(4):833–848

Asli M, Marcotte D (1995) Comparison of approaches to spatial estimation in a bivariate context. Math Geol 27(5):641–658

Burgess I, Webster R (1980) Optimal interpolation and isarithmic mapping of soil properties. I: the semivariogram and punctual kriging. J Soil Sci 31:315–331

Burrough PA, McDonnell RA (1998) Principles of geographical information systems, spatial information systems and geostatistics. Oxford University Press, New York

Cambardella CA, Moorman TB, Novak JM, Parkin TB, Karlen DL, Turco RF, Konopka AE (1994) Field-scale variability of soil properties in Central Iowa soil. Soil Sci Soc Am J 58:1501–1508

Chilés JP, Delfiner P (1999) Geostatistics. Modeling spatial uncertainty. Wiley series in probability and statistics. Wiley, New York, p 695

Gómez Hernández JJ, Cassiraga EF, Guardiola Albert C, Álvarez Rodríguez J (2001) Incorporating information from a digital elevation model for improving the areal estimation of rainfall. In: Monestiez P, Allard D, Kluwer RF (eds) geoENV III –geostatistics for environmental applications. Quantitative geology and geostatistics. Kluwer, Dordrecht, pp 67–78

Goovaerts P (1997) Geostatistics for natural resources evaluation. Applied geostatistics series. Oxford University Press, New York, p 483

Goovaerts P (2000) Geostatistical approaches for incorporating elevation into the spatial interpolation of rainfall. J Hydrol 228:113–129

Militino AF, Palacios MB, Ugarte MD (2001) Robust predictions of rainfall in Navarre, Spain. In: Monestiez P, Allard D, Kluwer RF (eds) geoENV III – geostatistics for environmental applications. Quantitative geology and geostatistics. Kluwer, Dordrecht, The Netherlands, pp 79–90

Mirás Avalos JM, Paz González A (2008) Improving the areal estimation of rainfall in Galicia (NW Spain) using digital elevation information. In: Soares A, Pereira MJ, Dimitrakopoulos R (eds) geoENV VI geostatistics for environmental applications. Quantitative geology and geostatistics. Springer, The Netherlands, pp 259–269

Mirás Avalos JM, Paz González A, Vidal Vázquez E, Sande Fouz P (2007) Mapping monthly rainfall data in Galicia (NW Spain) using inverse distances and geostatistical methods. Adv Geosci. The Netherlands, 10:51–57

Pebesma EJ (2000) GSTAT user's manual. Department of Physical Geography, Utrecht University, Utrecht, The Netherlands, p 100

Samper Calvete FJ, Carrera Ramírez J (1996) Geoestadística. Aplicaciones a la Hidrología Subterránea (2ª edición). Centro Internacional de Métodos Numéricos en Ingeniería. Barcelona, Spain, p 484

Stacey KF, Lark RM, Whitmore AP, Milne AE (2006) Using a process model and regression kriging to improve predictions of nitrous oxide emissions from soil. Geoderma 135:107–117

Thonon I, Paz González A (2004) A geostatistically interpolated digital elevation model of Galicia (NorthWest Spain). In: Sánchez Vila X, Carrera J, Gómez Hernández JJ (eds) geoENV IV – geostatistics for environmental applications. Quantitative geology and geostatistics. Kluwer, Dordrecht, The Netherlands, pp 532–533

Van Deursen WPA, Wesseling CG (1992) The PCRaster package. Vakgroep Fysische Geografie. Faculteit Ruimtelijke Wetenschappen. Universiteit Utrecht, Utrecht, The Netherlands, p 192

Watkins DW, Link GA, Johnson D (2005) Mapping regional precipitation intensity duration frequency estimates. J Am Water Resour Assoc 41(1):157–170

Webster R, Oliver MA (2001) Geostatistics for environmental scientists. Statistics in practice series. Wiley, New York

Geostatistics Applied to the City of Porto Urban Climatology

Joaquim Góis, Henrique Garcia Pereira, and Ana Rita Salgueiro

Abstract The Porto (Portugal) urban climatology was characterized by using of a set of data obtained daily during a 2 year temperature mobile monitoring campaign, performed by a measuring/recording appliance installed in a bus that maneuvered through a given path (established *a priori* in such a way that spatial variability within the city could be accounted for). In order to model such data by geostatistical techniques, a two step approach was adopted. The first step aims to obtain temperature probability density function (PDF) parameters for each sampled point in time. Using a flexible Weibull analytical model to interpolate the empirical histograms that represent the time PDF at each spatial station, two parameters (k – regarding form, and c – regarding scale) were obtained. In a second step, these parameters, viewed as regionalized variables, were used to obtain the corresponding kriged maps at any location in space. Based on these maps of the Weibull parameters, the time PDF is estimated at any unsampled point located in the nodes of a dense mesh that covers the city, allowing the calculation of the probability of exceeding certain thresholds (or the probability of maintaining temperature within a given range). The output of the methodology, which consists of temperature quantile or probability maps, was validated by expert knowledge on the particular climatology of the city, both in space and in time.

1 Introduction

Modern urban planning relies largely on the interaction between artificial systems that are to be built (or submitted to conservation procedures) in certain zones of a city and the environmental context prevailing in such zones. One of the

---

J. Góis (✉)
Engineering Faculty of Porto University, Mining Department – CIGAR,
Rua Dr. Roberto Frias. 4200-465 Porto, Portugal
e-mail: jgois@fe.up.pt

H.G. Pereira and A.R. Salgueiro
CERENA – Natural Resources and Environment
Center of IST, Lisboa, Portugal
e-mail: henrique.pereira@ist.utl.pt; rita.salgueiro@ist.utl.pt

P.M. Atkinson and C.D. Lloyd (eds.), *geoENV VII – Geostatistics for Environmental Applications*, Quantitative Geology and Geostatistics 16,
DOI 10.1007/978-90-481-2322-3_6,

relevant factors that affect quality of life in the above mentioned artificial systems is temperature. In order to develop construction schemes and devices that minimize energy waste for thermic control, a baseline model providing the temperature distribution in space and time is required. Such baseline consists of maps displaying expected values and corresponding variability of temperature across space and time, including the probability of extreme value occurrence.

To create a reliable baseline for urban planning in the Porto city in what temperature is concerned, a geostatistical methodology, based on measurements provided by a mobile monitoring apparatus that records the variable at selected stations, is presented and discussed. Such measurements are repeated in time, for 1 day slices spread over 2 years. The well known "urban heat island" phenomenon (Oke, 1982, 1987) was characterized in detail in this study, regarding spatial and temporal variability of temperature within the city, when compared with corresponding values in the surrounding rural areas.

2 State of the Art

In order to address the problem of estimating a Regionalized Variable whose structure depends jointly on two dimensions (space and time), the most common geostatistical technique is Space–Time Kriging (STK), as developed by Goovaerts (1977), and put into practice (using a particular Fortran software) by De Cesare et al. (2002).

This technique requires the extension of the usual 1D spatial autocovariance to a surface (a 2D function) that is intended to reflect, equally, spatial and temporal lags, as sketched in Fig. 1.

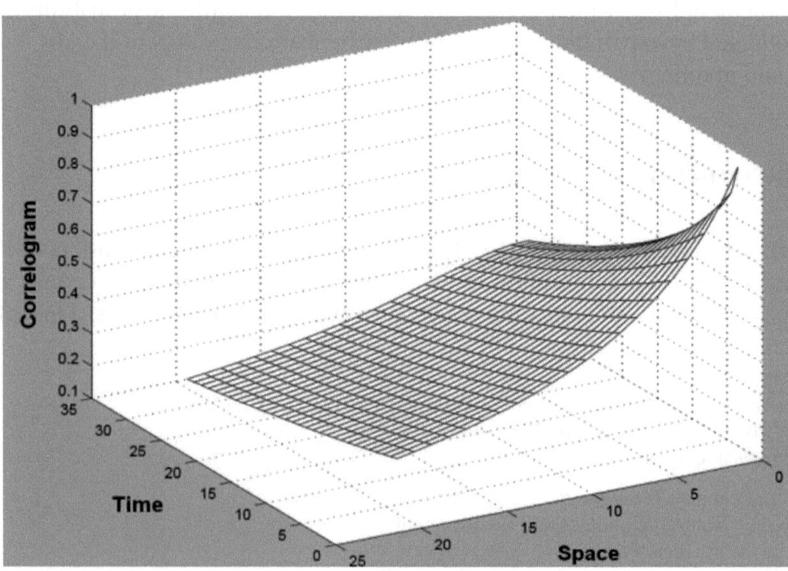

Fig. 1 Example of space time covariance surface (modified after Hjorth, 1999)

Obviously, the kriging system must be modified accordingly, taking into account the shortage of tractable covariance 2-D functions in space–time (Hjorth 1999).

Hence, STK applications are not straightforward (cf., for instance, Gething et al., 2006, referring to a health management system focused on malaria), since the two dimensions are difficult to model in what their joint development is concerned (apart from the fact that these dimensions, being essentially different in nature, are, consequently, hard to model under a common approach).

In addition, some applications lead to case studies where one of the dimensions is privileged over the other. In fact, for instance in the issue of plant disease propagation (where the spatial component prevails), the results obtained by STK are analogous to a visual comparison of successive maps deployed in time, each one of which is obtained by ordinary kriging in space (Alves et al., 2006). On the other hand, when a spatially sparse (but temporally rich) network of meteorological stations is available, the temporal component prevails, and satellite sensor imagery provides only some ancillary information on the spatial autocovariance structure (Spadavecchia and Wiliams, 2006).

Another view, which is out of the scope of 'pure' geostatistics, is given by Hoskin and Wallis (1997) for hydrological applications (stream flow vs. precipitation data, for instance). Under this view, the point is to fit a single frequency distribution in time, within a 'homogeneous' region in space. This method, based on L-moments (linear combinations of probability weighted moments of random variables), allows for quantile estimation at sites where no measurements are available. However, there are serious drawbacks in the practical treatment of urban climatology data using this kind of 'regional frequency analysis', since the criteria for defining 'homogeneous' regions in the context of a city are cumbersome, and spatial autocovariances are not accounted for.

3 Proposed Methodology

The geostatistical approach proposed here to handle space–time data referring to temperatures in Porto city relies on the review given in Kyriakidis and Journel (1999), specifically when the authors state their preference for a two step method: in the first step, time series at each point of space are modelled by 'conventional' techniques (such as ARMA or ARIMA); in the second step, parameters of such model are kriged in space, providing realizations of the underlying stochastic process, at unsampled locations.

In this paper, a new framework was developed along the above lines, detaching clearly the two dimensions of the problem (space and time). Since what is required for urban planning is not a set of time series, but a PDF generator for temperature at any point in space, the parameters of such PDFs were obtained by fitting Weibull theoretical distributions to the experimental histograms that refer to sampled locations. Then, in order to use kriging as the spatial estimation technique *par excellence*, variograms were computed for those parameters, viewed as Regionalized Variables that summarize temporal variability at each station. Finally, the PDF at any point of space was obtained by applying to the theoretical Weibull equation the kriged values of the above mentioned parameters, available all over the entire field.

4 Data Acquisition

The city of Porto (1 million inhabitants living in a 40 km^2 area) is located in the Northern region of Portugal, as displayed in Fig. 2.

A temperature monitoring plan was designed for the city, consistent with the guidelines given in Geiger et al. (1995). A path with a total length of 60 km was established, where 244 measuring stations were set up, according to the scheme shown in Fig. 3. During the period 1998/1999, the itinerary was replicated 57 times, covering the entire range of meteorological conditions that are considered by experts as significant for the region (Monteiro, 1994). Thus, a matrix of 244 lines in space by 57 columns in time is the data model for this study (Fig. 4).

The main point to be stressed in the sampling plan that gave rise to experimental data depicted symbolically in Fig. 3 is that it is representative both of spatial and temporal variability of temperature within the city. This condition was assured by expert knowledge on local climatology (Monteiro, 1994).

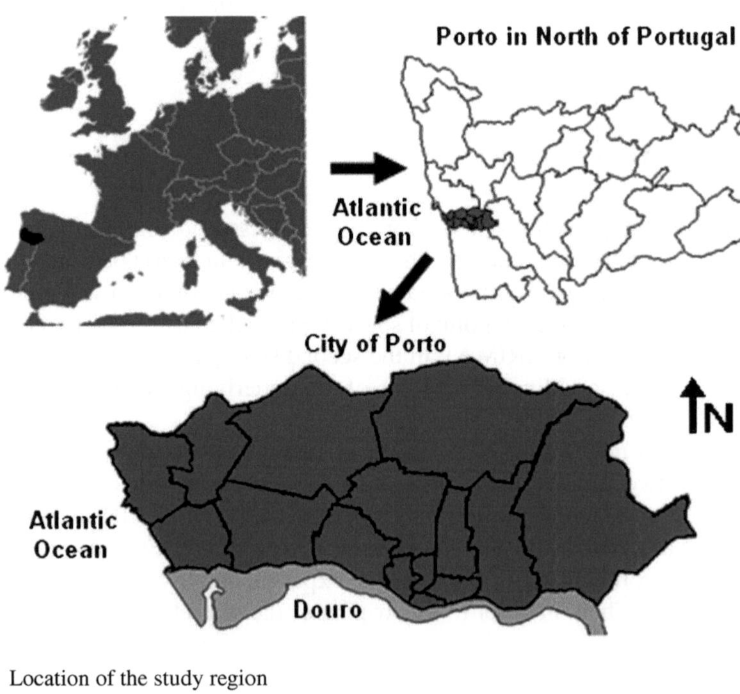

Fig. 2 Location of the study region

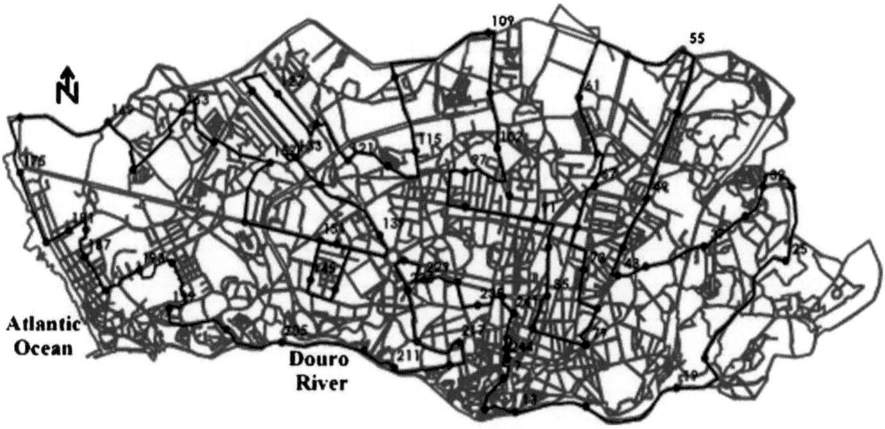

Fig. 3 Temperature monitoring itinerary for the Porto city

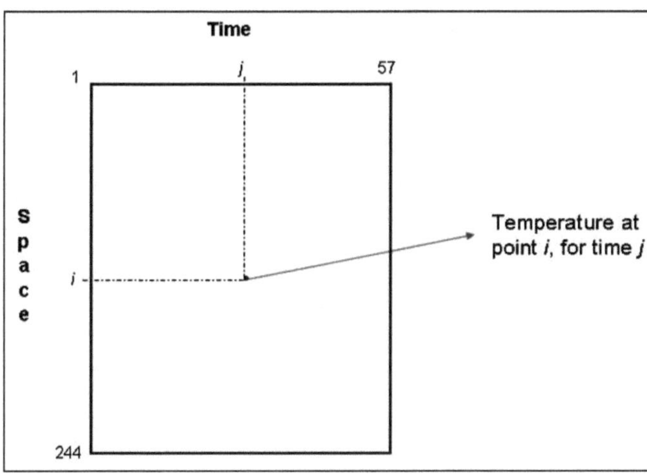

Fig. 4 Data model of recorded temperatures

5 Geostatistical Study

According to the above outlined proposed methodology, the first step is to take each line of Fig. 4 matrix, and construct 244 histograms of measures at each sample location. Such histograms were fitted by a two parameter Weibull probability density function (PDF), given by:

$$f(x,k,c) = \frac{k}{c^k} x^{k-1} e^{-\left(\frac{x}{c}\right)^k} \tag{1}$$

where k and c are the "form" and "scale" Weibull parameters, respectively.

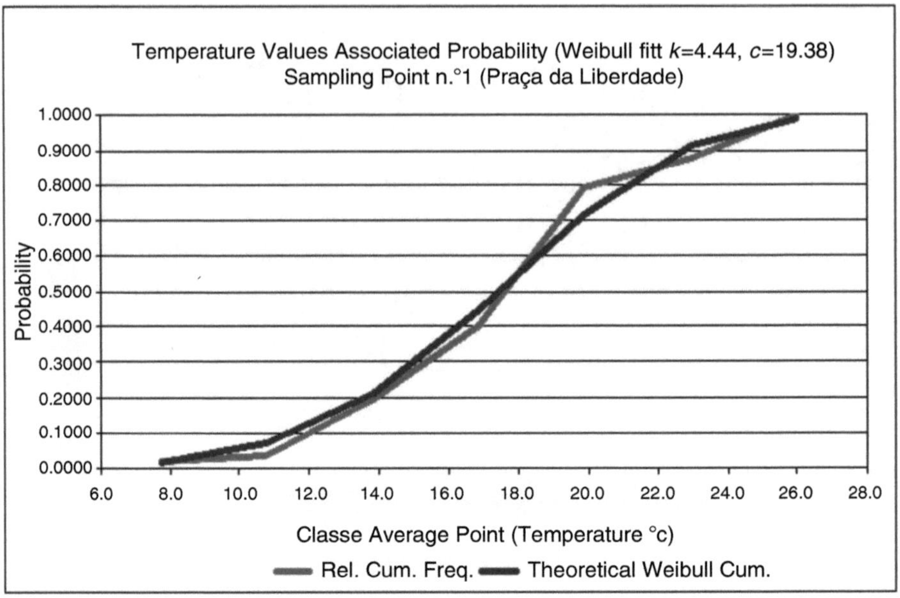

Fig. 5 Example of Kolmogorov–Smirnov model fitting

Table 1 Comparison of experimental and theoretical relative cumulative frequencies for Station n.° 1 – Praça da Liberdade

Absolute freq.	Relative freq.	Relative cum. Freq.	Theoretical Weibull Cum.	Expected vs. theoretical Difference
1	0.0175	0.0175	0.0161	0.0014
1	0.0175	0.0351	0.0723	0.0372
9	0.1579	0.1930	0.2102	0.0172
12	0.2105	0.4035	0.4455	0.0420
22	0.3860	0.7895	0.7179	**0.0911**
5	0.0877	0.8772	0.9125	0.0353
7	0.1228	1.0000	0.9867	0.0133
57	1	–		

An example of this fitting is given for a selected station. The cumulative histogram and the corresponding theoretical model are shown in Fig. 5 and Table 1. In Table 1, the maximum expected versus theoretical difference is 0.0911, which is lower than the allowed limit of the Kolmogorov–Smirnov test (K-S) for the 0.05 significance level (0.6342). Hence the Weibull distribution for parameters given in Fig. 5 is not rejected. By the same procedure, a set of 244 pairs of parameters are obtained. It is worth noting that the well known flexibility of the Weibull distribution gives rise to a minimum χ^2 statistic for all sets of data per station, when compared with other distributions laws that are equally not rejected by the K-S test (for instance, Normal, Lognormal, Erlang and Gamma).

Now, the two parameters (k and c) obtained by the above described procedure may be viewed as 'new' Regionalized Variables that summarize the local temporal variability. The spatial variograms for these new variables are given in Fig. 6. These variograms were built according to the usual Euclidean distance, since results are similar to those obtained by applying other city metrics (like the Manhattan distance, which is not suited to the irregular fabric of Porto's streets).

Hence, conditions are met to estimate by ordinary kriging these two parameters at every point of space located in the nodes of a dense mesh, providing the maps given in Figs. 7 and 8.

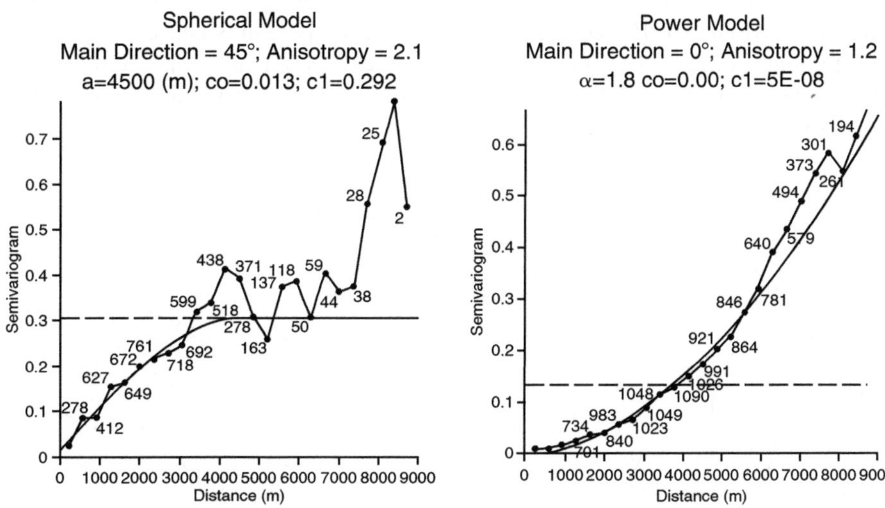

Fig. 6 Variograms of k and c Weibull parameters

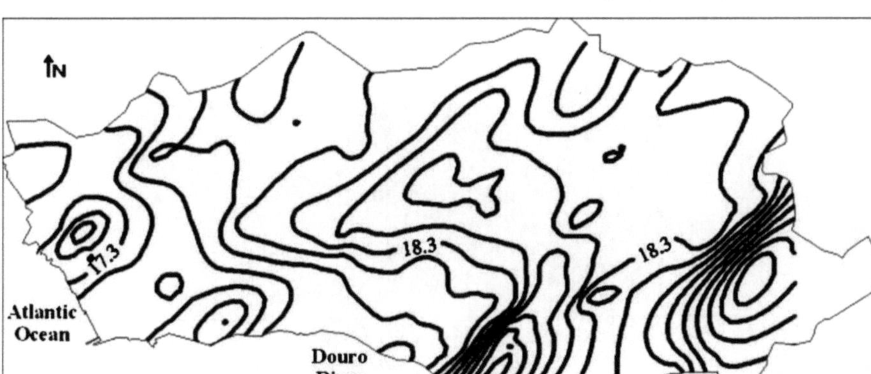

Fig. 7 Kriged map of k parameter

Based on values of k and c provided by the model underlying the maps given in Figs. 7 and 8, PDFs at any unsampled point are calculated by Eq. (1). Once obtained the PDFs at the nodes of the same mesh corresponding to parameters k and c, values for any quantile may be computed. For instance, medians of the temperature within the city are depicted in the map of Fig. 9, providing the 'general picture' of the most likely value of the variable to be controlled.

In addition to Fig. 9, such a control requires the identification of areas within the city where temperature drops below 10°, and others where it exceeds 25° (according to a consensus reached by climatologists and urban planners). These areas are given in Figs. 10 and 11, under a probabilistic form derived from the corresponding PDFs. On the grounds of maps of Figs. 10 and 11, whose reliability was assured by expert knowledge (Monteiro, 1994), regions of extreme temperatures can be spotted. To these regions, composed of sets of houses or other equipment, temperature control devices or specific construction systems are foreseen by urban planning, in order to reach prescribed comfort levels, based on minimum thermal stress.

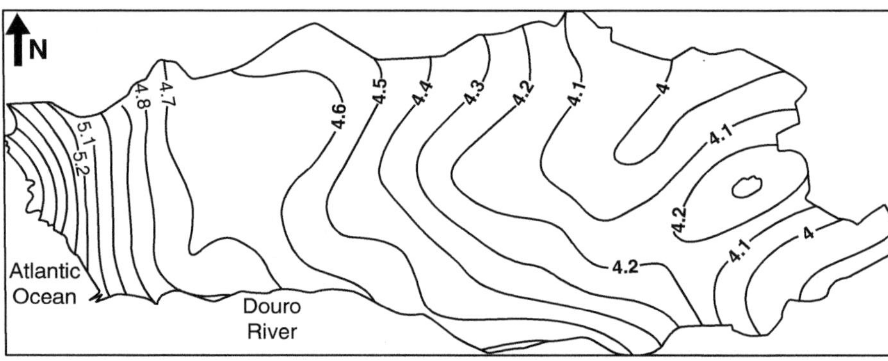

Fig. 8 Kriged map of c parameter

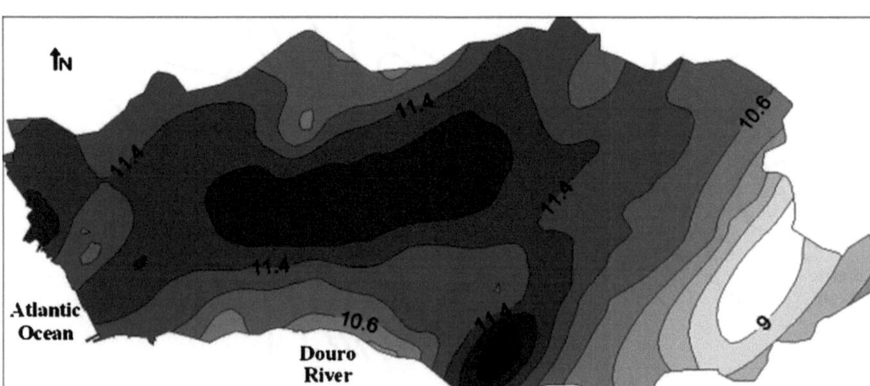

Fig. 9 Map showing the median temperatures in Porto city

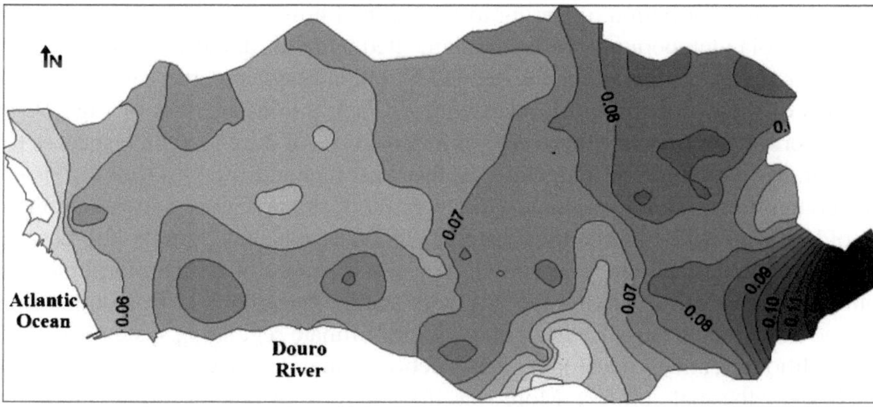

Fig. 10 Probability map of occurrence of temperatures below 10°

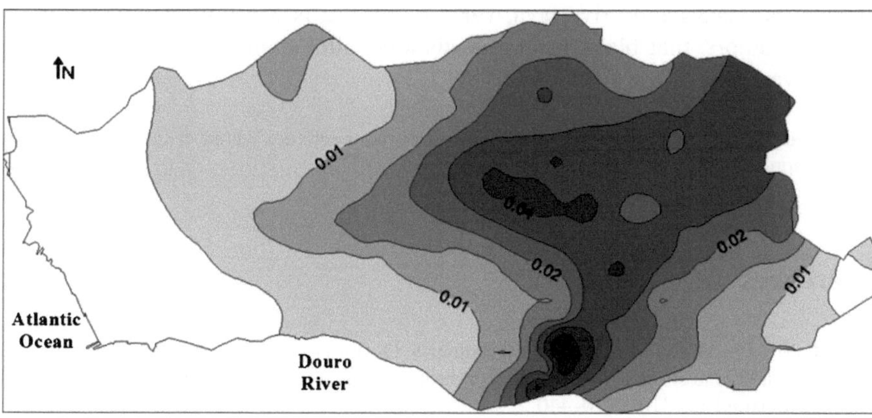

Fig. 11 Probability map of occurrence of temperatures above 25°

6 Discussion and Conclusions

The proposed approach addresses the space time geostatistical models in a simple manner, by separating the two dimensions that are ontologically different. Assuming that replications are representative of the temperature variability in time, and that ancillary variables like elevation are not relevant for characterizing the climatology of the city, advantage can be taken from the availability of a specific PDF for each point of space, obtained by substitution in Eq. (1) of the local values of k and c parameters (as displayed in Figs. 7 and 8). Hence, all results that can be acquired when an empirical PDF is at hand can be derived, in particular, the probability of exceeding a given threshold (without calling for the indicator formalism, which entails a raw data transformation, referring to pre-defined limits). In this case, maps displaying median temperature were complemented by plots, at the same scale, where zones of extreme temperatures (as defined by climatology experts) can be spotted, in terms of their probability of occurrence. In fact, a joint analysis of Figs. 9, 10 and 11 permits the accurate location of different zones of the city, with respect to the space–time distribution of temperatures: the West region displays a time-persistent temperate climate, due to oceanic influence; the Central region, corresponding to the old 'downtown', exhibits the expected "urban heat island"; finally, the East region is the coolest all the year round, as a consequence of its geomorphological configuration (a valley linking the city with semi-rural suburbs). This output may be used as a reliable basis for designing specific climate regulating systems, such as air-conditioners, and to developing construction or maintenance strategies aiming at maximizing thermal comfort. Moreover, if some form of block kriging is required (to estimate probabilities of extreme temperatures in a given spatial domain, for instance, a zone in the city, a borough, or a given quarter), there is no need to integrate point values, since the approach is valid for any support by modifying the second member of the ordinary kriging system, in order to account for point-block covariances (Isaaks and Srivastava, 1989, p. 325). It is worth noting, according to the same authors, that block kriging leads to smaller estimation errors than point kriging averages.

Acknowledgements Recognition is due to the anonymous referees whose criticism fueled a significant improvement of this paper.

References

Alves M, Pozza E, Machado J, Araújo D, Talamini V, Oliveira M (2006) Geostatistics as a methodology to study the space-time dynamics of diseases transmitted by seed-borne. Colletotrichum spp., Fitopatol Bras 31(6):430–440

De Cesare L, Myers D, Posa D (2002) FORTRAN programs for space-time modeling. Comput Geosci 28:205–212

Geiger R, Aron RH, Todhunter P (1995) The climate near the ground. Harvard University Press, Cambridge, MA

Gething P, Noor A, Gilkandi P, Ogara E, Hay S, Nixon M, Snow R, Atkinson P (2006) Improving imperfect data from health management information systems in Africa using space-time geostatistics. PLoS Med 3(6):825–831

Goovaerts P (1977) Geostatistics for natural resources evaluation. Oxford University Press, New York

Hjorth U (1999) On space-time covariance for geostatistical data. Preprint 64, Department of Mathematical Statistics, Chalmers/Gothenburg

Hoskin J, Wallis J (1997) Regional frequency analysis: an approach based on L-moments. Cambridge University Press, Cambridge

Isaaks EH, Srivastava RM (1989) An introduction to applied geostatistics. Oxford University Press, New York

Kyriakidis PC, Journel AG (1998) Geostatistical space-time models: a review. Math Geol 31:651–684

Monteiro A (1994) The changing urban climate of Oporto, Portugal, towards a sustainable future. In: Proceedings of the international conference on the environment "Promoting Sustainable Development", Manchester, UK
Oke TR (1982) The energetic basis of the urban heat island. Q J R Meteorol Soc 108:1–24
Oke TR (1987) Boundary layer climates. Routledge, London/New York
Spadavecchia L, Wiliams M (2006) Estimation of meteorological drivers for biosphere carbon models via space-time geostatistics. Geophys Res Abst 8:00477

Integrating Meteorological Dynamic Data and Historical Data into a Stochastic Model for Predicting Forest Fires Risk Maps

Rita Durão and Amílcar Soares

Abstract This paper couples a dynamic model of meteorological risk of forest fires with historical fire data in a stochastic model in order to predict forest fire risk maps. Daily Severity Rating (DSR), a meteorological risk of forest fire index, from the Canadian Forest Fire Weather Index System (CFFWIS), results from the transformation of daily weather observations into relatively simple indices that can be used to predict fire occurrence, behaviour and impact.

CFFWIS uses the daily weather observations or forecasts to calculate moisture of several fuel types and size classes, and combines them into indices of fire danger related to fire potential rate of spread, heat release, and fireline intensity.

The DSR index depends only on daily measurements of air temperature (°C), relative humidity (%), 10 m open wind speed (km/h) and 24 h accumulated precipitation (mm). DSR is extremely important for forest fire risk assessment but it is restricted to climatic factors.

DSR itself is an incomplete measure of seasonal fire activity because the latter is also dependent on the ignition pattern and the available control resources.

Durão proposed one Bayesian approach to calculate the local conditional probabilities of a forest fire occurring at any location $\mathbf{x}$, given the class $R(\mathbf{x})$ of predicted DSR for same location $\mathbf{x}$. Suppose an indicator variable $I(\mathbf{x})$ that takes the value 1 if a fire occurred in $\mathbf{x}$, otherwise $I(\mathbf{x}) = 0$. Let us call $R(\mathbf{x})$ as the classes of DSR predicted for control points and inferred by simulation for any location $\mathbf{x}$. In this paper, we calculate the probability of a forest fire occurring in $\mathbf{x}$, given $R(\mathbf{x})$ and the historical data of fires occurrence in $\mathbf{x}$, $D(\mathbf{x})$:

$$Prob\,\{I(\mathbf{x})|\ R(\mathbf{x}), D(\mathbf{x})\}$$

Both conditional probabilities Prob $\{I(\mathbf{x})|R(\mathbf{x})\}$ and Prob $\{I(\mathbf{x})|D(.)\}$ can be inferred at any location x. Hence conditional probability can be calculated with the method of Journel called *tau model*. Risk maps of forest fires can be driven from these conditional probabilities.

R. Durão (✉) and A. Soares
Instituto Superior Técnico, CERENA, Av. Rovisco Pais, 1049-001 Lisboa, Portugal
e-mail: rmdurao@ist.utl.pt; asoares@ist.utl.pt

P.M. Atkinson and C.D. Lloyd (eds.), *geoENV VII – Geostatistics for Environmental Applications*, Quantitative Geology and Geostatistics 16,
DOI 10.1007/978-90-481-2322-3_7,

A study was conducted for the period 2000–2005, but in this case study only the results for the 2 year period 2003–2004 of the Portuguese fire seasons are presented and discussed.

1 Introduction

Fire danger rating systems like the Canadian Forest Fire Weather Index System (CFFWIS) transform daily weather observations into relatively simple indices that can be used to predict fire occurrence, behaviour and impact (Stocks et al., 1989).

The CFFWIS uses the daily weather observations or forecasts to calculate moisture of several fuel types and size classes, and combines them into indices of fire danger related to fire potential rate of spread, heat release, and fireline intensity. CFFWIS's indices depend only on daily measurements of air temperature (°C), relative humidity (%), 10 m open wind speed (km/h) and 24 h accumulated precipitation (mm).

The Daily Severity Rating index (DSR) is an overall measure of the fire danger and can be understood as a numeric rating of the difficulty of controlling fire and is preferred for averaging meteorological risk of fire through time and across sites; therefore, it is very useful for regional scale studies.

The DSR is based on the Fire Weather Index (FWI), resulting from a deterministic model proposed firstly by Williams (1959) and modified later by Wagner (1970). It has the following expression:

$$\text{DSR} = 0.0272\,(\text{FWI})^{1.77} \tag{1}$$

However, the DSR itself is an incomplete measure of the fire's risk because it only accounts for meteorological factors while fire risk is also dependent on other factors, like the ignition pattern, fuel load type, topography, social factors and available control resources. DSR values are highly variable in space and time, conforming to different regional patterns. Consequently, for the same level of meteorological risk of fire, different regions over Portugal present different ranges of DSR values. For instance, in the northern part of Portugal, there are much lower DSR values than in the southern part, so the meteorological risk of fire is more severe in Southern Portugal; nevertheless, there are many more fire occurrences and burnt areas in the North than in the South. This means that this index has a great sensitivity to regional meteorological patterns, but depends also of the non-meteorological factors and characteristics of each Portuguese region, namely vegetation cover and human occupation which show dissimilar patterns in the south and north parts of the country. The DSR threshold can be interpreted as an indirect measure of the non-meteorological factors (anthropogenic, fuel load type, topography) that can contribute to fire risk. That is, the higher is the threshold the greater is the influence of other non-meteorological factors on the risk of fire.

The main idea of this present study is to use the DSR values and historical data of large forest fires to calculate and predict the risk of fire.

Conditional probability of a forest fire occurrence at a given location, given the meteorological risk predicted for the same location and the historical record of large fires of the region, is evaluated with the *tau model* proposed by Journel (2002). For assessing the spatial distribution of the fire risk conditional probabilities a geostatistical stochastic simulation, namely Direct Sequential Simulation (Soares, 2001), was used for the entire country in order to obtain several maps of forest fire risk driven from these conditional probabilities.

A study was conducted for the period 2000–2005, but here we present only and discuss the results for the 2003–2004 Fire Seasons and one prediction for a forecasted day, the 10th July 2006.

2 Materials and Methods

2.1 Meteorological and Forest Fire Data

The present analysis was applied to the so-called fire season in Portugal, defined here as starting on May 1st and ending on September 30th, for the 2 year period 2003–2004.

Meteorological data to calculate DSR values were obtained from 15 meteorological monitoring stations spatially distributed over Portugal where the fire occurrences were recorded (see Fig. 1).

The fire data used in this study were provided by the provided by the National Forestry Authority Detailed statistics for forest fires in Portugal are available since 1980 (http://www.afn.min-agricultura.pt/portal/dudf). The dataset includes information on fire location organized by district, municipality (LAU I) and civil parish (LAU II) levels, date and time of ignition and extinction, and burnt land cover type (forests, scrublands and agricultural crops).

2.2 Method for Forest Fire Risk Assessment

Durão (2006) proposed a method to obtain the Portuguese Fire Risk through Bayes Formalism.

Let us consider an indicator variable $I(\mathbf{x})$ that takes the value $I(\mathbf{x}) = 1$ if a fire occurs at a location $\mathbf{x}$ in a given day with a measurable value of burnt area greater than 1 ha; $I(\mathbf{x}) = 0$, otherwise.

The fire risk could be expressed as the conditional distribution $P[I(\mathbf{x})|R(\mathbf{x})]$, where $R(\mathbf{x})$ is a given class of the meteorological risk (DSR) at a municipality $\mathbf{x}$ which is usually divided in classes of severity, i.e. $R(\mathbf{x})$ is the DSR's class

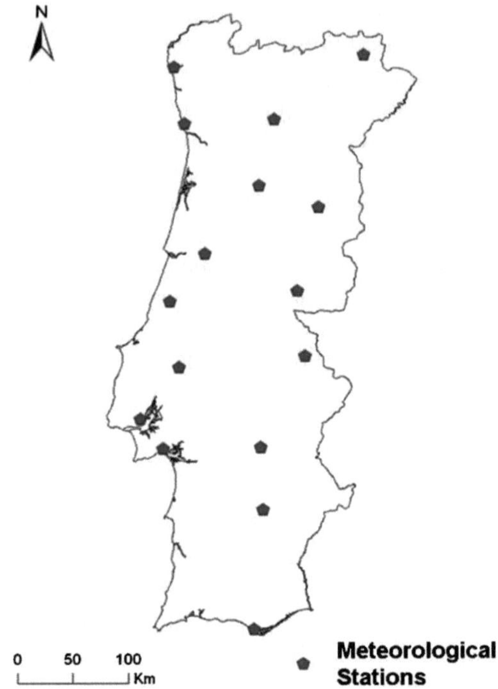

Fig. 1 Localization of the meteorological monitoring stations over Portugal

predicted for any municipality **x**. The marginal probability of fire $P[I(\mathbf{x})]$ at a given municipality was computed with the historical data (Durão et al., 2008) and the objective was to calculate the risk of fire, given the predicted meteorological risk DSR for a given time period. This *a priori* probability $P[I(\mathbf{x})]$ was obtained by averaging the output values of the function $I(\mathbf{x})$.

The proposed model (Durão, 2006) used DSR and fire data to up-date the *a priori* probability $P[I(\mathbf{x})]$ into *a posteriori* $P[I(\mathbf{x})|R(\mathbf{x})]$, through the Bayes formalism:

$$P\,[I(x)|R(x)] = \frac{P\,(R(x)|I(x))\,.\,P(I(x))}{P(R(x))} \tag{2}$$

The DSR values had been summarized in just two classes: High Risk (HR) and Low risk (LR) classes. The DSR's regional critical value that splits HR and LR must lead to high/moderate *a posteriori* risk of fire, such that:

$$P[I(\mathbf{x})|R(\mathbf{x})] >= 0.65. \tag{3}$$

and these regional thresholds can also be used to evaluate the dynamic evolution of local regions regarding the non-meteorological factors (Durão et al., 2008).

In this present work, the main goal is to calculate the probability of a forest fire occurrence in a municipality $\mathbf{x}$, given $R(\mathbf{x})$ and the historical data of fire occurrence, $D(\mathbf{x})$:

$$\text{Prob}\,\{I(\mathbf{x})|R(\mathbf{x}), D(\mathbf{x})\} \tag{4}$$

$D(\mathbf{x})$ is the relative number of times that a given location $\mathbf{x}$ has burnt in a past period. The conditional probability of expression (4) will be obtained through the *tau model* formalism.

2.3 Tau Model

Both conditional probabilities Prob$\{I(\mathbf{x})|R(\mathbf{x})\}$ and Prob$\{I(\mathbf{x})|D(\mathbf{x})\}$ can be inferred at any municipality $\mathbf{x}$. Hence the conditional probability (3) can be calculated using the *tau model* (Journel, 2002; Caers and Hoffman, 2006),

$$\text{Prob}\ \ (\text{A}|\text{B},\text{C}) = \frac{1}{(1+\text{X})} \tag{5}$$

with $\dfrac{x}{a} = \left(\dfrac{b}{a}\right)^{\tau 1}\left(\dfrac{c}{a}\right)^{\tau 2}$, and where,

$$b = (1 - \text{Prob}\,(\text{A} \mid \text{B}))/\text{Prob}\,(\text{A} \mid \text{B}),\ c = (1 - \text{Prob}\,(\text{A} \mid \text{C})/\text{Prob}\,(\text{A} \mid \text{C}),$$
$$a = (1 - \text{Prob}\,(\text{A}))/\text{Prob}\,(\text{A}).$$

Setting $\tau 1 = \tau 2$, Journel (2002) had shown that Eq. (5) is equivalent to the hypothesis of conditional independence:

$$\begin{aligned}\text{Prob}\,(\text{A} \mid \text{B},\text{C}) &= (\text{Prob}\,(\text{B} \mid \text{C},\text{A})\,\text{P}(\text{A}))/(\text{Prob}\,(\text{B} \mid \text{C})\\ &\sim ((\text{Prob}\,(\text{B} \mid \text{A})\,(\text{Prob}\,(\text{C} \mid \text{A})\,\text{Prob}\,(\text{A}))/\text{P}(\text{B},\text{C}))\end{aligned}$$

The τ-values in Eq. (5) allow modelling explicitly the dependency between the B and C data. These τ-values can be interpreted as "weights" given to each data type (Journel, 2002). Assuming conditional independence $\tau 1 = \tau 2 = 1$ results in a very particular dependency model, that must always be validated for each case study. In this present work we had considered conditional independence of meteorological factors $R(\mathbf{x})$ and the historical data of fires $D(\mathbf{x})$.

Hence the expression (5) becomes:

$$\text{Prob}\,(I(\mathbf{x}) \mid R(\mathbf{x}),\ D(\mathbf{x})) = 1/(1 + \text{x}) \tag{6}$$

where, $\text{x}/I(\mathbf{x}) = (R(\mathbf{x})/I(\mathbf{x}))(D(\mathbf{x})/I(\mathbf{x}))$.

2.4 *Spatial Pattern Assessment*

The conditional probabilities – Prob$\{I(\mathbf{x})|D(\mathbf{x})\}$, Prob $[I(\mathbf{x})|R(\mathbf{x})]$ and Prob $\{I(\mathbf{x})|R(\mathbf{x}),\ D(\mathbf{x})\}$ – were computed for each Fire Season. Then, the spatial correlation between meteorological monitoring stations was generalized in a correlation function of distance between any two points, the semivariogram, which summarizes the main spatial continuity patterns of all the obtained conditional probabilities. Afterwards, a stochastic simulation (direct sequential simulation) is used to evaluate the mean spatial pattern and the local variability of the conditional probabilities. The chosen geostatistical methodology follows two steps:

- In a first step DSR values and the conditional probabilities of the risk of fire were computed for each station and then space–time correlations (semivariograms) were obtained.
- In the second one, a stochastic simulation (direct sequential simulation) is used to evaluate the mean spatial pattern and the local variability of the DSR and of the conditional probabilities.

The direct sequential simulation approach (Soares, 2001) is used also to illustrate the results of the spatial patterns of those conditional probabilities. The spatial distributions of the fire's risk probabilities are visualised with maps generated by the simulation algorithm on a $1{,}000 \times 1{,}000$ m grid, using the spatial semivariogram models previously fitted for each fire season. The simulation algorithm generates a set of realisations of the spatial phenomena that roughly reproduces the *a priori* probability and the spatial covariances (variograms) of the computed *a posteriori* probabilities. In this case, for each fire season, 100 equiprobable simulated images were computed and, means and variances were calculated for each pixel generating new maps. Average maps (images) give a mean image of the conditional probabilities, per fire season, while the local variability maps enable quantification of spatial variability/homogeneity of each variable per fire season too.

3 Results and Discussion

The data for this study consisted on the series of maps from the historic fire occurrence events of the Portuguese fire database and the meteorological factor $R(\mathbf{x})$ calculated for each monitoring station.

For illustration purposes the following outputs for the 2003–2004 Fire Seasons and one prediction for a forecasted day, the 10th July 2006, are presented:

- Marginal probability of a forest fire occurrence given the historical data of fire occurrence, Prob$\{I(\mathbf{x})|D(\mathbf{x})\}$
- Conditional probability values, Prob$\{I(\mathbf{x})|R(\mathbf{x})\}$- probability of having a fire occurrence in $\mathbf{x}$ given *a priori* a High Risk class, calculated with Bayes's law (Durão et al., 2008)

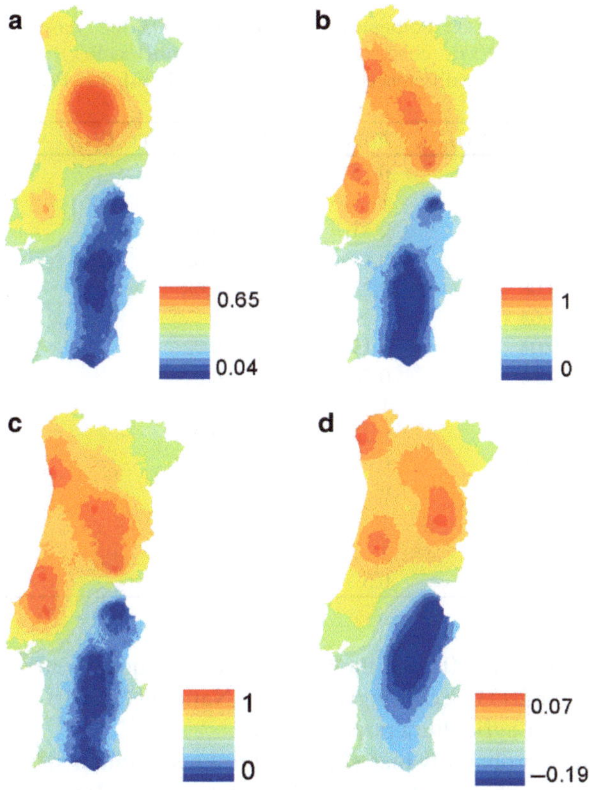

Fig. 2 Probability maps for 2003's fire season: (**a**) fire occurrence historical data, Prob{I(x)|D(x)}; (**b**) conditional probability, Prob{I(x)|R(x)}; (**c**) tau model's probability, Prob {I(x)|R(x), D(x)}; (**d**) difference between Prob{I(x)|R(x)} and Prob {I(x)|R(x), D(x)}

- Conditional probability values, Prob $\{I(\mathbf{x})|R(\mathbf{x}),\ D(\mathbf{x})\}$- probability of a forest fire occurrence in **x**, given $R(\mathbf{x})$ and the historical data of fire occurrence, $D(\mathbf{x})$, calculated with the tau model's formula
- Difference between the conditional probability Prob$\{I(\mathbf{x})|R(\mathbf{x})\}$ and Prob $\{I(\mathbf{x})|R(\mathbf{x}),\ D(\mathbf{x})\}$

The first output, Fig. 2, presents the Probability Maps of the Fire Season in 2003. The spatial distribution of the probability of a forest fire occurrence given the historical data (Fig. 2a) shows greater probability of fire occurrence in the Northern Portugal, where the majority of the forests and scrublands are located.

The second map, (Fig. 2b), shows the probability of fire occurrence given the predicted DSR High Risk class for all Portuguese municipalities, with Prob$\{I(\mathbf{x})|R(\mathbf{x})\}$ $>=$ 0.65 (Durão et al., 2008) and this spatial pattern shows the regions with higher probability of fire located mainly in the Northwest, in the centre of the country, having the main conditional probability of fire occurrence in the northern part of Portugal. The great majority of fires and burnt area took place

Table 1 Probabilities per each Portuguese municipality, 2003 fire season – fire occurrence historical data, Prob{I(x)|D(x)}; conditional probability, Prob{I(x)|R(x)}; tau model's probability, Prob {I(x)|R(x), D(x)}; difference between Prob{I(x)|R(x)} and Prob {I(x)|R(x), D(x)}, "Delta_Tau_Bayes"

Municipalities	*P(I(x)|D(x))*	*P(I(x)|R(x))*	*P(I(x)|R(x),D(x))*	*Delta_Tau_Bayes*
V. Castelo	0.39	0.75	0.82	0.07
Porto_PR	0.27	1	1	0
Coimbra	0.36	0.71	0.78	0.06
Faro	0.06	0.1	0.02	−0.08
Evora	0.08	0.27	0.08	−0.19
Viseu	0.65	1	1	0
Beja	0.04	0.13	0.02	−0.12
V. Real	0.34	0.76	0.81	0.04
C. Branco	0.33	1	1	0
Portalegre	0.06	0.23	0.04	−0.18
Bragança	0.26	0.62	0.59	−0.03
Guarda	0.42	0.83	0.9	0.07
Leiria	0.34	1	1	0
Santarém	0.44	1	1	0
Setúbal	0.24	0.5	0.44	−0.06

in Northern Portugal in 2003 and the model result fits quite well with results officially published for 2003, accordingly with the National Forestry Authority reports (http://www.afn.min-agricultura.pt/portal/dudf).

The *tau model's* probability map (Fig. 2c) presents a similar spatial pattern as the previous map with higher conditional probabilities of fire, Prob$\{I(\mathbf{x})|R(\mathbf{x}),\ D(\mathbf{x})\}$, located in the Northwest and the Central part of the country. In order to assess and compare the results between the conditional probability Prob$\{I(\mathbf{x})|R(\mathbf{x})\}$ and *tau-model's* probabilities differences between both maps were computed (Fig. 2d) and it shows that the *tau model* weights more the historical data than the conditional probability Prob$\{I(\mathbf{x})|R(\mathbf{x})\}$ (see Table 1).

Identical conclusions can be made with the 2004 fire season example (Fig. 3).

The spatial distribution of the probability of a forest fire occurrence given the historical data (Fig. 3a) shows greater probability of fire occurrence in Northern Portugal, where the majority of the forests and scrublands are located. The second map, (Fig. 3b), shows the probability of fire occurrence given the predicted DSR High Risk class for all Portuguese municipalities, and its pattern shows the regions with higher probability of fire located in the Northeast, Northwest, in the centre and South coast of the country, with the main conditional probability of fire occurrence in the northern part of Portugal.

The great majority of fires and burnt area took place North of the Tejo River and in Algarve (Southern Coast) in 2004 and besides having data for only 15 municipalities, the model result fits quite well with what happened and was officially published for 2004, accordingly with the National Forestry Authority reports (http://www.afn.min-agricultura.pt/portal/dudf).

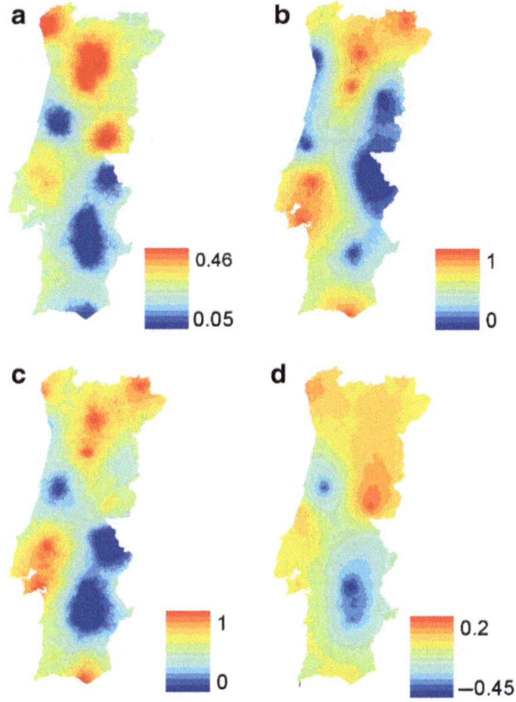

Fig. 3 Probability maps for 2004's fire season: (**a**) fire occurrence historical data, Prob{I(x)|D(x)}; (**b**) conditional probability, Prob{I(x)|R(x)}; (**c**) tau model's probability, Prob {I(x)|R(x), D(x)}; (**d**) difference between Prob{I(x)|R(x)} and Prob {I(x)|R(x), D(x)}

The *tau model* probability map (Fig. 3c) presents a similar spatial pattern as the previous map (Fig. 2b) with higher conditional probabilities of fire, Prob $\{I(\mathbf{x})|R(\mathbf{x}),\ D(\mathbf{x})\}$, located in the Northeast, Northwest, Centre and South coast of the country. In order to compare the results between the conditional probabilities Prob$\{I(\mathbf{x})|R(\mathbf{x})\}$ and $\{I(\mathbf{x})|R(\mathbf{x}),\ D(\mathbf{x})\}$ a difference map was computed (Fig. 3d).

The *tau model* combination results shows the influence of combining this additional information (historical data) into the *a posteriori* conditional probability Prob$\{I(\mathbf{x})|R(\mathbf{x})\}$ (see Table 2). If Prob$\{I(\mathbf{x})|R(\mathbf{x})\}$ is high (say 0.75), but the probability conditioned to historical data Prob$\{I(\mathbf{x})|D(\mathbf{x})\}$ is low (say 0.05) the *tau model* tends to give a low final result (0.31); otherwise if the probability conditioned, to historical data Prob$\{I(\mathbf{x})|D(\mathbf{x})\}$ is relatively high (say 0.41) the *tau model* tends to give a higher final result (0.74) than Prob$\{I(\mathbf{x})|R(\mathbf{x})\}$ (0.56). Hence the obtained results of the tau model show a very promising path for the assessment of risk of fires by providing a means to combine multiple sources of information.

Analogous risk maps are also presented for a forecasted day, the 10th July 2006 (Fig. 4, Table 3). For the forecasted day in 2006 historical data of 2003 and 2004 were used (Fig. 4a). The map of meteorological fire risk Prob $[I(\mathbf{x})|R(\mathbf{x})]$ (Fig. 4b) shows a quite similar pattern of the *tau model*'s probability map (Fig. 4c), with higher conditional probabilities of fire, Prob$\{I(\mathbf{x})|R(\mathbf{x}),\ D(\mathbf{x})\}$, around 100% (Fire forecast), located also in the northern coast of the country and in the Castelo

Table 2 Probabilities per each Portuguese municipality, 2004 fire season – fire occurrence historical data, Prob{I(x)|D(x)}; conditional probability, Prob{I(x)|R(x)}; tau model's probability, Prob {I(x)|R(x), D(x)}; difference between Prob{I(x)|R(x)} and Prob {I(x)|R(x), D(x)}, "Delta_Tau_Bayes"

Municipalities	*P(I(x)|D(x))*	*P(I(x)|R(x))*	*P(I(x)|R(x),D(x))*	*Delta_Tau_Bayes*
V. Castelo	0.42	0.82	0.91	0.1
Porto_PR	0.24	0.47	0.47	0
Coimbra	0.05	0.71	0.33	−0.38
Faro	0.11	1	1	0
Evora	0.05	0.75	0.31	−0.44
Viseu	0.38	1	1	0
Beja	0.05	0.48	0.12	0.36
V. Real	0.46	1	1	0
C. Branco	0.41	0.56	0.74	0.18
Portalegre	0.05	0.27	0.05	−0.22
Bragança	0.25	1	1	0
Guarda	0.29	0.45	0.51	0.06
Leiria	0.29	0.48	0.54	0.06
Santarém	0.33	1	1	0
Setúbal	0.24	1	1	0

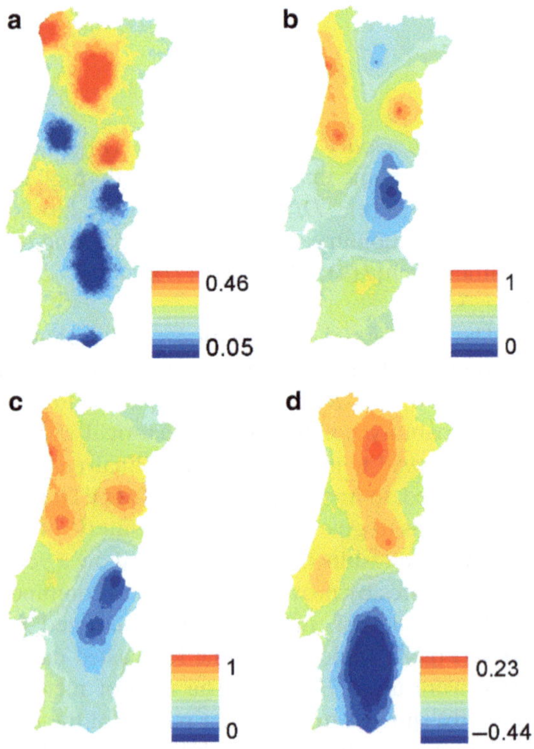

Fig. 4 Probability maps for the forecasted day, 10th July 2006: (**a**) fire occurrence historical data, Prob{I(x)|D(x)}; (**b**) conditional probability, Prob{I(x)|R(x)}; (**c**) tau model's probability, Prob {I(x)|R(x), D(x)}; (**d**) difference between Prob{I(x)|R(x)} and Prob {I(x)|R(x), D(x)}

Table 3 Probabilities per each Portuguese municipality for the forecasted day, 10th July 2006 – fire occurrence historical data, Prob{I(x)|D(x)}; conditional probability, Prob{I(x)|R(x)}; tau model's probability, Prob {I(x)|R(x), D(x)}; difference between Prob{I(x)|R(x)} and Prob {I(x)|R(x), D(x)}, "Delta_Tau_Bayes"

Municipalities	*P(I(x)|D(x))*	*P(I(x)|R(x))*	*P(I(x)|R(x), D(x))*	*Delta_Tau_Bayes*
V. Castelo	0.42	0.76	0.88	0.1
Porto_PR	0.24	1	1	0
Coimbra	0.05	1	1	−0.38
Faro	0.11	0.52	0.3	0
Evora	0.05	0.36	0.08	−0.44
Viseu	0.38	0.47	0.63	0
Beja	0.05	0.68	0.24	0.36
V. Real	0.46	0.28	0.51	0
C. Branco	0.41	0.25	0.43	0.18
Portalegre	0.05	0.10	0.02	−0.22
Bragança	0.25	0.43	0.44	0
Guarda	0.29	1	1	0.06
Leiria	0.29	0.47	0.54	0.06
Santarém	0.33	0.52	0.63	0
Setúbal	0.24	0.44	0.44	0

Branco's region and in the remaining areas of the country, the probabilities were again much lower, near 0% (No Fire forecast) in the same regions and also in the southern part of the country. The spatial distribution of the differences between the forecasted conditional probability Prob $[I(\mathbf{x})|R(\mathbf{x})]$ and *tau model*'s probabilities shows again that *tau model* weights more the historical data than meteorological risk given by the conditional probability Prob $[I(\mathbf{x})|R(\mathbf{x})]$ (Fig. 4d).

4 Conclusions

In the present work, we propose an approach forest fire risk assessment in Portugal by using the *tau model* formalism.

Firstly Bayes' law (Durão et al., 2008) was applied to get the risk of fire given by the conditional probabilities at each municipality **x**, and then the tau model combination (Journel, 2002) is applied to calculate the probability of a forest fire occurrence in **x**, given $R(\mathbf{x})$ and the historical data of fire occurrence, $D(\mathbf{x})$.

The *tau model* combination results had showed the influence of combining additional information, the historical data in this case study, into the *a posteriori* probability Prob$\{I(\mathbf{x})|R(\mathbf{x})\}$. The obtained results show a very promising path for assessment of the risk of fires by providing a means to combine multiple sources of information using this formalism. Therefore, the proposed model shows an improvement regarding the simple use of local conditional probabilities to meteorological risk of fire, DSR and regardless of having data from only 15 municipalities it reproduces reasonably what was officially published for 2003 and 2004 fire seasons, according to the National Forestry Authority report (http://www.afn.min-agricultura.pt/portal/dudf).

We intend to apply this methodology to all the municipalities of the 18 districts of Continental Portugal, in order to obtain more realistic and helpful risk maps which could be updated on a daily basis.

References

Durão RM (2006) Bayesian classification of the risk of fire risk in Continental Portugal. Master Thesis, IST

Durão RM, Soares A, Pereira JMC, Côrte-Real JA, Coelho MFES (2008) Bayesian classification of a meteorological risk index of forest fire (DSR). In: Soares A, Pereira MJ, Dimitrakopoulos R (eds) GeoENV 6th – geostatistics for environmental applications. Springer, pp 283–294

Caers J, Hoffman T (2006) The probability perturbation method: a new look at Bayesian inverse modeling. Math Geol 38(1):81–100

Journel AG (2002) Combining knowledge from diverse sources: an alternative to traditional data independence hypotheses. Math Geol 34(5):573–596

Stocks BJ, Lawson BD, Alexander ME, Van Wagner CE, McAlpine RS, Lynham TJ, Dube DE (1989) Canadian forest fire danger rating system: an overview. For Chron 65:258–265

Soares A (2001) Direct Sequential Simulation and Cosimulation. Math. Geology. 33(8):911–926

Using Geostatistical Methods in the Analysis of Public Health Data: The Final Frontier?

Linda J. Young and Carol A. Gotway

Abstract Geostatistical methods have been demonstrated to be very powerful analytical tools in a variety of disciplines, most notably in mining, agriculture, meteorology, hydrology, geology and environmental science. Unfortunately, their use in public health, medical geography, and spatial epidemiology has languished in favor of Bayesian methods or the analytical methods developed in geography and promoted via geographic information systems. In this presentation, we provide our views concerning the use of geostatistical methods for analyzing spatial public health data. We revisit the geostatistical paradigm in light of traditional analytical examples from public health. We discuss the challenges that need to be faced in applying geostatistical methods to the analysis of public health data as well as the opportunities for increasing the use of geostatistical methods in public health applications.

1 Introduction

Analysis of spatial data has come to be important for many studies in public health, medical geography and spatial epidemiology. Whereas geostatistical methods have been used extensively in a variety of disciplines, including mining, agriculture, meteorology, hydrology, geology and environmental science, they have found only limited application in health studies where Bayesian methods and analytical methods developed in geography and implemented in geographic information systems have dominated. Here, we consider some of the challenges encountered in our efforts to use geostatistical methods for analyzing spatial public health data and some of the solutions that have been proposed. This is not meant to be a comprehensive list, but one that reflects our experiences and identifies needs for additional research.

L.J. Young (✉)
Department of Statistics, 404 McCarty Hall C, P.O. Box 110339, University of Florida, Gainesville, FL 32611-0339, USA
e-mail: LJYoung@ufl.edu

C.A. Gotway
Centers for Disease Control and Prevention, Office of Workforce and Career Development, 1600 Clifton Rd, NE, MS E-94, Atlanta, GA 30333, USA
e-mail: cdg7@cdc.gov

P.M. Atkinson and C.D. Lloyd (eds.), *geoENV VII – Geostatistics for Environmental Applications*, Quantitative Geology and Geostatistics 16,
DOI 10.1007/978-90-481-2322-3_8,

2 Motivating Study

Our work with Florida's Environmental Public Health Tracking (EPHT) effort provides the motivating study (Young et al., 2008). Part of Florida's efforts to move toward implementation of EPHT is to develop models of the spatial and temporal association between myocardial infarctions (MIs) and the changing levels of ozone in outdoor air for Florida. To accomplish this, as with the majority of studies relating environmental changes to public health, especially those that are national or regional in scope, the analysis is based on pre-existing data. Florida's Department of Environmental Protection (FDEP) provided ozone measurements, recorded from a network of 48 air monitors placed throughout the state. Florida's Agency for Health Care Administration (AHCA), consistent with a data sharing agreement, provided all admissions to Florida's public and private hospitals where either the primary or secondary cause of admission was MI (*International Classification of Diseases*, 10th Revision (ICD-10) codes 410.0–414.0 [World Health Organization]). ACHA also provided both the zip code and county of residence for each patient's record and selected patient demographic information, including sex, age, and race/ethnicity. Selected sociodemographic data (age, race/ethnicity, sex, education) were obtained from the U.S. Census Bureau. Additional sociodemographic data were obtained from CDC's Behavioral Risk Factor Surveillance System (BRFSS). For March, 2001, the number of MI admissions per 10,000 population and the 48 ozone monitors functioning that month, are displayed in Fig. 1 (see Young et al., 2008 for full details).

3 Challenges for Public Health

3.1 *Spatial Support*

As illustrated in our Florida study, increasingly interest extends beyond the simple reporting of incidence or risk and turns to relating these responses to potential explanatory variables. As is also common, the variables used in each of these studies were collected from disparate sources and must be linked on a common set of spatial units for analysis. Moving from one set of spatial units to another can result in several challenging change of support problems (see Gotway and Young 2002 for a review). Most of the early geostatistical work on change of support problems was motivated by mining applications in which the inferential unit of interest was a block of ore. The rectangular shape of blocks made it possible to use a regular grid to discretize the blocks into points and approximate the integrals needed for block kriging using just a relatively few number of points. However, applications in the public health field call for a reassessment and extension to this and other geostatistical approaches.

First, the "blocks" are seldom rectangular in shape or consistent in size. As an example, note that the Florida counties (Fig. 1) vary considerably in size and are

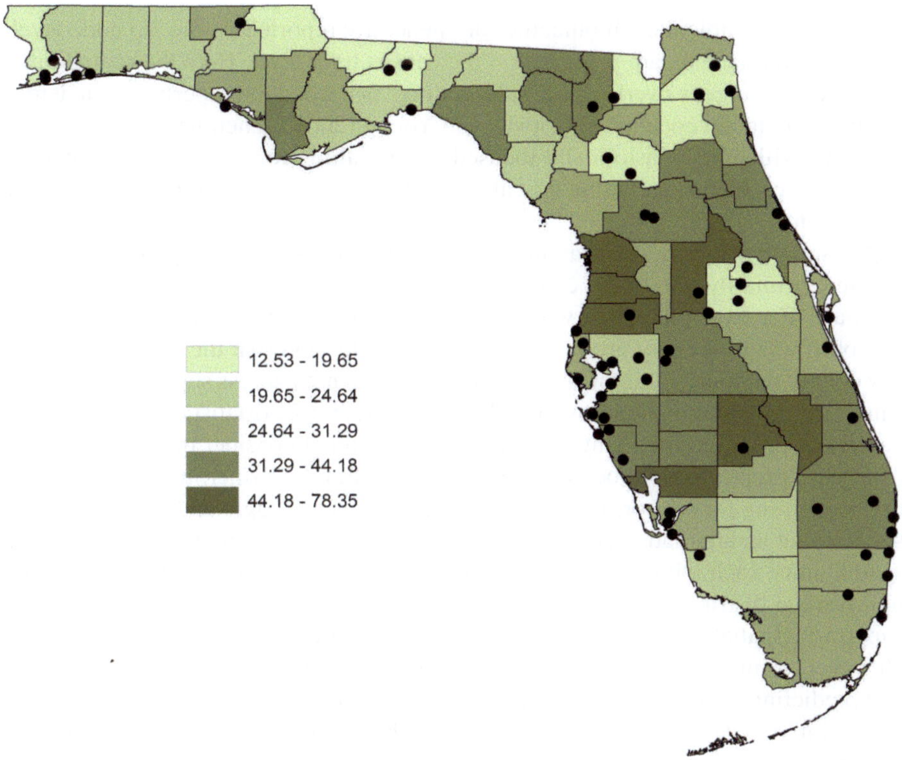

Fig. 1 The number of MI cases per 10,000 population recorded for each Florida county during March 2001 and the location of the ozone monitors functioning during that month

irregular in shape. This is typical of the postal codes, political boundaries, and census administration units often used in public health studies. Here we wanted to use the 48 ozone monitors functioning during March, 2001, to obtain an average maximum ozone value for each county. A regular grid was placed across all of Florida, and three counties did not have any grid points or monitors falling within them. One option was to make the grid very fine. This would have slowed computations tremendously and is inefficient because the larger counties would be over-characterized. Alternatively, we augmented the grid with a finer sub-grid for those three counties. Is this the best approach?

A second challenge results from the different change of support problems encountered in public health, often arising from data confidentiality concerns. Typically, the change of support problem is not one of upscaling (or aggregation). As an example, the incidence of low birth weight babies is available at the county level, but interest lies in incidence of low birth weight babies at the census tract level (Gotway and Young, 2007). This downscaling (or disaggregation) should ideally preserve the pycnophylactic property (Tobler, 1979) that the number of low birth weight babies from the census tracts within a county should equal the county total.

As another example, health outcomes are generally reported on the zip code level, but demographic data are provided on the census tract level. Here the spatial units from the two sources overlap and "side-scaling" is needed to properly assign demographic data to zip code units. Gotway and Young (2007) generalized geostatistical methods, which have historically focused on upscaling, for use in general change-of-support problems, including upscaling, down scaling, side-scaling, and intensity estimation.

We should acknowledge that, in many studies, the change-of-support problem is ignored, primarily due to the complexity of the solution proposed and the lack of software. For example, when working with the March ozone data, one suggested approach was to identify proxy monitors that would represent the ozone values for any county without a monitor, an approach that does not address support, but greatly simplifies the computational issues. Similarly, the non-geostatistical methods that have been proposed for change-of-support often do not consider the support of the data (e.g., proportional allocation, centroid smoothing). Further, instead of explicitly accounting for support in a geostatistical approach, Diggle and Robeiro (2007) suggest that an alternative approach is to partition the spatial region into n discrete spatial units, each with a response variable y_i, $i = 1, \ldots, n$, and then model the multivariate distribution for the random variable Y_i. Undoubtedly, accounting for support in spatial analysis is challenging, both theoretically and computationally. However, in mining, accounting for support was found to be critically important and predicting a spatial average is very different from simply predicting an average at a point. The lesson likely holds for public health as well, and we should learn from the mining experience where accurate block-grade predictions and inferences are critical to the profits of the industry.

3.2 Discrete Distributions

Geostatisticians working in public health and other application areas have responded to the need for new methods for discrete distributions, especially the Poisson and binomial distributions. Unlike the Gaussian distribution, the variance of any discrete probability distribution depends on the mean. For the Poisson, the mean and variance are equal; for the binomial, the variance is equal to the mean multiplied by a constant that is less than one. The Box-Cox family of transformations includes transformations which stabilize variance, and using the appropriate transformation from this family in a trans-Gaussian kriging (Schabenberger and Gotway, 2005, pp. 270–277) formulation may work well. However, models that explicitly account for the variance–mean relationships inherent in many discrete distributions are warranted for applications to other disciplines such as public health.

Poisson kriging was developed by Monestiez et al. (2005, 2006) for mapping the spatial distribution of fin whales and used to predict cancer mortality rates in a public health setting by Goovaerts (2005). In the public health context, we have $Z(\mathrm{B}_i)$, the count or total number of disease cases over the ith region with population $n(\mathrm{B}_i)$

at risk. Thus, $R(\mathrm{B}_i) = Z(\mathrm{B}_i)/n(\mathrm{B}_i)$ is the incidence proportion for region B_i, Assume that $\{\lambda(\mathbf{s})|\mathbf{s} \in D \subset \Re^2\}$ is an unobserved intensity process, with $\lambda(\mathbf{s}) \geq 0$ for all $\mathbf{s}$ in D. Assume this process has mean μ_λ and covariance function $C_{\lambda\lambda}(\mathbf{s_i}, \mathbf{s_j})$. Further assume that, conditional on this process, the observed frequencies (counts), $Z(\mathrm{B}_i)$, associated with an areal region B_i are independent Poisson random variables with means and variances both equal to $\lambda(\mathbf{s})n(\mathrm{B}_i)$. If we assume a linear prediction function for $\lambda(\mathbf{s})$, then the predictor of the intensity process at location $\mathbf{s}_0$ is

$$\hat{\lambda}(\mathbf{s}_0) = \sum_{i=1}^{N} w_i R(\mathrm{B}_i),$$

where N is the number of regions and optimal weights, w_i, can be obtained by solving

$$\sum_{k=1}^{N} w_k \left[C_{RR}(\mathrm{B}_i, \mathrm{B}_k) + \delta_{ik}\frac{\mu^*}{n(\mathrm{B}_i)}\right] + m = C_{\lambda R}(\mathbf{s}_0, \mathrm{B}_i), \qquad i = 1, \ldots, n$$
$$\sum_{k=1}^{N} w_k = 1.$$

Here $\delta_{ik} = 1$ if $\mathrm{B}_i = \mathrm{B}_k$ and 0 otherwise, μ^* is an estimate of the mean of $R(\cdot)$, and m is a Lagrange multiplier. A key to the estimation process is estimation of the point-support covariance function from which the cross-covariance function between the intensity process and the observed frequencies is determined. In an effort to adjust for heterogeneous variances, Monestiez et al. (2005, 2006) proposed weighting the difference pair by the corresponding population sizes. Extending the ideas of Mockus (1998), Goovaerts (2008) proposes an iterative deconvolution method. Here too is a change of support problem: $\lambda(\mathbf{s}_\mathrm{i})$ is assumed to be of point-support, but $R(\mathrm{B_i})$ is aggregated over areal regions. Binomial kriging (McNeill, 1991) has a similar derivation and leads to comparable challenges.

Gotway and Stroup (1997) developed models for generalized linear models, of which the Poisson and binomial are special cases. In an approach similar to that of trans-Gaussian kriging, they used Taylor series to linearize the problem so that the usual kriging predictor is optimal, but with variance-mean relationships built into models for spatial dependence. Gotway and Wolfinger (2003) compare these models to those conditioned on a latent process as in Poisson kriging, binomial kriging, and model-based geostatistics. Their results indicate that while conditionally-specified models can be used to build complicated, non-stationary models, they tended to under-predict both counts and rates and may severely over-estimate prediction uncertainty for data sets with moderate-to-large marginal spatial autocorrelation. The marginal models allow us to move away from any Gaussian assumptions and employ methods similar in form to least squares estimation. However, the estimation algorithm was not as stable for these models, and the predictions tended to vary more than those from the conditional model. Ordinary or universal kriging, with a semivariogram weighted inversely proportional to the assumed variance of the data (in this case, inversely proportional to $n(\mathrm{B_i})$) worked surprisingly well, demonstrating what most geostatistical practitioners have observed time and again: ordinary kriging is relatively robust to a variety of violations in assumptions. Although predictions may not be theoretically optimal, they are not grossly inaccurate either. Nevertheless, models that better describe the nature of the problem and the properties of the data are intuitively more appealing.

With both Poisson and binomial kriging, and marginal generalized linear models, two issues have yet to be fully addressed. One important issue is that, in geostatistical modelling, we are working with multivariate data and we need an underlying joint multivariate distribution for valid inference. Although this may appear to be a simple theoretical nuisance, the lack of such a multivariate distribution can cause difficulties, such as "covariance" matrices that are not positive definite, numerical instability, and order-relations problems, in some practical applications. Herein lies the problem with the non-parametric indicator approaches and Poisson, binomial, and generalized linear model approaches. A classic example is indicator kriging which predicts probabilities, which, theoretically, should by contained in [0,1]. However, any user of indicator methods has obtained predicted probabilities outside this range.

A number of the challenges arise in constructing non-Gaussian, multivariate distributions with specified correlation structure, marginal distributions, and conditional distributions (see Schabenberger and Gotway, 2004, pp. 192–195, for a full discussion). Constraints on the correlation exist for many multivariate distributions that are not constructed from an underlying multivariate Gaussian distribution. As an example, the multivariate binomial permits only negative correlations (Mardia 1970). For other models, no such multivariate distribution exists. For example, no multivariate distribution exists having both marginal and conditional distributions of Poisson form (Mardia, 1970).

Generating multivariate distributions sequentially from specified conditions overcomes some of these difficulties. In Bayesian hierarchical modeling, this sequential conditioning approach is used to generate fairly complex multivariate distributions, but the properties of the resulting distribution may not always be clear. As an example, suppose $Z_1(\mathbf{s})$ is a second-order stationary process with $\mathrm{E}[Z_1(\mathbf{s})] = 1$ and $\mathrm{Cov}[Z_1(\mathbf{u}),\ Z_1(\mathbf{u}+\mathbf{h})] = \sigma^2\rho_1(\mathbf{h})$. A simplified version of a common model used for modeling and inference with count data is obtained by conditioning $Z_2(s)$, a white noise process with mean and variance given by

$$E[Z_2(\mathbf{s})|Z_1(\mathbf{s})] = \exp\{\mathbf{x}(\mathbf{s})'\beta\}Z_1(\mathbf{s}) \equiv \mu(\mathbf{s}),\quad Var[Z_2(\mathbf{s})|Z_1(\mathbf{s})] = \mu(\mathbf{s})$$

on $Z_1(\mathbf{s})$. The marginal mean $\mathrm{E}[Z_2(\mathbf{s}) = \exp\{\mathbf{x}(\mathbf{s})'\boldsymbol{\beta}\}$, depends only on the unknown parameter $\boldsymbol{\beta}$, and the marginal variance, $\mathrm{Var}[Z_2(\mathbf{s})] = \mu(\mathbf{s}) + \sigma^2\mu(\mathbf{s})^2$, allows overdispersion in the data $Z_2(\mathbf{s})$, making the model attractive. Now, consider the marginal correlation of $Z_2(\mathbf{s})$

$$Corr[Z_2(\mathbf{s}), Z_2(\mathbf{s+h})] = \frac{\rho_1(\mathbf{h})}{\left[\left(1+\frac{1}{\sigma^2\mu(\mathbf{s})}\right)\left(1+\frac{1}{\sigma^2\mu(\mathbf{s+h})}\right)\right]^{1/2}}$$

If σ^2, $\mu(\mathbf{s})$, and $\mu(\mathbf{s}+\mathbf{h})$ are small, $\text{Corr}[Z_2(\mathbf{s}), Z_2(\mathbf{s}+\mathbf{h})] << \rho_1(\mathbf{h})$. Thus, while the conditioning induces both overdispersion and autocorrelation in the Z_2 process, the marginal correlation has a definite upper bound and so may not be a good model for highly correlated data. Most Bayesian models have a similar constraint built in, although it is often difficult to test either theoretically or empirically.

The second fundamental issue is that the marginal variance and the covariance function depends on $n(\mathrm{B}_i)$ (e.g., Goovaerts, 2005; Monestiez et al., 2005, 2006). Thus, neither Poisson nor binomial kriging are based on an intrinsically stationary process. Weighting the empirical semivariogram by factors that are inversely proportional to the standard deviation of the data (Goovaerts, 2005; Monestiez et al., 2005, 2006) ameliorates the problem. However, the semivariogram of the data process is only estimable (and arguably only defined) for intrinsically stationary processes. This problem of non-stationarity affects the validity of all the geostatistical tools such as measures of autocorrelation, spatial prediction, and geostatistical simulation methods. Moreover, covariates may not be spatially continuous and are often categorical. Thus, non-stationarity arises in two ways: differing populations and the need to adjust for covariates. Although most geostatistical tools are robust to departures from the assumption of stationarity, the lack of a more general paradigm may prevent their wide-spread adoption in public health.

More sophisticated models for prediction with discrete distribution have also been developed, including disjunctive kriging methods and isofactorial models (e.g., Rivoirard, 1994) and Bayesian methods (Diggle et al., 1998). Unfortunately, none of these approaches is ready for routine use, and the general Bayesian methods have yet to be extended to complex change of support problems.

Given the above discussion, the reasons for the popularity of the multivariate Gaussian distribution are evident. It has a closed form expression, permits pairwise correlations in $(-1, 1)$, each (Z_i, Z_j) has a bivariate Gaussian distribution, all marginal distributions are Gaussian, and all conditional distributions are Gaussian. Moreover, tractable multivariate distributions, such as the multivariate lognormal and the multivariate t-distribution can be derived from the multivariate Gaussian. The Gaussian distribution has truly earned its unique place in geostatistical theory. Thus, for our motivating study, instead of using methods developed for discrete distributions, the incidence of MI at the county level was indirectly standardized by age, sex, and education to the Florida population and the standardized event ratio (MI SER) computed. The MI SER was log-transformed (denoted by ln(SER) because the natural logarithm was taken) so that the assumptions of linear regression (normality and constant variance) would be more nearly met.

3.3 Spatial Regression

The traditional analytical approach, referred to here as *global regression*, is to conduct a multivariate linear regression analysis relating the health outcome to potential predictors with adjustments for sociodemographic variables (e.g., education, income, and percentage of smokers). For our study, a weighted regression was conducted with the weight being equal to the expected MI SER, and the coefficient on ozone was exponentiated to obtain the relative MI SER from the regression.

Just as ozone levels and the number of MI cases can vary over the state, the relative MI SER could also vary over the state. Hastie and Tibshirani (1993) introduced varying coefficient models, a class of regression and generalized regression functions in which the coefficients are allowed to vary as smooth functions of other variables. Müller (1998) adapted this idea to the spatial case and referred to the approach as *local regression*. Independently, Brunsdon et al. (1996) adapted the idea of varying coefficient models to the spatial case and called their method *geographically weighted regression*. More generally, when regression coefficients are assumed to vary smoothly over space, the models are referred to as spatially varying coefficient models (Gelfand et al., 2003).

To fit a local regression model, ideas from local smoothing and kernel regression are used to define spatial neighborhoods. The regression is performed by using only data in the spatial neighborhoods. As a consequence, the error terms are not necessarily constant for all locations. Further, because the spatial neighborhoods associated with different points in space overlap, the same data are used more than once to estimate all the spatial regression parameters. Local regression models are appealing because we expect risk to change over space as well as with time, and this can be an important outcome for public health studies. Yet this method has open questions. Because the same data are used more than once to estimate all the spatial regression parameters (βs), a correlation structure is induced among the βs. One consequence of this correlation might be overly smoothed predictions. In our motivating study, the estimated relative MI SERs are much smoother than either the MI SERs or the predicted ozone values. This phenomenon can be observed for other, similar local regression models for both frequentist (as presented here, see also Nakaya et al., 2005) and Bayesian analyses (e.g., Waller et al., 2007). As is often the case with Bayesian analyses, the local regression models are overparameterized, and assumptions (e.g., the form of the prior distributions) allow one to proceed with the analyses. In local regression, as in other analyses using overparameterized models, the impact of the assumptions is not fully evident.

Health outcomes are likely to depend on more than one environmental factor (e.g., the ozone levels considered here). This leads us to include other explanatory variables (e.g., PM2.5) in the models. Wheeler and Tiefelsdorf (2005) concluded that, for local regression, multicollinearity among the coefficients at a single location and the overall correlation between coefficients associated with two different explanatory variables (e.g., ozone and PM2.5) can make interpretation of the model coefficients problematic. Their results indicate that the collinearity among local regression coefficients might be present even if the process generating the explanatory variables leads them to be uncorrelated. This collinearity is likely caused by implicit conditions that are placed on the parameters during the estimation process. This is an open question worthy of further research, as is the more general concern of valid inference from all local regression models, because they were designed as exploratory smoothing methods and not inferential statistical tools.

4 Conclusions

Throughout this work, we have been critical of the existing methods as they related to public health studies. Our goal has been to emphasize the vast opportunities for research on important geostatistical issues. Here we want to take time to applaud the authors whose work we have critiqued. Although we have pointed out areas that need further development, we are encouraged that efforts are being made to address complex issues that arise.

Discrete and, more generally, non-Gaussian data are common in public health studies. Satisfactory multivariate non-Gaussian models have severe limitations. Either we do not get the marginal or conditional distributions that are desired or the choice of covariance structures is severely limited. Is the best solution to transform the data so that it is at least approximately normal and to then rely on the robustness of the standard geostatistical methods? Or, even with the disadvantages outlined here, is it better to use methods such as Poisson kriging? Is there a better approach? These are examples of the basic guidance that those working in public health need if geostatistical methods are to find broader application.

Acknowledgements The senior author was partially supported by the Florida Department of Health, Division of Environmental Health and Grant/Cooperative Agreement Number 5 U38 EH000177-02 from the Centers for Disease Control and Prevention (CDC). The findings and conclusions in this report are those of the authors and do not necessarily represent the views of the Centers for Disease Control and Prevention.

References

Brunsdon CF, Fotheringham AS, Charlton ME (1996) Geographically weighted regression: a method for exploring spatial nonstationarity. Geogr Anal 28:281–298

Centers for Disease Control and Prevention (CDC) (2006) Health risks in the United States: behavioral risk factor surveillance system 2006. US Department of Health and Human Services, CDC, Atlanta, GA

Diggle PJ, Robeiro PJ (2007) Model-based geostatistics. Springer, New York

Diggle PJ, Tawn JA, Moyeed RA (1998) Model based geostatistics. Appl Stat 47:299–350

Gelfand AE, Kim H-J, Sirmans CF, Banerjee S (2003) Spatial modeling with spatially varying coefficient processes. J Am Stat Assocn 98:387–396

Goovaerts P (2005) Geostatistical analysis of disease data: estimation of cancer mortality risk from empirical frequencies using Poisson kriging. Int J Health Geogr 4:31

Goovaerts P (2008) Geostatistical analysis of health data: state-of-the-art and perspective. In: Soares A, Pereira MJ, Dimitrakopoulos R (eds) geoENV VI –geostatistics for environmental applications. Springer, The Netherlands, pp 3–22.

Gotway CA, Stroup WW (1997) A generalized linear model approach to spatial data analysis and prediction. J Agric, Biol, Environ Stat 2:157–178

Gotway CA, Wolfinger RD (2003) Spatial prediction of counts and rates. Stat Med 22:1415–1432

Gotway CA, Young LJ (2002) Combining incompatible spatial data. J Am Stat Assoc 97:632–648

Gotway CA, Young LJ (2007) A geostatistical approach to linking spatially-aggregated data from different sources. J Comput Graph Stat 16:115–135

Hastie TJ, Tibshirani RJ (1993) Varying-coefficient models. J R Stat Soc B 55:757–796

Mardia KV (1970) Families of bivariate distributions. Hafner, Darienn, CT
McNeill L (1991) Interpolation and smoothing of binomial data for the Southern African Bird Atlas Project. S Afr J Stat 25:129–136
Mockus A (1998) Estimating dependencies from spatial averages. J Comput Graph Stat 7:501–513
Monestiez P, Dubroca L, Bonnin E, Durbec JP, Guinet C (2005) Comparison of model based geostatistical methods in ecology: application to fin whale spatial distribution in northwestern Mediterranean Sea. In: Leuangthong O, Deutsch CV (eds) Geostatistics Banf 2005, vol. 2. Kluwer, Dordrecht, The Netherlands, pp 777–786
Monestiez P, Dubroca L, Bonnin E, Durbec J-P, Guinet C (2006) Geostatistical modeling of spatial distribution of Balaenoptera physalus in the Northwestern Mediterranean Sea from sparse count data and heterogeneous observations efforts. Ecol Model 193:615–628
Müller WG (1998) Fundamentals of spatial statistics. In: Collecting spatial data: optimum design of experiments for random fields. Physica-Verlag, Heidelberg
Nakaya T, Fotheringham AS, Brunsdon C, Charlton M (2005) Geographically weighted Poisson regression for disease association mapping. Stat Med 24:2695–2717
Rivoirard J (1994) Introduction to disjunctive kriging and non-linear geostatistics. Clarendon, Oxford
Schabenberger O, Gotway CA (2005) Statistical methods for spatial data analysis. CRC Press, Boca Raton, FL
Tobler W (1979) Smooth pycnophylactic interpolation for geographical regions (with discussion). J Am Stat Assoc 74:519–536
U.S. Census Bureau (2001) Age: 2000. Economics and statistics administration. U.S. Department of Commerce, Washington, DC
Waller LA, Zhu L, Gotway CA, Gorman D, Gruenewald P (2007) Quantifying geographic variations in associations between alcohol distribution and violence: a comparison of geographically weighted regression and spatially varying coefficient models. Stoch Environ Res Risk Assess 21:573–588
Wheeler D, Tiefelsdorf M (2005) Multicollinearity and correlation among local regression coefficients in geographically weighted regression. J Geogr Syst 7:161–187
World Health Organization (2005) International classification of diseases and related health problems (ICD-10), 2nd edn. WHO
Young LJ, Gotway CA, Yang J, Kearney G, DuClos C (2008) Assessing the association between environmental impacts and health outcomes: a case study from Florida. Stat Med (in Press). doi: 10.1002/sim.3249

Second-Order Analysis of the Spatio-temporal Distribution of Human Campylobacteriosis in Preston, Lancashire

Edith Gabriel and Peter J. Diggle

Abstract We propose a method for analysing inhomogeneous spatio-temporal point process data by extending Baddeley's et al. (2000) inhomogeneous K-function to the spatio-temporal setting. We develop a non-parametric estimator of the space–time inhomogeneous K-function. We then apply the estimator to data on the spatio-temporal distribution of human campylobacteriosis cases in an area of north-west England and investigate evidence for spatio-temporal clustering and spatio-temporal interaction using tests based on this estimator.

1 Introduction

Campylobacter is the most commonly identified cause of bacterial gastro-enteritis in the developed world. Amongst the *campylobacter* species pathogenic to humans, 90% of disease is caused by *campylobacter jejuni* and most of the rest by *campylobacter coli*. Incidence of campylobacteriosis is typically sporadic, with a strong seasonal variation which rises sharply in spring and peaks in summer. In this paper, we shall analyse a data-set consisting of the locations and dates of notification of all cases of campylobacteriosis notified to the Preston Microbiology Services Laboratory in the Preston postcode district (Lancashire, England) between January 1st 2000 and December 31st 2002. These data can be considered as a single realisation of a spatio-temporal point process displaying a highly aggregated spatial distribution. As is common in epidemiological studies, the observed point pattern is spatially and temporally inhomogeneous, as the pattern of incidence of the disease reflects both the spatial distribution of the population at risk and systematic temporal variation

E. Gabriel (✉)
IUT STID – LANLG, Université d'Avignon, BP 1207, 84911 Avignon, France
e-mail: edith.gabriel@univ-avignon.fr

P.J. Diggle
Department of Medicine, Lancaster University, Lancaster LA1 4YF, UK
e-mail: p.diggle@lancaster.ac.uk

P.M. Atkinson and C.D. Lloyd (eds.), *geoENV VII – Geostatistics for Environmental Applications*, Quantitative Geology and Geostatistics 16,
DOI 10.1007/978-90-481-2322-3_9,

in risk. When analysing such spatio-temporal point patterns, a natural starting point is to investigate the nature of any stochastic interactions amongst the points of the process after adjusting for spatial and temporal inhomogeneity.

We shall use a method for analysing the second-order properties of inhomogeneous spatio-temporal point process data, based on the spatio-temporal inhomogeneous K-function (STIK-function) under the assumption of second-order intensity re-weighted stationarity. This extends to the spatio-temporal setting the inhomogeneous K-function proposed by Baddeley et al. (2000). We propose a non-parametric estimator for the STIK-function. Our pragmatic working assumption is that first-order effects are separable, meaning that the intensity can be factorised as the product of spatial and temporal intensities. We use Monte Carlo methods to assess the data for evidence of spatio-temporal clustering or spatio-temporal interaction. To test for clustering, the null hypothesis is that the underlying process is an inhomogeneous Poisson process. To test for spatio-temporal interaction, the null hypothesis is that the spatial and temporal component processes are stochastically independent. We then apply this methodology to our *campylobacter jejuni* data, from which we conclude that *campylobacter jejuni* cases exhibit both spatio-temporal clustering and spatio-temporal interaction.

2 The Space–Time Inhomogeneous K-Function

2.1 Definition

We consider an orderly point process, whose events define a countable set $\mathbf{x}_i = (\mathbf{s}_i, t_i) : i = 1, 2, \ldots$ in which $s_i \in \mathbb{R}^2$ is the spatial location of the ith event and $t_i \in \mathbb{R}$ its time of occurrence. Our data are a realisation of this process in $A = S \times T$, where $S \subset \mathbb{R}^2$ and $T \subset \mathbb{R}$. We denote by $Y(A)$ the number of events $\mathbf{x}_i \in A$.

The first-order properties of a point process are represented by the *(first-order) intensity function*,

$$\lambda(s,t) = \lim_{|d\mathbf{s}\times dt|\to 0} \frac{\mathbb{E}\left[Y(d\mathbf{s}\times dt)\right]}{|d\mathbf{s}\times dt|},$$

where $d\mathbf{s} \times dt$ defines a small region around the point $(\mathbf{s}, t)$ and $|d\mathbf{s} \times dt|$ is its volume. Informally, $\lambda(s,t)$ measures the mean number of events of the process per unit area per unit time in a neighbourhood of the point $(\mathbf{s}, t)$. Similarly, second-order properties can be represented by the *second-order intensity function*,

$$\lambda_2((\mathbf{s},t),(\mathbf{s}',t')) = \lim_{|d\mathbf{s}\times dt|,|d\mathbf{s}'\times dt'|\to 0} \frac{\mathbb{E}\left[Y(d\mathbf{s}\times dt)Y(d\mathbf{s}'\times dt')\right]}{|ds\times dt||d\mathbf{s}'\times dt'|},$$

or by a scaled version, the pair correlation function

$$g((\mathbf{s},t),(\mathbf{s}',t')) = \frac{\lambda_2((\mathbf{s},t),(\mathbf{s}',t'))}{\lambda(\mathbf{s},t)\lambda(\mathbf{s}',t')}.$$

First-order and second-order properties defined in this way can be considered as the point process analogues of the mean and covariance properties of a real-valued process. In particular, for any spatio-temporal Poisson process, $\lambda_2((\mathbf{s},t),(\mathbf{s}',t')) = \lambda(s,t)\lambda(s',t')$, hence $g((\mathbf{s},t),(\mathbf{s}',t')) = 1$. For this reason, the term "pair correlation" is perhaps confusing, since a value of 1 corresponds to the absence of second-order dependence. To add to the confusion, the function

$$\gamma((\mathbf{s},t),(\mathbf{s}',t')) = \lambda_2((\mathbf{s},t),(\mathbf{s}',t')) - \lambda(\mathbf{s},t)\lambda(\mathbf{s}',t'),$$

which is identically zero for a Poisson process, is sometimes called the *covariance density*.

Second-order stationarity of a point process holds when its first-order and second-order properties are invariant under translation, meaning that the intensity is constant and the second-order intensity only depends on the spatio-temporal difference vector. A point process is isotropic when its first-order and second-order properties are invariant under rotation. Hence, for a stationary, isotropic point process, we have $\lambda(\mathbf{s},t) = \lambda$ and $\lambda_2((\mathbf{s},t),(\mathbf{s}',t')) = \lambda_2(u,v)$, where $u = \|\mathbf{s}-\mathbf{s}'\|$ and $v = |t-t'|$. Second-order stationarity is too restrictive an assumption for most epidemiological applications. We therefore consider a weaker assumption, defined by Baddeley et al. (2000) and called *second-order intensity-reweighted stationarity*. This allows a non-constant intensity, but assumes that the pair correlation function depends only on the difference vector.

For a second-order, intensity reweighted stationary, isotropic spatio-temporal point process, we define the *space–time inhomogeneous K-function* (STIK-function) by

$$K^*_{ST}(u,v) = 2\pi \int_{-v}^{v}\int_{0}^{u} g(u',v')u'\,\mathrm{d}u'\,\mathrm{d}v', \tag{1}$$

where $g(u,v) = \lambda_2(u,v)/\left(\lambda(\mathbf{s},t)\lambda(\mathbf{s}',t')\right)$, $u = \|\mathbf{s}-\mathbf{s}'\|$ and $v = |t-t'|$. This definition extends Baddeley et al.'s definition of a second-order reweighted stationary isotropic spatial point processes to the spatio-temporal setting. Here, we restrict attention to future events only and define

$$K_{ST}(u,v) = 2\pi \int_{0}^{v}\int_{0}^{u} g(u',v')u'\,\mathrm{d}u'\,\mathrm{d}v'. \tag{2}$$

Note that definitions (1) and (2) only differ non-trivially because of the treatment of edge-effects when estimating these functions from data observed in a finite region A. In what follows, we focus on $K_{ST}(u,v)$.

The STIK function can be used as a measure of the spatio-temporal aggregation or regularity of clustering. Indeed, for any inhomogeneous spatio-temporal Poisson process with intensity bounded away from zero, $K_{ST}(u,v) = \pi u^2 v$. Values of $K_{ST}(u,v)$ greater than $\pi u^2 v$ indicate aggregation at spatial and temporal separations less than u and v, whilst $K_{ST}(u,v) < \pi u^2 v$ indicates regularity.

2.2 Non-parametric Estimation

For data $\mathbf{x}_i : i = 1, ..., n$ in a spatio-temporal region $A = S \times T$, where S is any polygonal region in $\mathbb{R}^2$ and $T = [T_0, T_1]$, we propose the following approximately unbiased estimator (see Gabriel and Diggle, 2009) of the STIK-function:

$$\widehat{K}_{ST}(u, v) = \frac{1}{|S \times T|} \frac{n}{n_v} \sum_{i=1}^{n_v} \sum_{j=1; j>i}^{n_v} \frac{1}{w_{ij}} \frac{1}{\lambda(\mathbf{x}_i)\lambda(\mathbf{x}_j)} \mathbf{1}_{\{u_{ij} \leq u\}} \mathbf{1}_{\{t_j - t_i \leq v\}}, \quad (3)$$

where n_v is the number of $t_i \leq T_1 - v$, $\lambda(\mathbf{x}_i)$ is the intensity at $\mathbf{x}_i = (\mathbf{s}_i, t_i)$, the $\mathbf{x}_i$ are ordered so that $t_i < t_{i+1}$ and w_{ij} is Ripley's (1976, 1977) spatial edge-correction, in which w_{ij} is the proportion of the circle centered on $\mathbf{s}_i$ and passing through $\mathbf{s}_j$, i.e. of radius $u_{ij} = \|\mathbf{s}_i - \mathbf{s}_j\|$, that lies inside S.

In practice the intensity is unknown and must be estimated. In general, this is an insoluble problem without additional assumptions; for example, a single realisation of a stationary Cox process with stochastic intensity $\Lambda(\mathbf{s}, t)$ is indistinguishable from a realisation of an inhomogeneous Poisson process whose first-order intensity function coincides with the unobserved realisation of $\Lambda(\mathbf{s}, t)$. In the current context, we resolve this ambiguity by making the pragmatic working assumption that first-order effects are separable, meaning that $\lambda(\mathbf{s}, t)$ can be factorised as

$$\lambda(\mathbf{s}, t) = m(\mathbf{s})\mu(t), \text{ for all } (\mathbf{s}, t) \in S \times T. \quad (4)$$

Under this assumption, any non-separable effects are interpreted as second-order, rather than first-order. Suitable estimates of the spatial intensity $m(\mathbf{s})$ and of the temporal intensity $\mu(t)$ in Equation (3) will depend on the characteristics of each application.

Assuming separability of the intensity (4) allows us to estimate both systematic (first-order) and stochastic (second-order) properties of the underlying point process, and hence to test for the existence of spatio-temporal clustering or interaction using the estimated STIK function. In particular, for a spatio-temporal Poisson process, representing the absence of spatio-temporal clustering, $K_{ST}(u, v) = \pi u^2 v$; whereas a process with no spatio-temporal interaction corresponds to the weaker assumption that $K_{ST}(u, v) = K_S(u)K_T(v)$, where $K_S(\cdot)$ and $K_T(\cdot)$ are spatial and temporal K-functions, respectively (Diggle et al., 1995).

3 Application

We apply our methodology to data on the locations (unit post-code) and notification dates of cases of human *campylobacter jejuni* infections reported from residential addresses in the Preston post-code sector, Lancashire, UK (Fig. 1a) over the 3 years

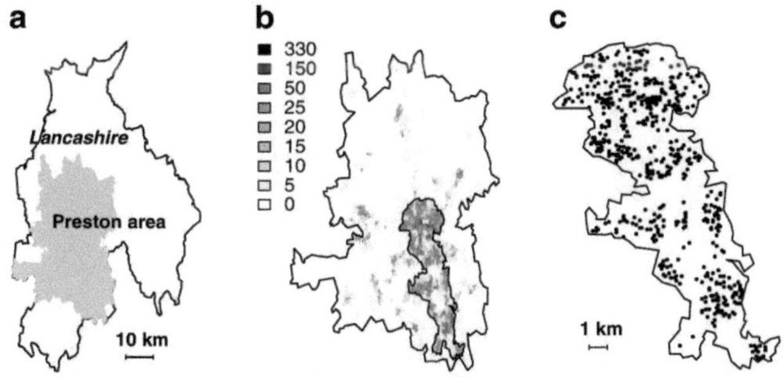

Fig. 1 (**a**) Preston post-code sector in Lancashire. (**b**) Population density in 2001 (number of people per hectare). (**c**) Locations of the 619 cases of *campylobacter jejuni* infections within the urban sub-region

2000 to 2002 inclusive. Figure 1b gives population density in this area in 2001. We restrict attention to the urban area, within which 619 cases have been recorded; Fig. 1c gives their locations.

3.1 Test for Spatio-temporal Clustering

To assess the data for evidence of spatio-temporal clustering, we follow common practice in constructing Monte Carlo tolerance envelopes around the estimator $\hat{K}_{ST}(u, v)$. The null hypothesis is that the underlying process is an inhomogeneous Poisson process with intensity $\hat{\lambda}(\mathbf{s}, t) = \hat{m}(\mathbf{s})\hat{\mu}(t)$, and the tolerance envelopes are therefore constructed from simulations of a Poisson process with an intensity of this form. We estimate the spatial intensity $m(s)$ using a Gaussian kernel with bandwidth chosen to minimize the estimated mean-square error of $\hat{m}(s)$, as suggested in Berman and Diggle (1989). To estimate $\mu(t)$ we use a Poisson log-linear regression model incorporating a time-trend, seasonal variation and day-of-the-week effects, hence

$$\log \mu(t) = \delta_{d(t)} + \sum_{k=1}^{3} \{\alpha_k \cos(k\omega t) + \beta_t \sin(k\omega t)\} + \gamma t,$$

where $\omega = 2\pi/365$ and $d(t)$ identifies the day of the week for day $t = 1, ..., 1{,}096$. Figure 2 shows the estimated spatial and temporal intensities.

3.2 Distribution of Cases Versus Population at Risk

Figure 1a and b show, unsurprisingly, that cases tend to be concentrated in areas of high population density. However, a test for spatial clustering (Diggle et al., 1995) showed that the spatial distribution of cases is more clustered than that of the population at risk.

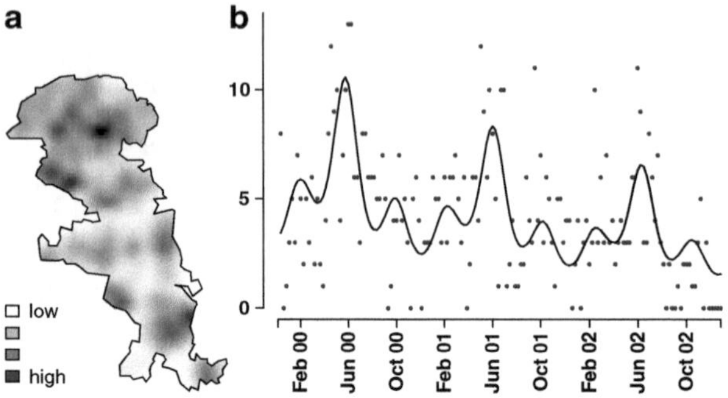

Fig. 2 (**a**) Kernel estimate of the spatial intensity; (**b**) weekly numbers (dots) of notified cases compared with fitted regression curve

To compare the spatio-temporal distribution of cases with the population at risk, we proceed as in Section 3.1, considering as null hypothesis that the process of *campylobacter jejuni* cases is an inhomogeneous Poisson process with intensity $\hat{\lambda}_0(\mathbf{s}, t) = \hat{m}_0(\mathbf{s})\hat{\mu}_0(t)$ proportional to the spatio-temporal intensity of the population at risk. The spatial intensity $m_0(\mathbf{s})$ is estimated as previously, whilst the temporal intensity $\mu_0(t)$ is assumed to depend only on day-of-the-week. To test for spatio-temporal clustering, we use a Monte Carlo approach based on the test statistic

$$Z = \int_0^{v_0} \int_0^{u_0} \left\{ \widehat{K}_{ST}(u, v) - E(u, v) \right\} \Big/ \; V(u, v)^{1/2} \, \mathrm{d}v \, \mathrm{d}u,$$

where $E(u, v)$ and $V(u, v)$ are the mean and variance of $\widehat{K}_{ST}(u, v)$ computed from 1,000 Poisson processes with intensity $\hat{\lambda}_0(s, t)$.

3.3 Test for Spatio-temporal Interaction

Separability of the STIK function into purely spatial and temporal components, $K_{ST}(u, v) = K_S(u) K_T(v)$, indicates absence of spatio-temporal interaction (Diggle et al., 1995). We use a Monte Carlo procedure to test for spatio-temporal interaction, where the null hypothesis is that the spatial and temporal component processes are independent, and construct tolerance envelopes by randomly permuting the observed spatial locations, $\mathbf{s}_i$, holding times t_i fixed.

3.4 Results

Figure 3a shows $\hat{K}_{ST}(u,v) - \pi u^2 v$ for the *campylobacter* data. The diagonal black hatching on Fig. 3b identifies those values of (u,v) for which the data-based estimate of $\hat{K}_{ST}(u,v) - \pi u^2 v$ lies above the 95*th* percentile of estimated calculated from 1,000 simulations of an inhomogeneous Poisson process with intensity $\hat{\lambda}(s,t)$. Similarly, the grey shading identifies those values of (u,v) for which $\hat{K}_{ST}(u,v) - \hat{K}_S(u)\hat{K}_T(v)$ lies above the 95*th* percentile envelopes calculated from 1,000 random permutations of the $\mathbf{s}_i$ holding the t_i fixed. The results suggest spatio-temporal clustering up to a distance of 300 m and a time-lag of 10 days, and spatio-temporal interaction at distances up to 400 m and time-lags up to 3 days. These findings are consistent with the infectious nature of the disease, leading to multiple cases from a common source that are relatively close both in space and in time. They also suggest the existence of stochastic structure that cannot be explained by $\hat{m}(\mathbf{s})\hat{\mu}(t)$. Note that the relatively large negative values of $\hat{K}_{ST}(u,v) - \pi u^2 v$ at large values of u and v are not significantly different from zero, because the sampling variance of $\hat{K}(u,v)$ increases with u and v. A test comparing the distribution of cases with that of the population at risk gives a significant p value, indicating that the distribution of cases is more spatio-temporally clustered than that of the population risk. Comparing $\hat{K}_{ST}(u,v) - E(u,v)$ with the 95*th* percentile envelopes calculated from 1,000 simulations of an inhomogeneous Poisson process with intensity $\hat{\lambda}_0(s,t)$ shows spatio-temporal clustering at distances up to 300 m and a time-lag of 10 days.

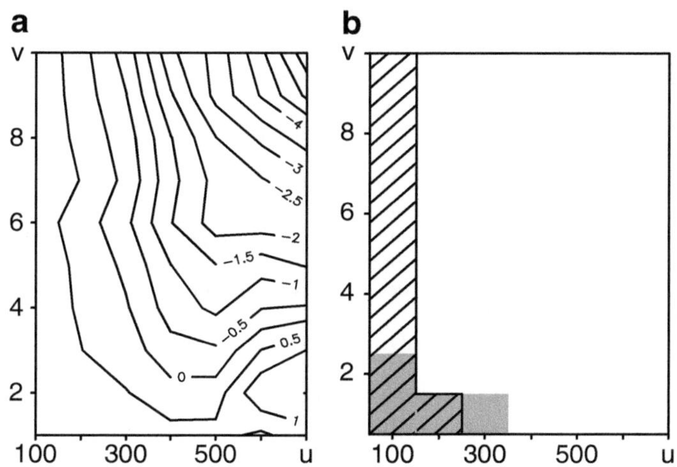

Fig. 3 (**a**) $\hat{K}_{ST}(u,v) - \pi u^2 v$ ($\times 10^6$). (**b**) Comparison between $\hat{K}_{ST}(u,v) - \pi u^2 v$ and tolerance envelopes indicating spatio-temporal clustering (*diagonal black hatching*) and comparison between $\hat{K}_{ST}(u,v) - \hat{K}_S(u)\hat{K}_T(v)$ and tolerance envelopes indicating spatio-temporal interaction (*grey shading*)

4 Conclusion

Our data show both spatial and temporal aggregation, which in general can arise through heterogeneity, clustering or a combination of the two. For example, in the current application we know that the spatial distribution of the population at risk is non-uniform, and that the risk of infection peaks each year in late spring. Our proposed methodology enables a pragmatic distinction between heterogeneity and clustering by identifying heterogeneity with separable first-order structure and clustering with residual second-order structure. We have proposed Monte Carlo tests for spatio-temporal clustering and for spatio-temporal interaction, based on the space–time inhomogeneous K function. Application of these tests to the *campylobacter jejuni* data suggests a combination of spatially and temporally localised variations in risk, and small-scale spatio-temporal clusters of cases. These empirical findings are consistent with there being both unmeasured socio-economic or environmental risk factors for the disease, and food-borne infections leading to multiple cases that are close in both space and time.

Acknowledgements We would like to thank Eric Bolton, John Cheesbrough, Paul Fearnhead, Andrew Fox, Steven Gee, Andrew J. H. Leatherbarrow, C. Anthony Hart and Daniel J. Wilson for stimulating discussion on various aspects of this work. This work was funded by the Department for Environment, Food and Rural Affairs as part of the Veterinary Training Research Initiative, and the National Centre for Zoonosis Research.

References

Baddeley AJ, Møller J, Waagpetersen R (2000) Non- and semi-parametric estimation of interaction in inhomogeneous point patterns. Stat Neerl 54:329–350

Berman M, Diggle PJ (1989) Estimating weighted integrals of the second-order intensity of a spatial point process. J R Stat Soc 51:81–92

Diggle PJ, Chetwynd AG, Haggkvist R, Morris S (1995) Second-order analysis of space-time clustering. Stat Methods Med Res 4:124–36

Diggle PJ, Zeng P, Durr P (2005) Nonparametric estimation of spatial segregation in a multivariate point process: bovine tuberculosis in Cornwall, UK. Appl Stat 54:645–658

Gabriel E, Diggle PJ (2009) Second-order analysis of inhomogeneous spatio-temporal point process data. Stat Neerl 63:43–51

Ripley BD (1976) The second-order analysis of stationary point processes. J Appl Probab 13: 255–266

Ripley BD (1977) Modelling spatial patterns (with discussion). J R Stat Soc 39:172–212

Application of Geostatistics in Cancer Studies

Pierre Goovaerts

Abstract This paper presents an overview of geostatistical methods available for the analysis of both areal and individual-level health data. The application of Poisson kriging and p-field simulation to lung cancer mortality rates recorded for white males in 688 US counties of the Southeast (1970–1994) allowed: (1) the creation of noise-filtered mortality maps at the county-level and over a fine grid (isopleth maps), (2) the detection of clusters of low or high mortality counties that are significantly correlated in space, and (3) the identification of areas where the local correlation of mortality rates is stronger for white males than for white females, revealing gender-specific factors such as occupational exposure. Then, indicator kriging is introduced as a way to map the risk for late stage breast cancer diagnosis using patient residences across Michigan.

1 Introduction

Cancer is a major public health problem in the United States and is currently the second leading cause of death. For cancer control activities and resource allocation, it is important to be able to compare incidence and survival rates, risk behaviors, screening patterns, diagnosis stage, and treatment methods across geographical and political boundaries and at as fine a spatial scale as possible. Although individual humans represent the basic unit of spatial analysis, the majority of cancer maps depict data in discrete or areal form. The so-called 'choropleth maps' are seen by many as an inferior representation of the basic data and their interpretation and analysis typically faces three major hurdles: (1) the presence of extreme unreliable rates that occur for sparsely populated areas and/or rare cancers, (2) the visual bias caused by the aggregation of health data within administrative units of widely different sizes and shapes, and (3) the mismatch of spatial supports for cancer rates and explanatory variables that prevents their direct use in correlation analysis (Goovaerts, 2009).

P. Goovaerts (✉)
BioMedware, Inc. 516 North State Street, Ann Arbor, MI, USA
e-mail: goovaerts@biomedware.com

P.M. Atkinson and C.D. Lloyd (eds.), *geoENV VII – Geostatistics for Environmental Applications*, Quantitative Geology and Geostatistics 16,
DOI 10.1007/978-90-481-2322-3_10,

Complex statistical techniques, usually involving Bayesian models, have recently been developed to increase the reliability of cancer risk maps. Yet, the estimation of model parameters requires iterative procedures, such as Markov Chain Monte Carlo methods, that are computer intensive and require fine-tuning, which makes their application and interpretation challenging for non-statisticians (Woodward, 2005). Furthermore, Bayesian algorithms are overwhelmingly used with the conditional auto-regressive (CAR) model for defining the random effect associated with spatial autocorrelation. The arbitrary neighborhood relationship underlying the CAR model is computationally convenient but is not well-suited to situations where geographical entities have different sizes and shapes and are not arranged in a regular pattern (Kelsall and Wakefield, 2002). Simulation studies have also demonstrated the strong smoothing effect of Bayesian disease-mapping models, in particular the BYM model (Besag et al., 1991), which limits their ability to detect localized increases in risk (Richardson et al., 2004).

Geostatistics provides a less cumbersome, powerful, yet still little known, model-based approach to disease mapping. Although it was introduced in the same year as the BYM model, the first initiative to tailor geostatistical tools to the analysis of disease rates (Lajaunie, 1991) went largely unnoticed. Rare applications of the method, known as binomial cokriging, include the study of the risk of childhood cancer in the West Midlands of England (Webster et al., 1994) and the mapping of lung cancer mortality in Long Island (Goovaerts, 2005a). A similar approach, named Poisson kriging, was developed more recently in the field of marine ecology (Monestiez et al., 2006) and generalized to the analysis of cancer mortality and cholera incidence data (Goovaerts, 2005b; Ali et al., 2006). Unlike the CAR model, the geometry of administrative units and the spatial repartition of the population at risk are accounted for in the geostatistical models (Goovaerts, 2006b), leading to more precise and accurate risk estimates than the Bayesian BYM model (Goovaerts and Gebreab, 2008). The BYM model also generates smoother risk surfaces, yielding much more false negatives than the geostatistical model in particular as the risk threshold raises. Last, area-to-point (ATP) Poisson kriging enables the creation of isopleth maps of mortality risk, which attenuates the visual bias associated with the interpretation of choropleth maps.

A limitation of all rate smoothers, including Poisson kriging, is that local details of the spatial variation of the risk are deleted from the maps. This smoothing has serious implications for local cluster analysis (LCA), since intuitively it should enlarge the size of clusters of low or high cancer risk while most spatial outliers would be filtered out. Static maps of estimated risk and kriging variance also fail to depict the spatial uncertainty attached to risk values and does not allow its propagation through multiple-point statistics such as local Moran's I in LCA. Goovaerts (2006a) proposed to combine Poisson kriging with a geostatistical simulation algorithm (p-field simulation) to generate multiple realizations of the spatial distribution of risk values. A set of simulated maps enables the quantification of how the spatial uncertainty about rates translates into uncertainty about the location of disease clusters (Goovaerts, 2006a), the presence of significant boundaries (Goovaerts, 2008a), or the relationship between health outcomes and putative risk factors (Goovaerts, 2009).

Another important contribution of geostatistical simulation is the generation of more realistic null hypotheses for statistical tests that are routinely performed by health scientists (e.g. to detect areas where mortality is significantly higher or lower than in adjacent geographical units). Most tests for spatial pattern are still based on the null hypothesis of spatial independence (SI) of observed rates and, provided the population sizes of areal units are fairly homogeneous, the assumption of constant or spatially uniform risk. The concept of "neutral model" (Goovaerts and Jacquez, 2004) allows the testing of more interesting hypotheses by accounting for spatial patterns and *a priori* information on the underlying risk in the formulation of null hypotheses. Geostatistical neutral models have been demonstrated to be useful for many types of applications, such as (1) the detection of significant clusters/outliers of breast cancer rates above and beyond a risk inferred from environmental covariates on Long Island, New York (Goovaerts, 2005a), (2) the identification of significant spatio-temporal changes in cervix cancer mortality rates above and beyond past spatial patterns (Goovaerts and Jacquez, 2005), (3) the assessment of significant clustering of residential histories in a case-control study of bladder cancer in Michigan (Jacquez et al., 2006), (4) the detection of significant differences in pancreatic cancer mortality between-county (boundary analysis, Goovaerts, 2008a), and (5) the study of the impact of demographic and economic factors on cervix cancer mortality in the Western US (Goovaerts, 2009).

This paper gives an overview of geostatistical methods for the analysis of both areal and individual-level health data, with applications to cancer studies.

2 Analysis of Areal Data

The analysis of areal data is illustrated for lung cancer which has been the leading cause of cancer deaths in the US for several decades. Figure 1a shows mortality rates for white males recorded over the period 1970–1994 for 688 counties of the Southeastern US. The population-weighted averaged mortality rate is 82.7 deaths per 100,000 person-years. The objectives of the study are threefold:

1. Create a reliable map of the spatial distribution of cancer mortality that accounts for small population sizes and the counties geography.
2. Identify groups of adjacent counties (i.e. local clusters) with significantly correlated low or high mortality rates.
3. Identify local clusters of mortality for males that exist above and beyond what should be expected based on female mortality rates that share similar environmental and socio-economic covariates.

2.1 Cancer Risk Mapping

The interpretation of cancer mortality maps is frequently biased by the large variation in the spatial support (e.g. county area) and level of confidence (small number

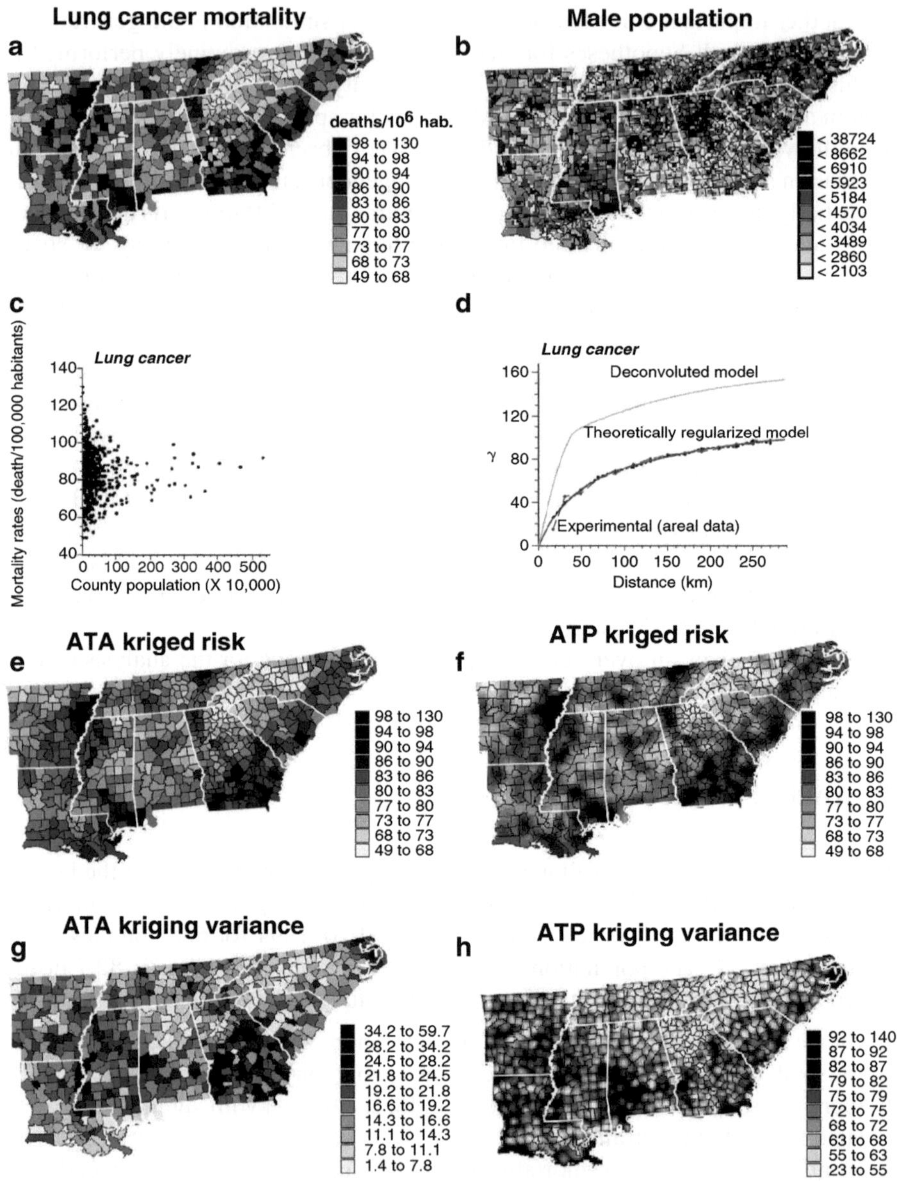

Fig. 1 (**a**) Lung cancer mortality rates recorded for white males (1970–1994) and (**b**) the population at risk assigned to 100 km^2 cells. (**c**) Scatterplot illustrates the larger variability of rates computed from sparsely populated counties. (**d**) Experimental semivariogram of the risk estimated from county-level rate, and the results of its deconvolution (*top curve*). The regularization of the point support model yields a curve (*short dashed line*) that is very close to the experimental one. The point-support model is then used to estimate lung cancer mortality risk (deaths/100,000 habitants) and associated prediction variance at the county level (ATA kriging) or at the nodes of a 10 km spacing grid (ATP kriging). Thick *white lines* delineate state boundaries

problem illustrated in Fig. 1c) of observation across the study area. For example, the areas and population sizes for counties in Fig. 1a vary by one and two orders of magnitude, respectively. Area-to-Area (ATA) and Area-to-Point (ATP) Poisson kriging allows the filtering of the spatially-varying noise while accounting for the heterogeneity in the shape, size and population repartition (Fig. 1b) among counties. The method, described in Goovaerts (2006b), proceeds as follows:

1. Compute the semivariogram of the risk from observed mortality rates (areal data), $z(v_\alpha)$, using the population-weighted estimator:

$$\hat{\gamma}_R(\mathbf{h}) = \frac{1}{2\sum_{\alpha,\beta}^{N(\mathbf{h})} \frac{n(v_\alpha)n(v_\beta)}{n(v_\alpha)+n(v_\beta)}} \sum_{\alpha,\beta}^{N(\mathbf{h})} \left\{ \frac{n(v_\alpha)n(v_\beta)}{n(v_\alpha)+n(v_\beta)} \left[z(v_\alpha) - z(v_\beta)\right]^2 - m^* \right\} \quad (1)$$

   where $N(\mathbf{h})$ is the number of pairs of areas (v_α, v_β) whose population-weighted centroids are separated by $\mathbf{h}$, and $n(v_\alpha)$ is the size of the population at risk.
2. Derive a point-support semivariogram model using an iterative deconvolution procedure (Goovaerts, 2008b) that seeks the point-support model that, once regularized, is the closest to the model fitted to areal data (Eq. (1)).
3. Estimate the noise-filtered mortality rate (mortality risk) and the associated kriging variance for the unit X using K neighboring rate data:

$$\hat{r}_{PK}(X) = \sum_{i=1}^{K} \lambda_i z(v_i) \qquad \sigma^2_{PK}(X) = C_R(0) - \sum_{i=1}^{K} \lambda_i \overline{C}_R(v_i, X) - \mu(X)$$

   where the unit X represents either an area v_α (ATA kriging) or a point $\mathbf{u}_s$ within that area (ATP kriging). The kriging weights and the Lagrange parameter $\mu(X)$ are computed by solving the "Poisson kriging" system:

$$\sum_{j=1}^{K} \lambda_j \left[\overline{C}_R(v_i, v_j) + \delta_{ij}\frac{m^*}{n(v_i)}\right] + \mu(X) = \overline{C}_R(v_i, X) \quad i = 1, \ldots, K$$

$$\sum_{j=1}^{K} \lambda_j = 1. \quad (2)$$

   where $\delta_{ij} = 1$ if $i = j$ and 0 otherwise, and m^* is the population-weighted mean of the N rates. The "error variance" term, $m^*/n(v_i)$, leads to smaller weights for less reliable data (i.e. rates measured over smaller populations). The area-to-area covariances $\overline{C}_R(v_i, v_j)$ and area-to-point covariances $\overline{C}_R(v_i, X = \mathbf{u}_s)$ are approximated as the average of the point support covariance $C(\mathbf{h})$ computed between any two locations discretizing the areas v_i and v_j, or v_i and $\mathbf{u}_s$.

Figure 1d shows the areal and point-support models inferred from 688 rate data. As expected, the point-support model (light gray curve) has a higher sill and its regularization (short dashed line) yields a semivariogram model that is close to the one fitted to experimental values, which validates the consistency of the deconvolution. This model was used to estimate mortality risk at the county level (ATA kriging) and to map the spatial distribution of risk within counties (ATP kriging). Both maps (Figs. 1e, f) are smoother than the map of raw rates since the noise due to small population sizes is filtered. High mortality is observed along the Mississippi valley (MS), the southern Atlantic (GA, SC) and across the Gulf Coast. Smoking patterns largely account for the regional variation in lung cancer mortality; for example, smoking habits, including the greater use of hand-rolled cigarettes, were found to contribute to the high rates in southern Louisiana (LA), especially in the Cajun population. In the 1970s and early 1980s, studies in coastal Georgia (GA), northeast Florida, and southern Louisiana (LA) also revealed an excess risk of lung cancer associated with work in shipyards, primarily during World War II (Devesa et al., 1999). North Carolina (NC) displays a clear East-West trend, with lower mortality in the more rural Western counties. By construction, aggregating the ATP kriging estimates within each county using the population density map (Fig. 1b) yields the ATA kriging map. The maps of kriging variance essentially reflect the lower confidence in risk estimated for sparsely populated counties and over smaller spatial support (i.e. ATP kriging).

2.2 Detection of Local Clusters of High and Low Mortality

A major goal of spatial analysis in public health is to detect local clusters (regions where adjacent areas have similar values) of high or low cancer mortality. Similarity between the rate measured within area v_α and those recorded in $J(v_\alpha)$ adjacent areas v_β (e.g. units sharing a common border or vertex with the kernel v_α) is often quantified by the local Moran's I statistic (Anselin, 1995) defined as:

$$I(v_\alpha) = \left[\frac{z(v_\alpha) - m}{s}\right] \times \left(\sum_{j=1}^{J(v_\alpha)} \frac{1}{J(v_\alpha)} \times \left[\frac{z(v_j) - m}{s}\right]\right) \tag{3}$$

where m and s are the mean and standard deviation of the set of N rates. This Local Indicator of Spatial Association (LISA) is simply the product of the kernel rate by the average of neighboring rates and can detect both positive and negative autocorrelations. It exceeds zero if the kernel and neighborhood averaged rates jointly exceed the global mean m (High-High, HH cluster) or are jointly below m (Low-Low, LL cluster). The uncertainty attached to mortality rates is propagated through the computation of the LISA statistic by replacing in Eq. (3) the rates $z(v_\alpha)$ by spatially correlated values computed as: $r^{(l)}(v_\alpha) = \hat{r}_{PK}(v_\alpha) + \sigma_{PK}(v_\alpha)w^{(l)}(v_\alpha)$, leading to a set of L simulated LISA values $\{I^{(l)}(v_\alpha), 1 = 1, \ldots L\}$. The L sets of random deviates, $\{w^{(l)}(v_\alpha), \alpha = 1, \ldots N\}$, are generated using non-conditional sequential

Gaussian simulation and the semivariogram of the risk, $\gamma_R(\mathbf{h})$, rescaled to a unit sill; see Goovaerts (2006a) for a detailed description of the p-field simulation algorithm.

To test whether any test statistic, $I^{(l)}(v_\alpha)$, is significantly greater than 0 (i.e. presence of spatial autocorrelation), one needs to know its probability distribution under the null hypothesis of spatial independence (SI). The common way to generate such reference distribution is to shuffle randomly the set of simulated rates, then use the shuffled values to compute the neighbourhood average in statistic (3) while the kernel rate remains the same. In other words, the value of the LISA statistic is computed for the scenario where the rates in adjacent areas are randomly distributed. This operation is repeated K times, i.e. $K = 999$ in this paper. Comparing the observed statistic (3) to the probability distribution enables the computation of the probability of not rejecting the null hypothesis of SI (p-value). The main drawback of this randomization procedure is that both the underlying mortality risk and population size are assumed uniform across the study area. To account for the population size, the random shuffling is replaced by the random sampling of a Poisson distribution $\text{Po}(\text{n}(v_j) \times m)$, where $\text{n}(v_j)$ is the size of the population at risk and m is the population-weighted average of rates.

The last step in the testing procedure is to compare the p-value to the significance level (e.g. 0.05 or 0.01) representing the risk of false positive (i.e. risk of rejecting the null hypothesis when it is true) that the user can tolerate. However, the repeated use of statistical tests (e.g. one for each county) increases the likelihood of false positives. For example, the independent testing of ten counties under a significance level of 0.05 will lead to a 0.4 probability that at least one test is significant even if none of the ten counties actually exhibits spatial autocorrelation with adjacent counties. In this paper, the multiple testing correction was conducted using the false discovery rate (FDR) approach that aims to control the expected proportion of true null hypotheses rejected out of the total number of rejections (Castro and Singer, 2006).

One hundred realizations (a number deemed reasonable for this application) of the spatial distribution of lung cancer mortality risk were generated by p-field simulation and underwent a local cluster analysis using a 0.01 significance level. Figure 2 shows the probability for each county to belong to a LL or HH cluster, which corresponds to the proportion of realizations for which the county falls within that category. Accounting for the population size in the randomization (Null hypothesis II) reduces the spread of the reference distribution resulting in smaller p-value, hence more significant tests in particular for heavily populated counties along the coast. For example, the analysis reveals clusters of high mortality around New Orleans, in coastal counties of North and South Carolina, and at the border of Tennessee (TN) and Georgia (GA) in Chattanooga that once held the unwelcome title of having the dirtiest air in the United States, a label provided by the federal government in 1969. Similarly, several LL clusters appear on the Null Hypothesis II maps; for example Benton County in the Northwest corner of Arkansas (AR) which has the second highest population and the lowest poverty rate of any county in the state.

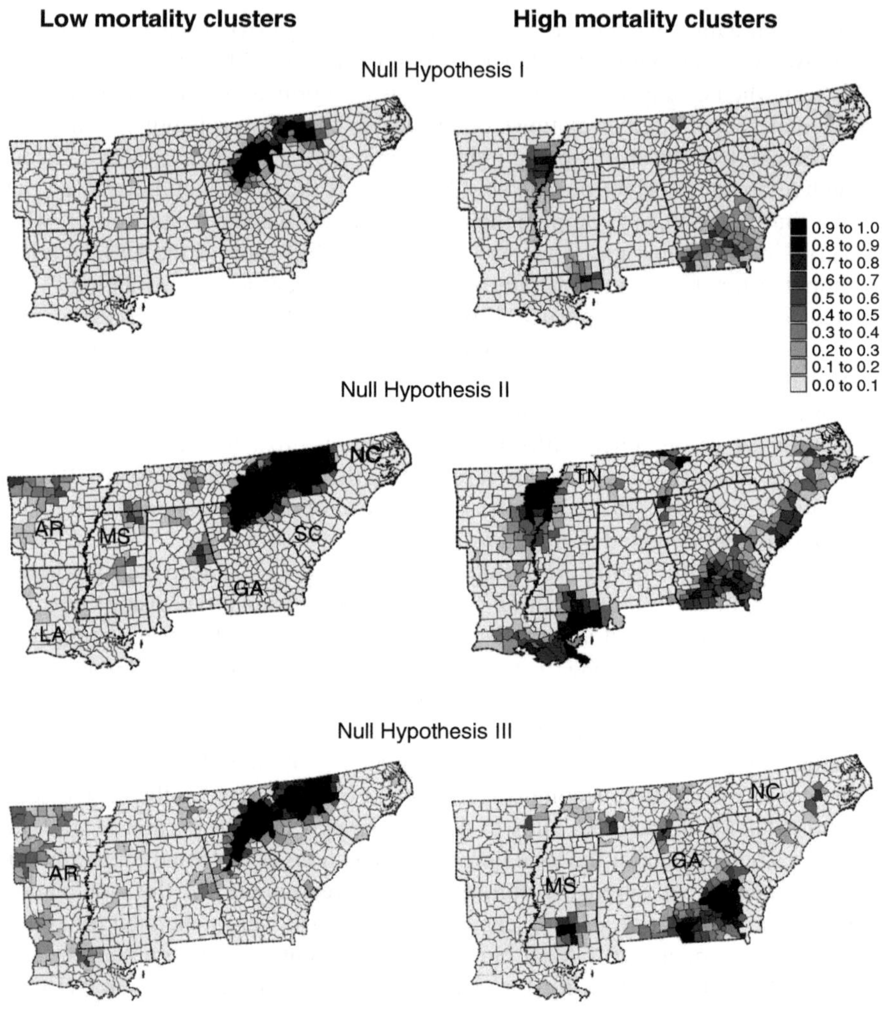

Fig. 2 Likelihood that a county belongs to a cluster of low or high cancer mortality computed from the local cluster analysis of 100 simulated risk maps. Three null hypotheses of increasing complexity are considered: uniform risk and population size (model I), uniform risk and heterogeneous population sizes (model II), heterogeneous risk (i.e. based on white female mortality) and population sizes (model III). Thick dashed lines delineate state boundaries

2.3 Tests of Hypothesis Using Spatial Neutral Models

Both null hypotheses I and II share the same assumption of uniform risk for lung cancer mortality. Yet, this risk clearly varies regionally as a result of changes in environmental exposure or other demographic, social, and economic factors (Devesa et al., 1999). The term "Neutral Model" captures the notion of a plausible system

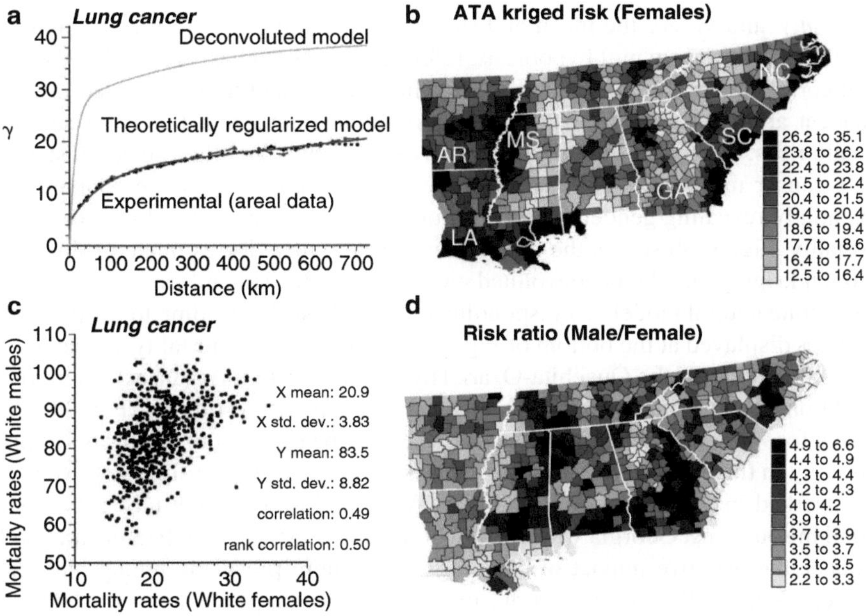

Fig. 3 (**a, b**) Cancer mortality risk estimated for white females using Poisson kriging and the deconvoluted semivariogram of the risk. (**c, d**) Scatterplot of risk estimates for males and females, and county-level map of their ratio

state that can be used as a reasonable null hypothesis. The problem then is to identify spatial patterns above and beyond that incorporated into the neutral model, enabling, for example, the identification of "hot spots" beyond background variation in a pollutant. Although lung cancer mortality is on average four times larger for males than females, the same environmental and socio-economic factors come into play, as exemplified by the 0.5 correlation computed between male and female risk estimates (Fig. 3c). Discrepancies between risk maps can be interpreted as signs of the local impact of gender-specific factors, such as occupational exposure to coal tar fumes in coal gasification and coke production, or asbestos in marine construction and repair, steel and iron mills, power generating stations, pulp and paper mills, and oil refineries.

Information on the spatial pattern of mortality risks for white females was incorporated into the generation of neutral models for the local cluster analysis. Specifically, the reference probability distribution of the local Moran's I statistic was inferred from $K = 999$ realizations of the spatial distribution of white female mortality risks, $\{y^{(k)}(v_\alpha), \alpha = 1, \ldots N\}$, generated by p-field simulation. The test statistic (3) is thus compared to the following set of simulated values:

$$I^{(k)}(v_\alpha) = \left[\frac{z(v_\alpha) - m}{s}\right] \times \left(\sum_{j=1}^{J(v_\alpha)} \frac{1}{J(v_\alpha)} \times \left[\frac{y^{(k)}(v_j) - m_Y}{s_Y}\right]\right) k = 1, \ldots, K \tag{4}$$

where m_Y and s_Y are the mean and standard deviation of the set of N simulated female rates. The new null hypothesis, referred to as Hypothesis III, is that the positive correlation between the male risk within area v_α and those recorded in J(v_α) adjacent areas v_j does not exceed the correlation found with the J(v_α) adjacent female risks. Rejecting the null hypothesis should thus highlight local clusters of male cancer mortality risks that are more similar than their female counterparts, potentially revealing gender-specific factors causing aggregates of lower or higher mortality. Figure 3b shows the map of female mortality risk estimated using ATA Poisson kriging and the deconvoluted model in Fig. 3a. This information was used to generate neutral models corresponding to Hypothesis III, leading to the probability maps displayed at the bottom of Fig. 2. Clusters of lower mortality are enhanced in the mountains of the Ouachita-Ozark Highlands (Western Arkansas) where rural conditions lead to smaller ratio of male versus female risk estimates (Fig. 3d). Adjusting for white female mortality leads to the disappearance of most clusters of high mortality in the coastal areas of Louisiana, South and North Carolina since females also worked in the shipbuilding industry. A large cluster of high mortality is still found in southwest Georgia where low income and poor access to health care likely enhance the negative impact of smoking that is more prevalent among men than women. A new HH cluster appears in central Mississippi, centered on Jefferson Davis County where the mortality risk for males is 6.44 times larger than for females (Fig. 3d). Another County with high male/female risk ratio (5.34) is Sampson County in North Carolina, which is adjacent to the two largest hog-producing counties in the United States.

3 Analysis of Individual-Level Data

Despite methodological advancements in the treatment of areal data, the degree of details in the isopleth risk maps will always be limited by the initial resolution of the choropleth map. Whenever possible, it is thus beneficial to avoid the tedious, arbitrary and inherently information-wasteful aggregation step and to process directly the point-based data. In addition to the greater accuracy in the location of health outcomes, the analysis of geocoded data can often capitalize on detailed information on residential history and a large number of potential risk factors.

Estimation and mapping of the spatial risk function requires the computation of the ratio of the case density to the population density. Using 'kernel density estimation methods', the number of cases and the total number of individuals at risk are simply summed within sliding windows and their ratio defines the rate assigned to the center (i.e. grid node) of that window (James et al., 2004). The operation is repeated for each grid node to create isopleth maps of, for example, late-stage cancer rates (ratio of number of late-stage cancer cases to total number of cancer patients). Unlike kernel density estimation, geostatistics takes into account the spatial support of the data and the pattern of spatial dependence (e.g. anisotropy, range of autocorrelation) in the computation of the weights assigned to neighboring data.

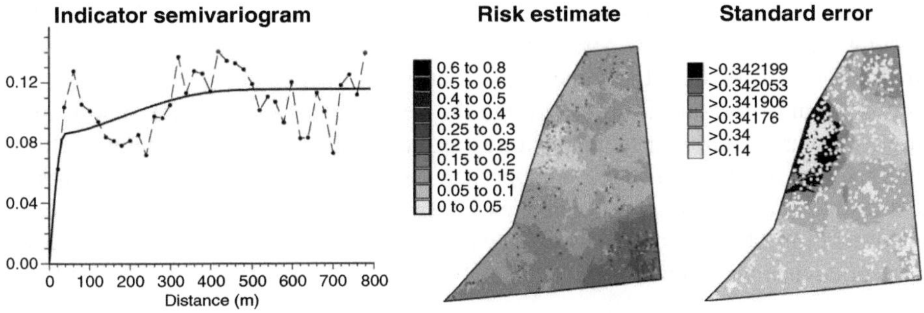

Fig. 4 Maps of late-stage breast cancer risk estimate, and the attached standard error, for white females in one Michigan County (1985–2002). The estimation was conducted at the nodes of a 200 m grid by indicator kriging of geocoded data

Each observation represents the probability (0 or 1) that the individual is a case (e.g. late stage cancer, birth defect), hence the analysis should be conducted using indicator kriging of categorical data. Since kriging is a non-convex interpolator, estimated probabilities can fall outside the range [0,1] and the faulty probabilities should be reset to the nearest bound, 0 or 1. This situation was not encountered in the present study.

Figure 4 illustrates the application of indicator geostatistics to the spatial distribution of late stage diagnosis of breast cancer within one Michigan County. The first step was to code each cancer case (N = 1,317) as 1 for late stage diagnosis and 0 otherwise (data not shown for confidentiality reasons). The indicator semivariogram (Fig. 4, left column) indicates that late detection cases do not occur randomly in space, yet individual-level factors such as age or family history generate a large variability over very short distances (first range = 40 m). The long-range structure (600 m) reflects the impact of contextual (i.e. neighborhood) factors, such as poverty and proximity to screening facilities. Capitalizing on this autocorrelation, indicator kriging (IK) was used to map the late-stage cancer risk and the standard error. These maps clearly illustrate a NW-SE increasing trend in the risk of late diagnosis, which should help selecting areas to be preferentially targeted for cancer screening and prevention activities.

4 Conclusions

The major difficulty in the analysis of health outcomes is that the patterns observed reflect the influence of a complex constellation of demographic, social, economic, cultural and environmental factors that likely change through time and space, and interact with the different types and scales of places where people live. It is thus primordial to incorporate the scale and spatial support of the data in their processing, as well as to account for the impact of population sizes on the reliability of rate estimates. Geostatistics provides a methodology to model the spatial correlation among rates measured over irregular geographic supports and to compute noise-free risk estimates over the same units or at much finer scales. It also enables the propagation of rate uncertainty through the delineation of areas with significantly higher/lower mortality or incidence rates, as well as the simulation of more realistic null hypotheses. In the future, the approach should be generalized to the multivariate case to analyze spatial relationships among diseases, which should facilitate the identification of common stressors, such as poverty level, lack of access to health care or environmental pollution.

Acknowledgments This research was funded by grant R44-CA132347-01 from the National Cancer Institute. The views stated in this publication are those of the author and do not necessarily represent the official views of the NCI.

References

Ali M, Goovaerts P, Nazia N, Haq MZ, Yunus M, Emch M (2006) Application of Poisson kriging to the mapping of cholera and dysentery incidence in an endemic area of Bangladesh. Int J Health Geogr 5:45

Anselin L (1995) Local indicators of spatial association – LISA. Geogr Anal 27:93–115

Besag J, York J, Mollie A (1991) Bayesian image restoration with two applications in spatial statistics. Ann Inst Stat Math 43:1–59

Castro MC, Singer BH (2006) Controlling the false discovery rate: a new application to account for multiple and dependent tests in local statistics of spatial association. Geogr Anal 38:180–208

Devesa SS, Grauman DJ, Blot WJ, Fraumeni JF Jr (1999) Cancer surveillance series: changing geographic patterns of lung cancer mortality in the United States, 1950 through 1994. J Natl Canc Inst 91(12):1040–1050

Goovaerts P (2005a) Detection of spatial clusters and outliers in cancer rates using geostatistical filters and spatial neutral models. In: Renard Ph, Demougeot-Renard H, Froidevaux R (eds) geoENV V – geostatistics for environmental applications. Springer, Berlin/Germany, pp 149–160

Goovaerts P (2005b) Geostatistical analysis of disease data: estimation of cancer mortality risk from empirical frequencies using Poisson kriging. Int J Health Geogr 4:31

Goovaerts P (2006a) Geostatistical analysis of disease data: visualization and propagation of spatial uncertainty in cancer mortality risk using Poisson kriging and p-field simulation. Int J Health Geogr 5:7

Goovaerts P (2006b) Geostatistical analysis of disease data: accounting for spatial support and population density in the isopleth mapping of cancer mortality risk using area-to-point Poisson kriging. Int J Health Geogr 5:52

Goovaerts P (2008a) Accounting for rate instability and spatial patterns in the boundary analysis of cancer mortality maps. Environ Ecol Stat 4:421–446

Goovaerts P (2008b) Kriging and semivariogram deconvolution in presence of irregular geographical units. Math Geosci 40:101–128

Goovaerts P, Gebreab S (2008) How does Poisson kriging compare to the popular BYM model for mapping disease risks? Int J Health Geogr 7:6

Goovaerts, P. 2009. Medical geography: a promising field of application for geostatistics. Math. Geosci 41(3):243–264

Goovaerts P, Jacquez GM (2004) Accounting for regional background and population size in the detection of spatial clusters and outliers using geostatistical filtering and spatial neutral models: the case of lung cancer in Long Island, New York. Int J Health Geogr 3:14

Goovaerts P, Jacquez GM (2005) Detection of temporal changes in the spatial distribution of cancer rates using LISA statistics and geostatistically simulated spatial neutral models. J Geogr Syst 7:137–159

Jacquez GM, Meliker JR, AvRuskin G, Goovaerts P, Kaufmann A, Wilson ML, Nriagu J (2006) Case-control geographic clustering for residential histories accounting for risk factors and co-variates. Int J Health Geogr 5:32

James L, Matthews I, Nix B (2004) Spatial contouring of risk: a tool for environmental epidemiology. Epidemiology 15:287–292

Kelsall J, Wakefield J (2002) Modeling spatial variation in disease risk: a geostatistical approach. J Am Stat Assoc 97(459):692–701

Lajaunie C (1991) Local risk estimation for a rare noncontagious disease based on observed frequencies. Note N-36/91/G. Centre de Géostatistique, Ecole des Mines de Paris

Monestiez P, Dubroca L, Bonnin E, Durbec JP, Guinet C (2006) Geostatistical modelling of spatial distribution of *Balenoptera physalus* in the Northwestern Mediterranean Sea from sparse count data and heterogeneous observation efforts. Ecol Model 193:615–628

Richardson S, Thomson A, Best N, Elliot P (2004) Interpreting posterior relative risk estimates in disease-mapping studies. Environ Health Perspect 112:1016–1025

Webster R, Oliver MA, Muir KR, Mann JR (1994) Kriging the local risk of a rare disease from a register of diagnoses. Geogr Anal 26:168–185

Woodward P (2005) BugsXLA: Bayes for the common man. J Stat Softw 14:5

Blocking Markov Chain Monte Carlo Schemes for Inverse Stochastic Hydrogeological Modeling

J. Jaime Gómez-Hernández and Jianlin Fu

Abstract Uncertainty characterization generally calls for a Monte Carlo analysis of many equally likely realizations that honor both direct information (e.g., conductivity data) and information about the state of the system (e.g., piezometric head or concentration data). The problem faced is how to generate multiple realizations conditioned (to parameter data) and inverse-conditioned (to dependent state data) over a large domain with high resolution. Traditional McMC methods face a big challenge in inverse-conditioning because of its slow convergence. In this study, we comment on several block updating schemes to improve the convergence performance of McMC.

1 Introduction

In the last decade, the problem of inverse conditional modeling has been recognized to be of paramount importance, especially in the hydrogeology and petroleum engineering fields, and a number of additional approaches have been developed for this purpose. Worth mentioning are the gradual deformation method (Hu, 2000), Markov chain Monte Carlo (Oliver et al., 1997), and the ensemble Kalman filter (Naevdal et al., 2003). The McMC methods do not see their wide application in engineering communities mainly because of their inefficiency in convergence velocity. In this paper, we follow the pioneering work of Oliver et al. (1997) in hydrogeology and propose several blocking Markov chain Monte Carlo (BMcMC) schemes to overcome the shortcomings of traditional McMC, e.g., to improve the convergence velocity and enhance the mixing speed.

J.J. Gómez-Hernández (✉) and J. Fu
Department of Hydraulic and Environmental Engineering, Universidad Politécnica de Valencia, 46071 Valencia, Spain
e-mail: jaime@dihma.upv.es; jianfu@dihma.upv.es

P.M. Atkinson and C.D. Lloyd (eds.), *geoENV VII – Geostatistics for Environmental Applications*, Quantitative Geology and Geostatistics 16,
DOI 10.1007/978-90-481-2322-3_11,

2 Bayesian Formulation

Consider a random function (RF) discretized at n grid nodes. Assume that there are k nonlinear state data, where the term "nonlinear" means that the dependent state data are a non-linear function of the model parameters (this non-linear function is implicitly given by the flow and transport partial differential equations). Specifically, let $x = (x_0, x_1, \ldots, x_{n-1})^T$ denote the RF, and $y = y_{\text{obs}} = (y_0, y_1, \ldots, y_{k-1})^T$ denote the k dependent state data. The joint prior probability density function (pdf) assuming a multi-Gaussian random field x is,

$$\pi(x) \propto \exp\left(-\frac{1}{2}(x - \mu_x)^T C_x^{-1}(x - \mu_x)\right) \tag{1}$$

where μ_x is the prior mean and C_x is the prior covariance. Assuming a multi-Gaussian error for the discrepancy between the observed state y and the state predictions resulting from the approximate solution of the state equations, the joint pdf of y conditional on a realization of the parameters x is given by,

$$\pi(y|x) \propto \exp\left(-\frac{1}{2}(y - g(x))^T C_y^{-1}(y - g(x))\right) \tag{2}$$

where $g(x)$ represents the state function (forward simulator) and C_y is the measurement error covariance, generally a diagonal matrix.

Following traditional McMC implementations (see, for instance, Oliver et al., 1997), after defining a transition distribution $q(x^*|x)$, a Markov chain can be built by drawing realizations from this transition distribution and retaining those that pass a Bernouilli trial with the following acceptance probability,

$$\alpha = \min\left(1, \frac{\pi(x^*)\pi(y|x^*)q(x|x^*)}{\pi(x)\pi(y|x)q(x^*|x)}\right) \tag{3}$$

The chain will converge to a series of realizations that is inverse conditional to the state data.

3 Blocking Markov Chain Monte Carlo

In this section, we will present several McMC schemes to improve the convergence velocity and mixing speed, which are critical for inverse stochastic hydrogeological modeling since the forward simulator $g(x)$ is usually very computationally demanding. What makes our method different from Oliver et al. (1997) is that the proposed member x^* is built from the previous member x by modifying a large block of grid nodes, instead of building the Markov chain modifying a node at a time. The first three schemes are essentially block updating McMC methods but have distinct convergence performance, while the other two are coupled McMC schemes that aim at combining together the best performances of the first three schemes.

The modification of the updating block is done by simulating the conductivity values within the block conditional to the values in the remainder of the previous realization. (For practical purposes, only the values in the skin next to the updating block, plus the prior conditioning data inside the block are retained.)

3.1 Scheme #1

If a field is small enough such that the inverse of the simple kriging covariance matrix C_x can be LU-decomposed, $C_x = LU$, in which case the acceptance rate in Eq. (3) can be easily computed using the method described by Davis (1987) and Alabert (1987) for the fast generation of conditional realizations via the LU decomposition, because this approach not only provides the values of the updating block but also the conditional covariance needed.

3.2 Scheme #2

If the field is too large such that the inverse of the simple kriging covariance matrix cannot be performed, we propose an approximation for the entire field by reducing the extent of the field to a smaller area centered in the block being updated, then the LU decomposition of the covariance matrix for this block can be performed.

3.3 Scheme #3

Finally, if the field is very large and the block being updated is also very large, we take the decision to use an independent proposal kernel in which the updating block is not made conditional on the previous realization but only on the prior conditioning data x_1.

3.4 Scheme #4

From our experience, it seems that a large updating block improves the convergence velocity of the Markov chain. It is also apparent that for the same blocking size, scheme #3 is the faster to converge. On the other hand, once the chain has converged, scheme #2 and a smaller updating block gives a better mixing of the chain. Thus, we propose a mixed scheme in which a large updating block and scheme #3 are used to drive the chain into the region of convergence, then, after a "burn-in" period after convergence, the proposal scheme is switched to a smaller updating block and scheme #2 to compute the acceptance rate.

3.5 *Scheme #5*

The scheme #4 can be modified to further improve mixing. Consider two separate chains that evolve in parallel: one is constructed by a large updating block and scheme #3, and the other with a small updating block and scheme #2. At each stage for a given number of iterations, the two chains exchange accepted members to form coupled Markov chains, which, in the end, should benefit from the strengths of each chain.

This coupled Markov chain concept leads us to propose a BMcMC similar to scheme #4 in which we switch back and forth between the two contributing chains. First, we run scheme #3, then we switch to scheme #2 but after generating a number of realizations, we switch back to scheme #3 to locate a new mode of the posterior distribution, and then switch to scheme #2 to generate realizations around this new mode, and so on, and so forth. This combination should produce a better and faster mixing.

4 A Numerical Experiment

A 2-D aquifer discretized on 32 by 32 cells with zero logconductivity mean and unit variance, and following an exponential covariance with practical range of 16 cells, was generated. In this aquifer a transient flow problem was modeled with piezometric heads measured at nine locations. The objective was to generate realizations of logconductivity with the same statistics and conditional to the nine transient piezometric head data.

Figure 1 shows the mismatch between predicted heads and measured heads as the chain progresses for different blocking schemes. Depending on the blocking schemes, the number of "runs" necessary to get a series of realizations that reproduce (to within measurement error) the piezometric head data, goes from 1,000 to 100,000. After this, which is termed the burn-in period in the McMC literature, realizations are properly conditioned. Although not presented here, scheme #2 almost has the same convergence velocity as scheme #1, indicating that using a subset to approximate the entire field is proper and efficient in computing the prior density by Eq. (1). Scheme #3 has a much better convergence velocity than scheme #1 and #2. Scheme #4 or #5 obviously has the same convergence velocity as scheme #3 but with a mixing speed enhanced. Figure 2 demonstrates such improvement in terms

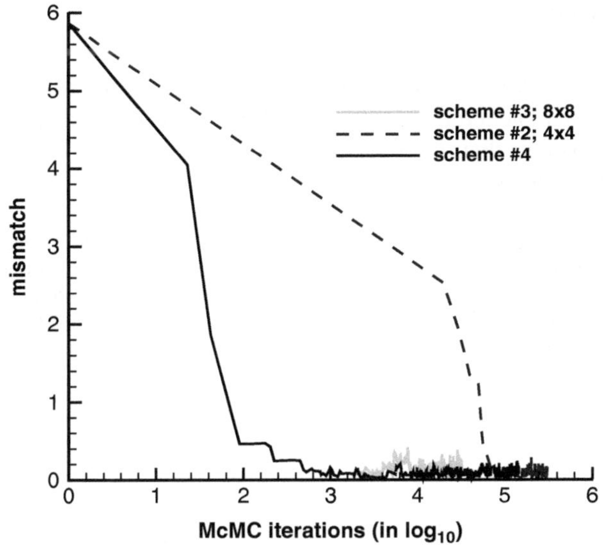

Fig. 1 A comparison on the convergence velocity of several blocking schemes

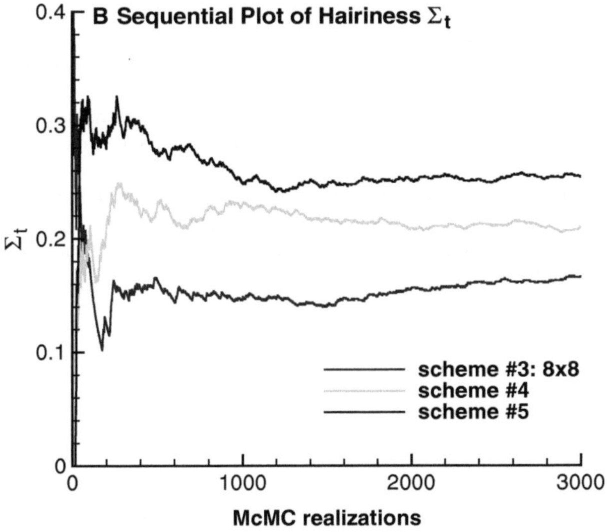

Fig. 2 A comparison on the mixing speed of several blocking schemes

of the mixing speed. The metric of mixing speed uses the so-called hairiness index that was originally developed by Brooks (1998) based on a method proposed by Yu and Mykland (1998). An ideal convergence sequence will have a hairiness index around 0.5.

5 Conclusions

Several blocking McMC schemes are presented to improve the convergence velocity of a Markov chain in generating inverse conditional realizations. A synthetic numerical experiment shows that the block updating schemes can efficiently improve the convergence performance.

Acknowledgements The first author thanks Universidad Politécnica de Valencia for a sabbatical grant during the preparation of this manuscript. The second author also thanks Universidad Politécnica de Valencia for a fellowship that supported him through his doctoral studies. The work on this manuscript also benefited from financial support from the Spanish Ministry of Education and Science through project CGL02004–2008, and from the European Commission through integrated project FI6W-516514.

References

Alabert F (1987) The practice of fast conditional simulations through the LU decomposition of the covariance matrix. Math Geol 19(5):369–386

Brooks SP (1998) Quantitative convergence assessment for Markov chain Monte Carlo via cusums. Stat Comput 8(3):267–274

Davis MW (1987) Production of conditional simulations via the LU triangular decomposition of the covariance matrix. Math Geol 19(2):91–98

Hu LY (2000) Gradual deformation and iterative calibration of gaussian-related stochastic models. Math Geol 32(1):87–108

Naevdal G, Johnsen M, Aanonsen SI, Vefring E (2003) Reservoir monitoring and continuous model updating using the ensemble Kalman filter, SPE Annual Technical Conference and Exhibition, SPE 84372

Oliver DS, Cunha LB, Reynolds AC (1997) Markov chain Monte Carlo methods for conditioning a log-permeability field to pressure data. Math Geol 29(1):61–91

Yu B, Mykland P (1998) Looking at Markov samplers through cusum path plots: a simple diagnostic idea. Stat Comput 8(3):275–286

Simulation of Fine-Scale Heterogeneity of Meandering River Aquifer Analogues: Comparing Different Approaches

Diana dell'Arciprete, Fabrizio Felletti, and Riccardo Bersezio

Abstract We compare different approaches to fine scale simulation of aquifer heterogeneity of meandering river depositional elements, based on the study of a 3-D quarry exposure of historical point bar-channel sediments of the Lambro River (Po plain, Northern Italy). The starting point is a sedimentological and hydrostratigraphic hierarchic model obtained after mapping of five quarry faces with centimeter-scale detail. The vertical facies maps show the shape and size of two superimposed composite bars, of their component unit bars and channel fills and the distribution of the individual facies within them. Textural and poro-perm analyses allowed the definition of the properties of four basic hydrofacies (Open Framework Gravels, Gravelly Sands and Sandy Gravels, Clean Sands, Sandy Silts and Clays), with permeability contrasts by at least one order of magnitude ($10^{-9} < K < 10^{-1}$). The correlation of hydrofacies has been quantified after discretization of the maps with square cells (side 0.05 m), by both transition-probability geostatistics and variographic analysis, to support 3-D pixel-oriented simulation of the volume. We found a high level of correspondence between the semivariogram ranges and the experimental transition probabilities computed on the entire dataset. Several realizations of 3-D conditioned simulations, that honour the vertical facies maps, were computed using Sequential Indicator Simulation (SIS) and T-Progs (transition-probability geostatistics software). Both methods yield more realistic results if the highest rank depositional elements are simulated separately than if the sedimentary volume is simulated on the whole. Image analyses on random sections through selected realizations shows that, in this specific case, SIS yields the most realistic simulations. However, both techniques are not capable of accounting for trends of depositional features that determine a non-stationary behaviour at the facies scale.

D. dell'Arciprete (✉), F. Felletti, and R. Bersezio
Dipartimento Scienze della Terra – Università di Milano, via Mangiagalli 34, 20133 I-Milano, Italy
e-mail: diana.dellarciprete@unimi.it; fabrizio.felletti@unimi.it; riccardo.bersezio@unimi.it

P.M. Atkinson and C.D. Lloyd (eds.), *geoENV VII – Geostatistics for Environmental Applications*, Quantitative Geology and Geostatistics 16,
DOI 10.1007/978-90-481-2322-3_12,

1 Introduction

The hydrogeological properties of fluvial sediments are determined by textural variations within the hierarchic arrangement of depositional units, from individual strata to depositional systems, and by the geometry of these units, at different scales (Jordan and Prior, 1992; Lunt et al., 2004; Bridge and Lunt, 2006; Rubin et al., 2006). This complex heterogeneity, which is characterized by multiple scale lengths, affects groundwater flow and contaminant transport. Correlation models of this spatial variability are obtained from various geostatistical tools, such as indicator variograms and transition probabilities. The variogram describes the degree of correlation between two points over a range of separation distances. The form of the experimental semivariogram yields information about the correlation length scales (Johnson and Dreiss 1989, Rubin and Journel, 1991; Johnson, 1995; Ritzi et al., 2004; Dai et al., 2004). Transition probabilities (Carle and Fogg, 1996) describe the spatial correlation structure of a sediment volume in a similar way, computing the probability that a transition from one class to another (e.g., from a facies to another) occurs over a range of separation distances. These tools have been applied extensively to describe braided river aquifers (Rubin et al., 2006 with references). Relatively few examples of statistical description, simulation and modelling of meandering river aquifers have been presented so far (Bierkens and Weerts, 1994; Kostic and Aigner, 2007).

In this work we compare the results yielded by two different techniques of geostatistical description and simulation (SIS, see for example Goovaerts [1997] with references, and T-Progs, Carle and Fogg, 1996) applied to an outcrop aquifer analogue exposing composite point bar and channel systems of a monocursal meandering river. The main goal is evaluating to what extent pixel-oriented techniques can reproduce complexity, at the fine scale, in the extremely heterogeneous case of meandering river sediments.

2 Case History and Methods

Excavation of gravel and sand in the Po alluvial plain (northern Italy, Fig. 1a) offers several ephemeral exposures of different types of fluvial aquifer analogues. For this study we had the opportunity to investigate the historical sediments of the Lambro River at a quarry site just south of Milan (Fig. 1a). In this sector the Lambro River is a monocursal, meandering river, flowing since the post-glacial age within a narrow valley encased into the Upper Pleistocene sandur of the Lecco glacial amphitheatre. The quarry site (Fig. 1b) exposes two superimposed depositional units, formed by sands and gravels, that could be attributed to an historical age after discovery of Roman to Middle Age and Renaissance Age artifacts (bricks, tiles, ceramics), imbricated within dunes and bars (Bersezio et al., 2007). The two units correspond to the exposed parts of two composite point bars with minor channel fills on top. We named them respectively unit A (the lower, with Roman-Middle Age artefacts) and

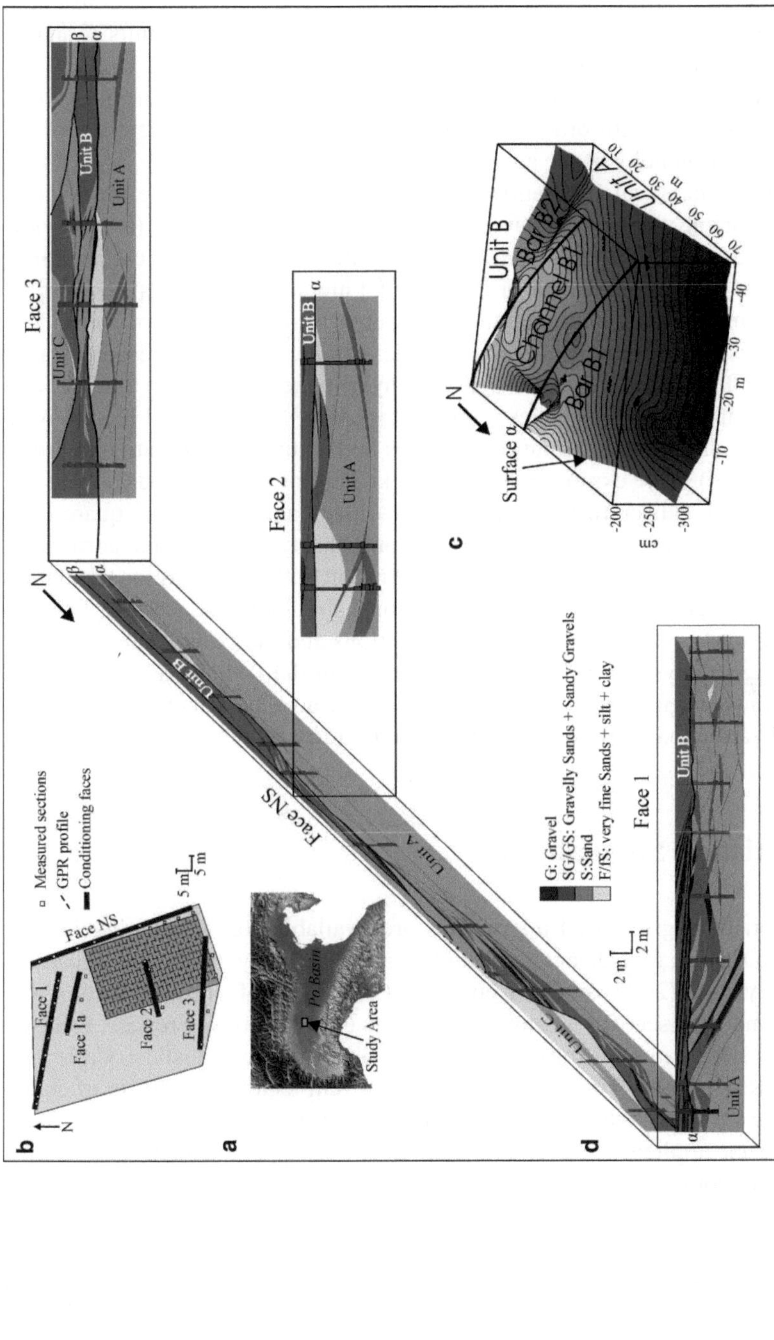

Fig. 1 **(a)** Location of the study area. **(b)** Scheme of the study site. The location and orientation of the pit wall exposures (Face NS, Face 1, 2 and 3) and of the GPR profiles is drawn. **(c)** Kriging of the erosion surface α between unit A and B. **(d)** Vertical facies maps of Face 1 and Face NS that include the shape and hierarchy of the depositional units, the distribution of the four operative hydrofacies

unit B (the upper, with Renaissance Age artefacts). Unit A shows a composite point bar to main channel fill lateral transition, unit B is mostly represented by a composite point bar, with chute channel scour and fills on top. The erosion surface between them, tapered by lag deposits, is the α surface of Fig. 1c. A later channel (unit C, bounded by erosion surface β) removed part of unit B, and will not be considered here, because of the very poor observation available. Together with units A and B, it is cut by the modern and present-day courses of the Lambro river. Units A and B are formed by a hierarchic arrangement of depositional units, from the second order of bedsets to the fifth order of the bar/channel systems, that determines the architectural heterogeneity of the aquifer analogue.

To obtain geostatistical simulations comparable with the available observation of the real sediment volume, we developed the following procedure: (i) plan-view and vertical mapping of morphology and sedimentary facies, taking care to faithfully reproduce geometry and size of the different hierarchic elements; (ii) vertical logging with cm-scale resolution, textural and petrophysical analysis of facies. The products of these steps consist of the hierarchic classification and interpretation of the sediments, definition of the four operative hydrofacies reported in Table 1, and representation of five vertical facies maps that include the shape and hierarchy of the depositional units, the distribution of facies and of the four operative hydrofacies (Fig. 1d); (iii) GPR exploration, to assist kriging of the α boundary between A and B units; (iv) discretization of the facies maps and logs with square cells (0.05×0.05 m); (v) variographic and transition probability description of correlation of hydrofacies within the vertical maps; (vi) definition of the domains for simulation (units A and B, α and β surfaces); (vii) conditional simulation by SIS and T-Progs; (viii) comparisons between the different realizations and the geological model, assisted by image analysis of selected sections through the simulated volumes; (ix) conclusive considerations. One final product includes also the selection of the most realistic realization that will be used for flow and transport numerical experiments.

3 Geostatistical Simulations

Many geostatistical grid-based approaches are available for distributing heterogeneities in 3D space; for a discussion about their applicability in practical situations see de Marsily et al. (2005) and Falivene et al. (2007). Lithofacies distribution was simulated using SIS: see, for example, Goovaerts (1997) with references; Deutsch and Journel (1992) and T-Progs (Transition-probability geostatistics: Carle and Fogg, 1996; Ritzi, 2000) which simulates the different facies in the form of coded indicator-type variables, where each value corresponds to a given facies.

SIS has been applied at different scales in a variety of depositional settings such as fluvial (Journel et al., 1998; Seifert and Jensen, 1999; Zappa et al., 2006; Felletti et al., 2006; Falivene et al., 2007), deltaic (Cabello et al., 2007), aeolian (Sweet et al., 1996), and turbidite settings (Journel and Gómez-Hernández, 1993; Falivene et al., 2007). Transition-probability geostatistics (T-Progs) has been used

Table 1 Facies classification adopted in this study, correlative hydrofacies and permeability values used for simulation

Facies Class	Facies	Interpretation	Adopted K values (m/s)	Operative hydrofacies
F	Fm,	Clay plug, mud balls	1×10^{-9}	F/fS
	Fl	Clay drapes	÷	
S	Sh	Low-relief bedwaves, upper flow regime	1×10^{-4}	
	Sm	Channel fills, lower flow regime		
	St	3D sand dunes		
	Sp	2D sand dunes	5×10^{-4}	S
	Sl	Sand drape		
	Sr	Ripples		
SG	SGm	Avalanching (scroll bars and channel fills)	5×10^{-3}	SG-GS
	SGt	3D gravelly sand dunes		
	SGp	2D gravelly sand dunes		
	SGh	Traction carpet, upper flow regime		
	SGl	Bedload sheets		
GS	GSm	Avalanching (scroll bars and channel fills)		
	GSt	3D gravelly sand dunes		
	GSp	2D gravelly sand dunes		
	GSh	Traction carpets, upper flow regime		
	GSl	Bedload sheets		
G	Gm	Avalanching (scroll bars and channel fills)	5×10^{-2}	G
	Gt	Migration of 3D gravel dunes		
	Gp	Migration of 2D gravel dunes		
	Gh	Bedload sheets, upper flow regime		

to model facies distribution in braided river (Felletti et al., 2006) and in alluvial fans (Fogg et al., 1998; Carle et al., 1998; Weissmann et al., 1999; Weissmann and Fogg, 1999).

In this study, four operative hydrofacies (F, S, SG and G; Table 1) have been utilized. In Fig. 2 there is an example of the computed semivariograms and transition probabilities graphs. The full 3D facies simulation was run on a volume of approximately $47 \times 75 \times 8.6$ m, including the five vertical facies maps used for conditioning.

We have reconstructed the entire volume in different ways: first we simulated the entire volume of units A and B; successively we simulated separately the same units, followed by merging the simulations through the kriged α boundary between A and B (Fig. 1c).

Semivariogram computation and SIS were performed with the geostatistical library Isatis v.3 (Bleines et al., 2000). Markov chains and transition-probability geostatistics were computed with T-Progs (Carle, 1999).

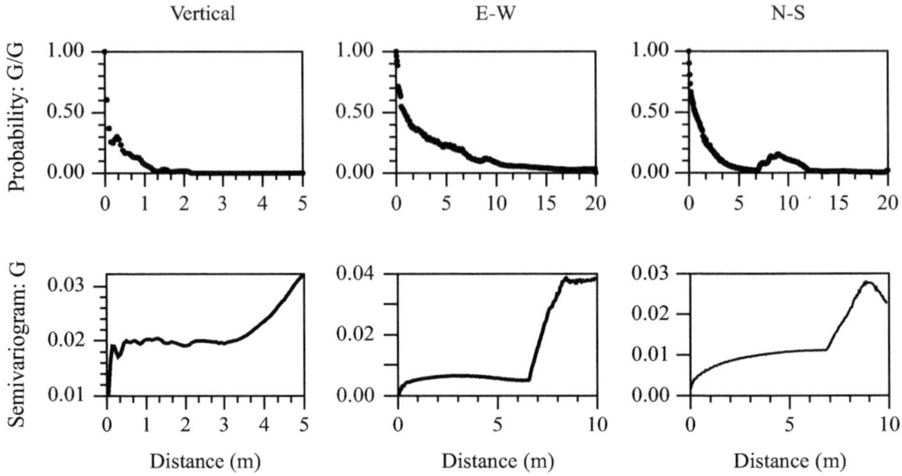

Fig. 2 Example of transitional probabilities and semivariograms computed in different directions for the hydrofacies G (gravel)

4 Discussion of Results

We studied one simulation among several equiprobable realizations (Fig. 3). The comparison between the geological model and the simulations was made by visual inspection and image analysis on the vertical facies maps and on sections cut through the simulated volume (Fig. 4 and Table 2). For this analysis we have considered the following criteria: (i) statistical parameters such as facies proportions, variogram ranges and anisotropy axes, (ii) existing facies trends and (iii) geometry of architectural elements. The following results can be highlighted:

1. Concerning the statistical analysis of correlation of the four hydrofacies we found a high degree of correspondence between the semivariogram ranges and the experimental transition probabilities computed on the entire dataset. Moreover, no significant difference on proportions, orientation of the anisotropy axes, geometry and pattern of facies distribution was observed comparing semivariograms and transition probability curves. This correspondence occurs because the points on the variograms are obtained from several thousand observations, with sampling continuously along the vertical and horizontal directions in our indicator database, which consists of the five vertical lithofacies maps and 31 logs. This dataset does not include the multiple sources of error typical of the databases consisting only of borehole logs (bias in estimates of facies proportion and spurious lateral indicator correlation, respectively due to clustering and sparse and non-random distribution of logs).
2. Concerning simulations, we preliminarily observe that, as expected, both SIS and T-Progs yielded unrealistic results for the undivided volume of units A and B. The realizations obtained by separate simulation of the two units were by far

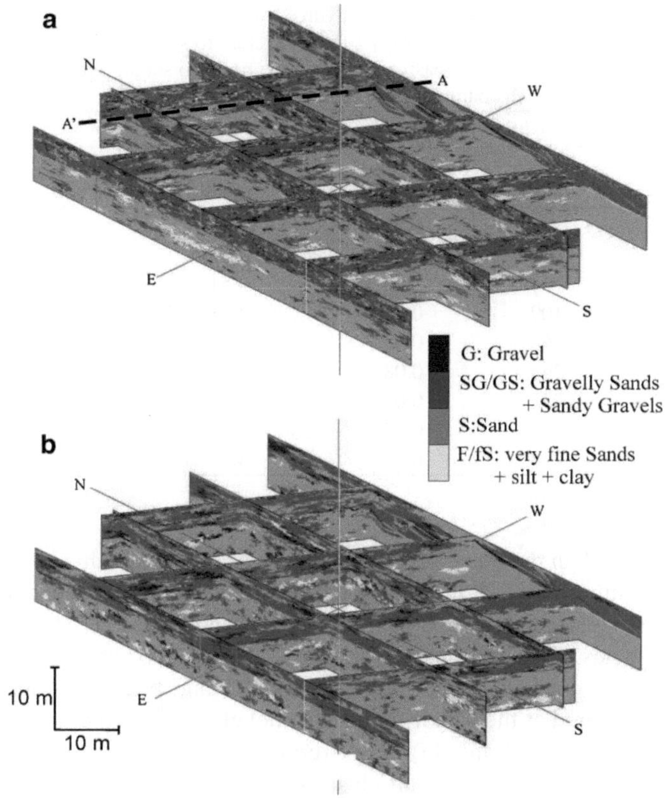

Fig. 3 Fence diagram of the simulated volume. (**a**) Simulation obtained with T-Progs and (**b**) simulation obtain with SIS. AA′ is the trace of the cross-section presented in Fig. 4b and c

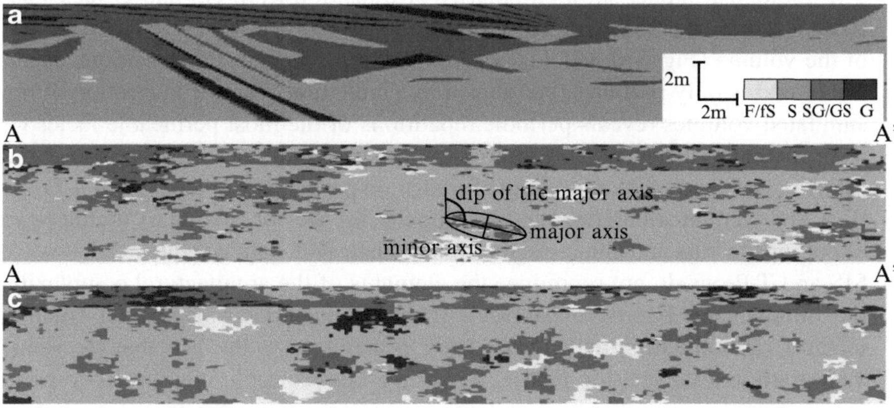

Fig. 4 (**a**) Discretized facies map of the Face 1 (location in Fig. 1b). (**b** and **c**) Sections cut into T-Progs and SIS simulations 5 m south of face 1 (location in Fig. 3). (**b**) Illustrates also the parameters considered for the image analysis

Table 2 Example of results of image analysis (hydrofacies G). Location of conditioning Face 1 is in Fig. 1b. Labels SIS 1 to 10 m and T-Progs 1 to 10 m refers to sections cut trough the simulated volumes parallel to Face 1, with increasing separation to north to south. See Face 1, SIS 5 m and T-Progs 5 m in Fig. 4 where the parameters here considered are shown

Image analysis – Operative hydrofacies G

	n° objects	Area of objects (m^2)	Anisotropy	Major axis (m)	Major axis dip (°)
Conditioning Face 1	59	0.33	19.77	2.16	94.70
SIS 1 m	77	0.41	4.15	0.92	91.57
SIS 3 m	64	0.53	4.51	1.02	92.05
SIS 5 m	109	0.20	4.58	0.69	92.36
SIS 10 m	63	0.32	4.57	0.86	93.31
T-Progs 1 m	268	0.08	4.38	0.46	90.81
T-Progs 3 m	235	0.06	4.14	0.42	90.60
T-Progs 5 m	232	0.05	4.28	0.40	91.23
T-Progs 10 m	246	0.05	4.23	0.38	89.89

more realistic than the outcomes of the previous attempt. In this second case, the analysis of simulations shows that both techniques underestimate continuity and size of the low-rank geological elements (facies and bed-sets). The critical distribution of facies G (open framework gravels along the lower part of the inclined bed-sets of the composite bars; Figs. 1d and 4a) and F (m-sized lenses of very fine sand and mud at the top of minor channel elements and dm-size mud clasts at their base; Figs. 1d and 4a) is not reproduced by T-Progs simulations, that yield a scattered pattern of small clusters, sparse in a "matrix" of facies S (the most abundant one; Figs. 3 and 4). Simulations by SIS reproduced more effectively than T-Progs the size, shape, distribution and orientation (sloping features of lateral and frontal accreted elements) of these low-hierarchy elements.

3. The geological model shows a polarity of transition from GS and G facies association to S and less abundant F facies towards the western and southern part of the volume (Figs. 1d and 4a), where the bar to channel-fill transitions occur. This trend is only partially reproduced by simulations. Visual inspection of the simulated volumes reveals periodic repetitions of the most permeable facies G, at a separation distance that is multiple of the variogram range in the case of the SIS (Fig. 4c), and of the minimum of transition probability in the case of T-Progs (Fig. 4b). Neither simulation method accounts for the non-stationary architecture of composite bars and channels, thus losing their real spatial trends.
4. SIS and T-Progs do not reproduce the elements of the architectural complexity, like minor channels, erosion bases, etc. This problem affects many pixel-oriented methods of simulation and, in our case, it arises from the fact that the semi-variogram and correlation matrix are a bivariate isotropic measure (two-point autocorrelation), and therefore any non-linear correlation structure (e.g., curved surfaces) cannot be reproduced. In principle, these difficulties could be overcome by pixel-oriented methods based on higher-order correlations (Liu et al. 2002), which are nevertheless rarely used because of their practical complexity.

Moreover, vertical tendencies at the scale of the bed-sets and bed-set groups (2–4 m), which are evident in the cross-variogram and in the off-diagonal vertical transition-probability plots of the facies maps, are partially lost in the 3D simulation. The representation of such non-stationary periodicities is still an open issue and cannot be resolved using "classical stationary" semivariogram or Markov chain models.

5. To perform 2-D image analysis at the hydrofacies scale, we considered five parameters: (a) number of objects (clusters of connected pixels belonging to the same hydrofacies), (b) average area of objects, (c) average major axis of objects, (d) average anisotropy ratio between maximum and minimum axis of objects, computed as shown in Fig. 4b and e dip of the major axis (Fig. 4b). These parameters were computed for A and B units separately and together over the conditioning faces 1 and NS (Fig. 1b) and on four sections cut into the SIS- and T-Progs- simulated volumes, parallel to the conditioning faces at increasing separation distances. As an example in Table 2 we present the parameters computed for hydrofacies G referred to the E-W sections.

   Visual inspection and image analysis show major differences between the conditioning faces and the simulated sections when studying units A and B together and separately.

   In general, the number of objects is overestimated by both simulation techniques and consequently their length, area and continuity are underestimated (see Table 2, hydrofacies G). However, SIS yields a more realistic number of objects, with comparable average area with respect to the conditioning faces, than T-Progs (Table 2). The same observations hold also for the estimated major axis of the individual objects.

   The best simulation obtained with SIS shows firstly increasing (1–5 m) then decreasing (5–10 m) fragmentation of objects. This behaviour is linked to the correlation length of semivariograms (10 m in a direction perpendicular to Face 1).

   Concerning the anisotropy, we observed that only SIS yields reasonably satisfactory estimates of the average anisotropy angle (Table 2). On the contrary both techniques underestimate seriously the anisotropy ratio. This fact is a direct consequence of the averaging effect of the variogram and of the transition probabilities with respect to the variable orientation of the objects through space.

6. At last, visual inspection of the simulated sections shows that the elements of the architecture with the rank of bed-sets (hydrofacies associations) have been reproduced most well by SIS, with shapes, sizes, spatial orientation and arrangement comparable to the geological model.

5 Conclusions

1. Simulations of the undivided volume, obtained by both SIS and T-Progs, are unrealistic because units A and B are characterized by very different statistical properties (frequency and correlation of hydrofacies). In order to realize

realistic simulations, in fact, it is necessary that statistical properties do not vary significantly throughout the studied domain (Falivene et al., 2007).

2. Pixel-oriented simulation of fine-scale heterogeneity of aquifer analogues characterized by high textural and structural complexity is possible, but realistic results are far too difficult to obtain at present. Our attempts yielded realizations that show many similarities with the geological model. SIS results suggested a greater capacity to reproduce size, continuity and shape of the low-rank elements of the sedimentary architecture (bed-sets) than was the case for T-Progs.
3. Both the studied composite bar and channel systems are characterized by facies trends that introduce non-stationarity. Both SIS and T-Progs reproduce non-stationary features only in an indirect way, accounting for facies proportions of the conditioning faces. How to account for depositional trends that are associated with periodicities at different scales, as is the case of point-bar complexes, looks to be an open problem, specifically due to the huge computational loads that arise with the existing mathematical solutions (Liu et al., 2002).
4. The finest scale heterogeneities can be simulated accounting for number and size of the individual facies units, particularly by SIS. On the contrary, it was impossible to reproduce adequately their anisotropy. Non-stationary anisotropic features, derived from quantification of the geological model, should be introduced in the simulation as a conditioning parameter.

References

Bersezio R, Giudici M, Mele M (2007) Combining sedimentological and geophysical data for high resolution 3-d mapping of fluvial architectural elements in the Quaternary Po plain (Italy). Sediment Geol 202:230–247

Bierkens MFP, Weerts HJT (1994) Block hydraulic conductivity of cross-bedded fluvial sediments. Water Resour Res 30:2665–2678

Bleines P, Deraisme J, Geffroy F, Perseval S, Rambert F, Renard D, Touffait Y (2000) Isatis Software Manual. Geovariances, Avon Cedex

Bridge JS, Lunt IA (2006) Depositional models of braided rivers. In: Sambrook Smith GS et al. (eds) Braided rivers, Spec Publ Int Assoc Sedimentol 36:11–50

Cabello P, Cuevas JL, Ramos E (2007) 3D modelling of grain size distribution in quaternay in deltaic plain deposits (Llobregat Delta, NE Spain). Geologica Acta 5:231–244

Carle SF (1999) T-PROGS: Transition Probability Geostatistical Software version 2.1. University of California, Davis

Carle SF, Fogg GE (1996) Transition probability-based indicator geostatistics. Math Geol 28:453–477

Carle SF, Labolle EM, Weissmann GS, Van Brocklin D, Fogg GE (1998) Conditional simulation of hydrofacies architecture: a transition probability/Markov approach. In: Fraser GS, Davis JM (eds) Hydrogeologic models of sedimentary aquifers. SEPM Special Publication, Concepts in Hydrogeology and Environmental Geology, pp. 147–170

Dai Z, Ritzi RW Jr, Huang C, Rubin Y, Dominic DF (2004) Transport in heterogeneous sediments with multimodal conductivity and hierarchical organization across scales. J Hydrol 294:68–86

de Marsily G, Delay F, Gonçalvès PRJ, Teles V, Violette S (2005) Dealing with spatial heterogeneity. Hydrogeol J 13:161–183

Deutsch CV, Journel AG (1992) GSLIB—Geostatistical Software Library and Users Guide: New York, Oxford University Press, 340 p

Falivene O, Cabrera L, Muñoz JA, Arbués P, Fernández O, Sáez A (2007) Statistical grid-based facies reconstruction and modelling for sedimentary bodies. Alluvial-palustrine and turbiditic examples. Geol Acta 5:199–230

Felletti F, Bersezio R, Giudici M (2006) Geostatistical simulation and numerical upscaling to model groundwater flow in a sandy-gravel, braided river aquifer analogue. J Sediment Res 76 (11):1215–1229

Fogg GE, Noyes CD, Carle SF (1998) Geologically-based model of heterogeneous hydraulic conductivity in an alluvial setting. J Hydrogeol 6:131–143

Goovaerts P (1997) Geostatistics for natural resources evaluation. Oxford University Press, New York

Johnson NM (1995) Characterization of alluvial hydrostratigraphy with indicator variograms. Water Resour Res 31:3217–3227

Johnson NM, Dreiss SJ (1989) Hydrostratigraphic interpretation using indicator geostatistics. Water Resour Res 25:2501–2510

Jordan DW, Pryor WA (1992) Hierarchical levels of heterogeneity in a Mississippi River meander belt and application to reservoir systems. AAPG Bull 76:1601–1624

Journel AG, Gómez-Hernández JJ (1993) Stochastic Imaging of the Wilmington Clastic Sequence. SPE paper 19857. Society of Petroleum Engineers Formation Evaluation March, pp. 33–40

Journel AG, Gundeso R, Gringarten E, Yao T (1998) Stochastic modelling of a fluvial reservoir: a comparative review of algorithms. J Petroleum Sci Eng 21:95–121

Kostic B, Aigner T (2007) Sedimentary architecture and 3-D ground penetrating radar analysis of gravelly meandering river deposits (Neckar valley, SW Germany). Sedimentology 54:789–808

Liu X, Srinisivan S, Wong DW (2002) Geological characterization of naturally fractured reservoir using multiple point geostatistics. Society of Petroleum Engineers 75246, SPE/DOE Symposium on Improved Oil Recovery

Lunt IA, Bridge JS, Tye RS (2004) Development of a 3-D depositional model of braided river gravels and sands to improve aquifer characterization. In Aquifer Characterization, Bridge JS, Hyndman D (eds) Spec Publ SEPM Soc Sediment Geol 80:139–169

Ritzi RW (2000) Behavior of indicator variogram and transition probabilities in relation to the variance in lengths of hydrofacies. Water Resour Res 36:3375–3381

Ritzi RW, Dai Z, Dominic DF (2004) Spatial correlation of permeability in cross-stratified sediment with hierarchical architecture. Water Resour Res 40, w03513, doi:10.1029/2003WR002420

Rubin Y, Journel AG (1991) Simulation of non-Gaussian space random functions for modeling transport in groundwater. Water Resour Res 27:1711–1721

Rubin Y, Lunt IA, Bridge JS (2006) Spatial variability in river sediments and its link with river channel geometry. Water Resour Res 42, w06d16, doi:10.1029/2005wr004853

Seifert D, Jensen JL (1999) Using sequential indicator simulation as a tool in reservoir description: issues and uncertainties. Math Geol 31:527–550

Sweet ML, Blewden CJ, Carter AM, Mills CA (1996) Modeling heterogeneity in a low permeability gas reservoir using geostatistical techniques, Hyde Field. American Association of Petroleum Geologists Bulletin, 80:1719–1735

Weissmann GS, Fogg GE (1999) Multi-scale alluvial fan heterogeneity modeled with transition probability geostatistics in a sequence stratigraphic framework. J Hydrol 226:48–65

Weissmann GS, Carle SF, Fogg GE (1999) Three-dimensional hydrofacies modeling based on soil surveys and transition probability geostatistics. Water Resour Res 35:1761–1770

Zappa G, Bersezio R, Felletti F, Giudici M (2006) Modeling aquifer heterogeneity at the facies scale in gravel-sand braided stream deposits. J Hydrol 325:134–153

Application of Multiple-Point Geostatistics on Modelling Groundwater Flow and Transport in a Cross-Bedded Aquifer

Marijke Huysmans and Alain Dassargues

Abstract In this work, the problem of modelling groundwater flow and transport in a heterogeneous environment with complex geological structures using multiple-point geostatistics is addressed. This study demonstrates how a training image can be constructed based on geological and hydrogeological field data and how multiple-point geostatistics can be applied to determine the impact of complex geological heterogeneity on groundwater flow and transport in a real aquifer. Application of the proposed approach of a hypothetical contaminant case in Brussels Sands (Belgium) shows that the type of heterogeneity encountered in the Brussels Sands may have a significant effect on contaminant transport and should be taken into account in groundwater contamination studies.

1 Introduction

Sedimentological and erosional processes often result in a complex three-dimensional subsurface architecture of sedimentary structures and facies types. Such complex sedimentological heterogeneity may induce a highly heterogeneous spatial distribution of hydrogeological parameter values in porous media at different scales (Klingbeil et al., 1999) and may consequently greatly influence subsurface fluid flow and solute migration (Koltermann and Gorelick, 1996). Because of the limited access to the relevant hydraulic properties, deterministic models often fall short in characterizing the subsurface heterogeneity and its inherent uncertainty. In recent decades, numerous stochastic approaches have been developed to overcome this problem. Most of these methods employ a variogram to characterize

M. Huysmans (✉) and A. Dassargues
Katholieke Universiteit Leuven, Applied Geology and Mineralogy, Celestijnenlaan 200 E - Bus 2408, 3001 Heverlee, Belgium,
e-mail: marijke.huysmans@ees.kuleuven.be

A. Dassargues
Université de Liège, Hydrogeology and Environmental Geology, Department of Architecture, Geology, Environment, and Civil Engineering (ArGEnCo)
e-mail: alain.dassargues@geo.kuleuven.ac.be

P.M. Atkinson and C.D. Lloyd (eds.), *geoENV VII – Geostatistics for Environmental Applications*, Quantitative Geology and Geostatistics 16,
DOI 10.1007/978-90-481-2322-3_13,

the heterogeneity of the hydraulic parameters. Variograms are calculated based on two-point correlations only and therefore have some important limitations. Variograms are not able to describe realistic heterogeneity in complex geological environments. Complex geological patterns including sedimentary structures, multi-facies deposits, structures with large connectivity, curvi-linear structures, etc. cannot be characterized using only two-point statistics (Koltermann and Gorelick, 1996; Fogg et al., 1998). Moreover, variograms, as a limited and parsimonious mathematical tool, cannot take full advantage of the possibly rich amount of geological information from outcrops (Caers and Zhang, 2004). Multiple-point geostatistics (Strebelle, 2000, 2002; Caers and Zhang 2004; Feyen and Caers 2006) aims to overcome the limitations of the variogram. The premise of multiple-point geostatistics is to move beyond two-point correlations between variables and to obtain (cross) correlation moments at multiple locations at a time (Strebelle and Journel, 2001). Because of the limited direct information from the subsurface, such statistical information cannot directly be obtained from samples. Instead, "training images" are used to characterize the patterns of geological heterogeneity. A training image is a conceptual explicit representation of the expected spatial distribution of hydraulic properties or facies types. The main idea is to borrow geological patterns from these training images and anchor them to the subsurface data domain. This study demonstrates how multiple-point geostatistics can be applied to determine the impact of complex geological heterogeneity on groundwater flow and transport in a real aquifer. More precisely, multiple-point geostatistics is used in this study to investigate the effect of complex small-scale sedimentary heterogeneity on the short term migration of a contaminant plume and its uncertainty. This paper also shows how a training image can be constructed based on geological and hydrogeological field data.

2 Materials and Method

2.1 Geological Setting

The aquifer of interest is the Brussels Sands formation in Belgium. Approximately 29,000,000 m^3 of groundwater per year is pumped from this aquifer. The Brussels Sands display a complex geological heterogeneity and anisotropy that complicates pumping test interpretation, groundwater modeling and prediction of pollutant transport. The Brussels Sands formation is an early Middle-Eocene shallow marine sand deposit in Central Belgium (Fig. 1). The depositional environment of the Brussels Sands is studied in detail by Houthuys (1990) based on field studies and descriptions of approximately 90 outcrops and hundreds of boreholes. The Brussels Sands are a tidal sandbar deposit, deposited at the beginning of an important transgression at the southern border of the Eocene North Sea. The Brussels Sands display several features and sedimentary structures typical for tidal deposits, such as important grain size variations, cross-bedding, bottomsets, foresets, mud drapes and unidirectional

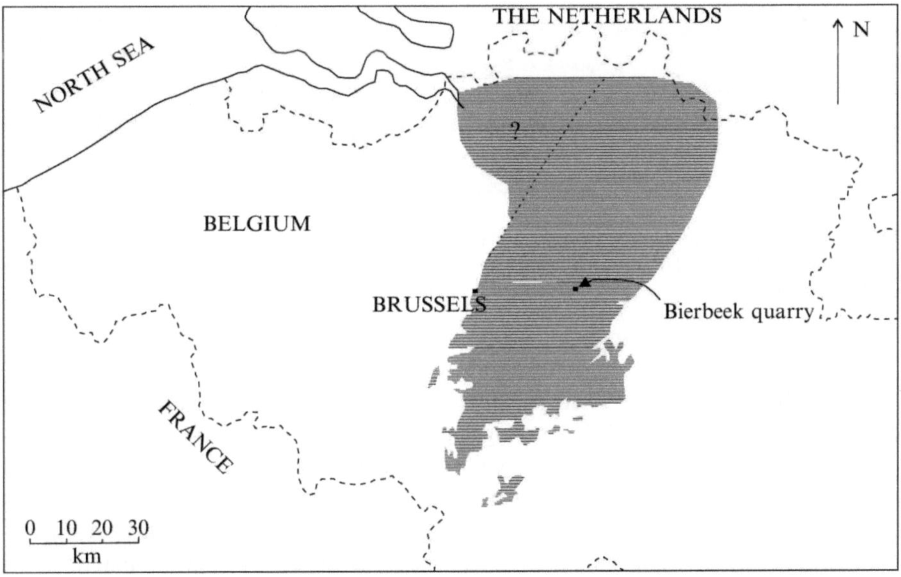

Fig. 1 Map of Belgium showing Brussels Sands outcrop and subcrop area (*shaded part*) and the location of the Bierbeek quarry (modified after Houthuys, 1990)

reactivation surfaces. Bottomset beds are approximately horizontal beds consisting of finer grained sediment and form the base of most cross-bedded beds. Mud drapes are thin layers of mud within the cross-bedded beds.

2.2 Field Measurements

An extensive field campaign is carried out consisting of field observations of the sedimentary structures and 2,750 small-scale in situ measurements of air permeability in the Brussels Sands. The results and conclusions of this field campaign are summarized in this section. More details about this field campaign can be found in Huysmans et al. (2008). A representative Brussels Sands outcrop (Bierbeek quarry near Leuven, Belgium) is mapped in detail with regard to the spatial distribution of sedimentary structures and lithologies. Geological sketches and digital photographs from all faces of the quarry are made. A visual distinction between sand-rich and clay-rich zones, hereafter called the sand facies and the silt facies respectively, is made in situ based on sediment characteristics. Figure 2 shows an interpreted photomosaic of one of the outcrops of the vertical quarry walls, corrected for perspective distortion. Thickness and dip measurements of several sedimentary features are made at various locations in the quarry and analyzed statistically. Histograms of bottomset thicknesses, set thicknesses and lamination dipping angles measured during this measurement campaign and from Houthuys (1990) are calculated.

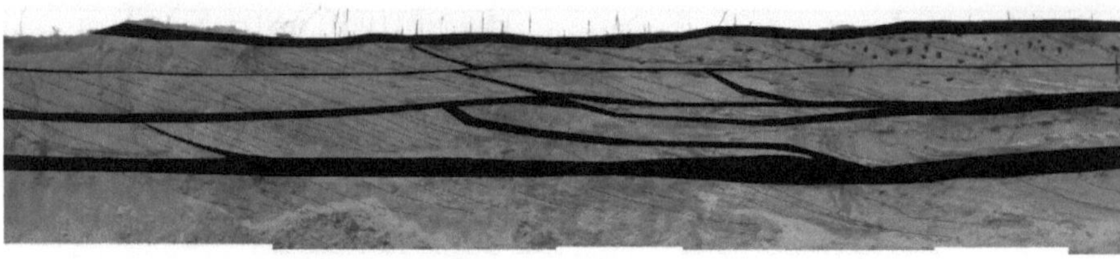

Fig. 2 Interpreted photomosaic of quarry wall showing the silt facies consisting of clay-rich bottomsets and distinct mud drapes in black. Height of quarry wall is approximately 4–5 m (Huysmans et al., submitted)

Additionally, a total of 2,750 air permeability measurements at centimeter-scale are carried out in situ. Permeability histograms and variograms of the sand and silt facies are calculated. Analysis of the spatial distribution of sedimentary structures and permeability shows that silt facies consisting of clay-rich sedimentary features such as bottomsets and distinct mud drapes exhibit a different statistical and geostatistical permeability distribution compared to the sand facies. Variogram map analysis of the air permeability data shows that permeability anisotropy in the cross-bedded lithofacies is dominated by the foreset lamination orientation. The results show that small-scale sedimentary heterogeneity has a dominant control on the spatial distribution of the hydraulic properties and induces permeability heterogeneity and anisotropy.

2.3 Training Image Construction

To demonstrate the need for "training images" in multiple-point geostatistics, this section first briefly recalls the mathematical basis behind multiple-point geostatistics. The remainder of this section describes the training image construction process for this study. Consider an attribute S, taking J possible states $\{s_j, j = 1 \ldots J\}$. S can be a categorical property, e.g. facies, or a continuous value such as permeability, with its interval of variability discretized into J classes. A data event d_n of size n centered at location $\mathbf{u}$ is constituted by (1) the data geometry defined by the n vectors $\{\mathbf{h}_\alpha, \alpha = 1 \ldots n\}$ and (2) the n data values $\{s(\mathbf{u} + \mathbf{h}_\alpha), \alpha = 1 \ldots n\}$. A data template τ_n comprises only the previous data geometry. The categorical transform of the variable S at location $\mathbf{u}$ is defined as:

$$I\,(\mathbf{u}; j) = \begin{cases} 0 & \text{if } S\,(\mathbf{u}) = s_j \\ 1 & \text{if } S\,(\mathbf{u}) \neq s_j \end{cases}$$

The multiple-point statistics are probabilities of occurrence of the data events $d_n = \{S(\mathbf{u}_\alpha) = s_{j,\alpha}, \alpha = 1 \ldots n\}$, i.e. probabilities that the n values $s(\mathbf{u}_1) \ldots s(\mathbf{u}_j)$ are jointly in the respective states $s_{j,1} \ldots s_{j,n}$. For any data event d_n, that probability is also the expected value of the product of the n corresponding indicator data:

$$\text{Prob}\{d_n\} = \text{Prob}\left\{S(\mathbf{u}_\alpha) = s_{j,\alpha}; \alpha = 1 \ldots n\right\} = \text{E}\left[\prod_{\alpha=1}^{n} I\left(\mathbf{u}_\alpha, j_\alpha\right)\right]$$

Such multiple-point statistics or probabilities cannot be inferred from sparse field data. Their inference requires a training image depicting the expected patterns of geological heterogeneities. Training images can be obtained from observations of outcrops, geological reconstructions and geophysical data (Strebelle and Journel, 2001). In this study, training images are constructed based on observations of outcrops. 2D vertical training images of clay and sand occurrence in different orientations are constructed based on field photographs and observations of the geometry and dimensions of the sedimentary structures. The 2D training images are composite sketches composed of smaller scale photographs and field sketches conditioned by the histograms of set thicknesses, bottomset thicknesses and lamination angles. The training image size is 30 by 30 m. To capture the thin clay drapes, a small grid cell size of 0.05 by 0.05 m is adopted so that the training image consists of 360,000 grid nodes. Figure 3 shows the 2D training images in the N40°E direction and the approximately perpendicular N45°W direction. These training images show that the facies distribution in the N40°E direction is rather complex while almost horizontal layering is observed in the perpendicular direction. Since the facies changes in the N45°W direction are so limited compared to the other direction, 2D analyses are carried out in the remainder of this paper only considering the training image shown in Fig. 3a.

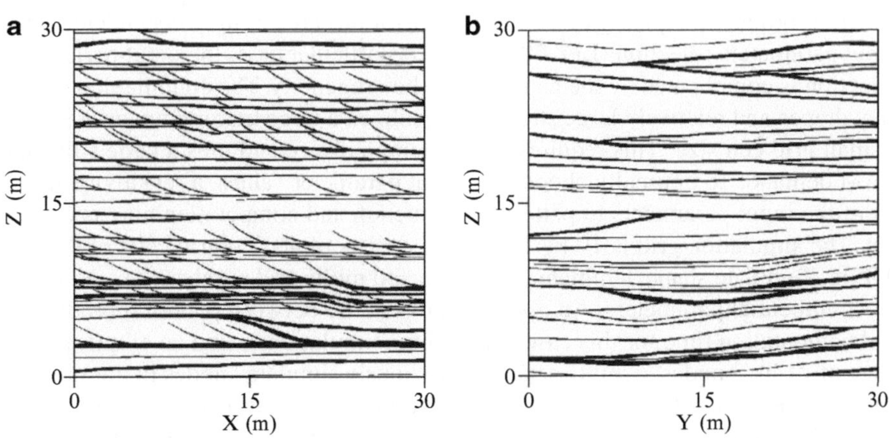

Fig. 3 Vertical 2D training image of 30 × 30 m in (**a**) N40 °E direction and (**b**) N45 °W direction (white = sand facies, black = silt facies)

2.4 Multiple-Point Geostatistical Facies Realizations

Multiple-point statistics are borrowed from the training image to simulate multiple realizations of silt and sand facies occurrence using the single normal equation simulation (SNESIM) algorithm (Strebelle, 2002). Snesim is a pixel-based sequential simulation algorithm that obtains multiple-point statistics from the training image, exports it to the geostatistical numerical model and anchors it to the actual subsurface hard and soft data. For each location along a random path the data event d_n consisting of the set of local data values and their spatial configuration is recorded. The training image is scanned for replicates that match this event to determine the local conditional probability that the unknown attribute S(u) takes any of the J possible states given the data event d_n, as

$$\text{Prob}\left\{S(\mathbf{u}) = s_j \,|d_n\right\} = \frac{\text{Prob}\left\{S(\mathbf{u}) = s_j \text{ and } S(\mathbf{u}_\alpha) = s_{j,\alpha}; \alpha = 1 \ldots n\right\}}{\text{Prob}\left\{S(\mathbf{u}_\varepsilon) = s_{j,\alpha}; \alpha = 1 \ldots n\right\}}$$

The denominator can be inferred by counting the number of replicates of the conditioning data event found in the training image. The numerator can be obtained by counting the number of those replicates associated to a central value $S(\mathbf{u})$ equal to s_k. A maximum data search template is defined to limit the geometric extent of those data events. SNESIM makes reasonable CPU demands by scanning the training image prior to simulation and storing the conditional probabilities in a dynamic data structure, called the search tree. The theory and algorithm behind SNESIM are described in Strebelle (2002). Descriptions of SNESIM parameters are in Liu (2006), Strebelle and Remy (2005) and Strebelle (2003). The computation time and pattern reproduction quality of SNESIM realizations are strongly dependent on the input parameters selection (Liu, 2006). In this particular case, the input parameters selection is complicated by the nature of the heterogeneity. The combination of thin clay drapes and relatively large structures results in a large training image size with a small grid cell size. This requires a large template size and thus a large CPU and RAM demand. To optimally choose the input parameter values, a sensitivity analysis of the input parameters to pattern reproduction and computation time is carried out. The simulation grid is 10 by 10 m and consists of 40,000 grid cells of 0.05 by 0.05 m. Template shape, template dimension and multi-grid number prove to be the most influential parameters. An optimal compromise between pattern reproduction and computation time for this case is found for simulations using an elliptical template of 21 by 3 nodes, 6 multi-grids, 48 previously simulated nodes in the sub-grid approach, a re-simulation threshold of 50 and 6 re-simulations iterations. A total of 150 SNESIM realizations of 10 by 10 m are simulated using the optimal input parameter selection. Figure 4a shows three example SNESIM facies realizations.

2.5 *Intrafacies Permeability Simulation*

Intrafacies permeability variability within the sand and silt facies is simulated using conventional variogram-based geostatistical methods based on histograms and variograms obtained from the in situ air permeability measurements. The simulation algorithm used in this study is direct sequential simulation with histogram reproduction (Oz et al., 2003). The input statistics and variogram parameters of permeability for both facies are presented in Table 1. Air permeability realizations are converted into hydraulic conductivity realizations to serve as input to a local groundwater flow model. In this way, intrafacies hydraulic conductivity of the 150 facies realizations is simulated. Figure 4b shows the hydraulic conductivity realizations of the facies realizations of Fig. 4a. The silt facies are visible in the hydraulic conductivity realizations as areas with lower hydraulic conductivity. The low conductivity zones are, however, not continuous flow barriers, since the sand and silt permeability distributions are overlapping.

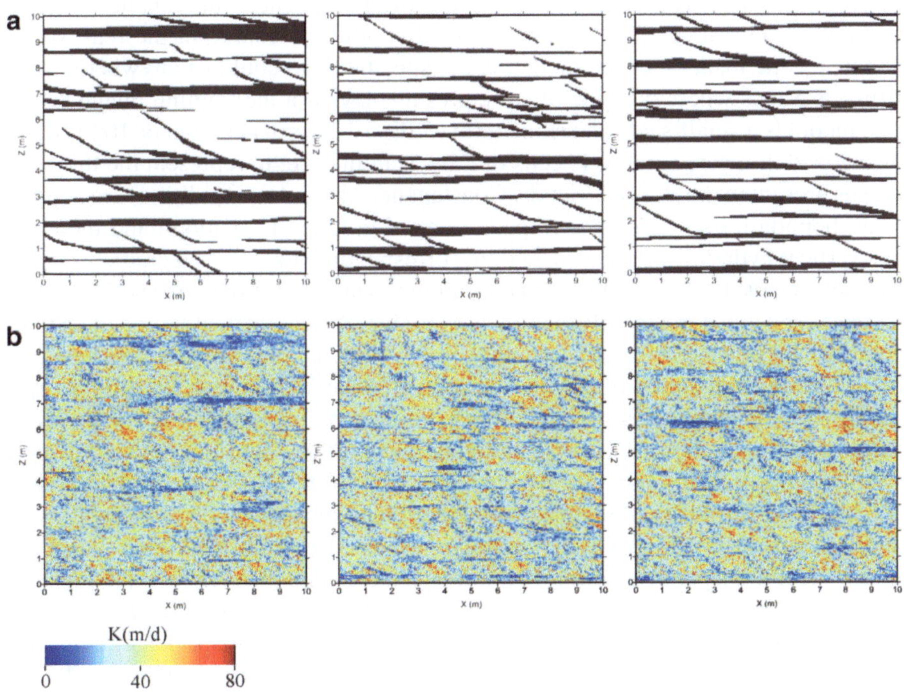

Fig. 4 (**a**) Three example 2D vertical SNESIM facies realizations (white = sand facies, black = silt facies); (**b**) Corresponding hydraulic conductivity (m/d) realizations

Table 1 Statistical and variogram parameters of permeability in mD (milliDarcy) for the sand and silt facies (values from Huysmans et al., 2008)

	Sand facies	Silt facies
Mean k (mD)	58,700	42,200
Variance k (mD)^2	3.6×10^8	2.55×10^8
Variogram type k	Spherical	Spherical
Nugget (mD)^2	2.09×10^8	1.03×10^8
Sill (mD)^2	1.51×10^8	1.52×10^8
Dip angle of major axis of anisotropy	26°	0° (horizontal)
Lamina parallel range (m)	0.6	1.9
Lamina perpendicular range (m)	0.3	0.4

2.6 Groundwater Flow and Transport Model

The simulated hydraulic conductivity realizations are used as input to a groundwater flow and transport model to investigate the effect of the small-scale sedimentary heterogeneity on early contaminant plume migration. The contaminant source is a hypothetical source. The location of this hypothetical source in the real world, and hence the location of the model, is not specified and could be anywhere in the Brussels Sands where the type of structures displayed in the training image occur. The model is a small-scale and short-term (3-day) 2D vertical model of 10 by 10 m, discretized into very small grid cells of 5 by 5 cm in order to represent the thin clay drapes present in the Brussels Sands. Constant head boundary conditions are applied to all boundaries so that the average horizontal gradient is 10 m/km and the average vertical hydraulic gradient is 5 m/km corresponding to observed gradients in the Brussels Sands. Porosity of the sand and silt facies are both assumed to be 30% since no facies specific porosity information is available. A hypothetical source of an inert contaminant is assumed at the surface at $x = 2$ with an arbitrarily chosen flow rate of 1,000 l/day and an arbitrarily chosen source concentration of 1,000 mg/l. Corresponding to the very small grid cell dimension, a very low longitudinal dispersivity value of 0.01 m is chosen based on extrapolation of the relationships between dispersivity and the scale of observation from Gelhar et al. (1992). Transverse dispersivity is taken to be one order of magnitude smaller than longitudinal dispersivity (Zheng and Bennett, 1995). Dispersivity values are assumed equal in both facies since no facies specific dispersivity information is available. The differential equations describing groundwater flow are solved by MODFLOW (McDonald and Harbaugh, 1988), a block-centered finite-difference method based software package. Transport by advection and dispersion is simulated with MT3DMS (Zheng and Wang, 1999), using the high-order finite-volume TVD solver. The Courant number used for determination of the time step size for transport calculations is 0.75. This groundwater flow and transport model is run 150 times for the 150 simulated hydraulic conductivity realizations. The distributions and uncertainty of the following three relevant output parameters are calculated and studied: (1) the maximum solute concentration after 3 days, (2) the maximum depth where a concentration of 1 mg/l is reached after

3 days and (3) the maximum horizontal distance to the source where a concentration of 1 mg/l is reached after 3 days. The convergence of the output parameter statistics in terms of the number of simulations is also studied in order to assess whether 150 simulations are sufficient.

3 Results and Discussion

Figure 5 focuses on the calculated contaminant plume for the three hydraulic conductivity realizations of Fig. 4 and shows simulated hydraulic head contours and contaminant concentrations for $t = 3$ days. These figures show a different plume shape and extent and different maximum concentrations for the different hydraulic conductivity realizations. Figure 6 shows histograms of the three relevant output parameters defined in the previous section. The maximum simulated solute concentration for $t = 3$ days varies between 6.3 and 22.0 mg/l and shows a slightly skewed distribution with a mean of 10.7 mg/l and a standard deviation of 2.7 mg/l. The maximum depth with a concentration of 1 mg/l for $t = 3$ days varies between 1.3 and 1.9 m and shows a symmetric distribution with a mean of 1.6 m and a standard deviation of 0.1 m. The maximum horizontal distance to the source with a concentration of 1 mg/l for $t = 3$ days varies between 4.3 and 5.6 m and shows a slightly skewed distribution with a mean of 5.2 m and a standard deviation of 0.2 m. The contaminant plumes of different realizations thus have significantly different characteristics. The largest maximum simulated solute concentration is more than three times larger than the smallest maximum simulated solute concentration. The largest maximum depth with $c = 1\,\mathrm{mg/l}$ is almost 50% larger than the smallest maximum depth with $c = 1\,\mathrm{mg/l}$ and the largest maximum horizontal distance with $c = 1\,\mathrm{mg/l}$ is 30% larger than the smallest maximum horizontal distance with $c = 1\,\mathrm{mg/l}$. These results show that the uncertainty on the spatial facies distribution and intrafacies hydraulic conductivity distribution results in a significant uncertainty on the calculated concentration distribution. In particular, the maximum simulated concentration value can vary strongly among the different input hydraulic conductivity realizations.

4 Conclusions

This study applies multiple-point geostatistics in the field of hydrogeology on a real aquifer. This study demonstrates how a training image can be constructed based on geological and hydrogeological field data and how multiple-point geostatistics can be applied to determine the impact of complex geological heterogeneity on groundwater flow and transport in a real aquifer. Application of the proposed approach to a hypothetical contaminant case in Brussels Sands shows that the uncertainty on the spatial facies distribution and intrafacies hydraulic conductivity distribution

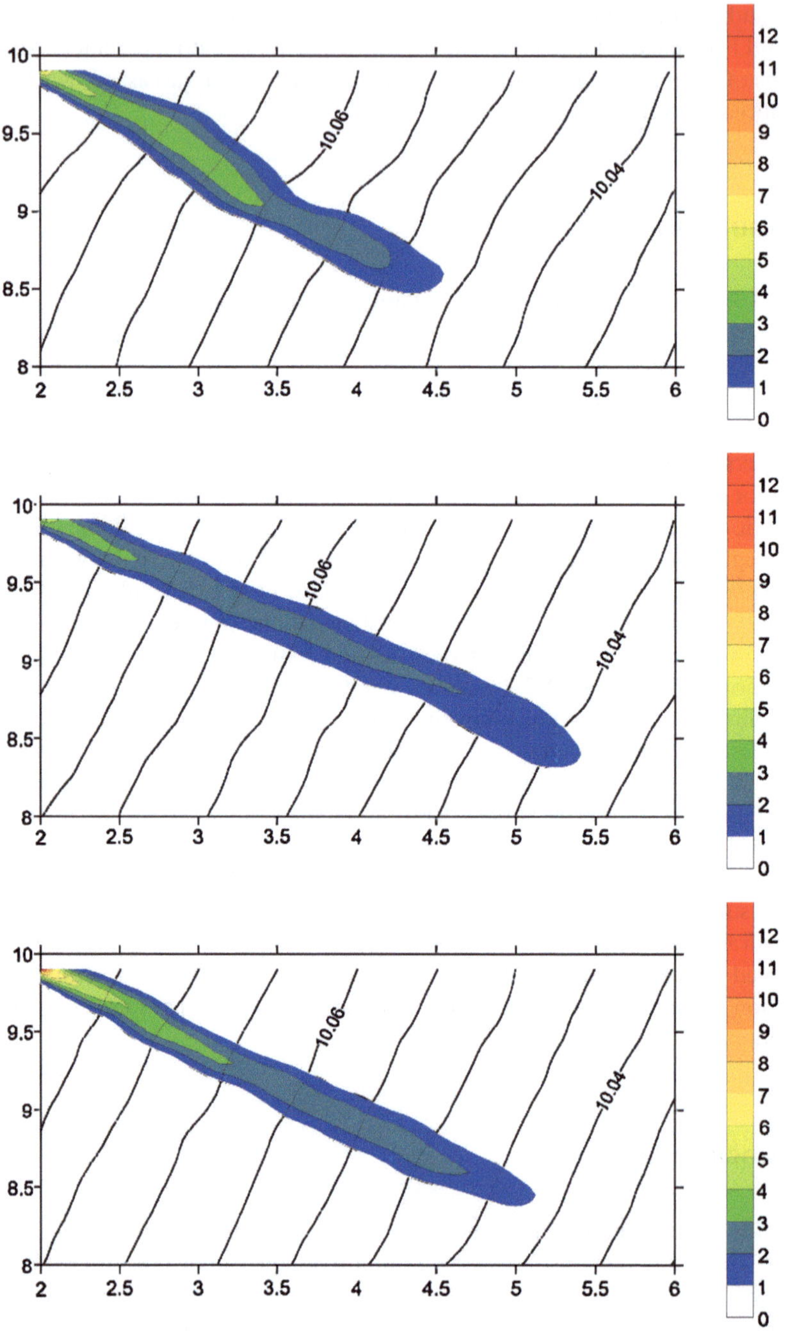

Fig. 5 Simulated hydraulic head contours and contaminant concentrations for $t = 3$ days for the three realizations of Fig. 4

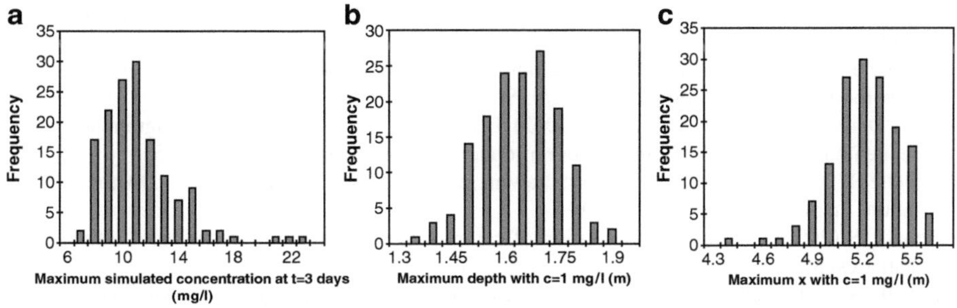

Fig. 6 Histograms of (**a**) maximum solute concentration after 3 days, (**b**) maximum depth where a concentration of 1 mg/l is reached after 3 days and (**c**) maximum horizontal distance to the source where a concentration of 1 mg/l is reached after 3 days

results in a significant uncertainty on the calculated concentration distribution. The small-scale sedimentary heterogeneity in the Brussels Sands has a significant effect on the calculated concentration distribution and using a homogeneous model instead of a heterogeneous model could lead to significant error in the prediction of contaminant plume migration and concentrations. This shows that the type of heterogeneity encountered in the Brussels Sands may have a significant effect on contaminant transport and should be taken into account in groundwater contamination studies.

Acknowledgements The authors wish to acknowledge the Fund for Scientific Research – Flanders for providing a Postdoctoral Fellowship to the first author.

References

Caers J, Zhang T (2004) Multiple-point geostatistics: a quantitative vehicle for integrating geologic analogs into multiple reservoir models. In: Integration of outcrop and modern analog data in reservoir models, AAPG memoir 80. Tulsa, pp. 383–394

Feyen L, Caers J (2006) Quantifying geological uncertainty for f low and transport modeling in multi-modal heterogeneous formations. Adv Water Resour 29(6):912–929

Fogg GE, Noyes CD, Carle SF (1998) Geologically based model of heterogeneous hydraulic conductivity in an alluvial setting. Hydrogeol J 6(1):131–143

Gelhar LW, Welty C, Rehfeldt KR (1992) A critical review of data on field-scale dispersion in aquifers. Water Resour. Res. 28(7):1955–1974

Houthuys R (1990) Vergelijkende studie van de afzettingsstruktuur van getijdenzanden uit het Eoceen en van de huidige Vlaamse banken. Aard Mededelingen 5, Leuven Univ Press, p 137

Huysmans M, Peeters L, Moermans G, Dassargues A (2008) Relating small-scale sedimentary structures and permeability in a cross-bedded aquifer. J Hydrol 361(1–2):41–51

Koltermann CE, Gorelick S (1996) Heterogeneity in sedimentary deposits: a review of structure imitating, process-imitation, and descriptive approaches. Water Resour Res 32(9):2617–2658

Klingbeil R, Kleineidam S, Asprion U, Aigner T, Teutsch G (1999) Relating lithofacies to hydrofacies: outcrop-based hydrogeological characterisation of quaternary gravel deposits. Sediment Geol 129(3–4):299–310

Liu Y (2006) Using the Snesim program for multiple-point statistical simulation. Comput Geosci 32(10):1544–1563

McDonald MG, Harbaugh AW (1988) A modular three-dimensional finite-difference ground-water flow model. Technical report, U.S. Geol. Survey, Reston, VA

Oz B, Deutsch CV, Tran TT, Xie Y (2003) DSSIM-HR: A FORTRAN 90 program for direct sequential simulation with histogram reproduction. Comput Geosci 29(1):39–51

Strebelle S (2000) Sequential simulation drawing structures from training images, Doctoral dissertation, Stanford University, USA, 164 p

Strebelle S (2002) Conditional simulation of complex geological structures using multiple-point statistics. Math Geol 34:1–22

Strebelle S (2003) New multiple-point statistics simulation implementation to reduce memory and CPU-demand. Proceedings to the IAMG 2003, Portsmouth, UK, September 7–12

Strebelle S, Journel A (2001) Reservoir modeling using multiple-point statistics: SPE 71324 presented at the 2001 SPE Annual Technical Conference and Exhibition, New Orleans, September 30–October 3

Strebelle S, Remy N (2005) Post-processing of multiple-point geostatistical models to improve reproduction of training patterns. In: Leuangthong O, Deutsch CV (eds) Geostatistics Banff 2004, vol. 2: Springer, Dordrecht, pp. 979–987

Zheng C, Bennett GD (1995) Applied contaminant transport modeling, theory and practice. Wiley, New York, 433 pp.

Zheng C, Wang PP (1999) MT3DMS, a modular three-dimensional multi-species transport model for simulation of advection, dispersion and chemical reactions of contaminants in groundwater systems. Documentation and user's guide. US Army Engineer Research and Development Center Contract Report SERDP-99–1, Vicksburg, MS

Assessment of the Impact of Pollution by Arsenic in the Vicinity of Panasqueira Mine (Portugal)

Ana Rita Salgueiro, Paula Helena Ávila, Henrique Garcia Pereira, and Eduardo Ferreira da Silva

Abstract The mining and beneficiation processes at Panasqueira mine have given rise, during a long production period, to a large amount of sulphide-rich waste, contained in several tailing ponds, two of them located near a small village. Among the pollutant elements that occur in the surrounding area, arsenic (AS) was selected to illustrate a geostatistics based methodology aiming at combining land use with the spatial distribution of the contaminant concentration in soils, by taking the former as an external drift to estimate the latter. Since land use is an ordinal variable, its combination, via the external drift algorithm, with As concentration requires its prior transformation into a real number. The proposed transformation relies on the Correspondence Analysis (CA) of the contingency table crossing classes of As concentration with classes of land use. The co-ordinates of samples projection onto the CA first axis turned out to be a reliable proxy of the interaction between As concentration and land use, providing the required real variable to be used as external drift. Hence, 'raw' As concentration maps were 'corrected' through the external drift algorithm, leading to an increase where land use is more 'valuable' (populated areas) and to a decrease where land use is less 'valuable' (barren soil). Obviously, the 'corrected' maps are a more realistic basis for reclamation planning than the 'raw' ones.

A.R. Salgueiro (✉) and H.G. Pereira
CERENA – Natural Resources and Environment Center of IST, Lisboa, Portugal
e-mail: rita.salgueiro@ist.utl.pt; henrique.pereira@ist.utl.pt

P.H. Ávila
LNEG – S. Mamede Infesta Laboratory, Porto, Portugal
and
GeoBioTec – GeoBiosciences, Technologies and Engineering Research Unit, Aveiro, Portugal
e-mail: paula.avila@ineti.pt

E.F. da Silva
GeoBioTec – GeoBiosciences, Technologies and Engineering Research Unit,
Campus Universitário de Santiago 3810-193, Aveiro, Portugal
e-mail: eafsilva@ua.pt

P.M. Atkinson and C.D. Lloyd (eds.), *geoENV VII – Geostatistics for Environmental Applications*, Quantitative Geology and Geostatistics 16,
DOI 10.1007/978-90-481-2322-3_14,

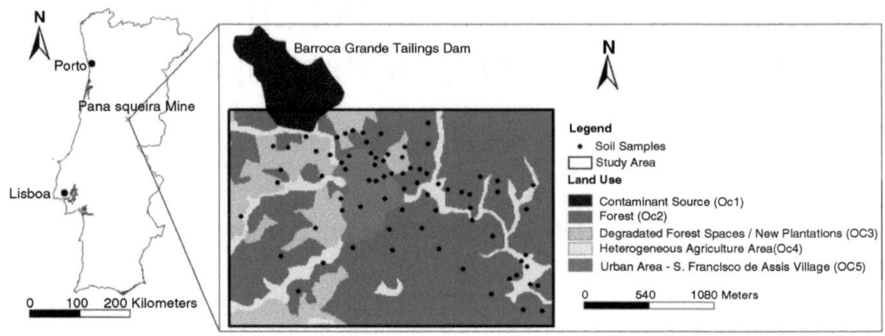

Fig. 1 Study area, including sample location and land use codes

1 Introduction

A major environmental issue associated with mining and beneficiation wastes is the release of heavy metals and arsenic into the environment. Since sulphur is often present in the tailings where such wastes are dumped, exposure to the atmosphere in the presence of water leads to the production of acid mine drainage, a common type of pollution in mining areas that results from the oxidation of sulphide minerals leading to generation of free acidity and soluble metal species. The consequences of the contamination of the surrounding topsoil may become particularly worrisome when mining and ore treatment operations occur in populated areas.

The Panasqueira wolframite ore deposit is the biggest of Western Europe, which has been in operation from 1896 to the present date (with periods of higher and lower extraction rate, according to W prices in the international market). This long exploitation history gave rise to, among others, a huge tailing (7,000,000 m^3) and two mud dams (see Fig. 1), which are the source of pollution in the vicinity of the mining area.

Arsenic (As) was chosen as the target of this case study due, mainly, to the following reasons:

- Arsenopyrite is the most common sulphide that is present in Panasqueira complex paragenesis (Breiter, 2001; Correa and Naique, 1998; Noronha et al., 1992).
- The effect of mining and arsenic release from acid mine waters to groundwater and the related arsenic accumulation in soils (Ávila et al., 2008).
- Soils at S. Francisco de Assis village, downstream Barroca Grande tailing, are a major repository for arsenic released by the Panasqueira mining activities (see Fig. 1).
- Plants absorb arsenic rather easily, so that high-ranking concentrations may be present in food whenever As rich soil is used for agriculture purposes (Walsh et al. 1977).

Arsenic is considered as a "priority pollutant element" (Glanzman and Closs, 1993), harmful both for humans and for ecosystems above certain thresholds (678 and 40 mg/kg, respectively, as given by Swartjes [1999]). In particular, epidemiological data analysis has shown a link between environmental As exposure and an increased risk of cancer in human populations (ATSDR, 1993).

In order to assess the As pollution derived from the contamination source displayed in Fig. 1 (as Oc1), aiming at making available the guidelines for an eventual remediation procedure that avoids the above mentioned harmful effects, a soil sampling campaign was performed (see also Fig. 1 for location of the set of 76 samples in their land use environment). The samples were oven dried before dry sieving at 40°C, mixed, homogenized and sieved through a <200 mesh screen for chemical analysis. For trace metal analysis, a 0.5 g split was leached in hot (95°C) aqua regia ($HCl - HNO_3 - H_2O$) for 1 h. After dilution to 10 ml with deionized water, the solutions were analyzed for As and other 11 elements.

It is worth noting that concentrations of As as high as 29,000 mg/kg were found (the sampling average is 497.7 mg/kg and the 95% percentile is 429 mg/kg, which compares with the corresponding values of the background of 13.6 and 28.3 mg/kg, respectively).

Based on the As analytical results given by the sample campaign, a geostatistical methodology combining pollutant concentration with land use was applied. Obviously, it is less risky to handle high As concentrations in a barren land than in an urban area; conversely, the later land use magnifies low As concentrations.

2 Methodology

The proposed methodology to evaluate As distribution in the study area (accounting, jointly, for concentration thresholds and land use categories, arranged in ascending order from Oc2 to Oc5) addresses the problem of balancing the intensity of contamination with the socio-economic importance of the zones where such contamination occurs.

In particular, a single gradation combining the real variable (As concentration) with the ordinal one (land use) is to be obtained, ranging from the less harmful extreme (low As in forest zones) to the more damaging one (high As in urban zones). The external drift technique (Maréchal, 1984) allows the production of an output depicting such a gradation, provided that land use vulnerability can be put under a quantitative form.

In order to apply the proposed methodology, the following steps were considered:

1. Disregarding for eventual remediation purposes the areas given by indicator kriging where As concentration is lower than 40 mg/kg(limit for ecosystems vulnerability)
2. Estimation of As concentration in a regular mesh by ordinary kriging
3. Submission to Correspondence Analysis (CA) the cross-tabulation of As classes by land use categories, in order to obtain a quantitative variable referring to each sample and accounting for As *vs.* land use interaction
4. Application of the variable provided by 3. as an external drift to produce a kriged map where 'raw' As concentration is balanced with its harmfulness for each land use class

The proposed methodology was developed as an extension of the usual geostatistical techniques for pollution evaluation in soils, based on kriging or stochastic simulation (Goovaerts, 1999). These techniques provide, as a final output, a map of contaminant concentrations, which is not a very helpful tool when remediation planning is the aim of the study, since high pollution spots are not linked with the social-economical context where they occur. On the other hand, when ancillary information is available (as nominal soil characteristics, in Jeannée and Fouquet, 2003, or remote sensing data, in Choe et al., 2008), this information is used to improve the estimation of pollution levels by increasing its precision. To the best of our knowledge, the ultimate product of such studies remains a (more or less precise) contaminant concentration map.

3 Results

To accomplish the first step of the proposed methodology, an indicator variable was established, taking the value 1 if the As concentration is higher than 40 mg/kg and 0, otherwise. The omnidirectional variogram of such a variable is given in Fig. 2a, together with the fitted spherical model parameters.

Based on the variogram model of Fig. 2a, the indicator variable was estimated in a 10×10 m regular mesh, providing the map exhibited in Fig. 2b when an average based cut-off is applied to the kriged indicator values. The white zone of Fig. 2b is to be disregarded, since concentrations in that area are likely to be lower than the lowest limit for ecosystem vulnerability in what As is concerned.

The second step of the methodology requires, as a prolegomenon, the construction of the histogram of Fig. 3a, since the examination of the entire set of raw data shows a high skewed distribution containing some outliers, as described in the Introduction section. In fact such a distribution is suited to a logarithmic transformation, as shown in Fig. 3b.

The variogram of log As concentrations is given in Fig. 4a (together with fitted model parameters), and the kriged map for the same 10×10 m regular mesh is given in Fig. 4b.

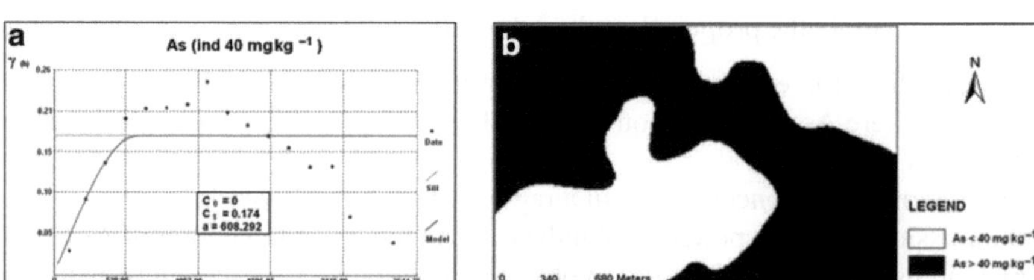

Fig. 2 (**a**) Variogram of the indicator variable (threshold in As concentration of 40 mg/kg) and (**b**) Map with the limits of the zone where As concentration <40 mg/kg

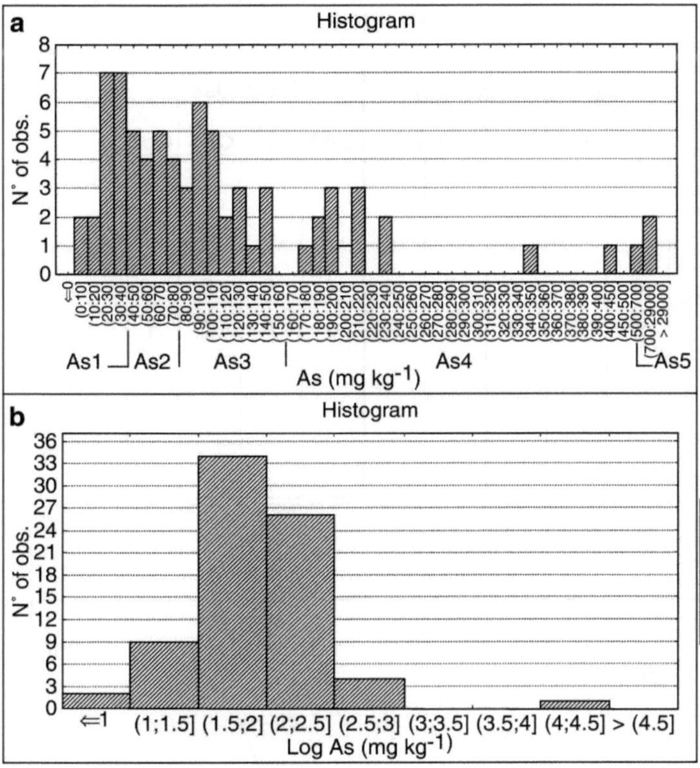

	As
Minimum	9.00
Lower Quartile	44.00
Mean	497.66
Median	85.50
Upper Quartile	143.00
Maximum	29000.00
Variance	10993694.87
Std.Dev	3315.67
Skewness	8.70
Kurtosis	75.75

	LogAs
Minimum	0.95
Lower Quartile	1.64
Mean	1.93
Median	1.93
Upper Quartile	2.16
Maximum	4.46
Variance	0.24
Std.Dev	0.49
Skewness	1.80
Kurtosis	8.87

Fig. 3 Histograms of (**a**) As concentrations and (**b**) log As concentrations

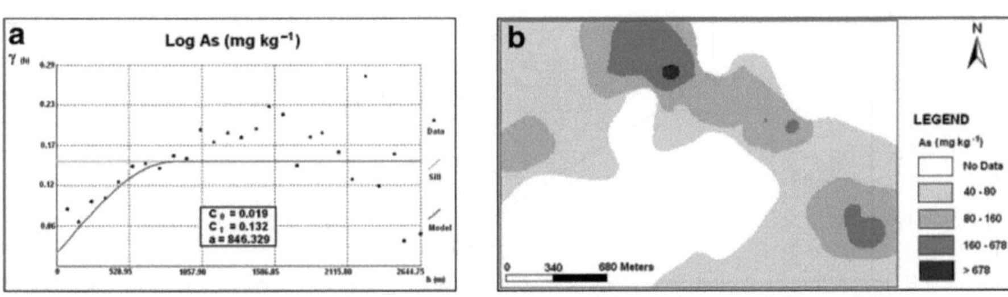

Fig. 4 (**a**) Variogram of log As concentrations and (**b**) estimated map by lognormal kriging

The third step of the methodology consists of applying the Correspondence Analysis algorithm to the data model of Fig. 5, where samples are defined by one category of land use (excluding Oc1, which refers to the unsampled pollution source), and by one class of As concentration. These classes were established on the grounds of the 'natural' splitting of the histogram of Fig. 3a, taking into account the thresholds provided in the Introduction section.

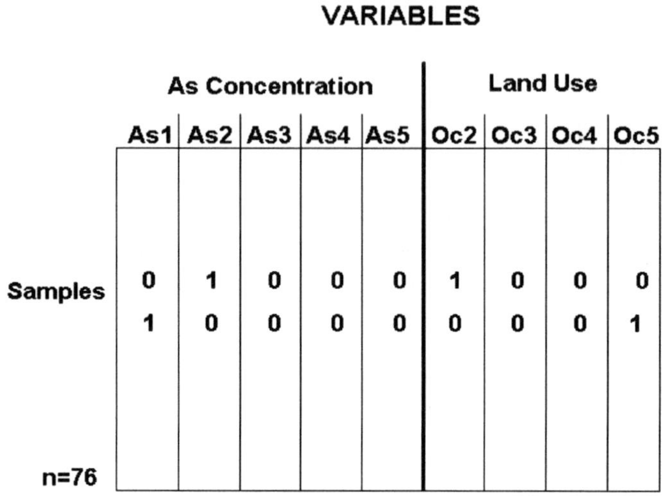

Fig. 5 Data model for correspondence analysis

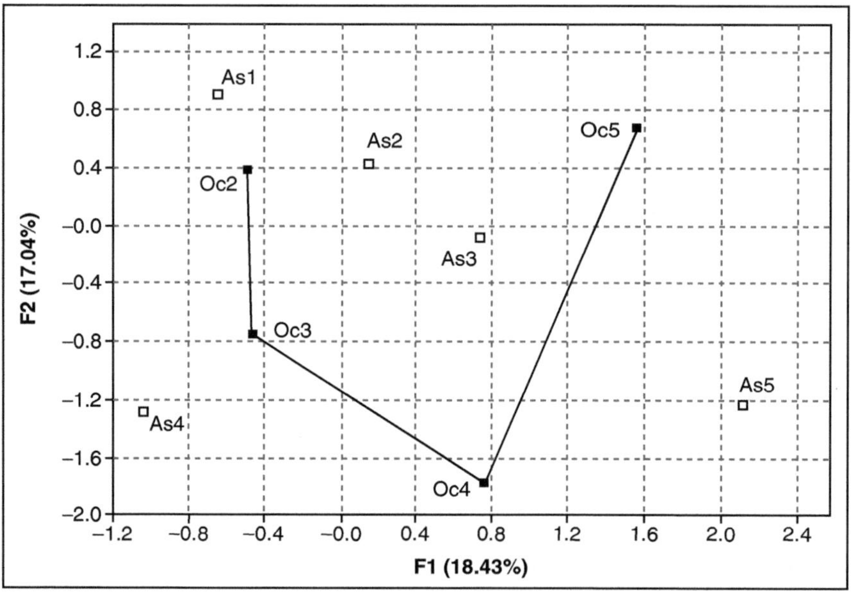

Fig. 6 First factorial plane of correspondence analysis showing modalities projections

Results of CA are given in Fig. 6, showing the projections of variable modalities onto the first factorial plane. It is worth noting, in the graph of Fig. 6, the "horseshoe effect" (Greenacre, 1984) referring to the sequence of categories of the ordinal land use variable, whose projections grow along F1. Hence, given the "barycentric property" (Greenacre, 1984), which allows the 'correspondence' of variables and

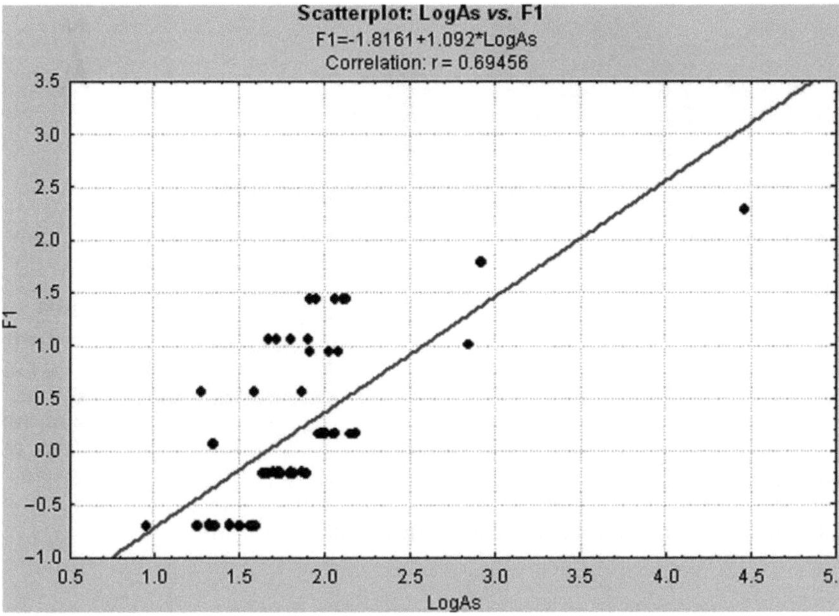

Fig. 7 Bi-plot of As concentrations versus projection of samples onto the first CA axis (excluding As4 class)

samples in the same graph, sample co-ordinates in F1 may be viewed as the quantitative variable that 'substitutes' the ordinal land use attribute, in what its interaction with As concentration is concerned.

An inspection of Fig. 6 in regard to As classes sequence reveals that As4 class is not linked to axis 1 (its contribution to axis 2 is higher). On the other hand, the histogram of Fig. 3a shows a very scattered (and heterogeneous) allocation of frequencies within As4 class. Hence, it was decided to disregard values belonging to such a class, generating the bi-plot of Fig. 7 (that exhibits a reasonable correlation between F1 – to be used as the drift – and log As).

Finally, the real variable 'projection of sample onto F1' was used as a surrogate of land using classes for the application of the external drift technique, which produces the map of Fig. 8.

4 Discussion and Conclusions

Since it is not available any quantification of land use (in monetary units) and the harmfulness to humans can not be put in the same scale as to ecosystems, a remediation cost–benefit analysis is not feasible, at this stage. However, a quantitative scenario for As concentration interdependence with land use is provided by the

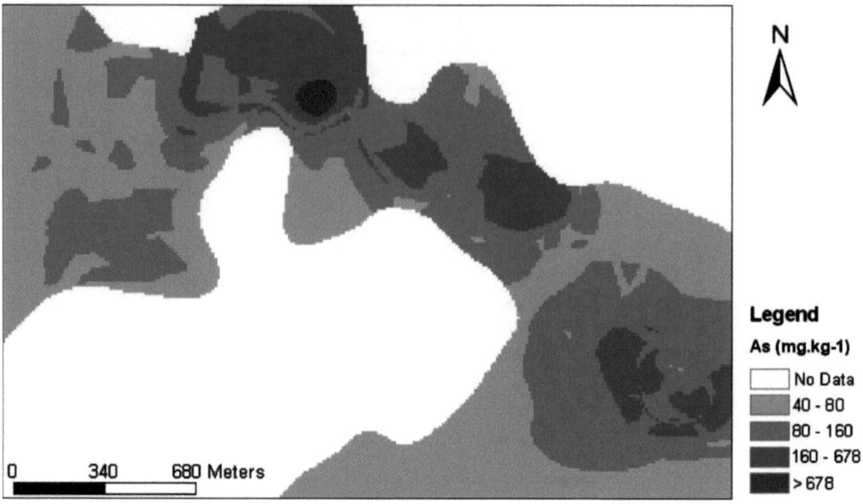

Fig. 8 Estimated map of As concentrations by kriging with external drift

proposed methodology, whose crucial point is the application of a proxy of land use, transforming the corresponding ordinal variable into a real number though Correspondence Analysis. For creating such a quantitative scenario, the external drift technique was adopted, rather than the soft cokriging method for categorical information (Goovaerts, 1997), because advantage may be taken by introducing the ordinal character of the land use attribute.

The practical results of the study allow to select areas where an eventual remediation procedure may be foreseen; in such areas, the 'raw' As concentrations were 'blended' with land use, producing a more realistic map which flattens the innocuous zones and highlights the dangerous ones, according to As concentration versus land use interaction.

Acknowledgments This research was funded by the European Commission through the e-Ecorisk Project (# EVG1–2002–25 0068), "A regional enterprise network decision-support system for environmental risk and disaster management of large-scale industrial spills".

References

ATSDR (1993) Toxicological profile for arsenic. US Department of Health and Human Services, Public Health Service. Agency for toxic substances and diseases registry, Atlanta, Georgia

Ávila P, Ferreira da Silva E, Salgueiro AR, Farinha JA (2008) Geochemistry and mineralogy of mill tailings impoundments from the Panasqueira mine (Portugal): implications for the surrounding environment. Mine Water Environ doi:10.1007/s10230-008-0046-4

Breiter K (2001) Report about Laboratory Investigations of Rock Samples from the Panasqueira Mine and Recommendations for Future Exploration, Nov 2001

Choe E, Van de Meer F, Ruitenbeek F, Van der Werff H, Smeth B, Kim K (2008) Mapping of heavy metal pollution in stream sediments using combined geochemistry, field spectroscopy and hyperspectral remote sensing: a case study of the Rodalquilar mining area, SE Spain. Remote Sensing Env 112:3222–3233

Correa A, Naique RA (1998) Minas Panasqueira, 100 Years of Mining History. Proceedings of the 1998 conference of the International Tungsten Industry Association (ITIA)

Glanzman RK, Closs LG (1993) Quality assurance and control guidelines for exploration and environmental geochemistry investigation. Explore 78:1–6

Goovaerts P (1997) Geostatistics for natural resources characterization. Oxford University Press, New York

Goovaerts P (1999) Geostatistics in soil science: state-of-the-art and perspectives. Geoderma 89(1–2):1–45

Greenacre MJ (1984) Theory and applications of correspondence analysis. Academic Press, London/Orlando, FL

Jeannée N, Fouquet C (2003) Contribution of auxiliary information for the estimation of grades in heterogeneous media: example in soil pollution. C R Geosci 335:441–449

Maréchal A (1884) Kriging seismic data in presence of faults. In: Verly G, David M, Journel A, Maréchal A (eds) Geostatistics for natural resource characterization. Reidel, Dordrecht

Noronha F, Doria A, Dubessy J, Charoy B (1992) Characterization and timing of the different types of fluids present in the barren and ore veins of the W-Sn deposit of Panasqueira, Central Portugal. Miner Deposita 27:72–79

Swartjes F (1999) Risk-based assessment of soil and groundwater quality in the Netherlands: standards and remediation urgency. Risk Anal 19:1235–1249

Walsh L, Sumner M, Keeney D (1977) Occurrence and distribution of arsenic in soils and plants. Environ Health Perspect 19:67–71

Simulation of Continuous Variables at Meander Structures: Application to Contaminated Sediments of a Lagoon

Ana Horta, Maria Helena Caeiro, Ruben Nunes, and Amílcar Soares

Abstract Simulation of continuous variables conditioned to meander structures is an important tool in the context of soil contamination assessment, namely, when the contamination is related with depositional sediments in water channels. Hence, this paper proposes using bi-point statistics stochastic simulation with local anisotropy trends to simulate continuous variables inside predefined channels. To accomplish this objective, the Direct Sequential Simulation (DSS) algorithm was modified to account for local anisotropy when searching for the simulation node. This methodological approach was applied to the spatial characterization of polluted sediments in a coastal lagoon located in the North of Portugal (Barrinha de Esmoriz).

1 Using Geostatistics for the Characterization of Meander Structures

Modelling curvilinear or meander structures can help to differentiate between different geological media and/or to condition the estimation/simulation of data to those structures. For petroleum applications, to recognize the shape of the structures can be a first step whilst for hydrological or environmental applications, meander forms can be visual and numerically recognized. In this situation, the issue will be to assess spatial distribution limited to those shapes.

One of the first attempts to model the morphology of geological curvilinear structures using geostatistics was made by Soares (1990), who proposed the use of local anisotropy directions to estimate (using morphological kriging) folded geological strata. This result was particularly important for petroleum applications, since it provided the possibility to identify different structures for the numerical modelling of reservoirs. This idea was used by Luis and Almeida (1997) and Xu (1997) to condition sequential simulation procedures for the characterization of sand channels

A. Horta (✉), M.H. Caeiro, R. Nunes, and A. Soares
Centro de Recursos Naturais e Ambiente, Instituto Superior Técnico,
Universidade Técnica de Lisboa, Lisbon, Portugal
e-mail: ahorta@ist.utl.pt; helena.caeiro@ist.utl.pt; nunesrfm@gmail.com; asoares@ist.utl.pt

P.M. Atkinson and C.D. Lloyd (eds.), *geoENV VII – Geostatistics for Environmental Applications*, Quantitative Geology and Geostatistics 16,
DOI 10.1007/978-90-481-2322-3_15,

geometry in a fluvial reservoir. Their work presented a pixel-based approach to simulate the geometry of sand channels taking into account morphological information and local continuity directions. When compared to object-based algorithms, which are an alternative way to reproduce curvilinear shapes, these pixel-based algorithms accounting for directional information were better suited to reservoir characterization due to the possibility to incorporate local field data. A first application of this concept to an environmental problem was presented by Caetano et al. (2004) who used wind directions as local anisotropy information to condition the estimation (kriging) of atmospheric pollutant distribution. Another example is the work presented by Stroet and Snepvangers (2005) that uses local anisotropy kriging to interpolate bathymetric data. These applications are based on two point statistics by using a kriging algorithm. Recently, multiple-point statistics (MPS) has been proposed for the characterization of meander structures and further variable simulation (see Strebelle, 2002). In the context of petroleum applications, simulation of meander structures with MPS consists of extracting patterns from training images and then reproducing those patterns conditioned to local field data (Strebelle, 2007). Also relying on a pixel-based sequential approach, MPS can be used for the simulation of categorical and continuous variables. However, modelling of continuous properties implies a discretization into a small number of classes to process simulation and a discrete-to-continuous transformation afterwards (Strebelle, 2007).

Thus, considering the present state-of-the-art, this paper aims to provide a solution for the simulation of continuous variables conditioned to meander structures. To achieve this goal, a pixel-based sequential algorithm (Direct Sequential Simulation; Soares, 2001) was used to reproduce bi-point statistics plus local anisotropy information (local directions and ratios).

2 Objectives

The aim of this paper is to present an application of Direct Sequential Simulation (DSS) to the characterization of a continuous variable with a spatial distribution conditioned to a meander structure, i.e., the algorithm had to be modified to account for local anisotropy information (direction of maximum continuity and anisotropy ratio). The problem of conditioning simulation to a specific curvilinear form was raised in the context of an environmental application related to the assessment of sediment contamination in a coastal lagoon, with a permanent water/sediment flow due to effluent water channels and the sea. A rationale was established to better approach the problem:

(i) Pollutant contamination patterns in sediments usually follow preferential main flow paths. Hence, it is not advisable to simulate a pollutant concentration ignoring a preferential transport/accumulation path.
(ii) Knowing the water flow regime enables us to determine a main flow direction (and thus the main direction of continuity for the dispersion contaminant path) and flow velocity can be related to the degree of anisotropy of such patterns.

Hence, once the main flow trends in the meanders have been defined, local anisotropy parameters can be estimated. For the presented case study, a satellite image was used to define main water flow paths and compute local directions and ratios.

3 Simulation of Continuous Variables Conditioned to Meander Structures

To determine the spatial distribution of a certain attribute conditioned to a curvilinear (or meander) structure, the use of stochastic simulation is a reliable option. Simulation algorithms not only allow for spatial assessment of an attribute but also provide information about the spatial uncertainty involved on that evaluation. DSS had been used for the spatial characterization of continuous variables related with several environmental problems such as air pollution (Soares and Pereira, 2007; Russo et al., 2008) or soil quality assessment (Franco et al., 2006; Horta et al., 2008). In these examples, the spatial correlation is evaluated across a Euclidean space, without differentiating sample locations (for example, samples exposed to different wind conditions or samples collected in different soil types). Thus, DSS was performed with global variogram parameters (direction, range and ratio of anisotropy), assumed to be representative for the entire the study area. Therefore, when it comes to conditioning the simulation to a meander structure – typical non-stationary situation – DSS is not able to reproduce the curvilinear shapes. One solution is to introduce local spatial trends representing local anisotropy variations that will reproduce the meander aspect of the structure where the variable is to be simulated.

3.1 Introducing Local Anisotropy in the DSS Algorithm

Let us consider the continuous variable $\underline{Z}(\mathbf{x})Z(x)$ with a global cumulative distribution function (cdf) $F_z(z) = \text{Prob}\{Z(\mathbf{x}) \leq \mathbf{z}\}$. The main sequence of methodological steps of DSS can be summarized as follows:

(i) Define a random path over the entire grid of nodes $x_u(u = 1, \ldots, N)$ to be simulated.
(ii) Estimate the local mean and variance of z(x), identified, respectively, with the simple kriging estimate $z^*SK(x)$ and variance $\sigma^2SK(x)$, conditioned to the original data z(x) and the previous simulated values $z^l(x)$.
(iii) Define the interval Fz(z) to be sampled (defined by the local mean and variance of z(x)).
(iv) Draw a value $z^l(x)$ from the cdf Fz(z).
(v) Loop until all N nodes have been visited and simulated.

To solve the simple kriging system (step ii), experimental samples are selected with an elliptical search radius which is defined using global variogram parameters, as

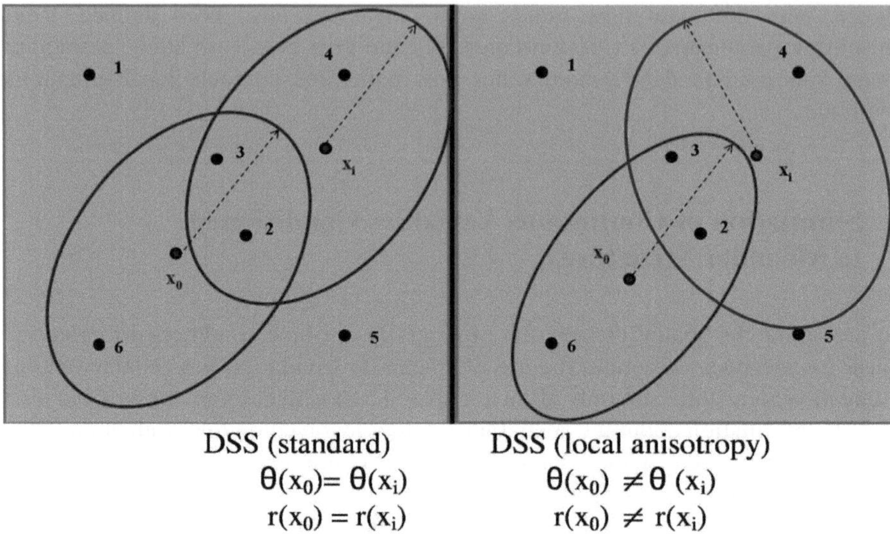

Fig. 1 Representation of search radius definition for standard DSS and DSS with local anisotropy

illustrated in Fig. 1. In practical terms, accounting for local anisotropy parameters, namely, direction of maximum continuity (given by azimuth θ) and anisotropy ratio (r), means changing the search radius from node to node to be simulated, as illustrated in Fig. 1. Thus, in step (ii), the matrix of data-to-data covariances and the vector of data-to-unknown covariances are calculated with corrected local covariances $C_{\theta,r}(\mathbf{h})$ by the local values of $\theta(\mathbf{x})$ and $r(\mathbf{x})$. The simple kriging estimate of local mean becomes a function of $\theta(\mathbf{x})$ and $r(\mathbf{x})$. Note that to estimate a local cdf at given location $\mathbf{x}_\mathrm{u}$ only the local angle of $\mathbf{x}_u$ is retained.

The practical application of this idea raised other issues such as choosing the range of maximum continuity (search ellipse major axis a_θ). For this paper purpose, it was assumed that $a_\theta$remained constant and equal to the range of the global variogram. Only the minor range of the search ellipse was conditioned to the width of the meander structure in each simulated node. Thus, changes in anisotropy parameters determine that the variogram model is non-stationary.

4 Application

4.1 Study Area

The methodology presented in this paper was developed within a project framework which aims to characterize soil/sediment contamination using state-of-the-art geostatistical models to assess spatial uncertainty. The study area is a coastal lagoon,

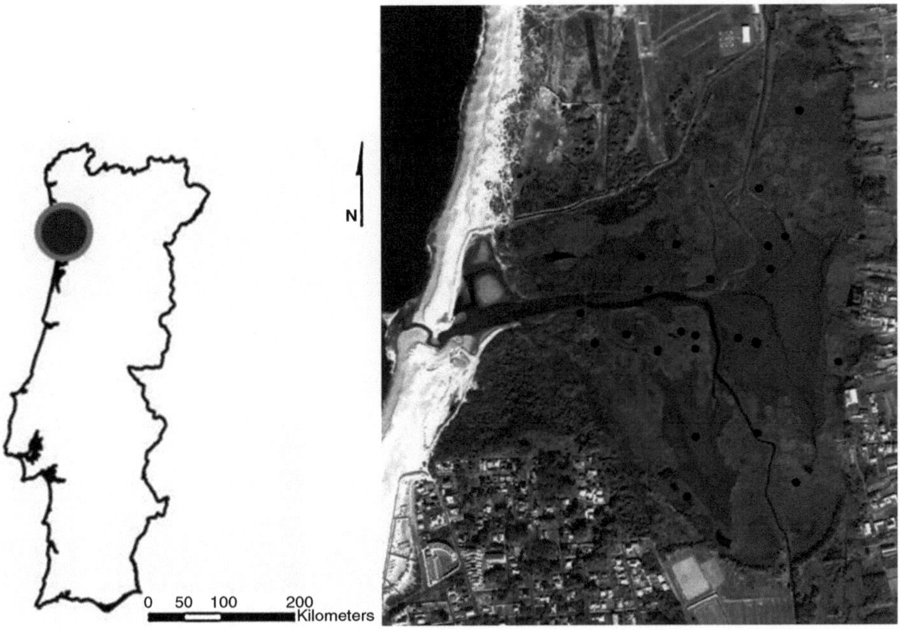

Fig. 2 Study area (Barrinha de Esmoriz) and Sampling point distribution

located in the Portuguese Northern Region, named Barrinha de Esmoriz (Fig. 2). In terms of its ecological value, the Barrinha de Esmoriz was included in the list of the natural sites to be integrated in the Natura Network 2000. The lagoon is about 1,500 length and 700 m width, surrounded by dense vegetation (reeds and scrubs) and bordered by the dune. The sea is about 400 m distance and it connects with the lagoon through a 50 m width channel. Also, two water ditches flow into the lagoon, coming from the North and from the South, using the lagoon as a discharge point from the water basin.

A sedimentation process has been taking place in the last few decades, reducing lagoon's area and water depth. Also, there have been reports of serious pollution discharges from the Northern ditch, mainly industrial water discharges coming from the industrial sites located in the Northern part of the water basin. Evidence of this pollution has already been reported in a previous soil contamination assessment.

4.2 Soil Contamination Data

A previous soil contamination report (DHVFBO, 2001), developed to evaluate the degree and the extent of contamination in the lagoon's sediments, contained information about heavy metal concentrations at 25 sampling points, distributed as

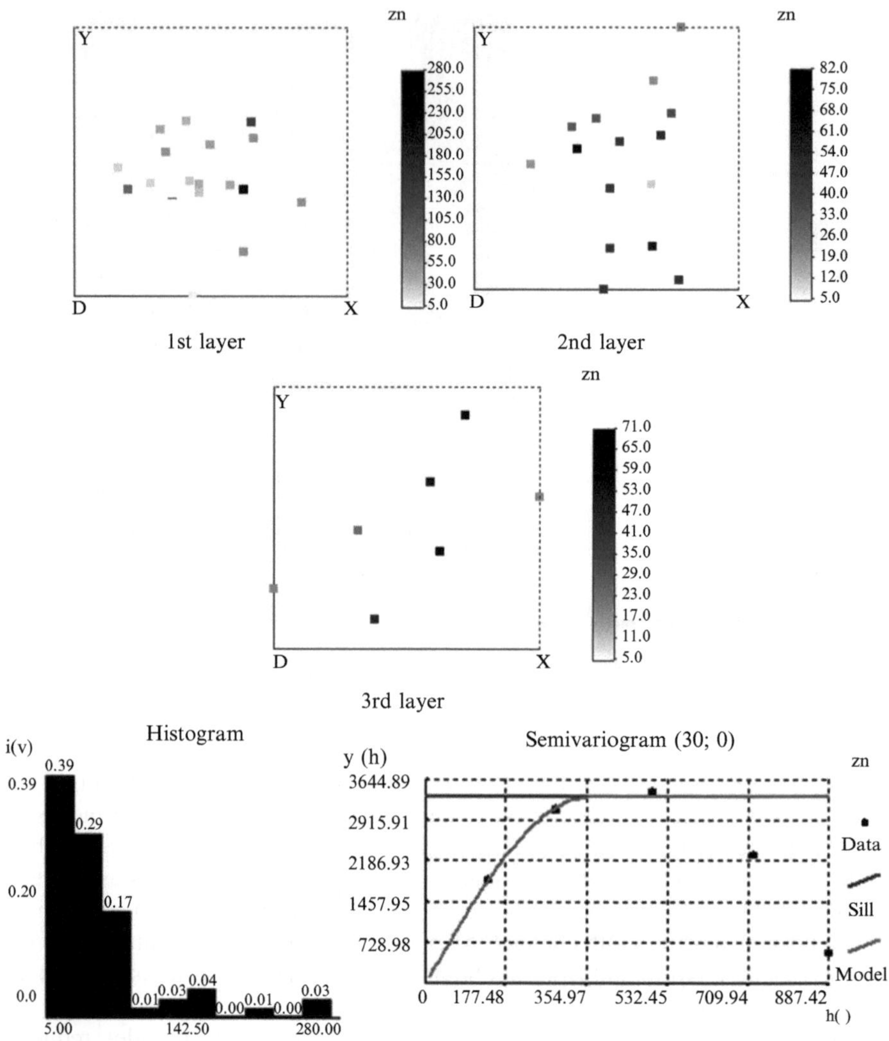

Fig. 3 Zn spatial distribution, histogram, and mean variogram (spherical model, angular tolerance of 20°)

shown in Fig. 2. The samples were collected in the first 1, 2 and 3 m, depending on field conditions. For simulation purposes, 64 data values were used, obtained for Arsenic (As), Copper (Cu), Cromium (Cr), Niquel (Ni) and Zinc (Zn). From this set, 39 values correspond to concentrations in the upper sediment layer. As an example, Zn concentration distribution in the three sampled layers is presented in Fig. 3. Also, Fig. 3 shows the sample locations at different layers, and the global histogram and mean variogram.

4.3 Model Implementation and Results

For the assessment of sediment contamination with Zn, the following methodological steps were performed:

1. Flow Direction Assessment: using a Quickbird satellite image (2006) the main trends of water flow channels were visually recognized and used to define flow direction vectors.
2. Computing of Local Anisotropy: estimation of direction of maximum continuity (θ) and anisotropy ratio (r), using a kriging algorithm (Fig. 4).
3. Contamination Assessment: simulation using DSS with correction for local anisotropies and uncertainty evaluation based on the variance of simulated images (Figs. 5 and 6).

Regarding the practical implementation of DSS with local anisotropy to this case study, besides the modification introduced to account for local direction and ratio, also a connected sequential simulation path was imposed to improve the calculation of contaminant concentration. Instead of choosing one point x_u in the random path to be simulated, a set of connected points $\left[x_u, x_1, ..x_{np}\right]$ was chosen to be simulated in a row (Yao, 2007). Each point x_u is in the direction if $x_u + 1$ defined by the angle $\theta(x_u)$: arctg $(x_u - x_u + 1) = \theta(x_i)$. The number np of connected points to be simulated is randomly defined at each sequential step.

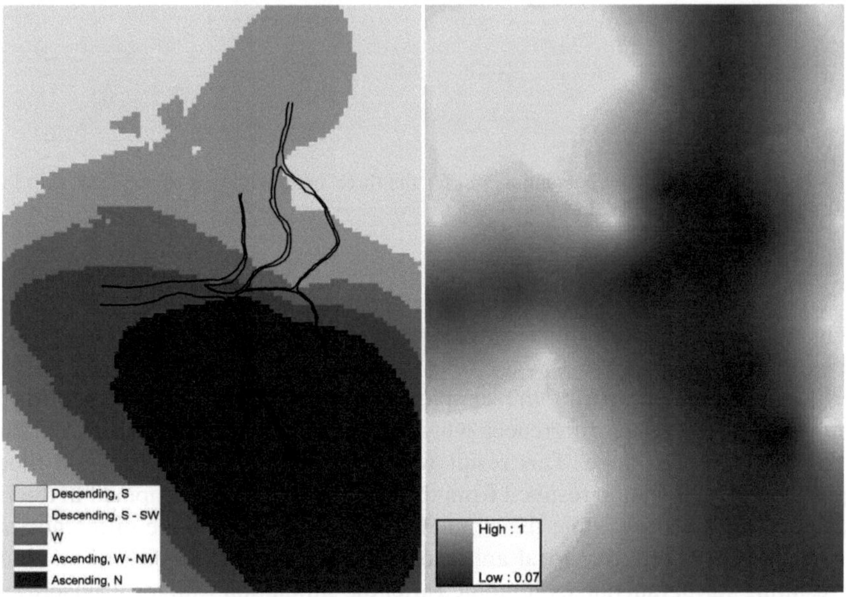

Fig. 4 Local Anisotropy, from *left* to *right*: (**a**) Flow main directions (**b**) Anisotropy ratio

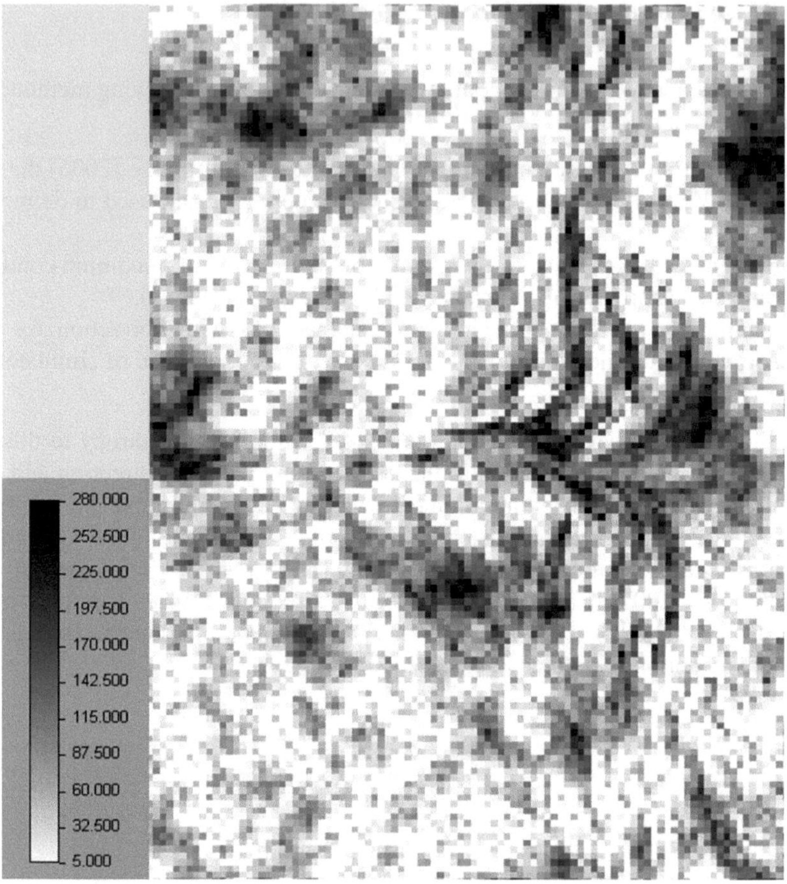

Fig. 5 One realization of Zn contamination, for the three sampled layers (using DSS with local anisotropies and connectivity flow path)

To check the quality of simulation performance, histogram and variogram reproduction in the simulated images were verified and produce generally the results in Fig. 7.

However, when comparing the sample variogram and the one obtained for the simulated images, some differences where detected (Fig. 8), mainly in what concerns the computed range. This result was expected since the variogram model imposed to the simulation resulted from the sample variogram computed in the Euclidean reference space while the simulated values result from the different water flow channels i.e. different local anisotropy relations and main directions. Hence the resulting variogram ranges computed after the simulation with local anisotropies tend to be smaller than the imposed model.

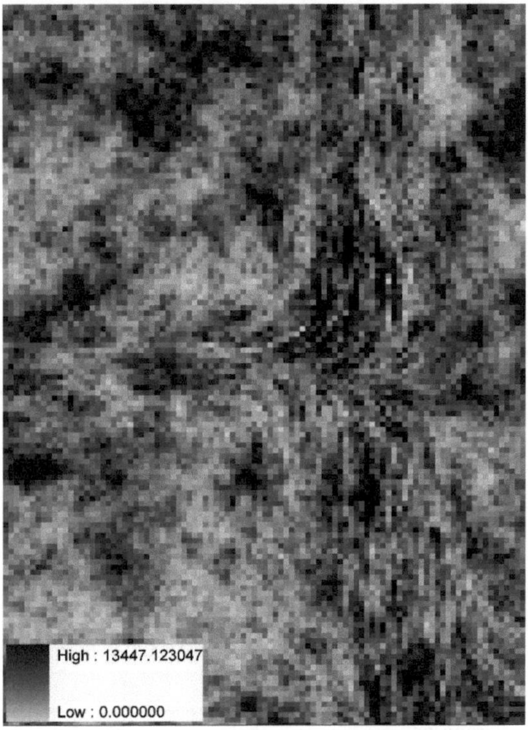

Fig. 6 Uncertainty evaluation for the first layer

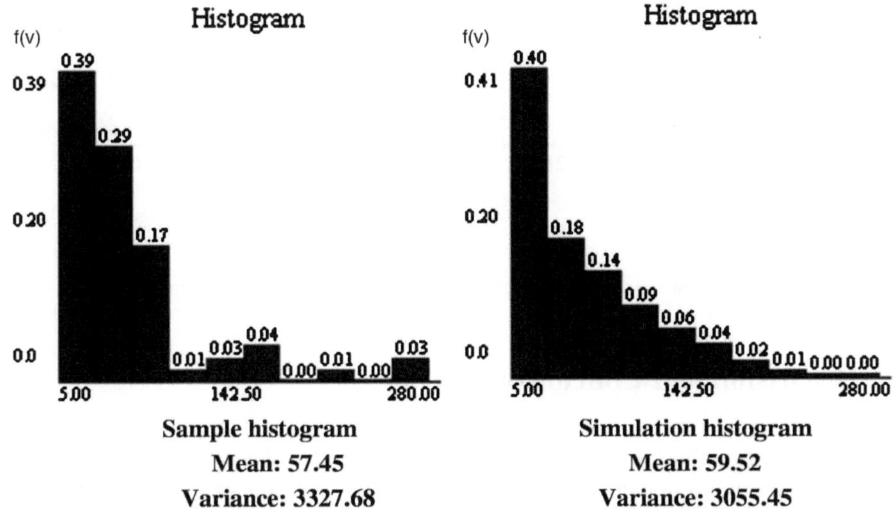

Fig. 7 Comparison between sample histogram and simulated images histogram

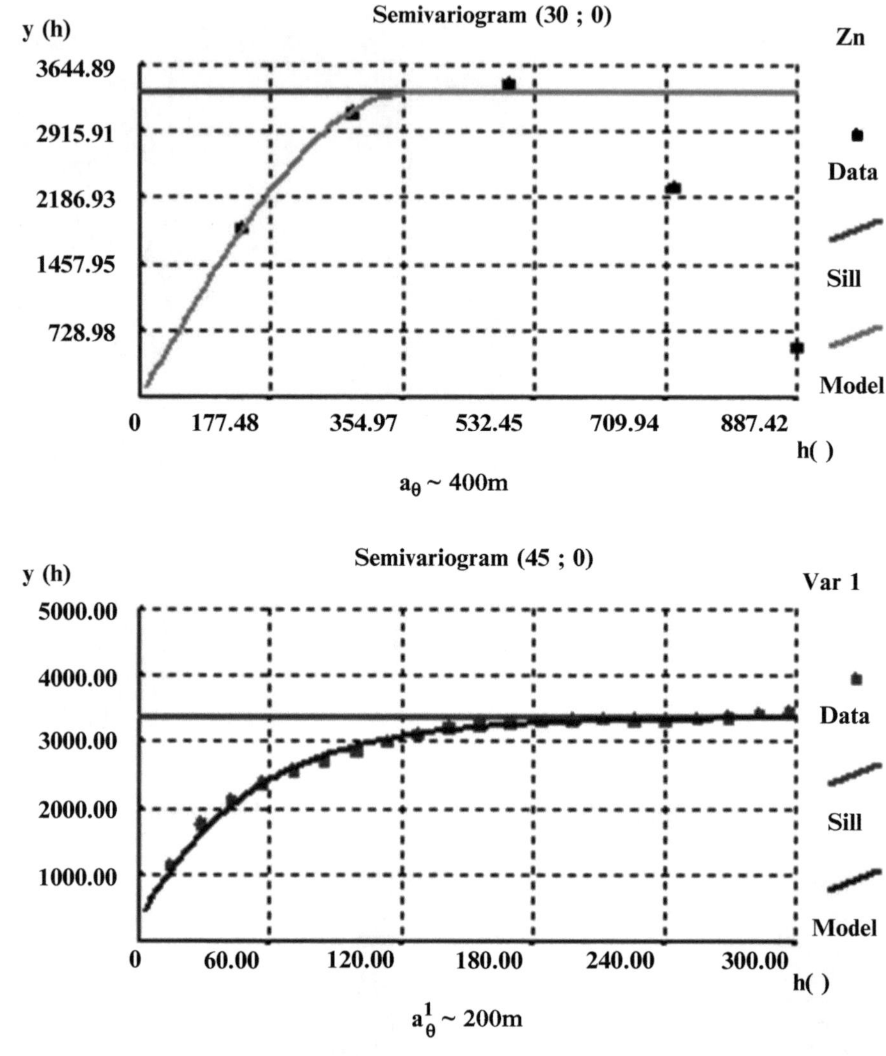

Fig. 8 Comparison between sample variogram and simulated images variogram

5 Discussion and Conclusion

The presented method refers to the application of Direct Sequential Simulation to the characterization of a continuous variable with a spatial distribution conditioned to a meander structure. With this purpose, the DSS algorithm had to be modified to account for local anisotropy information (direction of maximum continuity and anisotropy ratio).The proposed methodology has shown quite promising results for the Barrinha de Esmoriz case study. It was possible to obtain a set of probable images for contamination dispersion along the lagoon channels thus identifying hot spots. However, uncertainty evaluation as presented in Fig. 6 shows high values for variance for the concentrations calculated along some parts of the channels (especially in the Northern part of the lagoon). This may be due to the lack of hard contaminant data along the channel paths. This information will be used to define an improved sampling campaign for the Barrinha de Esmoriz project.

Regarding further developments in the application of DSS using local anisotropy, this method can be generalized to the application to other fields, namely, the characterization of internal properties of reservoirs inside channel boundaries previously simulated by MP statistics.

Finally, a crucial point of this methodology is the determination of local directions and ratios of anisotropy. For the presented case study, as the main channel trends were visible in aerial photos, those parameters were directly inferred by the shape of meander structures. The main vectors defining main flow directions were first identified in the channels and, afterwards, they were populated for a regular grid of points covering the entire set of channels, using a kriging algorithm. Note that, instead of kriging, these main flow directions parameters could also be simulated (Luis and Almeida, 1997; Xu, 1997), principally when there is a high uncertainty about the meander's shape and location.

Acknowledgments This paper was produced in the context of Project "Soil contamination risk assessment" (PTDC/CTE-SPA/69127/2006) (financed by FEDER through the national Operational Science an Innovation Program 2010 with the support of the Foundation for Science and Technology).

References

Caetano H, Pereira MJ, Guimarães C (2004) Use of factorial kriging to incorporate meteorological information in estimation of air pollutants. In: Sanchez-Vila X, Carrera J, Gómez-Hernández J (eds) geoENV IV – Geostatistics for environmental applications. Kluwer, The Netherlands, pp 55–66

DHVFBO (2001) Technical Report "Soil and Groundwater Contamination Assessment in Barrinha de Esmoriz"

Franco C, Soares A, Delgado J (2006) Geostatistical modelling of heavy metal contamination in the topsoil of Guadiamar river margins (S Spain) using a stochastic simulation technique. Geoderma 136(3–4):852–864

Horta A, Carvalho J, Soares A (2008) Assessing the quality of the soil by stochastic simulation. In: Soares A, Pereira MJ, Dimitrakopoulos R (eds) geoENV VI – geostatistics for environmental applications. Springer, Berlin, pp 385–396

Luis JJ, Almeida JA (1997) Stochastic characterisation of fluvial sand channels. In: Baafi EY, Schofield NA (eds) Geostatistics Wollongong 96, vol. 1. Kluwer, The Netherlands, pp 477–488

Russo A, Trigo RM, Soares A (2008) Stochastic modelling applied to air quality space-time characterization. In: Soares A, Pereira MJ, Dimitrakopoulos R (eds) geoENV VI – geostatistics for environmental applications. Springer, Berlin, pp 83–93

Soares A (1990) Geostatistical estimation of orebody geometry: morphology kriging. Math Geol 22(7):787–802

Soares A (2001) Direct sequential simulation. Math Geol 33(8):911–926

Soares A, Pereira MJ (2007) Space–time modelling of air quality for environmental-risk maps: a case study in south Portugal. Comput Geosci 33(10):1327–1336

Strebelle S (2002) Conditional simulation of complex geological structures using multi-point statistics. Math Geol 34(1):1–21

Strebelle S (2007) Simulation of petrophysical property trends within facies geobodies. In: EAGE (ed) Petroleum Geostatistics 2007

Stroet C, Snepvangers J (2005) Mapping curvilinear structures with local anisotropy kriging. Math Geol 37(6):635–649

Xu W (1997) Conditional curvilinear stochastic simulation using pixel-based algorithms. In: Baafi EY, Schofield NA (eds) Geostatistics Wollongong 96, vol. 1. Kluwer, The Netherlands, pp 454–464

Yao T, Calvert C, Jones T, Foreman L, Bishop G (2007) Conditioning geologic models to local continuity azimuth in spectral simulation. Math Geol 39:349–354

Joint Space–Time Geostatistical Model for Air Quality Surveillance/Monitoring System

Ana Russo, Amílcar Soares, Maria João Pereira, and Ricardo M. Trigo

Abstract Air quality is usually driven by a complex combination of factors where meteorology, physical obstacles and interaction between pollutants play significant roles. The use of models that are able to characterize space–time dispersion of pollutants at fine scales in urban areas (e.g. stochastic and neural networks models) is becoming a common practice. The main objective of this work is to produce an integrated air quality model designed to monitor Lisbon's metropolitan area. This model, which allows forecasting critical concentration episodes of a certain pollutant by means of a hybrid approach, is based on the combined use of neural network models and stochastic simulations. A stochastic simulation of the spatial component with a space–time trend model is proposed to characterize critical situations at a given present period or for a very near future period, taking into account data from the past and a space–time trend from the recent past. To identify critical episodes in the near future period $t + 1$, predicted values from neural networks are used at each monitoring station. The neural network model was developed taking into account historical data of pollutants' concentrations and meteorological conditions measured and also predicted for each monitoring station. First, a joint space–time model is used to build the trend model based on historical data (period $\leq t$). Afterwards, stochastic simulation is performed to predict the period $t + 1$ at any location x, allowing for the local conditional distribution functions characterization and spatial uncertainty assessment. As this approach is performed sequentially in the time domain, the space–time trend is sequentially updated for every new period $t + i$, $i = 1\ldots, N$. This spatial-temporal model has been developed and applied to the urban area of Lisbon. An application to the prediction of mean daily NO_2 concentration is presented in this paper.

A. Russo (✉), A. Soares, and M.J. Pereira
CERENA, Instituto Superior Técnico, Av. Rovisco Pais, 1. 1049-001 Lisboa, Portugal
e-mail: arusso@ist.utl.pt

R.M. Trigo
Centro de Geofísica da Universidade de Lisboa, Ed. C8, Campo Grande,
1749-016 Lisboa, Portugal

P.M. Atkinson and C.D. Lloyd (eds.), *geoENV VII – Geostatistics for Environmental Applications*, Quantitative Geology and Geostatistics 16,
DOI 10.1007/978-90-481-2322-3_16,

1 Introduction

The industrial and urban development that took place in the last few centuries led to a generalized increase of atmospheric pollutant emissions. As a result, atmospheric pollution is nowadays considered to be a problem, especially in major cities. Given that it is a problem that threatens peoples' health (Seinfeld, 1986; Cobourn et al., 2000; Kolehmainen et al., 2000), it is mandatory to develop tools that are able to identify and predict harmful situations, in order to take measures destined to its prevention, mitigation and risk assessment. Providentially, during the last few decades there has been a marked increase in research activity focusing on modelling and simulating air quality. This growing interest was also motivated by the large amount of data from monitoring activities and by the necessity of answering important environmental problems such as the ones mentioned above.

Air quality, as for most natural phenomena, can be seen as a space–time process. However, the simultaneous integration of space and time is not an easy task to achieve (Nunes and Soares, 2005). The difficulty of a simultaneous integration results from the fact that space and time relationships have usually quite different characteristics and levels of uncertainty (Nunes and Soares, 2005). Usually, the selection of a model is based on several issues, such as, data availability, purpose of the study or computational cost. However, another well-known characteristic common to most air quality monitoring networks can present itself as a problem: high density of sample values in time collected at just a few spatial locations. This can be a serious limitation if one wishes to evaluate impact costs or carry out an environmental risk analysis of the emissions for public health, the different land uses, eco-systems and natural resources of a region (Russo et al., 2005). Spatial-temporal geostatistical models can constitute an alternative to other types of modelling techniques (Kyriakidis and Journel, 1999; Nunes and Soares, 2005) because they allow the characterization of uncertainty, supplying images of a probable reality that reproduces patterns of spatial continuity quantified by the observations available. In Portugal the major impacts of air pollution tend to be registered generally in areas with large urban concentration and/or in the presence of large industrial units.

The main objective of this work is to produce an integrated air quality surveillance/monitoring system, which allows forecasting critical concentration episodes of a certain pollutant by means of a hybrid approach, based on neural network models and stochastic simulations. Nowcast and forecast spatial-temporal air quality models are developed, including information regarding different time and spatial scales. A space–time model system, taking into account meteorological conditions for the characterization of the spatial and temporal distribution of the pollutants is proposed. After the completion of the main objective of this work it will be possible to foresee critical air quality situations. Real time forecasts provide air quality alerts, allowing sustainable management of environmental risks for public health, thus, supporting the decision of the responsible entities for environmental and health management.

2 Methodology

The proposed surveillance/monitoring methodology is based on a space–time stochastic simulation model. This methodology includes short term predictions by means of a neural network model fitted for each monitoring station, and also the past space–time trend obtained from historical data.

2.1 Prediction Model Formulation – Neural Network Modelling

Presently, neural network (NN) models constitute the best technique that is able to identify complex non-linear relations between variables – inputs and outputs – without previous integral understanding of the phenomenon's nature (Haykin, 1994; Beale and Demuth, 1998). A number of other methods (e.g. Box-Jenkins time series models) have been applied to time-series for air pollutants (Simpson and Layton, 1983; Ziomas et al., 1995; Shi and Harrison, 1997), including comparisons with neural network methods (Yi and Prybutok, 1996; Comrie, 1997; Gardner and Dorling, 1999; Cobourn et al., 2000; Kolehmainen et al., 2000). Most of the comparative studies concluded that ANN generally provides as accurate as or more accurate results than linear methods. The NN model used was trained and tested using air quality and meteorological data and was processed by a multiple layers neural network with feed-forward propagation (feed-forward multi-layer perceptron) trained by a back-propagation algorithm. The prediction of the daily average pollutant's concentration at the monitoring station α for day $t+1$, $Z^*(x_\alpha, t+1)$, is achieved based on air quality data and meteorological data of the previous day t and from the early hours of day $t+1$, by running the NN model fitted.

2.2 Space–Time Model

For this study, a stochastic simulation of spatial component with a space–time trend model, taking into account the predicted data of "next day" and a space–time trend from the recent past, is proposed. This model was conceived in order to deal with spatial non-stationary situations. Consider an attribute $Z(x,t)$ defined at a spatial location x, $x \in D$ at day $t \in T$. In this model, the attribute value z is decomposed into a trend $M(x,t)$ and a residual $R(x,t)$:

$$Z(x,t) = M(x,t) + R(x,t) \tag{1}$$

The proposed approach can be summarized in three basic iterative steps:

(i) Characterization of the space–time trend for day t (the present time) $M(x,t)$ based on data from previous days, $Z(x_\alpha, t+1-i)$, $i = 1, \ldots, N_\mathrm{d}$

(ii) Simulation of residuals, taking into account the predictions $Z^*(x_\alpha, t+1)$ for future day $t+1$ at monitoring station α (c.f. Section 2.1) and the space–time trend $M(x,t)$
(iii) Update the space–time trend by adding data of $Z(x_\alpha, t+1)$ to the condition data set and return to step i

2.2.1 Characterization of a Space–Time Trend with Joint Space–Time Models

For the first image of the space–time trend $M(x,t)$, the objective is to weight the different periods or spatial location of experimental data according the proximity to the location (x,t). Host et al. (1995) decomposed the trend in spatial and time components interpreted as spatial and temporal random fields. In another application (ecological resources) Santos et al. (2000) used a different approach to weight uneven dispersed monitoring stations with the estimation (kriging) variance.

In this study, we propose to characterize the first image of a space–time trend $M(x,t)$ with a space–time simulation – direct sequential simulation of $Z(.)$ (Soares, 2002) – based on observed data at the monitoring stations on previous time periods $t' < t$. A stationary space–time covariance model is adopted for this first trend image (Soares, 2002; Kyriakidis et al., 1999). At the location (x,t), the trend $M(x,t)$ is calculated by averaging N simulated realizations $z^i(x,t)$:

$$M(x,t) = \sum_{i=1}^{N} z^i(x,t) \tag{2}$$

2.2.2 Simulation of $Z(x, t+1)$ Based on NN Predictions

For the day $t+1$, one has the predicted values $Z^*(x_\alpha, t+1)$ (obtained by NN modelling). The corresponding residuals $R(x_\alpha, t+1)$ are first calculated at the location $(x_\alpha, t+1)$:

$$R(x_\alpha, t+1) = Z^*(x_\alpha, t+1) - M(x_\alpha, t) \tag{3}$$

and simulated at any spatial location x, $r^i(x, t+1)$. The simulated values of pollutant $z^i(x, t+1)$ are obtained by adding the simulated residuals to the space–time trend:

$$z^i(x, t+1) = M(x,t) + r^i(x, t+1) \tag{4}$$

This is equivalent to performing the simulation (Direct Sequential Simulation – DSS) of $Z(x,t)$ with local means given by the space–time trend.

2.2.3 Update of the Space–Time Trend for the Next Day $t + 1$

The space time trend is updated by the inclusion of the new measured data $z(x_\alpha,$, $t+1)$ in the next day $t+1$ in the conditioning data set. Simulated values (DSS with local means) are obtained as predicted data in the previous step. At the location $(x, t+1)$, the trend $M(x, t+1)$ is calculated by averaging N simulated realizations $z^i(x, t+1)$:

$$M(x, t+1) = \sum_{i=1}^{N} z^i(x, t+1) \quad (5)$$

Then, step 2.2.2 is repeated, where $M(x_\alpha, t)$ refers to the updated $M(x, t+1)$.

2.3 *Spatial and Temporal Uncertainty Assessment: Critical Areas and Time Periods Characterization*

For any day $t + 1$ and spatial location x, one can assess the local conditional distribution functions of $Z(x, t)$ with the N simulated realizations $z^i(x, t)$. Spatial and temporal uncertainty can be evaluated and, consequently, critical situations can be identified by the joint probability of a set of points (x_j, t_i) to be greater than a given threshold. Note that at this stage the evaluated uncertainty does not account for the prediction errors of the neural network. In other words, temporal uncertainty of the neural network prediction is not included in those uncertainty maps calculated with simulated images.

To account for the uncertainty of temporal prediction of neural networks, one suggest the use of historical bivariate probability distribution $\mathrm{F}(\mathrm{Z}(.), \mathrm{Z}^*(.))$ between the predicted values $\mathrm{Z}^*(x, t)$ and observed values $\mathrm{Z}(x, t)$ at the same space–time location (x, t). For each predicted value $\mathrm{Z}^*(x_0, t_0)$ at time t_0 at a specific spatial location x_0, one can calculate the conditional distribution $\mathrm{F}(\mathrm{Z}(x, t) \,|\mathrm{Z}^*(x, t)) = \mathrm{z}^*(x_0, t_0))$ from the historical bi-distribution. Measures of uncertainty (variance, inter-quartiles) can be calculated from these conditional distributions for each monitoring station.

3 Case Study

The proposed methodology intends to characterize critical concentration episodes of NO_2. In order to test the proposed methodology a case study is presented: the city of Lisbon and its surroundings (Fig. 1). A long NO_2 spatial-temporal data sample was collected at 22 air quality monitoring stations (Fig. 1) on an hourly basis for a period of 11 years (from 1/1/1995 to 31/12/2005). Afterwards, the original hourly air quality data series was converted to daily averages. Meteorological data – temperature and radiance on an hourly basis – were also collected.

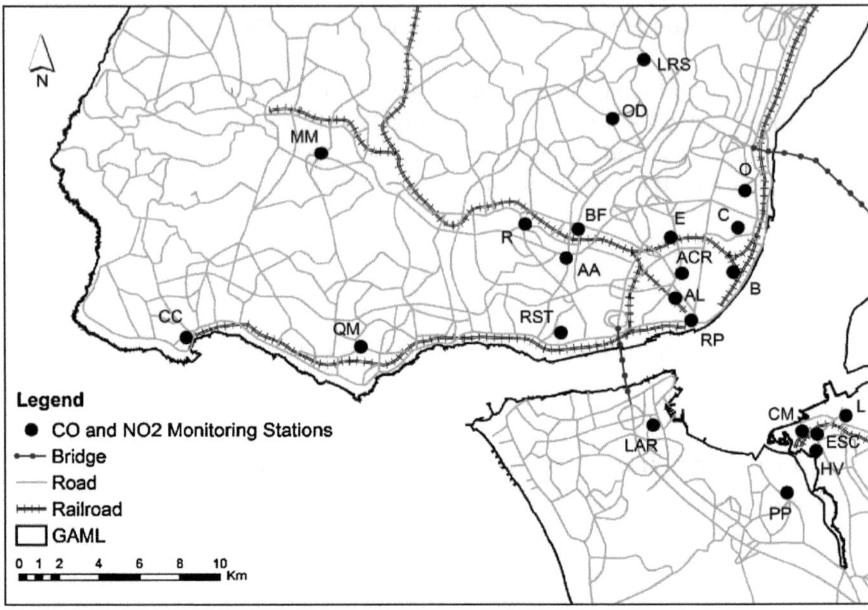

Fig. 1 Location of the monitoring stations at the metropolitan area of Lisbon

4 Results and Discussion

4.1 Exploratory NO_2 *Data Analysis*

The monitors are located in a mix of urban, suburban and rural sites in the vicinity of Lisbon. The hourly average NO_2 concentrations were obtained from the website of the Portuguese Environmental Agency (www.qualar.org) and meteorological data were obtained from the Instituto de Geofísica do Infante D. Luiz at the University of Lisbon (38°43′ N; 09°09′ W; 77 m). The basic statistics of mean daily NO_2 concentrations at each monitoring station are shown on Table 1. The mean daily NO_2 concentrations exhibit a positive asymmetric distribution (Fig. 2a) and cyclical behavior over time dependent on the season of the year (Fig. 2b). As expected, mean hourly NO_2 concentrations (Fig. 3a) are higher during peak traffic emissions (8–9 a.m. and 7–8 p.m.), as it is the case of most cities, such as London and Hong-Kong (Chit-Ming et al., 2002). Figure 3b illustrates the number of exceeded values registered by the 22 monitoring stations for each hour of the day. The majority of hourly NO_2 concentrations exceeding the legal limit occur also during peak traffic emissions.

Spatial and temporal NO_2 variograms (Fig. 4) were determined for the NO_2 daily averages calculated from the original 22 monitoring station data sets. As the 22 monitoring stations are the only available spatial data, it is assumed that the variogram

Table 1 Basic statistics of daily mean NO_2 concentrations ($\mu\ g/m^3$)

	Mean	Minimum	Maximum	Std. dev.	Coef. variation	Skewness
Entrecampos (E)	44.1	0.0	150.4	22.6	0.51	0.7
Olivais (O)	31.7	0.0	180.8	20.8	0.66	1.7
Chelas (C)	36.0	4.3	152.9	20.7	0.58	1.0
Beato (B)	24.6	1.1	110.5	14.5	0.59	1.3
Av. Liberdade (AL)	64.7	5.2	259.4	32.1	0.50	1.1
Benfica (BF)	50.8	3.0	191.1	25.8	0.51	0.9
R. Prata (RP)	57.9	9.2	154.8	22.4	0.39	0.6
Av. Casal Ribeiro (ACR)	56.9	0.4	250.8	34.9	0.61	1.5
Hospital Velho (HV)	27.1	0.0	194.0	14.2	0.52	1.3
Lavradio (L)	39.7	3.4	175.6	24.1	0.61	1.7
Paio Pires (PP)	21.8	0.0	114.8	14.0	0.64	1.2
C. Municipal (CM)	26.4	2.2	108.6	15.4	0.58	0.9
Escavadeira (ESC)	24.7	0.0	100.6	14.3	0.58	1.4
Reboleira (R)	24.2	0.4	114.0	17.4	0.72	1.3
Laranjeiro (LAR)	28.5	3.8	104.2	15.2	0.53	1.0
Loures (LRS)	20.4	0.1	96.4	14.0	0.69	1.2
Alfragide-Amadora (AA)	40.5	1.7	255.8	26.8	0.66	1.8
Restelo (RST)	21.7	1.4	104.4	14.3	0.66	1.3
Cascais (CC)	37.6	7.3	113.9	14.1	0.38	1.4
Q. Marques (QM)	18.7	0.0	81.1	13.8	0.74	1.2
Mem-Martins (MM)	15.5	0.8	72.2	12.0	0.77	1.4
Odivelas (OD)	28.0	1.8	98.3	16.3	0.58	1.1

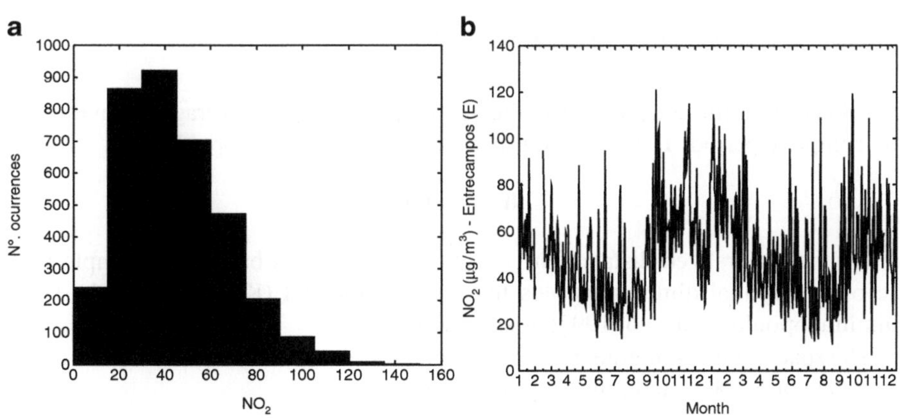

Fig. 2 Mean daily NO_2 concentration at Entrecampos monitoring station: **(a)** histogram; **(b)** time series

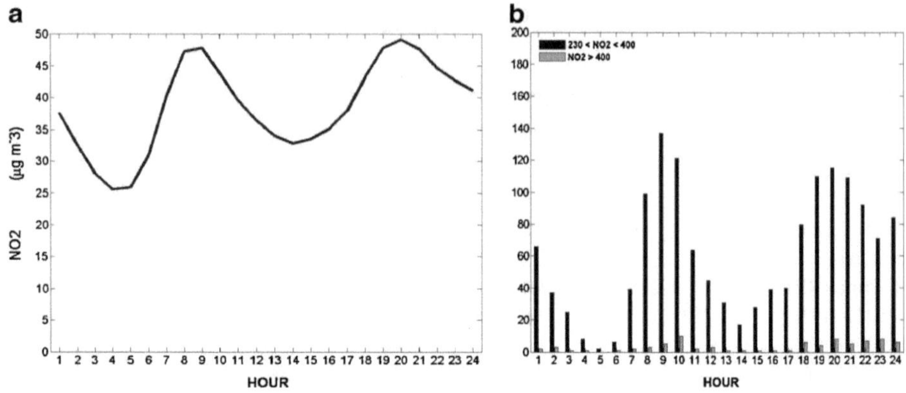

Fig. 3 Hourly NO_2 concentrations: **(a)** mean hourly concentrations for 22 monitoring stations for the period between 1995 and 2005; **(b)** number of mean hourly NO_2 concentration values exceeding the legal limit

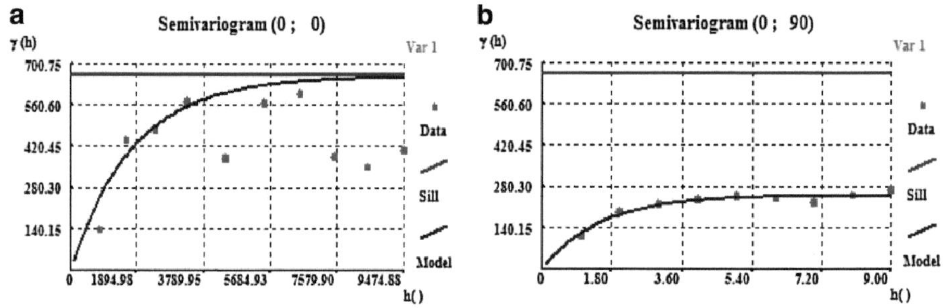

Fig. 4 NO_2 spatial **(a)** and temporal **(b)** variograms

calculated with these data reflects the spatial pattern of the average behaviour for the complete period. The following space–time model was adopted:

$$C(\mathbf{h}, t) = C(|\mathbf{h}|) \tag{6}$$

where the generalized distance $|\mathbf{h}| = \sqrt{x^2 + y^2 + t^2}$ is based on a simple metric of two spatial dimensions plus the time component (Kyriakidis et al., 1999; Dimitrakopoulos and Luo, 1993; Soares, 2002).

The space–time variogram model is an anisotropic exponential model with a spatial range of 5,300 m and time range of 4.5 h.

4.2 Predictive Neural Network Model

First, the neural network (NN) model described previously was properly calibrated and validated for a 5 year period (2001–2005). Considering that the available data

Table 2 NN inputs and respective lags

Input	Lag (day)
NO_2	t, t + 1 at 5 a.m.
Maximum temperature	t, t + 1
Sin ($2\pi t/365$) and Cos ($2\pi t/365$)	t + 1
Sin ($2\pi t/7$) and Cos ($2\pi t/7$)	t + 1
CO	t
Radiance	t

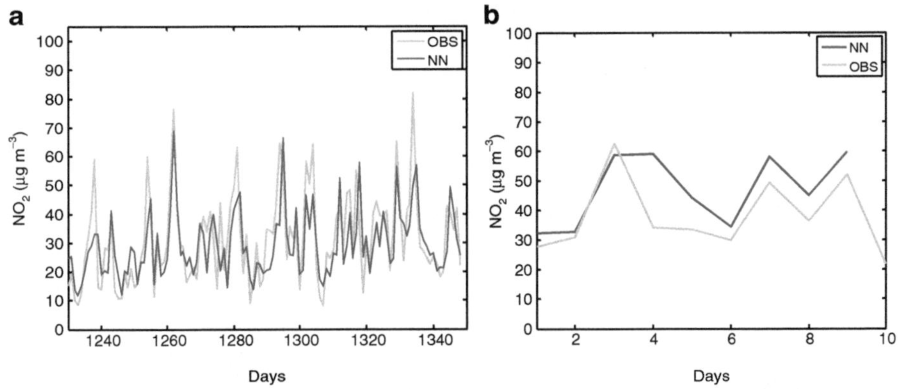

Fig. 5 NN results versus actual observed NO_2 values at the Laranjeiro monitoring station for part of the calibration/validation period **(a)** and for 24/4/2006 – 3/5/2006 **(b)**

sets correspond to a long data set of daily averages, cross-validation was used. The weights were randomly initialised and the NN had only one hidden layer. The best set of inputs used in order to reach the target (NO_2 for the next day) is represented in Table 2.

Afterwards, the model was used to produce daily NO_2 forecasts for the monitoring stations, for a period of 100 days. The dataset used was an independent NO_2 sample, not used previously for calibration and validation. The results attained by the NN model were then compared with the actual observed NO_2 values at the monitoring stations (Figs. 5 and 6).

The results obtained by the NN model are reasonable (mean correlation coefficient of 82% and mean skill against persistence of 52%) and indicate that a useful spatial-temporal model can be developed in forecast mode.

4.3 Stochastic Simulation Model

The sequential methodology previously described was applied to perform a joint space–time modelling procedure in order to obtain the first trend model through the use of DSS. The first trend refers to the period (23rd April 2006) (Fig. 7a).

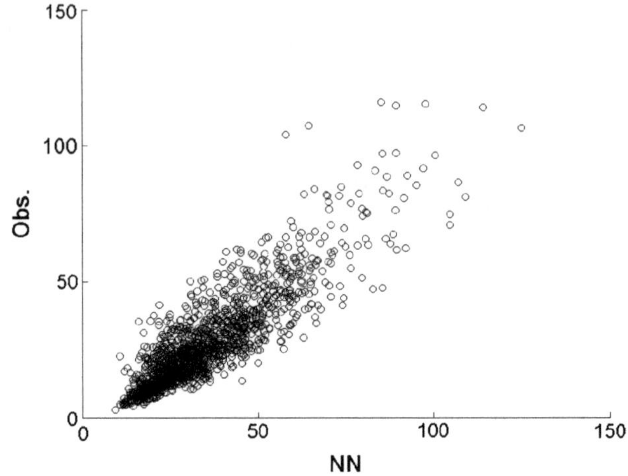

Fig. 6 Bi-plot of 100 daily NO_2 predictions by NN versus actual observed NO_2 for all the monitoring stations

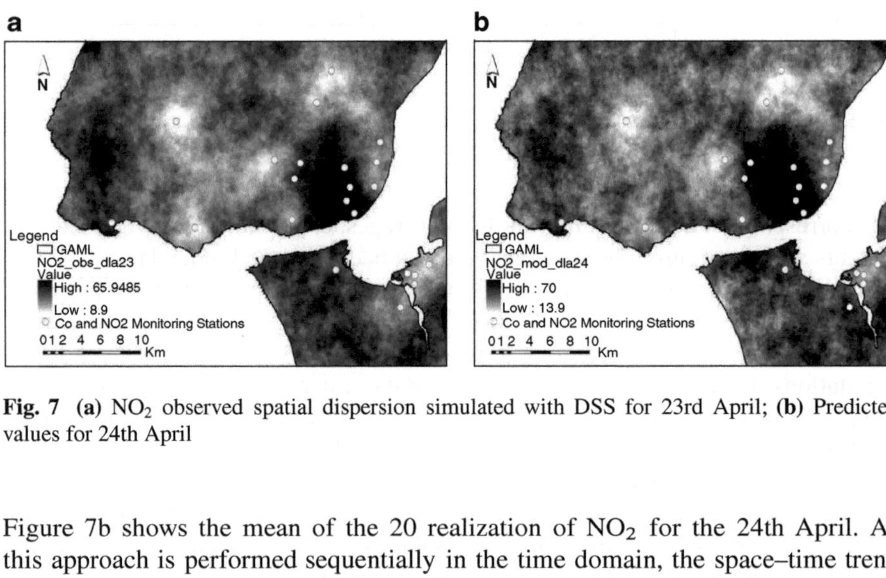

Fig. 7 **(a)** NO_2 observed spatial dispersion simulated with DSS for 23rd April; **(b)** Predicted values for 24th April

Figure 7b shows the mean of the 20 realization of NO_2 for the 24th April. As this approach is performed sequentially in the time domain, the space–time trend is sequentially updated everyday. Subsequently, this procedure was repeated for the NO_2 daily measured data for the next day (24th April). An average image, shown in Fig. 8a), will be used in the next iteration as the trend for 24th April. The difference between the average maps based on real observations (Fig. 8a) and predicted values (Fig. 7b) is shown on Fig. 8b.

The spatial uncertainty can be achieved by the set of simulated images of NO_2 – variance of NO_2 (Fig. 9).

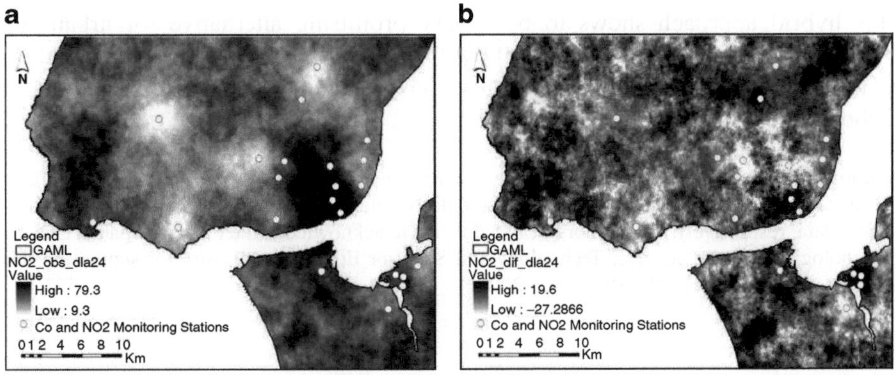

Fig. 8 **(a)** NO_2 observed spatial dispersion simulated with DSS for 24th April; **(b)** Difference between Figs. 8b and 9a

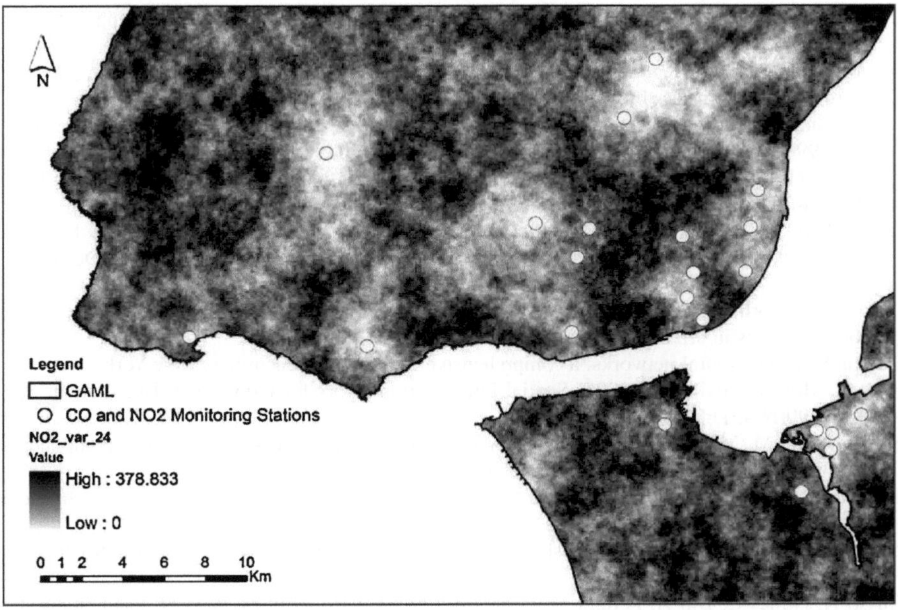

Fig. 9 NO_2 variance for April, 24th

5 Conclusions

The proposed methodology aims at the development of an AQ model which allows forecasting critical concentration episodes of a certain pollutant by means of a hybrid approach, combining iteratively the use of two efficient space–time modelling techniques: neural network models and geostatistical stochastic simulations.

This hybrid approach shows to be a very promising alternative for urban AQ characterization. These results will allow further developments in order to produce an integrated air quality and health surveillance/monitoring system in the area of Lisbon.

Acknowledgments The authors would like to acknowledge the Instituto de Geofísica do Infante D. Luiz at the University of Lisbon and Instituto do Ambiente for the meteorological and environmental data, respectively. The authors would also like to acknowledge the Fundação para a Ciência e Tecnologia from the Science, Technology and Superior Education Ministry for supporting this research through grant SFRH/BD/27765/2006.

References

Beale MH, Demuth HB (1998) Neural network toolbox for use with MATLAB, User's Guide, version 3. The MathWorks, Inc

Chit-Ming W, Atkinson R, Ross Anderson H, Hedley A, Ma S, Chau YK, Tai-Hing L (2002) A tale of two cities: effects of air pollution on hospital admissions in Hong Kong and London compared. Environ Health Persp 110(1):1999–2009

Cobourn WG, Dolcine L, French M, Hubbard MC (2000) A comparison of nonlinear regression and neural network models for ground-level ozone forecasting. J Air Waste Manag Assoc 50:1999–2009

Comrie A (1997) Comparing neural networks and regression models for ozone forecasting. J Air Waste Manag Assoc 47:653–663

Dimitrakopoulos R, Luo X (1994) Spatiotemporal modelling: covariances and ordinary kringing systems. In: Dimitrakopoulos R (ed) Geostatistics for the next century. Kluwer, Dordrecht, pp 88–93

Gardner M, Dorling S (1999) Neural network modelling and prediction of hourly Nox and NO_2 concentrations in urban air in London. Atmos Environ 33:709–719

Haykin S (1994) Neural networks: a comprehensive foundation. Macmillan, New York

Host G, More H, Switzer P (1995) Spatial interpolation errors for monitoring data. J Amer Stat Assoc 90(N-431):853–861

Kolehmainen M, Martikainen H, Ruuskanen J (2000) Neural networks and periodic components used in air quality forecasting. Atmos Environ 35:815–825

Kyriakidis P, Journel A (1999) Geostatistical space–time models: a review. Math Geol 31(6):651–685

Nunes C, Soares A (2005) Geostatistical space–time simulation model. Environmetrics 16:393–404

Russo A, Nunes C, Bio A, Pereira M, Soares A (2005) Air quality assessment using stochastic simulation and neural networks. In: Leuangthong O, Deutsch CV (eds) Geostatistics Banff. Springer, pp 797–807

Santos E, Almeida J, Soares A (2000) Geostatistical Characterization of the Migration Patterns and Pathways of the Wood Pigeon in Portugal. In Proceedings of the 6th International Geostatistics Congress. Cape Town

Seinfeld JH (1986) Atmospheric chemistry and physics of air pollution. Wiley, New York

Shi J, Harrison R (1997) Regression modelling of hourly NO_x and NO_2 concentrations in urban air in London. Atmos Environ 31:4081–4094

Simpson RW, Layton AP (1983) Forecasting peak ozone levels. Atmos Environ 17:1649–1654

Soares A (2002) Stochastic Modelling of Spatio-Temporal Phenomena in Earth Sciences. Geoinformatics. Atkinsson P (ed) Encyclopedia of Life Suport Systems Chapter (EOLSS). UNESCO, Eolss Publishers, Oxford, UK [http://www.eolss.net]

Yi J, Prybutok V (1996) A neural network model forecasting for prediction of daily maximum ozone concentration in an industrialized urban area. Environ Pollut 92:349–357
Ziomas I, Melas D, Zerefos CS, Bais AF, Paliatsos AG (1995) Forecasting peak pollutant levels from meteorological variables. Atmos Environ 29:3703–3711

Geostatistical Methods for Polluted Sites Characterization

Amílcar Soares

Abstract In the last 2 decades, Geostatistics had a quite significant increase in methodological developments and applications in the environment field, in particular related with natural resources management and anthropogenic pollution characterization. Contaminated sites, air pollution and polluted water characterization are among the most serious environmental problems for which characterization and management, geostatistical methods play an important role. However, geostatistical methods are still far from the main stream of industrial applications; in particular the industry of contaminated site remediation and air pollution control and management. In most cases, compliance with legislation is the only environmental requirement for industries. Hence, less attention is paid to the development of better assessment tools. But as for the lack of water resources, the migration and concentration of world population in large urban areas, soil degradation and environmental health become of main concern in developed countries, this scenario will probably change as air, water and soil will turn into more valuable resources and geostatistics continues to provide robust and accurate answers to the solution of monitoring and characterization of these resources. This paper intends to present some methods as potential paths for geostatistical models to approach the pollution problem of environmental systems in this new framework:

- Direct sequential simulation with joint distributions for environmental impact and risk assessment in polluted soil sites
- The use of hybrid models coupling deterministic fluid dispersion models with stochastic simulation with locally varying anisotropy for contaminated river water and sediment characterization
- Space–time modelling of air quality in urban areas

A. Soares (✉)
Centro de Recursos Naturais e Ambiente, Instituto Superior Técnico,
Universidade Técnica de Lisboa, Lisbon, Portugal
e-mail: asoares@ist.utl.pt

P.M. Atkinson and C.D. Lloyd (eds.), *geoENV VII – Geostatistics for Environmental Applications*, Quantitative Geology and Geostatistics 16,
DOI 10.1007/978-90-481-2322-3_17,

1 Introduction

Since the 1980s, environmental problems have led the geostatistical community to approach problems of air quality, soil contamination, hydrology, natural resources management, with identical methods that had been, until then, mainly focused on mining studies. The Tahoe conference (Geostatistics for Natural Resources Characterization [1983]) marks the beginning of environmental applications. But still the main innovations in geostatistical methods and algorithms were driven by petroleum and mining, the fields with significant investments in research. By then environmental studies were mostly dominated by qualitative judgments with a clear predominance of ecological concerns.

With the increase of global social concerns about environmental problems – ozone hole, greenhouse effect, global warming, desertification, droughts, floods – geostatistics began to be viewed as a potential tool for the quantification of those problems, mainly in observation, characterization and management of physical phenomena of natural resources and polluted areas. Hence, one can consider the decade of 1990s as the mature phase of the discipline. The conferences Geostatistics Troia'92, in Troia (Soares, 1993), Geostatistics for Environmental and Geotechnical Applications in Phoenix (Rouhani et al., 1996), and geoENV series of conferences in Lisbon (Soares et al., 1997), Valencia (Gómez-Hernandez et al., 1999), Avignon (Monestiez et al., 2001), Barcelona, (geoEN IV, 2003), Neuchatel (Renard et al., 2006) and in Rhodes (Soares et al., 2008) are reference milestones of the geostatistical applications in the environmental field (Sanchez-Villa et al., 2004).

Among the diversity of environmental applications those related with anthropogenic pollution have particular characteristics – complexity and space–time heterogeneity of natural phenomena (air, soil, water) and the dynamic of different contaminations – that put them in a specific position to be approached by innovative geostatistical tools. This note will only focus on soil and air quality. In contaminated sites, mapping of pollutants concentration have been treated by classical estimators of kriging (Goovaerts and Van Meirveinne, 2001; Atkinson and Lloyd, 2001), stochastic simulation (Goovaerts, 1997) Markov random fields (Cressie et al., 1999) and co-simulation of several contaminants (Franco et al., 2004). Integration of secondary information to characterize a main pollutant have been approached by several authors with the use of secondary information such as geophysical data (Garcia and Froidevaux, 1997) and soil type (Pereira et al., 2001). Air quality characterization introduced the time component to the usual framework of geostatistics (Kyriakidis and Journel, 1999; De Iacco et al., 2001; Nunes et al., 2004) or in BME framework (Serre et al., 2003).

With this paper I intend to present some methodological trends of geostatistics for the applications of air pollution and soil contamination characterization: polluted soil sites characterization with direct sequential simulation with joint distributions; the use of hybrid models coupling deterministic fluid dispersion models with stochastic simulation with local varying anisotropy for contaminated river water and sediments characterization; real time monitoring and modeling of air quality in urban areas. These are illustrated with examples of real case studies.

2 Direct Sequential Simulation with Joint Distributions for Contaminated Site Characterization

Among the sequential algorithms of stochastic simulation, one advantage of direct sequential simulation and co-simulation (Soares, 2001) is precisely the use of original variables instead of the transformed Gaussian (sequential Gaussian simulation) or indicator (sequential indicator simulation). Direct sequential simulation and co-simulation have been applied in several soil and air quality characterization studies, with promising results.

The use of original (non-transformed) variables and the method of generating a simulated value by re-sampling the global probability distribution function (pdf), opened the door to new ways of this resampling approach: Carvalho et al. (2006) proposed to resample local distributions taken from a secondary image (instead of the global pdf), in an application of data fusion of satellite images; Horta et al. (2008a) proposed the resampling of joint distributions for co-simulation of a set of variables, for soil quality evaluation. This new algorithm of DSS with joint distributions can be summarized in the following sequel.

Consider the direct sequential co-simulation (Soares, 2001) of just two variables, Z_1 and Z_2, with the following implementation:

(i) Direct sequential simulation of $Z_1(_x)$ that reproduces the marginal distribution $F_{Z1}(Z)$ and variogram $\gamma_{Z1}(\mathrm{h})$.
(ii) Co-simulation of $Z_2(_x)$ with $Z_2(_{x\alpha})$ data and previously simulated $Z_1{}^l(_x)$ as secondary variable, by using co-located co-kriging to estimate local means and variances. Marginal distributions of Z_2, $F_{Z2}(Z)$, variogram $\gamma_{Z2}(\boldsymbol{h})$ and the joint spatial pattern characterized by the co-variograms $\gamma_{Z1,Z2}(\boldsymbol{h})$ are reproduced at the final results.

This is usually sufficient to characterize both spatial random functions Z_1 and Z_2 for the most in environment and Earth sciences applications. However, in some situations one wishes that the bi-distributions would be reproduced or, at least, the final results do not violate the experimental bi-histogram boundaries. For example, suppose that one intends to jointly simulate two pollutant concentrations, relatively highly correlated, in a contaminated site. For a particular class of values of $Z_1(_x) = Z_1$, one would like to impose limits to the prob $\{Z_2(_x) < Z|Z_1(_x) = Z_1\}$. In other words, one would wish that the simulated values $Z_2{}^1(_x)$ must not exceed the limits found in the experimental conditional histogram of $Z_2(_x)$ given $Z_1(_x) = Z_1$. As this is not imposed, in any sequential co-simulation algorithms (sGs or co-DSS), it is not reproduced in the final results. In fact, after the conditional mean and variance are estimated, $Z_1(_x)$ and $Z_2(_x)$ are drawn from the global marginal distributions (DSS) or from its local Gaussian transform (sGs). Although the marginal histograms are reproduced, at the very beginning of the sequential simulation process, high conditional variances can drive simulated values out of the boundaries of conditional histograms. This can lead to erroneous conclusions when non-linear cost functions are applied to the pairs of co-simulated values.

In short, sequential co-simulation procedure does not guarantee that the conditional distributions $F[Z_2(_x)|Z_1(_x)]$ are reproduced.

Hence the new approach of direct sequential co simulation (Horta et al., 2008b) is based on the very simple idea of resampling $Z_2(_x)$ from the joint distribution $F_{Z1,Z2}(z_1, z_2)$ which can be summarized very shortly in the following sequence:

(i) Estimate the global bi-distributions from experimental data. Smooth algorithm (Deutsch and Journel, 1998) can eventually be used when there is a lack of experimental data.
(ii) First covariate $Z_1(_x)$ is simulated using Direct Sequential Simulation. Realizations of $Z_1{}^1(_x)$ reproduce the variogram $\gamma Z1(h)$ and marginal pdf $FZ(Z1(_x))$.
(iii) Co-simulation of $Z_2(_x)$: at each location $_{x0}$, estimate local mean and variance, identified with estimated simple collocated co-kriging and corresponding estimation variance: $[Z_2(_{x0})]_{\text{sck}}$ and $\sigma^2{}_{\text{sck}}(_{x0})$.
(iv) Based on previously simulated $Z_1{}^1(_x0)$, conditional pdf $F_Z[Z_2(_x)|Z_1(_x) = Z_1{}^1(_{x0})]$ are calculated from the bi distribution $F_{Z1,Z2}[Z_1(_x), Z_2(_x)]$.
(v) Simulated value $Z^s{}_2(_{x0})$ is re-sampled from the conditional pdf $FZ[Z_2(_x)| Z_1(_x) = Z_1{}^1(_{x0})]$ (as in the usual direct sequential procedure with marginal pdfs [Soares, 2001]).

This new approach of direct sequential co-simulation with joint distributions was applied to a contaminated site, the Guadiamar River, South of Spain, after a spill of a mining dam. Figure 1a presents the experimental bi-histogram of two soil contaminants (Arsenic [As] and Copper [Cu]) and Fig. 1b refers to the resulting co-simulated values (Franco et al., 2004) using classical DSS (Soares, 2001). Although

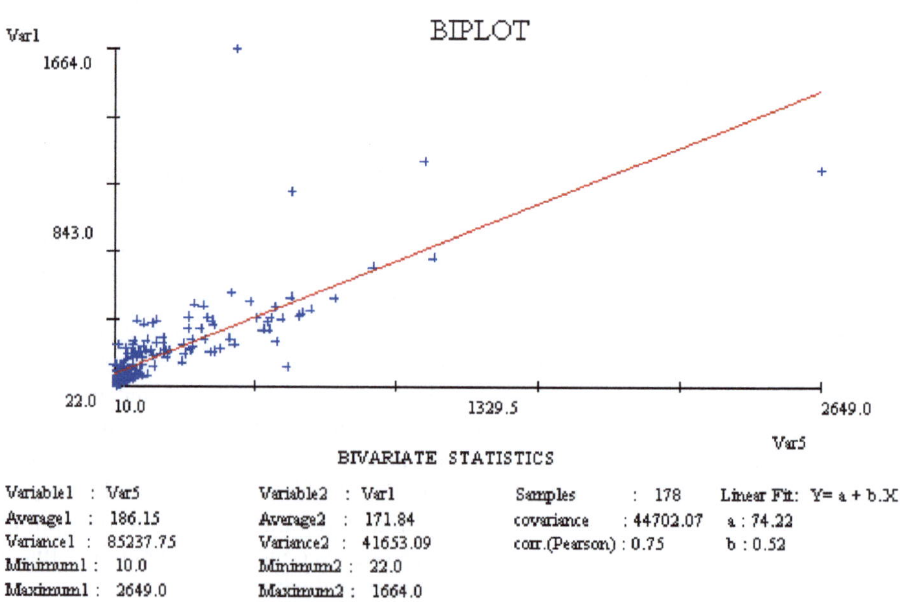

Fig. 1 Experimental bi-histogram of two soil contaminants (Arsenium [As] and Copper [Cu])

the marginal histograms, variograms and correlation coefficient are very well reproduced, the conditional histograms are not. Looking to the values of As for the class of Cu < 500 ppm in the bi-plot of Fig. 1a, a clear bias occurs in this conditional histogram: there are 15% of values with As > 500 ppm and Cu < 500 ppm that do not exist in the experimental data. This can lead to a misevaluated risk and cost functions calculation based on the joint probability of exceedences (Franco et al., 2004).

For comparison purposes, Fig. 2 shows the bi-histogram of simulated covariates with the new approach of DSS with joint distributions. The marginal histograms and the correlation coefficient are fairly well reproduced as in the traditional co-DSS (Fig. 1a). But, the bi-histograms of simulated covariates with both methods have some differences which are reflected in the simulated images of As (Fig. 3). The bias previously reported in the class of As > 500 ppm and Cu < 500 ppm is not found in the bi-plot of Fig. 2 using co-DSS with joint distributions.

This simple example shows the clear advantage of using co-DSS with joint distributions when one wishes to impose conditional distributions to the final covariate realizations. Another example, but in a different application (air pollution) of this new approach is presented in a following section (Section 3.2).

3 Hybrid Models for Air Quality and Soil Contamination Characterization

When pollution is a result of spatial and temporal dispersion of a contaminant driven by physical phenomenon with high dynamic behavior like, for example, air pollution dispersion or percolation of a pollutant through different types of soil, geostatistical models can be enriched by integration of those dynamic characteristics through another deterministic or stochastic model. These hybrid models have been quite widely used in hydrogeology or petroleum applications, by coupling the dynamic simulation models with geostatistical models to integrate dynamic responses known at the wells (Hendricks-Franssen et al., 1999; Meier et al., 1999).

When the dynamic behavior is so complex that hardly can be modeled by physical laws that govern deterministic dispersion models like, for example, air pollution in urban areas, non-linear classifiers (neural networks) can be used to integrate the temporal dynamics.

Two examples of hybrid models are presented to illustrate the paths of geostatistical modelling in this field.

3.1 Contaminated Sediment Characterization by a Hybrid Model

Let us consider a situation where contaminants are being accumulated in a sediment bed naturally created by the river flow. Characterization of such pollutant

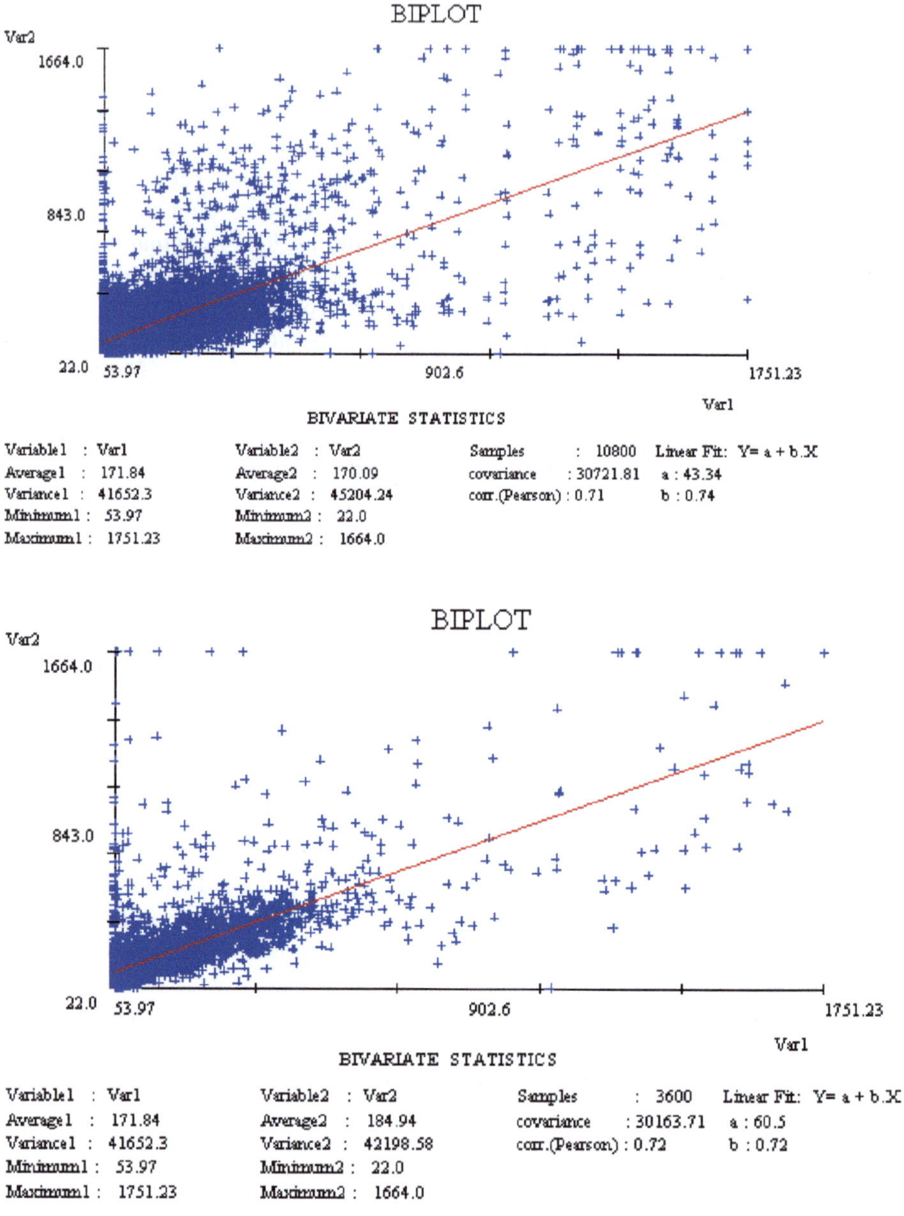

Fig. 2 Bi-histogram of co-simulated (with joint distributions) soil contaminants (As and Cu)

concentration by a geostatistical model must account of preferential sediment deposition along different channel morphology. Taking this into account, Horta et al. (2008a) proposed a simulation of continuous variables conditioned to meander structures. Basically, a bi-point statistics stochastic simulation with local anisotropy

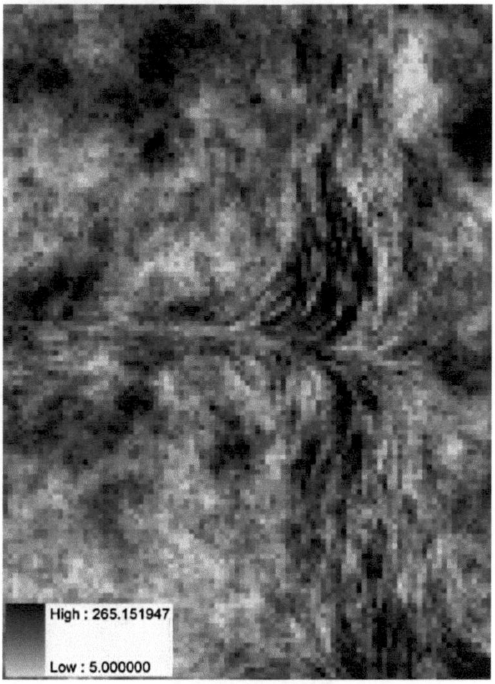

Fig. 3 Mean of 30 simulated maps of Zinc

trends is used to characterize continuous variables inside pre-defined channels. To accomplish this, a new version of the DSS algorithm is proposed to account for local anisotropy directions and ratios.

Let us assume $\theta(_x)$ and $r(_x)$, the local angle of anisotropy and the ratio of anisotropy, are known at spatial location $_x$. The main sequence of methodological steps of this version of DSS can be summarized in the following:

(i) Choose a given location $\boldsymbol{x}_0$ in a random path of a regular grid to be simulated.
(ii) Local means and variances of $z(_{x0})$ are estimated – simple kriging – with corrected local covariances $C_{\theta,r}(\boldsymbol{h},)$ by the local values of $\theta(_x)$ and $r(_x)$. Hence the simple kriging estimate of local mean becomes a function of $\theta(_x)$ and $r(_x)$.
(iii) Draw a simulated value at $_{x0}$ by re-sampling the global histogram and return to (i) until all nodes of the regular grid have been simulated.

This method was applied to the assessment of sediment contamination in a coastal lagoon where serious pollution discharges have been reported coming from the industrial sites located in the North part of the water basin (Horta et al., 2008a). After a sediment monitoring campaign, the previous method of DSS with local directions and ratios of anisotropy was applied to characterize the spatial dispersion of different pollutants. In Fig. 3, the final mean of 30 simulated maps of Zinc is presented, where the influence of meanders structures is quite visible.

An important and crucial point of this methodology is precisely the determination of local directions and ratios of anisotropy. As the channels were visible in

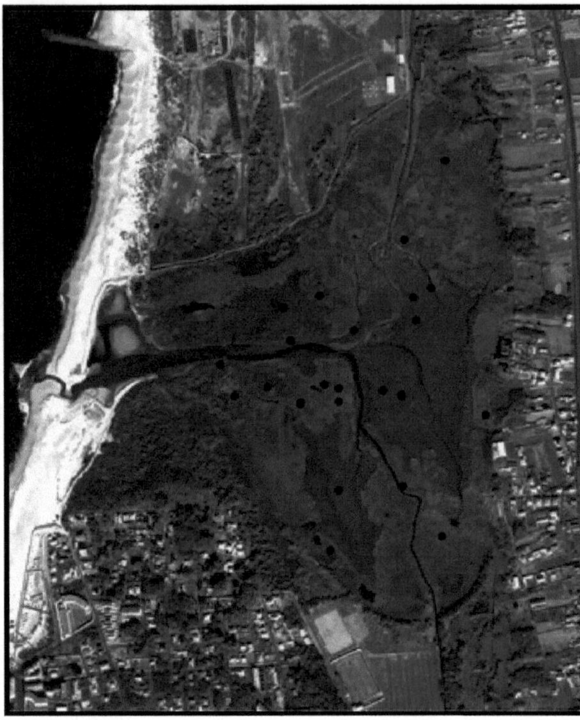

Fig. 4 Satellite sensor image (Quickbird, 3 m spatial resolution)

a satellite image (Quickbird, 3 m spatial resolution), Horta et al. 2008a) inferred those parameters by the shape of meanders structures delineated from the image (Fig. 4). However, as directions and ratios of anisotropy are related with the fluid flow characteristics, those can be automatically inferred with a dynamic dispersion model response. In Fig. 5a one can see only the main meander bathymetry, and the response of the dynamic dispersion model MOHID (www.mohid.com; Fig. 5b). Local anisotropy angles and ratios can be inferred (linearly related) by the velocity vectors of the fluid.

3.2 *Real Time Monitoring and Modelling of Air Quality in Urban Areas*

An example of a hybrid model for characterization of air pollution in industrial areas by integrating dynamic deterministic modelling (Gaussian plume) with stochastic simulation has been proposed by Pereira et al. (1997). In most of the industrial areas, contamination plume sources are identified, which makes use of such physical models easy to implement. However, in urban areas the dynamic of the pollutants depend

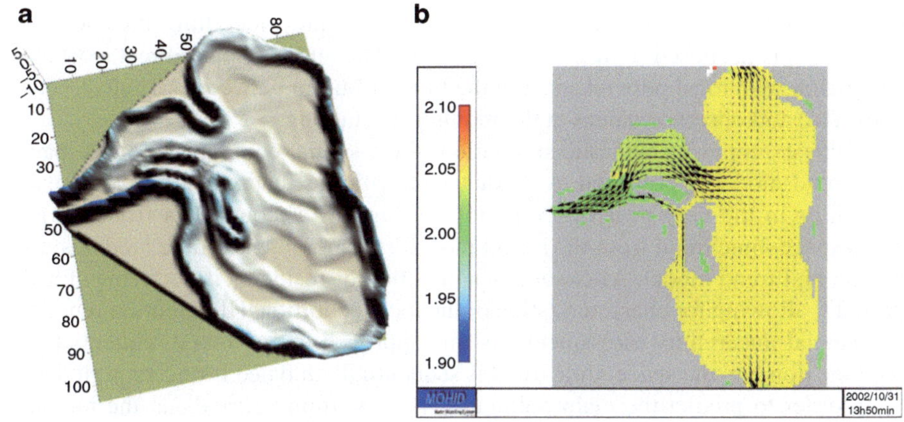

Fig. 5 (**a**) Main meander bathymetry. (**b**) Velocity vectors of the fluid given by the dynamic dispersion model

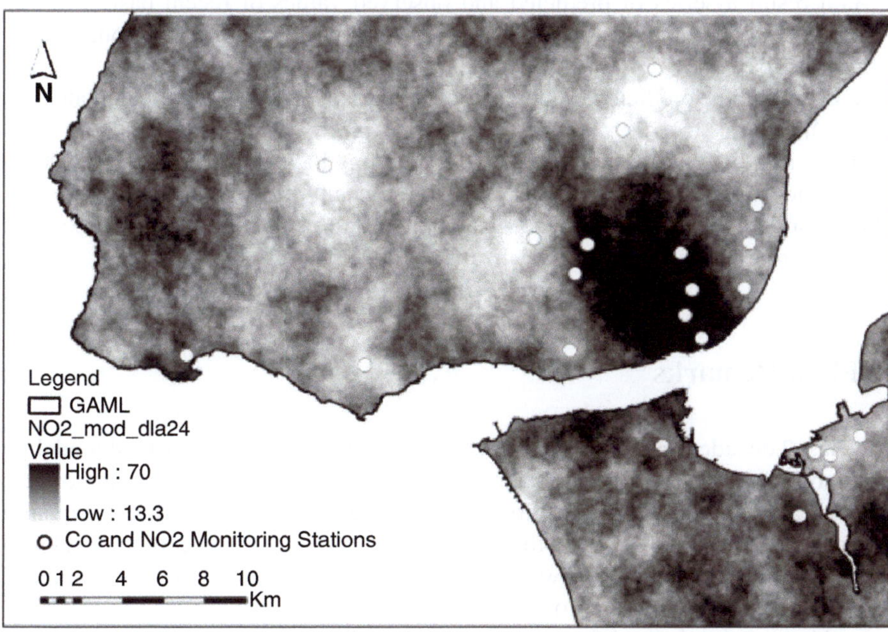

Fig. 6 Mean of a set of simulated realizations for NO_2 in Lisbon

on a complex set of factors related with meteorological conditions, and buildings structure, but mostly on the diffuse source of the contamination. This extremely complex behavior makes the implementation of deterministic dispersion models very simplistic and most of the time useless. For these situations a hybrid model by coupling the use of neural networks for short-term prediction in monitoring stations and stochastic simulations for space–time modelling the pollutant dispersion, is proposed by Russo et al. (2008). The authors apply direct stochastic simulation with local pdfs taken from the bi-distribution between predicted (neural networks) and observed values at the monitoring stations.

In this example, a stochastic simulation of the spatial component with a space–time trend model is proposed to characterize critical situations at a given present period, or in the very near future period, taking into account data from the past and a space–time trend from the recent past. First, a joint space-time model is used for the first trend model. Afterwards a simulation of residuals is performed for the period t, allowing for characterization of the local conditional distribution functions and spatial uncertainty assessment. As this approach is performed sequentially in the time domain, the space–time trend is sequentially updated for every period t.

In order to predict the main pollutants for near future periods, at the monitoring stations, a neural network was developed taking into account the pollutants concentration at past periods and the meteorological conditions measured and also predicted for each monitoring station.

As a result a series of predicted and observed values of recent past gave rise to bi-distributions at the monitoring stations location. Hence conditional distributions of real values $Z(_{x\alpha}|\mathrm{t}_i)$ at the monitoring station $_{x\alpha}$ and period t_i, given the corresponding predicted value (neural networks) $Z(_{x\alpha}|\mathrm{t}_i)^*$ are calculated: $F(Z(_{xa}\mathrm{t}_i)|Z(_{xa}\mathrm{t}_i)^*)$.

DSS with joint distributions and with re-sampling local pdfs (Horta et al., 2008a) was applied to simulate the next day period concentration $\mathrm{Z}(_{x\alpha}|\mathrm{t}_i)$. Russo et al. (2008) applied this methodology to Lisbon city. Figure 5 is represented an example of the mean of a set of simulated realizations for NO_2.

4 Final Remarks

In the last 2 decades, in spite of the increased performance of geostatistical modeling for the monitoring and characterization of polluted sites, its use is still far from the main stream of industrial applications (contaminated sites remediation, air quality control). However, it is expected that external factors of main concern to developed countries, such as the lack of water resources, desertification and environmental health, will change this scenario.

In this context, deterministic models of air pollution dispersion in urban areas (where most of the population is concentrated) just give qualitative (decorative) indicators and are almost useless for short-term prediction purposes. On the other hand, there is a health cost issue directly associated with risk assessment concerning either poor air quality or contaminated soil sites. Geostatistical modelling for polluted sites characterization will tend to play a more important role in this new framework of environmental problems.

This paper presented some methods which are potentially innovative and efficient tools for approaching problems of pollution within the geostatistical framework:

- The use of direct sequential simulation and co-simulation with joint distributions for simulation of covariates in contaminated soil sites or in space–time air quality short-term prediction
- The use of hybrid models coupling deterministic and stochastic simulation to integrate the dynamic component of complex phenomena.

References

Atkinson PM, Lloyd CD (2001) Ordinary and Indicator kriging of monthly mean nitrogen dioxide concentrations in the United kingdom. In: Monestiez P et al. (eds) geoENVIII – geostatistics for environmental applications. Kluwer, Dordrecht, pp 33–44

Carvalho J, Delgado GJ, Soares A (2006) Merging Landsat and SPOT digital data using stochastic simulation with reference images. 2006 Spatial Accuracy – The Seventh International Symposium on Spatial Accuracy Assessment in Natural Resources and Environmental Sciences

Cressie N, Keiser M, Daniels M, Lee J, Lahiri S, Cox L (1999) Spatial analysis of particulate matter in an urban environment. In: Gómez-Hernandez J et al. (eds) geoENVII – geostatistics for environmental applications. Kluwer, Dordrecht, pp 41–52

Deutsch CV, Journel AG (1998) Gslib geostatistical software library and user's guide, Oxford University Press, New York

Franco C, Soares A, Delgado-Garcia J (2004) Characterization of environmental hazard maps of metal contamination in Guadimar river margins. In: Sanchez-Villa X et al. (eds) geoENV IV – geostatistics for environmental applications. Kluwer, Dordrecht, pp 425–2436

Garcia M, Froidevaux R (1997) Application of geostatistics to 3D modeling of contaminant sites: a case study. In: Soares A et al. (eds) GeoENVI – geostatistics for environmental applications. Kluwer, Dordrecht, pp 309–326

Gómez-Hernandez J et al. (eds) (1999) GeoENVII – geostatistics for environmental applications. Kluwer, Dordrecht

Goovaerts P (1997) Kriging vs stochastic simulation for risk analysis in soil contamination. In: Soares A et al. (eds) GeoENVI – geostatistics for environmental applications. Kluwer, Dordrecht, pp 247–258

Goovaerts P, Van Meirveinne M (2001) Deliniation of hazardeous areas and additional sampling strategy in presence of a location-specific threshold. In Monestiez P et al. (eds) GeoENVIII – geostatistics for environmental applications. Kluwer, Dordrecht, pp 125–136

Hendricks-Frassen H, Cassiraga J, Gomez-Hernandez J, Sahuquillo J, Capilla J (1999) Inverse modeling of groundwater flow in a 3D fractures media. In: Gómez-Hernandez J et al. (eds) GeoENVII – geostatistics for environmental applications. Kluwer, Dordrecht, pp 283–294

Horta A, Caeiro M, Nunes R, Soares A (2008a) Simulation of continuous variables at meander structures: application to contaminated sediments of a lagoon. GeoENV2008, Southampton

Horta A, Soares A (2008b) Data integration model for soil degradation risk assessment. In Proceedings of the Eight International Geostatistics Congress, Ortiz JM, Emery X (eds.) – Santiago 2008, pp 931–940

Kyriakidis P, Journel A (1999) Stochastic modelling of spatiotemporal distributions: application to sulphate deposition trends over Europe. In: Gómez-Hernandez J et al. (eds) GeoENVII – geostatistics for environmental applications. Kluwer, Dordrecht, pp 89–100

Meier P, Medina A, Carrera J (1999) Inverse geostatistical modeling of pumping and tracer tests within a shear zone in granite. In: Gómez-Hernandez J et al. (eds) GeoENVII – geostatistics for environmental applications. Kluwer, Dordrecht, pp 295–306

Monestiez P et al. (eds) (2001) GeoENVIII – geostatistics for environmental applications. Kluwer, Dordrecht
Nunes C, Soares A (2004) Geostatistical space-time simulation model for characterization of air quality. In: Sanchez-Villa X et al. (eds) GeoENV IV – geostatistics for environmental applications. Kluwer, Dordrecht, pp 103–114
Pereira M, Soares A, Branquinho C (1997) Stochastic simulation of fugitive dust emissions. In: Baafi EY et al. (eds) Geostatistics Wollongong'96, vol 2. Kluwer, Dordrecht, pp 1055–1065
Pereira MJ, Almeida J, Costa C, Soares A (2001) Accounting for soft information in mapping soil contamination with TPH at an oil storage site. In: Monestiez P et al. (eds) GeoENVIII – geostatistics for environmental applications. Kluwer, Dordrecht, pp 475–486
Renard P et al. (eds) (2005) Geostatistics for environmental applications. Springer, Berlin
Rouhani S et al. (ed) (1996) Environmental and geotechnical applications. STP 1283, ASTM Pub
Russo A, Soares A, Trigo R (2008) Geostatistical model for air quality surveillance/monitoring system. GeoENV2008. Porthmouth, UK
Sanchez-Villa X et al. (eds) (2004) GeoENV IV – geostatistics for environmental applications. Kluwer, Dordrecht
Serre M, Christakos G, Lee S (2003) Soft data space-timemaping of coarse particulate matter annual arithmetic average over the U.S. In: Sanchez-Villa X et al. (eds) GeoENV IV – geostatistics for environmental applications. Kluwer, Dordrecht, pp 115–126
Soares A (ed) (1993) Geostatistics TROIA'92, vols 1 and 2. Kluwer, Dordrecht
Soares A (2001) Direct Sequential simulation and co-simulation. Math Geol 33(8):911–926
Soares A et al. (eds) (1997) GeoENVI – geostatistics for environmental applications. Kluwer, Dordrecht
Verly G et al. (eds) (1984) Geostatistics for natural resources characterization, vols 1 and 2, D. Reidel Pub., Dordrecht

Geostatistical Mapping of Outfall Plume Dispersion Data Gathered with an Autonomous Underwater Vehicle

Maurici Monego, Patrícia Ramos, and Mário V. Neves

Abstract The main purpose of this study was to examine the applicability of geostatistical modeling to obtain valuable information for assessing the environmental impact of sewage outfall discharges. The data set used was obtained in a monitoring campaign to *S. Jacinto* outfall, located off the Portuguese west coast near *Aveiro* region, using an AUV. The Matheron's classical estimator was used the compute the experimental semivariogram, which was fitted to three theoretical models: spherical, exponential and Gaussian. The cross-validation procedure suggested the best semivariogram model and ordinary kriging was used to obtain the predictions of salinity at unknown locations. The generated map shows clearly the plume dispersion in the studied area, indicating that the effluent does not reach the nearby beaches. Our study suggests that an optimal design for the AUV sampling trajectory from a geostatistical prediction point of view, can help to compute more precise predictions and hence to quantify more accurately dilution. Moreover, since accurate measurements of plume's dilution are rare, these studies might be very helpful in the future for validation of dispersion models.

1 Introduction

Outfalls are designed to promote the natural assimilative capacity of the oceans to dispose of wastewaters with minimal environmental impact. This is accomplished through the vigorous initial mixing that is followed by oceanic dispersion within spatially and temporally varying currents. Usually, those mixing processes,

M. Monego (✉), P. Ramos, and M.V. Neves
Faculty of Engineering of University of Porto, Rua Dr. Roberto Frias, 4200-465 Porto, Portugal
e-mail: mdmonego@fe.up.pt; patricia@fe.up.pt; mjneves@fe.up.pt

P. Ramos
Institute of Accountancy and Administration of Porto, Department of Mathematics,
R. Jaime Lopes Amorim, 4465-004 S. M. Infesta, Portugal
e-mail: patricia@fe.up.pt

P.M. Atkinson and C.D. Lloyd (eds.), *geoENV VII – Geostatistics for Environmental Applications*, Quantitative Geology and Geostatistics 16,
DOI 10.1007/978-90-481-2322-3_18,

in conjunction to bacterial mortality, result in rapid reductions in the concentrations of contaminants and organisms present in the wastewater to near background levels. However, coastal physical, chemical and biological processes, very dynamic and complex, and intimately coupled to the concentration and content of wastewater, are in most instances, poorly understood. Consequently, how sewage disperses and how effluent modifies and is modified by coastal environments remain in many aspects unknown and unpredictable. The impacts of discharged wastewaters on human beings may include direct contact (e.g., by swimmers, surfers, beachgoers) with chemical contaminants or pathogens, and indirect effects through the consumption of contaminated food suppliers (e.g., fish, shellfish). Much effort has been devoted recently to improve the means to monitor and characterize effluent plumes under a variety of oceanographic conditions, on relevant temporal and spatial scales. However, effluent plume dispersion is still a difficult problem to study *in situ*. The difficulties in conducting field studies arise from the rapid spatial and temporal variations in physical, chemical and biological processes and oceanographic conditions that can occur in coastal waters. Additional logistical difficulties that include variability of discharge flowrate, high costs, and large area extent to be monitored, make reliable field measurements of coastal outfall plumes rare.

Autonomous Underwater Vehicles (AUVs) have already been demonstrated to be appropriate for high-resolution surveys of small features such as outfall plumes (Ramos, 2005). Some of the advantages of these platforms include: easier field logistics, low cost per deployment, good spatial coverage, sampling over repeated sections, and capability of feature-based or adaptive sampling. Demands for more reliable model predictions, and predictions of quantities that have received little attention in the past are now increasing. These are driven by increasing environmental awareness, more stringent environmental standards, and application of diffusion theory in new areas. While the gross properties of the plume can be reasonably predicted by the most commonly used marine discharge models, there remain many aspects which cannot be, particularly the patchy nature of the wastefield. This patchiness, which has been observed in field studies, is not incorporated into any of those models. They implicitly assume properties to vary smoothly in space, an assumption that is true only for time-averaged plumes. If we want to calibrate these models with real data we have to be able to quantify spatial correlations and other related characteristics.

In this paper, we use geostatistics in the spatial analysis of environmental data gathered with an autonomous underwater vehicle (AUV) in a monitoring campaign targeted to a sea outfall, aiming: (i) to distinguish the effluent plume from the receiving water; (ii) to estimate the salinity value at unknown locations and map its distribution by kriging interpolation, motivated by environmental impact assessment for decision-making and (iii) to validate predictions of plume dispersion models.

Geostatistical modeling has been used to analyze and characterize the spatial variability of soil properties (Saby et al., 2006; Wei et al., 2007), to obtain information for assessing water and wind resources (Shoji, 2006; Shoji and Kitaura, 2006), to design sampling strategies for estuarine sediment collection (Caeiro et al., 2003), to study the thickness of effluent-affected sediment in the vicinity of wastewater discharges (Murray et al., 2002), and to obtain information about the spatial distribution of sewage pollution in coastal sediments (Poon et al., 2000), among many others.

Although very chaotic, due to turbulent diffusion, plume's dispersion process tends to a natural variability mode when the plume stops rising and the intensity of turbulent fluctuations approaches to zero (Roberts, 1996). This region is called the end of "near field" or "initial mixing region". After the end of the near field the established wastefield spreads laterally, drifting with the ocean current diffused by oceanic turbulence. In the near field the dilution increases rapidly with downstream distance, due to the turbulent kinetic energy generated by the buoyancy and momentum of the discharge. However, after the end of the near field the rate of increase of dilution is much lower. Dilution is then usually evaluated, for risk assessment purposes, at the end of the near field. It is likely that after the end of the near field pollutant concentrations are spatially correlated. In this sense, geostatistics appears to be an appropriate technique to estimate dilution and map the plume dispersion.

In this work we conduct a geostatistical study of salinity measurements, obtained in the vicinity of an outfall discharge, using ordinary kriging interpolation. In a first step the spatial structure of the observations was inspected thought a descriptive statistical analysis. Then, the degree of spatial correlation among data in the study area as function of the distance and direction was expressed in terms of the semivariogram. Finally, ordinary kriging was used to estimate salinity at unknown locations, and a map of this parameter distribution in the field was generated. Cross-validation indicators and additional model parameters helped to choose the most appropriate model.

2 Geostatistical Analysis

The data set used in this analysis was obtained in a monitoring campaign of *S. Jacinto* outfall, located off the Portuguese west coast near *Aveiro* region, using the AUV of the Underwater Systems and Technology Laboratory of University of Porto. A rectangular area of $200 \times 100\,\text{m}^2$ starting 20 m downstream from the middle point of the outfall diffuser was covered. As planned, the vehicle performed six horizontal trajectories at 2, 4, 6, 8, 10 and 12 m depth. In each horizontal section the vehicle described six parallel transects, perpendicular to the current direction, of 200 m length and spaced at 20 m. While navigating at a constant velocity of approximately 1 m/s, CTD (conductivity, temperature, depth) data were collected and recorded at a rate of 2.4 Hz. Consecutive measurements at horizontal sections were then distanced at about 0.4 m.

In this study, we analyse salinity data (computed from conductivity, temperature and depth) from the horizontal section at 2 m depth, where the effluent plume was found established and dispersing horizontally. The trajectory of the AUV at this section is shown in Fig. 1.

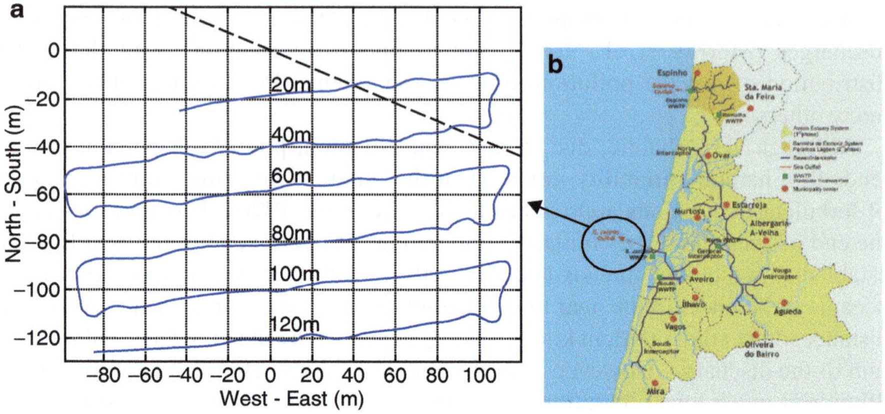

Fig. 1 (**a**) AUV sampling trajectory at 2 m depth. (**b**) Study area off the Portuguese west coast near *Aveiro* region

2.1 Exploratory Analysis

Table 1 gives the summary statistics of the salinity data set (2,470 measurements). The salinity ranged from 35.152 to 35.607 psu. The mean value of the data set was 35.451 psu, being close to the median value that was 35.463 psu. As in conventional statistics, a normal distribution for the variable under study is desirable in linear geostatistics (Wackernagel, 2003).

It can be seen from Table 1 that both skewness and kurtosis values are low indicating an approximated normal distribution of the raw data.

Figure 2 shows the frequency distribution of the salinity data set. The left tail of the histogram shows a lightly negatively skewed distribution, which is in accordance with the negative value of the skewness parameter in Table 1. This can be justified by the sampling strategy adopted. Since transects were all perpendicular to the current direction (and not parallel to the outfall diffuser), the ones closer to the diffuser still caught the plume ascending giving much lower values of salinity.

2.2 Semivariogram

Geostatistical methodology uses the semivariogram to quantify the spatial variation of the variable in study (Cressie, 1993; Isaaks and Srivastava, 1989). The semivariogram measures the mean variability between two data points as a function of their distance. Matheron's classical estimator of the semivariogram was used in this study, whose computing equation is (Matheron, 1965):

$$\gamma(\mathbf{h}) = \frac{1}{2N(\mathbf{h})} \sum_{i=1}^{N(\mathbf{h})} [Z(\mathbf{x}_i) - Z(\mathbf{x}_i + \mathbf{h})]^2 \tag{1}$$

Table 1 Summary statistics of the salinity data set

Summary statistics of the salinity data set	
Number of data	2,470
Minimum	35.152 psu
Mean	35.451 psu
Median	35.463 psu
Maximum	35.607 psu
Variance	0.004
Standard deviation	0.067
Skewness	−0.52
Kurtosis	0.006

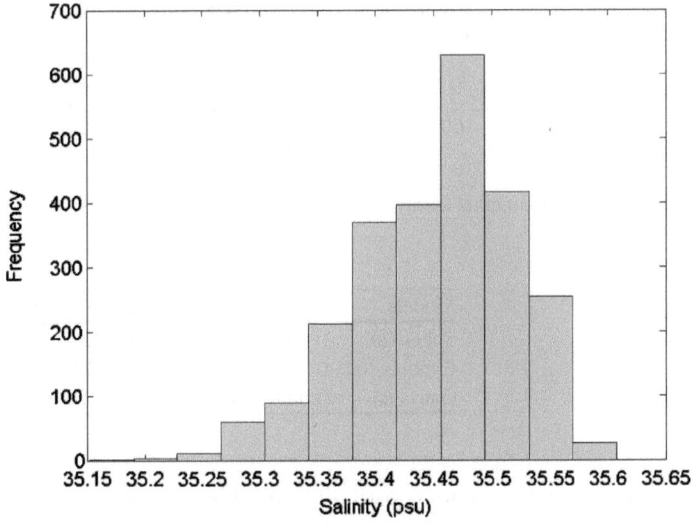

Fig. 2 Frequency distribution of the salinity data set

where $\gamma(\mathbf{h})$ is the semivariogram, $Z(\mathbf{x}_i)$ is the salinity value measured at location $\mathbf{x}_i$, $\mathbf{h}$ is the lag distance and $N(\mathbf{h})$ is the number of pairs of measurements which are $\mathbf{h}$ distance apart. The experimental semivariogram is calculated for several lag distances. Once the experimental semivariogram is computed, the next step is to fit it into a theoretical model. This model gives information about the structure of the spatial variation being also used for the spatial prediction by kriging. The most commonly used theoretical models are circular, spherical, exponential and Gaussian (Kitanidis, 1997).

Figure 3 shows the omnidirectional experimental semivariogram of the salinity data set and the models spherical, exponential and Gaussian fitted.

Estimation of semivariances was carried out using a lag distance of 10 m. Anisotropy was investigated by calculation of semivariogram for several directions. However, no effect of anisotropy could be shown. The nugget, sill, and range parameters of the three fitted models are shown in Table 2.

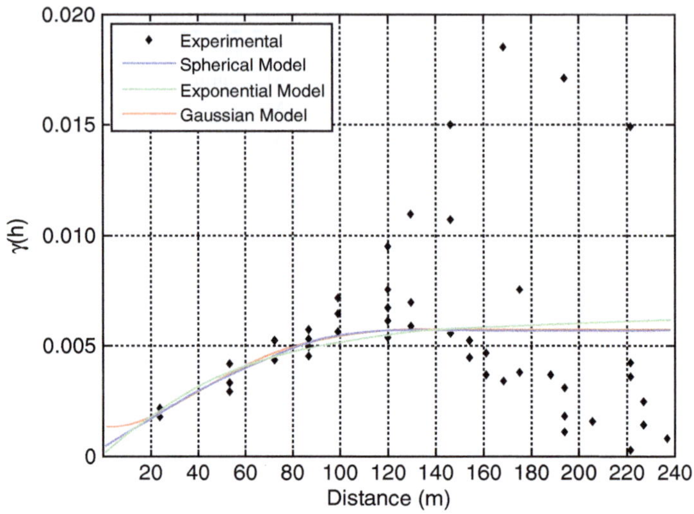

Fig. 3 Omnidirectional experimental semivariogram and fitted models

Table 2 Parameters of the semivariogram models

Models	Nugget	Sill	Range	Nugget/Sill (%)
Spherical	0.00021	0.00555	109.772	3.9
Exponential	0	0.00492	109.772	0
Gaussian	0.00093	0.00608	109.772	15.3

The degree of spatial dependence of the variable in study can be evaluated through the nugget/sill ratio. According to Wei et al. (2007), nugget/sill ratios less than 25% suggest that the variable has a strong spatial dependence; nugget/sill ratios between 25% and 75% suggest that the variable has a moderate spatial dependence; and nugget/sill ratios above 75% suggest that the variable has low spatial dependence. As can be observed in Table 2, the nugget/sill ratios of salinity for all the semivariogram models are low and less then 25%, suggesting that the variable has a strong spatial dependence and that probably local variations could be captured, as expected.

2.3 *Cross-Validation*

Cross-validation was used to compare the prediction performances of the three semivariogram models. In this procedure, each sample is eliminated in turn and the remaining samples are used by kriging to predict the eliminated observation. The differences between the observations and the predictions are then evaluated using the mean error (ME), the root mean squared error (RMSE), and the root mean

Table 3 Cross-validation parameters for the semivariogram models

Models	ME	RMSE	RMSSE
Spherical	-3.8×10^{-5}	0.01476	0.8077
Exponential	0.29×10^{-5}	0.01409	1.6310
Gaussian	-29.9×10^{-5}	0.02495	0.7461

squared standardized error (RMSSE), computed respectively according to the following equations:

$$\mathrm{ME} = \frac{1}{N}\sum_{i=1}^{N}\left[\hat{Z}(\mathbf{x}_i) - Z(\mathbf{x}_i)\right] \tag{2}$$

$$\mathrm{RMSE} = \sqrt{\frac{1}{N}\sum_{i=1}^{N}\left[\hat{Z}(\mathbf{x}_i) - Z(\mathbf{x}_i)\right]^2} \tag{3}$$

$$\mathrm{RMSSE} = \sqrt{\frac{1}{N}\sum_{i=1}^{N}\left[\frac{\hat{Z}(\mathbf{x}_i) - Z(\mathbf{x}_i)}{\sigma^2(\mathbf{x}_i)}\right]^2} \tag{4}$$

where $\hat{Z}(\mathbf{x}_i)$ is the predicted value at cross-validation point $\mathbf{x}_i$, $Z(\mathbf{x}_i)$ is the actual (measured) value at point $\mathbf{x}_i$, N is the number of measurements of the data set, and $\sigma^2(\mathbf{x}_i)$ is the kriging variance at cross-validation point $\mathbf{x}_i$. Table 3 shows these indicators for the spherical, exponential and Gaussian models that helped to choose the best semivariogram model among these candidates.

For a model that provides accurate predictions the ME should be close to zero, indicating that the predictions are unbiased. The RMSE should be as small as possible, indicating that the predictions are close to the measured values. If the kriging variances are accurate, then the RMSSE should be close to 1 (Wackernagel, 2003). If it is higher, the kriging predictions are too optimistic about the variability of the estimates. The results given by Table 2 and Table 3 suggest that the spherical model should be used to estimate salinity over the studied area.

2.4 Ordinary Kriging

After selecting a variogram model, kriging was applied to estimate the value of the variable at unsampled locations, using data from surrounding sampled points. The estimation is also based on the semivariogram model, and therefore, takes into account the spatial variability of the variable in study.

The kriging method belongs to the best linear unbiased estimators (BLUE) family. It is said to be linear because the estimated value is a linear combination of the measurements, being written in the form of:

$$\hat{Z}(\mathbf{x}_0) = \sum_{i=1}^{M}\alpha_i Z(\mathbf{x}_i) \tag{5}$$

where $\hat{Z}(\mathbf{x}_0)$ is the estimated value for location $\mathbf{x}_0$, M is the number of observations in the neighborhood of $\mathbf{x}_0$ used in the estimative, and α_i are the correspondent weights.

Ordinary kriging is used when the mean value of the variable in study is unknown. For this estimator to be unbiased, for any value of the mean, it is required that $\sum_{i=1}^{M} \alpha_i = 1$. The estimated value is obtained by minimizing the kriging variance with the help of the Lagrange multipliers, in order to impose the unbiased condition (Cressie, 1993; Kitanidis, 1997).

3 Results

The kriged maps of salinity of the horizontal section at 2 m depth using the spherical, exponential and Gaussian models are shown in Fig. 4. All maps show clearly the spatial variation of salinity in the studied area. From these maps it is possible to identify unambiguously the effluent plume and its dispersion downstream in the current direction. It appears as a region of lower salinity compared to the surrounding ocean waters at the same depth. It is also possible to observe the plume edges since the wastefield width is shorter than the survey width. We may say that the results obtained with the three semivariogram models are quite similar. However, in the prediction using the Gaussian model some small local variations were not captured. Figure 5 shows the prediction error map when using the spherical model. It can be seen, as expected, that the prediction error is smaller the closer the prediction to the trajectory of the vehicle.

Salinity differences compared to the surrounding waters at 2 m depth started to be about 0.455 psu in the first two transects (20 and 40 m), decreasing to about 0.293 psu in the third transect (60 m), to about 0.215 psu in the forth transect (80 m), to about 0.176 psu in the fifth transect (100 m), ending almost equally to background waters at 120 m distance, with a difference of about 0.071 psu. Washburn et al. (1992) observed salinity differences compared with the surrounding waters of the order of 0.1 psu, while Petrenko et al. (1998) found differences of the order of 0.2 psu.

A sharp difference in salinity at the effluent plume lateral edges is clearly visible, being the wastefield spreading almost centered in the survey area. This indicates that the sampling strategy designed was successful, even for a surfacing plume which can be considered as the most complicated case in terms of natural tracer tracking.

The plume exhibits a considerably more complex structure than the compact shape of the classical picture of the buoyant plume, but not so patchy as in previous studies, maybe due to the increase in horizontal resolution and also possibly due to the kriging results.

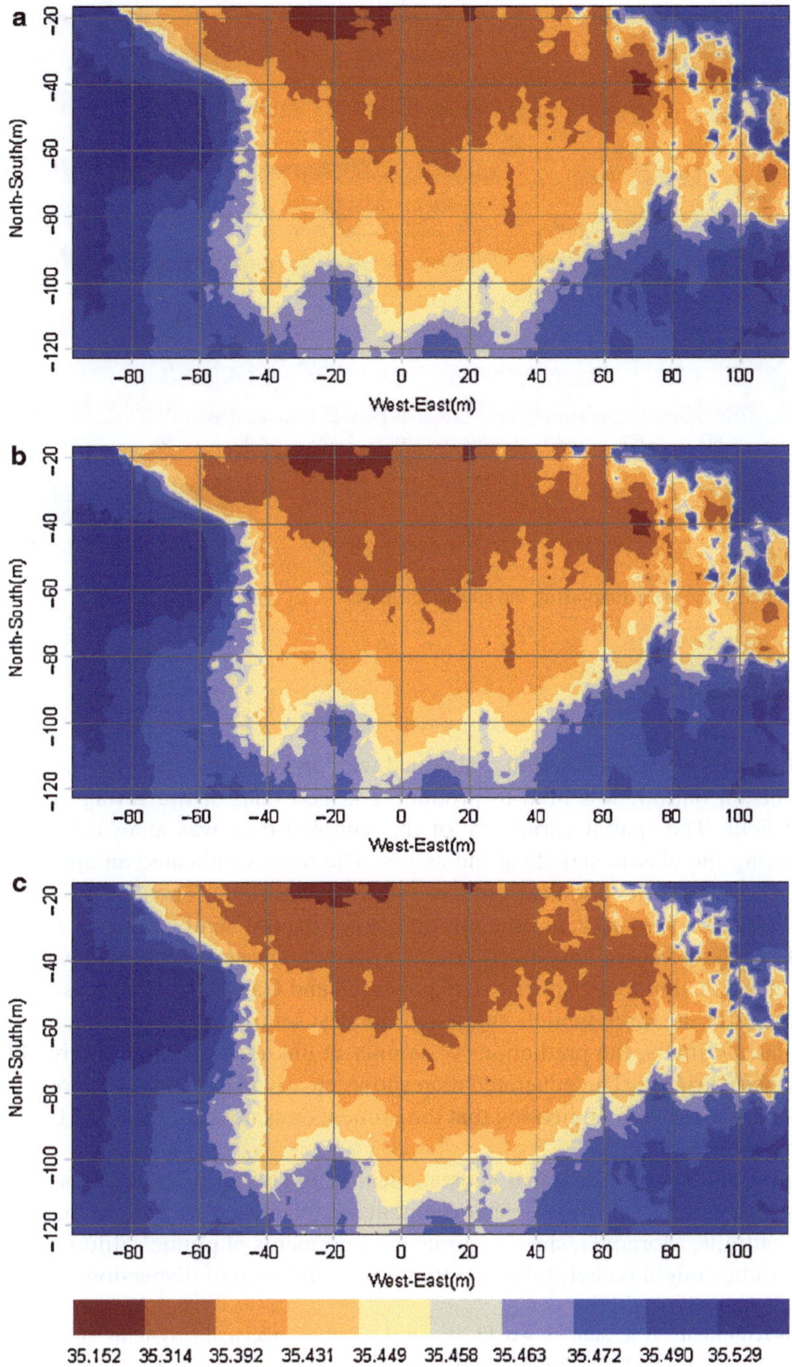

Fig. 4 Prediction maps of salinity distribution using the: (**a**) spherical model. (**b**) Exponential model. (**c**) Gaussian model

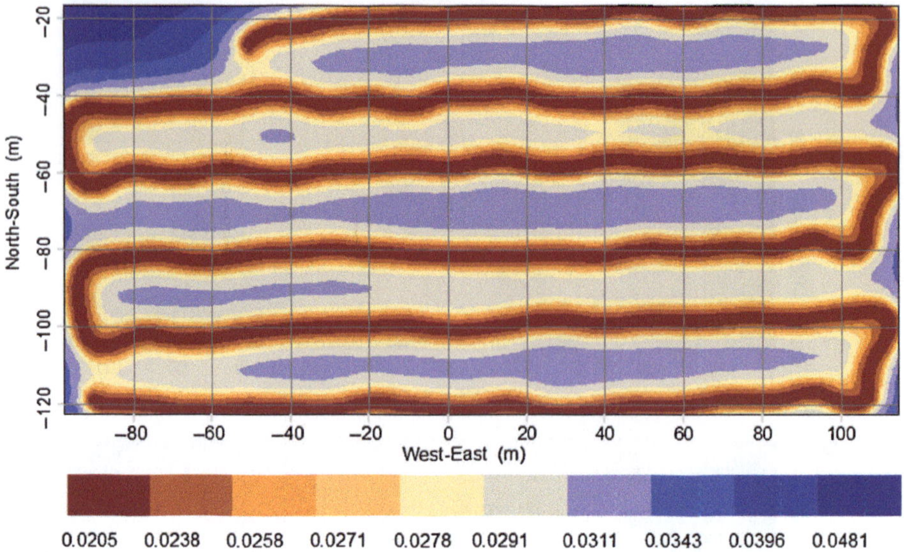

Fig. 5 Prediction error map using the spherical model

4 Conclusions

Geostatistical analysis of salinity, obtained with an AUV in a monitoring campaign to an ocean outfall, was able to produce a kriged map of the sewage dispersion in the field. The spatial variability of the sampled data was analysed previously calculating the classic statistical indicators. The results indicated an approximated normal distribution of the data samples, which is desirable. Then, Matheron's classical estimator was used to compute the experimental semivariogram for several directions. No effect of anisotropy could be shown. The semivariogram was fitted to three theoretical models: spherical, exponential and Gaussian. The cross-validation indicators for the three models suggested the best semivariogram model among the candidates. Finally, the predictions of salinity at unknown locations were obtained by ordinary kriging. The generated map shows clearly the spatial variation of salinity in the studied area, indicating that the effluent does not reach the nearby beaches distanced about 3 km.

Our study demonstrates that geostatistical analysis can provide estimates of effluents dispersion, valuable for environmental impact assessment and management of sea outfalls. Moreover, since accurate measurements of plume's dilution are rare, these studies might be helpful in the future for validation of dispersion models.

Acknowledgments The authors would like to thank the Underwater Systems and Technology Laboratory of University of Porto for the data set used in this analysis.

References

Caeiro S, Painho M, Goovaerts P, Costa H, Sousa S (2003) Spatial sampling design for sediment quality assessment in estuaries. Environ Model Softw 18:853–859

Cressie N (1993) Statistics for spatial data. Wiley Interscience, New York

Isaaks EH, Srivastava RM (1989) Applied geostatistics. Oxford University Press, New York

Kitanidis P (1997) Introduction to geostatistics: applications in hydrogeology. Cambridge University Press, New York

Matheron G (1965) Les variables régionalisées et leur estimation: une application de la théorie des fonctions aléatoires aux sciences de la nature. Masson, Paris, France

Murray CJ, Leeb HJ, Hampton MA (2002) Geostatistical mapping of effluent-affected sediment distribution on the Palos Verdes shelf. Cont Shelf Res 22:881–897

Petrenko AA, Jones BH, Dickey TD (1998) Shape and Initial Dilution of Sand Island, Hawaii Sewage Plume. J Hydraul Eng, ASCE 124:565–571

Poon KF, Wong RWH, Lam MHW, Yeung HY, Chiu TKT (2000). Geostatistical modelling of the spatial distribution of sewage pollution in coastal sediments. Water Res 34:99–108

Ramos P (2005) Advanced mathematical modeling for outfall plume tracking and management using autonomous underwater vehicles based systems. Ph.D. thesis, Faculty of Engineering of University of Porto

Roberts PJW (1996) Environmental hydraulics. In: Singh VP, Hager WH (eds) Sea outfalls. Kluwer, The Netherlands

Saby N, Arrouays D, Boulonne L, Jolivet C, Pochot A (2006) Geostatistical assessment of Pb in soil around Paris, France. Sci Total Environ 367:212–221

Shoji T (2006) Statistical and geostatistical analysis of wind. A case study of direction statistics. Comput Geosci 32:1025–1039

Shoji T, Kitaura H (2006) Statistical and geostatistical analysis of rainfall in central Japan. Comput Geosci 32:1007–1024

Wackernagel H (2003) Multivariate geostatistics: an introduction with applications. Berlin, Springer

Washburn L, Jones BH, Bratkovich A, Dickey TD, Chen M (1992) Mixing, Dispersion, and Resuspension in Vicinity of Ocean Wastewater Plume. J Hydraul Eng, ASCE 118:38–58

Wei H, Dai L, Wang L (2007) Spatial distribution and risk assessment of radionuclides in soils around a coal-fired power plant: a case study from the city of Baoji, China. Environ Res 104:201–208

References

Change of the *A Priori* Stochastic Structure in the Conditional Simulation of Transmissivity Fields

Carlos Llopis-Albert and José Esteban Capilla Romá

Abstract The development of methods for the stochastic simulation of transmissivity (T) fields has progressed, allowing simulations that are conditional not only to T measurements but to piezometric head and solute concentration data. Some methods are even able to honour secondary data and travel time information. However, most of these methods require an *a priori* definition of the stochastic structure of T fields that is inferred only from T measurements. Thus, the additional conditioning data, that implicitly integrate information not captured by T data, might lead to changes in the *a priori* model. Different simulation methods will allow different degrees of structure adaptation to the whole set of data. This paper illustrates the application of a new stochastic simulation method, the Gradual Conditioning (GC) method, to two different sets of data, both non-multiGaussian, one based on a 2D synthetic aquifer and another on a 3D real case (MADE site). We have studied how additional data change the *a priori* model. Results show how the GC method honours the a priori model in the synthetic case, showing fluctuations around it for the different simulated fields. However, in the 3D real case study, it is shown how the a priori structure is slightly modified not following just fluctuations but possibly the effect of the additional information on T, implicit in piezometric and concentration data. Thus, we consider that implementing inversion methods able to yield *a posteriori* structures that incorporate more data might be of great importance in real cases.

1 The GC Method

The GC method (Llopis-Albert, 2008; Capilla and Llopis-Albert, 2009; Llopis-Albert and Capilla, 2009) presents a new stochastic inverse modeling technique for the simulation of transmissivity (T) fields which has been developed to overcome

C. Llopis-Albert (✉) and J.E.C. Romá
Instituto de Ingeniería del Agua y Medio Ambiente, Universidad Politécnica de Valencia, Camino de Vera s/n, 46071-Valencia, Spain
e-mail: cllopisa@gmail.com; jcapilla@upv.es

P.M. Atkinson and C.D. Lloyd (eds.), *geoENV VII – Geostatistics for Environmental Applications*, Quantitative Geology and Geostatistics 16,
DOI 10.1007/978-90-481-2322-3_19,

several of the limitations found in the already existing techniques. It uses an iterative optimization procedure to simulate T fields honoring T measurements, secondary information obtained from expert judgment or geophysical surveys, transient piezometric head (h) data and concentration (c) measurements. Travel time data can also be considered by means of a backward-in-time probabilistic model (Neupauer and Wilson, 1999), which extends the applications of the method to the characterization of sources of groundwater contamination. The formulation of the method does not require assuming the classical multiGaussian hypothesis allowing the reproduction of strings of extreme values of T that often take place in nature, these being formation features crucial in order to obtain realistic and safe estimations of mass transport predictions. The method has been developed using a modified version of the gradual deformation technique (Hu, 2000), and applying a Lagrangian approach to solve the mass transport equation. This allows avoiding numerical dispersion usually found in Eulerian approaches. The new algorithm has been implemented for 3D transient flow problems under variable density flow conditions, considering the dispersion as a tensorial magnitude, and a first-order mass transfer approach. Performing a Bernoulli trial on the appropriate phase transition probabilities, the particle distribution between the mobile domain and the immobile domain can be determined (see Salamon et al., 2006).

1.1 Iterative Optimization Process

The iterative optimization process for constraining stochastic simulations to data is carried out by doing non-linear combinations of seed conditional realizations. These seed conductivity (K) fields are already conditional to K and secondary data, and are generated by sequential indicator simulation. The *a priori* stochastic structure of these K seed fields is defined by means of the local cumulative density functions (ccdf's) and the indicator variograms, thus allowing the GC method to adopt any Random Function (RF) model. As a first step, the GC method builds linear sequential combinations of multiGaussian K fields that honour K data:

$$K^m = \alpha_1 K^{m-1} + \alpha_2 K_{2m} + \alpha_3 K_{2m+1} \text{ with } K^0 = K_1 \tag{1}$$

where subscripts stand for seed fields and superscripts for conditional fields resulting from a previous linear combination That is, at m iteration, the field K^{m-1}, from the previous iteration, is combined with two new independent realizations K_{2m} and K_{2m+1}. The procedure requires combining at least three conditional realizations at a time to ensure the preservation of mean, variance, variogram and K data in the linearly combined field. The coefficients have also to fulfill the constraints in Equation (2):

$$\begin{cases} \alpha_1 + \alpha_2 + \alpha_3 = 1 \\ (\alpha_1)^2 + (\alpha_2)^2 + (\alpha_3)^2 = 1 \end{cases} \tag{2}$$

being the parameterization ofα_i given by Equation (3):

$$\begin{cases} \alpha_1 = \frac{1}{3} + \frac{2}{3}\cos\theta \\ \alpha_2 = \frac{1}{3} + \frac{2}{3}\sin(-\frac{\pi}{6} + \theta) \\ \alpha_3 = \frac{1}{3} + \frac{2}{3}\sin(-\frac{\pi}{6} - \theta) \end{cases} \quad \text{with} \quad \theta \in [-\pi, \pi] \tag{3}$$

The α_i coefficients are different in every iteration m, and correspond to a unique parameter θ; note the one to one correspondence between the parameter and the combined realization K^m.

Because the linear combination of independent non-Gaussian random functions does not preserve the non-Gaussian distribution, although the variogram is preserved, a transformation between Gaussian to the non-Gaussian fields (and vice versa) is required. This transformation is performed through the probability fields (see Capilla and Llopis-Albert, 2009 for more details). Finally, at each iteration m of the method the parameter θ is determined by minimizing an objective function that penalizes deviations among computed and measured data. This way of operating has been successfully applied in both synthetic and real cases (see Llopis-Albert and Capilla, 2007, 2009).

2 Application to a 2D Synthetic Data Set

The flow domain has a size of 226.4×246.4 m and is discretized in 37×34 square blocks of size 6.6 m with prescribed head boundary conditions and transient flow conditions with three stress periods of length 31.7, 761 and 2,378.27 years, respectively. A pumping well with a rate of 1.8 l/s is activated during the second stress period. Other parameter values are defined as: porosity of 0.35, longitudinal dispersion of 0.3 m, transversal dispersion of 0.03 m, specific storage of $2.5 \cdot 10^{-4}$ 1/m and a number of 4,900 particles are used to solve the transport equation. T seed fields (only conditioned to T data) have been generated by sequential indicator simulation, code ISIM3D (Gómez-Hernández and Srivastava, 1990). A mosaic variography has been chosen, which is spherical, with equal ranges in all directions of 40 m, 0.04 of nugget effect, and sill of 0.22. The *a priori* conditional cumulative density function (ccdf) displays a highly asymmetrical distribution with a long lower tail. Figure 1 shows additional information about well and initial particle locations as well as the T reference field (selected between the generated seed fields). Figure 2 shows the spatial location for the following sets of conditioning data: sixteen regularly spaced T measurements, 16 regularly spaced piezometric head measurements at the end of the three time steps considered and 40 solute concentration measurements distributed in time and space to capture the shape and extension of the plume (three snapshots at time 412.22, 792.74 and 1,902.58 years).

Ensemble variograms, presented in Fig. 3, illustrate that the method tends to preserve the *a priori* spatial structure for every threshold, at time that reduces the uncertainty when conditioning to all available information. Moreover, this is true

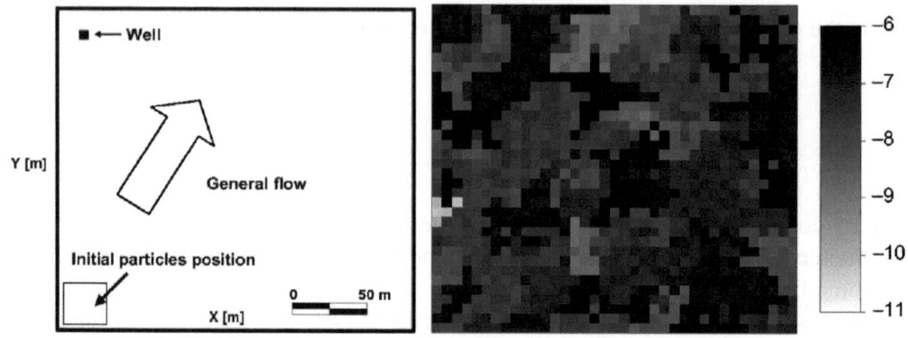

Fig. 1 Geometry, boundary conditions, and initial particles and well position (*left*). Log T (m/s) of the reference filed (*right*)

Fig. 2 Piezometric head fields at the end of the three stress periods and spatial location for T data, c data at three snapshots and h data at three periods (from *left* to *right*)

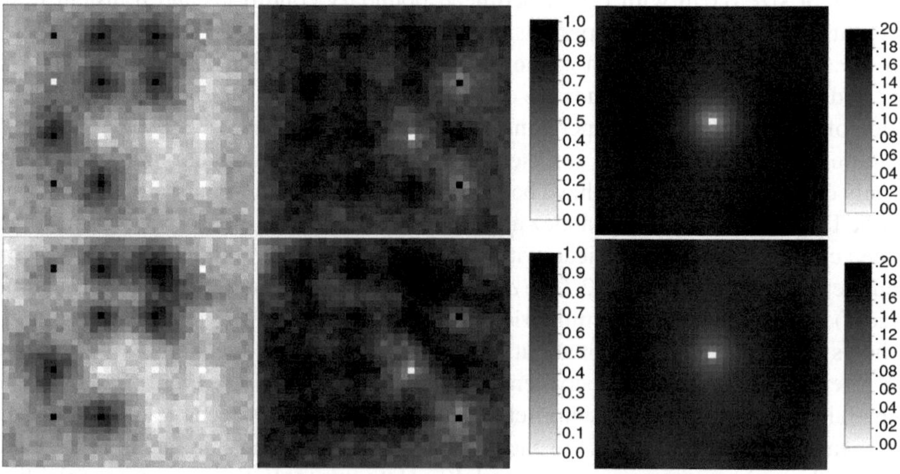

Fig. 3 Probability maps (local ccdfs for deciles 1 and 7) and variogram map (decile 2) for an ensemble of 100 seed fields (*above*) and 100 fields conditioned to T and h data at iteration 50 (*below*)

regardless of the iteration number of the optimization process or the conditioning information used. Results show the normal ergodic fluctuations for different simulated fields, which disappear at ensemble variograms. However, the method can slightly modify the variogram in a certain way to better reproduce the conditioning information and get closer to the unknown reality. This can also be seen in the local cumulative density functions (ccdfs), which display a great similarity for all deciles, leading to a random function model preservation, while the non-multiGaussian feature is also retained. Again, probability maps show that the method is able to produce interconnected zones (reflected in the modified local ccdfs) that were not captured by the data used in the *a priori* stochastic structure, at time that avoids a homogenization of the T field, since the inverse model is able to increase or reduce the T values where necessary.

3 Application to a 3D Real Case (MADE Site)

The GC method was successfully applied in a highly heterogeneous aquifer ($\sigma^2_{lnK} \approx 0.5$) at the Macrodispersion Experiment (MADE-2) site on Columbus Air Force Base in Mississippi (Llopis-Albert and Capilla, 2007), since a good agreement between data and simulated mass distribution at $t = 328$ days, including the non-Gaussian plume behaviour, was reached. Furthermore, MonteCarlo simulations, using seed fields, showed the existence of a high uncertainty when not using all available information and the need to condition to as much information as possible. For the sake of conciseness, the reader may be referred to Llopis-Albert and Capilla (2007) for details concerning the modeling approach. The *a priori* random function modelling is based on a similar indicator geostatistical analysis as presented, for the MADE site, by Salamon et al. (2007), although they used a flowmeter measurement support scale. We assume that depositional structures in the aquifer are approximately horizontal, as argued by various authors (e.g., Salamon et al., 2006). Hence, spatial continuity is only analyzed in the completely horizontal and vertical direction. No significantly higher spatial continuity in the extreme thresholds was detected in the horizontal plane, in spite of the fact that preferential flow pathways have a significant effect on the anomalous tracer plume spreading at the MADE site (e.g., Llopis-Albert and Capilla, 2007). In addition, modeller decisions are needed to complete the indicators variography definition due to the fact that there are not enough K data within each threshold. We have adopted a higher spatial continuity definition for the extreme thresholds, thus allowing the reproduction of existing preferential flow pathways (consistent with results obtained from indicator variography). The iterative optimization process is carried out by combining alternately seed T fields generated with this variography with others without this higher spatial continuity definition. Once again, it is clear from Fig. 4 that the *a priori* structure is slightly modified according with the additional conditioning information.

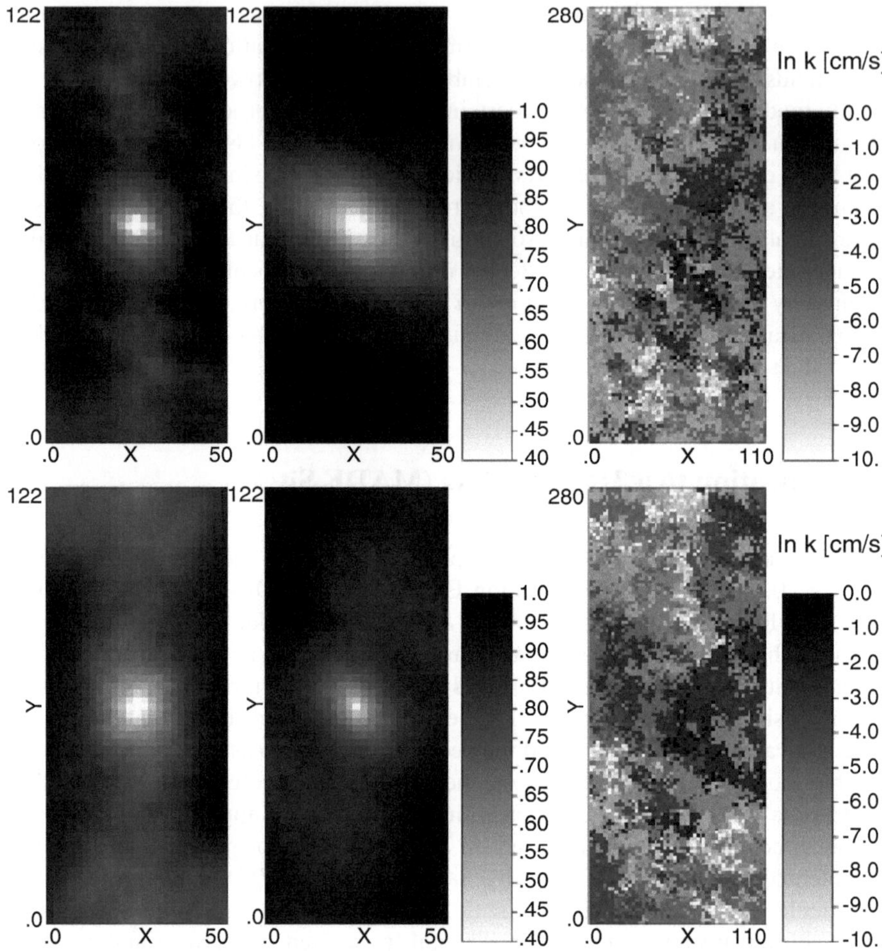

Fig. 4 Standardized indicator variograms (decile 1 and 5 of layer 10) and horizontal slices (layer 10) of ln conductivity fields (cm/s) for a seed field (*above*) and a field conditioned to K, h and c data (*below*)

4 Conclusions

As shown, the method tends to preserve the *a priori* spatial structure of the stochastic process during the iterative optimization procedure of non-linear combinations of T fields, although some modifications can take place to better reproduce the conditioning information and become closer to the unknown reality, at least in terms of honoring other available information. This can be seen in the 2D synthetic case, in which the GC method honours the *a priori* model (which is shared by all seed fields), although showing the normal ergodic fluctuations for the different simulated fields. Besides, additional conditioning data, that implicitly integrate information not captured by T data, lead to changes in the *a priori* stochastic model, as shown for example when reproducing interconnected zones, which might provide safer estimations of mass transport predictions. However, in the 3D real case study, results show how the *a priori* structure is modified not obeying just fluctuations but possibly also the effect of the additional information on T, implicit in h and c data. Finally, both 2D and 3D cases retain the non-Gaussian feature in the conditioned fields, since the variogram is kept for all deciles.

Acknowledgements Financial support from the Spanish Ministry of Science and Education (Ref.REN2003-06989) is gratefully acknowledged.

References

Capilla JE, Llopis-Albert C (2009) Stochastic inverse modeling of non multiGaussian transmissivity fields conditional to flow, mass transport and secondary data. 1 Theory. J Hydrol doi:10.1016/j.jhydrol.2009.03.015

Gómez-Hernández JJ, Srivastava RM (1990) ISIM3D: An ANSI-C three dimensional multiple indicator conditional simulation program. Comput Geosci 16(4):395–440

Hu LY (2000) Gradual deformation and iterative calibration of gaussian-related stochastic models. Math Geol 32(1)87–108

Llopis-Albert C (2008) Modelación inversa estocástica no multigaussiana condicionada a datos de flujo y transporte. Ph.D. Thesis, Technical University of Valencia, 274 pp. ISBN: 978-84-691-9796-7

Llopis-Albert C, Capilla JE (2007) A new approach for the stochastic inversion of flow and transport data: Application to the macrodispersion experiment (MADE-2) site. Calibration and Reliability in Groundwater Modelling: Credibility of Modelling (Proceedings of ModelCARE 2007 Conference, held in Denmark, September 2007). IAHS Publ. 320, Copenhagen

Llopis-Albert C, Capilla JE (2009) Stochastic inverse modeling of non multiGaussian transmissivity fields conditional to flow, mass transport and secondary data. 2 Demonstration on a synthetic aquifer. J Hydrol doi:10.1016/j.jhydrol.2009.03.014

Neupauer RM, Wilson JL (1999) Adjoint method for obtaining backward-in-time location and travel time probabilities of a conservative groundwater contaminant. Water Resour Res 35(11):3389–3398

Salamon P, Fernàndez-Garcia D, Gómez-Hernández JJ (2006) Modeling mass transfer processes using random walk particle tracking. Water Resour Res 42:W11417, doi:10.1029/2006WR004927

Salamon P, Fernàndez-Garcia D, Gómez-Hernández JJ (2007) Modeling tracer transport at the MADE site: the importance of heterogeneity. Water Resour Res 43:W08404, doi:10.1029/2006WR005522

Geostatistical Interpolation of Soil Properties in Boom Clay in Flanders

Annelies Govaerts and André Vervoort

Abstract This contribution examines the applicability of ordinary kriging to interpolate the results of cone penetration tests or CPTs. The advantages of geostatistics are studied for two datasets of CPTs measured in a typical soil encountered in Flanders, Belgium: the Boom clay. Firstly the unit pile base resistance of a pile with a diameter of 0.4 m is studied, for a specific depth separately. Secondly the characteristic resistance of an axially loaded pile is considered. In that part of the study the whole profile is estimated at once, instead of estimating the value at a specific depth. From the study one can conclude that it is possible to estimate the bearing capacity accurately in a point if sufficient measurements are carried out in the immediacy of that point and that ordinary kriging can be a good method to make an estimation of the characteristic resistance of an axially loaded pile.

1 Introduction

A detailed soil investigation program is an important step that cannot be omitted in the planning phase of many environmental and geotechnical projects like e.g. site remediation, waste management operations, and civil engineering projects. Soils are naturally formed in different depositional environments; therefore their physical properties vary from point to point (in a horizontal as well as vertical plane). This variation can even exist in an apparently homogeneous soil unit (Jones et al., 2003). In most cases the relevant soil parameters are spatially correlated and, hence a geostatistical approach is advisable. This is also the case when one investigates the bearing capacity of the soil, using the in-situ cone penetration test, or CPT (Govaerts, 2006). This test gives information as a function of the depth at only one particular position. Hence, interpolation techniques are necessary to evaluate the properties between the CPTs. Poorly guided inter- and extrapolation of

A. Govaerts (✉) and A. Vervoort
Research unit mining, K.U. Leuven, Kasteelpark Arenberg 40 bus 2448, B-3001 Leuven, Belgium
e-mail: annelies.govaerts@bwk.kuleuven.be; andre.vervoort@bwk.kuleuven.be

P.M. Atkinson and C.D. Lloyd (eds.), *geoENV VII – Geostatistics for Environmental Applications*, Quantitative Geology and Geostatistics 16,
DOI 10.1007/978-90-481-2322-3_20,

the data gathered can be the cause of large problems (financially and others). This contribution examines the applicability of geostatistical techniques (ordinary kriging) to interpolate the results of CPTs.

2 Data

2.1 The Cone Penetration Test

The CPT is a fast and cheap test to explore the soil. It is one of the most widely used methods for soil investigation worldwide. The test is performed using a cylindrical penetrometer with a conical tip (cone) penetrating the soil at a constant rate. During the penetration, measurements are made of the resistance to penetration of the cone. The total force acting on the cone, divided by the protected area of the cone, produces the cone resistance q_c. A cone resistance profile consists of q_c-values at certain depth intervals, e.g. every 0.2 m (see Fig. 1 as an example). The CPT has three main applications in the site investigation process: (1) to determine the sub-surface stratigraphy and identify materials present, (2) to estimate geotechnical parameters and (3) to provide results for direct geotechnical design (Lunne et al., 1997). The geotechnical parameters that can be deduced are for example shear strength, density, elastic modulus and rate of consolidation (Brouwer, 2007). In practice one could be interested in an estimation of the whole cone resistance profile, but also in the value

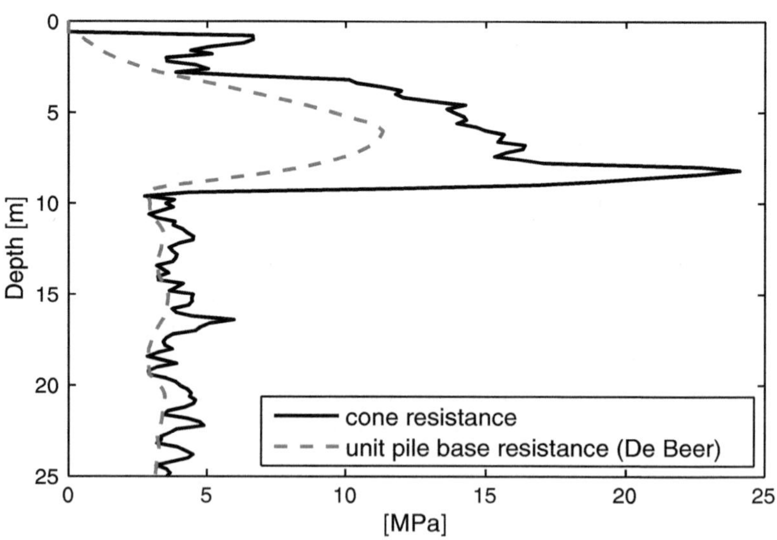

Fig. 1 An example of a cone resistance profile and the deduced unit pile base resistance profile

of one of these parameters at a certain depth. The depth one is interested in could for example be the depth one wants to design a foundation for. This depth is chosen based on the measured, and possibly the estimated, profiles.

2.2 Two Datasets

In this study, firstly, the unit pile base resistance of a pile with a diameter of 0.4 m is studied. This unit pile base resistance is deduced from the q_c-profile, based on the method of De Beer (1971). These profiles do not have the high frequent fluctuations of the q_c-profiles (Fig. 1). In fact these fluctuations do not hold much information for practical foundation calculation and are not spatially correlated. To estimate the unit pile base resistance at a certain depth in a certain point, most of the time one cannot interpolate from CPT-values above this point of interest (e.g. from a CPT-test which is stopped above the horizon of interest). Hence, one has to estimate the pile base resistance in an unsampled point from neighbouring CPTs. If the site is not too outstretched the resistance-values at the same depth, relatively to the top or bottom of the soil layer, are the most correlated. Therefore it is decided to execute the estimations in a horizontal plane parallel to the top of the studied soil layer.

In this study, secondly, the characteristic resistance of an axially loaded pile is considered. In this part the entire profile is estimated instead of estimating the resistance at individual depth values.

The advantages of geostatistics are studied for CPT-datasets measured in a typical soil encountered in Flanders, Belgium: the Tertiary overconsolidated Boom clay. This clay is a marine deposit of Middle Oligocene age (35 million years) (Schittekat, 2001). The total thickness of the original clay formation could have been well above 100 m. Later erosion has removed part of the clay to leave a thickness of 70 m in, for example, Antwerp and surroundings. Furthermore there is evidence that it was covered by thicker deposits than those left today. The removed overburden in the Antwerp area is estimated to be 90 m. Consequently, the Boom clay may be considered as an overconsolidated clay (Schittekat, 2001). Afterwards these clay sediments are covered by quaternary formations.

In this study two different datasets are used (Fig. 2). Both datasets consist of a set of electric CPTs and are measured in the region of Antwerp, Belgium. The first test site, with an area of 2,000 × 5,500 m, is located at Kruibeke. There are 73 CPTs conducted at this site. The 25 CPTs of the second dataset are situated in the river Schelde near the ‘Sint-Anna-strand’ on a smaller area, 100 × 450 m. At both sites the subsoil consists of a couple of meters of Quaternary layers followed by the Boom clay. Only that part of the profiles that is describing the upper 7 m of the Boom clay is studied. The part through the Quaternary layers is omitted.

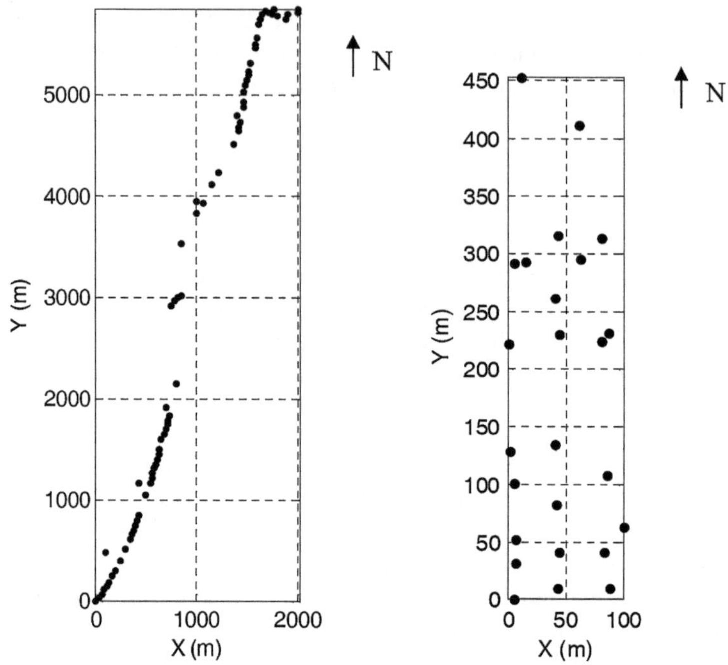

Fig. 2 Relative position of the 73 CPT's at the site near Kruibeke (*left*) and of the 25 CPT's at the site at Sint-Anna (*right*)

3 Unit Pile Base Resistance at a Specific Depth

3.1 Semivariogram

Firstly the unit pile base resistance at a specific depth[1] is studied in the horizontal plane. One can determine a semivariogram at every depth for both sites. In Fig. 3 the semivariograms at respectively 5 m below the top of the Boom clay at the site in Kruibeke and at 3.4 m below the top at the site in Sint-Anna are shown. In Table 1 the parameters of the spherical semivariograms at several depths are given for both sites. In Kruibeke the ranges are situated between 250 and 500 m. The smaller ranges are found at the larger depths. The nugget effect varies from 0% to 35% of the total sill value. The smallest nugget effects are also found at the larger depths. At the second site some of the semivariograms are modelled as a pure nugget effect. For the other depths the range and nugget effect vary between respectively 50 and 120 m, and 0% and 22% of the total sill value. Most of the semivariogram models do not include a nugget effect.

[1] Depth = distance below the top of the Boom clay.

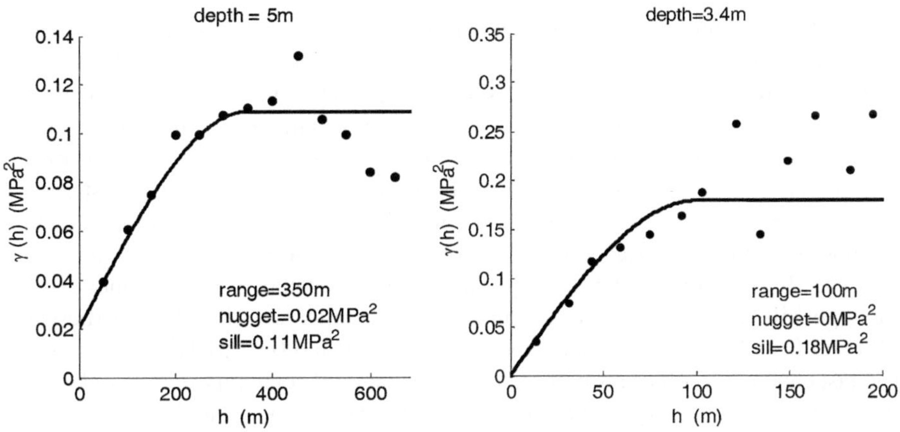

Fig. 3 Semivariogram of the unit pile base resistance value at a depth of 5 m in Kruibeke (*left*) and at a depth of 3.4 m in Sint-Anna (*right*)

Table 1 Parameters describing the spherical semivariogram models of the unit pile base resistance value at different depths in Kruibeke and Sint-Anna

	Kruibeke				Sint-Anna			
Depth (m)	Range (m)	Nugget (MPa^2)	Sill (MPa^2)	Nugget effect (%)	Range (m)	Nugget (MPa^2)	Sill (MPa^2)	Nugget effect (%)
0.0	450	0.03	0.12	26	p.n.[a]	p.n.[a]	0.16	100
0.6	480	0.03	0.11	28	50	0.00	0.11	100
1.0	500	0.02	0.10	21	60	0.01	0.12	8
1.4	500	0.02	0.09	22	80	0.03	0.13	22
1.8	500	0.02	0.09	23	p.n.[a]	p.n.[a]	0.13	100
2.2	500	0.02	0.08	24	p.n.[a]	p.n.[a]	0.12	100
2.6	500	0.02	0.10	21	70	0.00	0.12	0
3.0	400	0.04	0.11	35	80	0.00	0.14	0
3.4	400	0.04	0.12	33	100	0.00	0.18	0
3.8	400	0.03	0.14	22	120	0.00	0.22	0
4.2	400	0.04	0.13	33	120	0.00	0.28	0
4.6	350	0.04	0.12	18	90	0.00	0.39	0
5.0	350	0.02	0.11	10	80	0.00	0.40	0
5.4	300	0.01	0.10	14	60	0.00	0.41	0
5.8	300	0.01	0.11	5	60	0.00	0.42	0
6.2	250	0.01	0.10	0	70	0.00	0.41	0
6.6	250	0.00	0.09	6	80	0.00	0.39	0
7.0	250	0.01	0.09	26	70	0.00	0.37	0

[a] p.n. = pure nugget model.

Although the two datasets are recorded in the same Boom clay one can clearly see a difference between the parameters at both sites. This can be because of the scale differences: the overall area (=extent) of the site at Kruibeke is larger and also the spacing between the datapoints is larger at this site. On average the distance between one point and its closest neighbour is 66 m in Kruibeke and 31 m at Sint-Anna. These scale differences influence the semivariogram parameters. For the ideal case of very small spacings and very large extents the estimated range and the estimated variance are close to their true values. However, as the spacing increase or the extent decreases these estimated parameters differ more from the true values. Western and Blöschl (1999) showed that the apparent range increases for larger spacing or extent, and that the apparent variance increases for larger extent, but does not change with spacing. This explains the larger ranges at the site in Kruibeke. But it cannot explain the larger sill values, thus larger variances, at the smaller site of Sint-Anna (between 0.08 and 0.14 MPa^2 in Kruibeke and between 0.11 and 0.42 MPa^2 in Sint-Anna). The last can be a result of a small scale process that did act at this site but did not act at the site in Kruibeke (possibly due to the influence of the Schelde river). But the mean values of the unit pile bearing resistance at every depth are also larger for the dataset at Sint-Anna (between 1.0 and 2.3 MPa in Kruibeke and between 3.4 and 3.9 MPa in Sint-Anna). When the dimensionless coefficient of variation is studied, both datasets show similar values (between 0.13 and 0.32 in Kruibeke and between 0.10 and 0.17 in Sint-Anna). Probably there is also an effect of anisotropy. One of the causes is likely the river Schelde. At the site in Kruibeke it flows approximately south-north, east of the datapoints. At Sint-Anna it runs from east to west and the datapoints are situated in the river. But in Kruibeke all data points are approximately situated in the same direction. Therefore the calculated semivariogram is a unidirectional one and it is impossible to determine the semivariogram in other directions to detect the anisotropy. This is not the case for the site in Sint-Anna, but at this site there are not enough data points to determine the directional semivariograms. Based on this study, one can conclude that a semivariogram based on data of one site in the Boom clay cannot easily be extrapolated to another site with data recorded in the same Boom clay.

3.2 Estimation

To evaluate the estimation techniques cross-validation is used. This means that values are kriged at each sampled location, assuming that the corresponding sample is missing. The semivariograms are used to estimate the unit pile base resistance at a specific depth at all measured locations, based on the values in the five nearest data points. A different semivariogram is used for each depth (Table 1). Several indices are suitable to evaluate the performance of an estimation technique. These indices are all a measure of the estimation error, which is the difference between the estimated value and the true value. The true value is in this case the unit pile base resistance calculated based on the CPT-measurement in a certain location. The estimated value is the unit pile base resistance that was estimated based on the unit pile bearing resistance in the five nearest locations. The more similar both values are, the

Table 2 Minimum, maximum and mean of the estimation error, absolute value of the estimation error and slope of the linear fit through the estimated unit pile base resistance versus calculated (De Beer) pile base resistance plot for both datasets: comparison of the kriging results and the arithmetic mean

	Kruibeke						Sint-Anna					
	Kriging			Mean			Kriging			Mean		
	Min	Max	Mean	Min	Max	Mean	Min	Max	Mean	Min	Max	Mean
Error (MPa)	−0.96	1.05	0.01	−0.84	1.04	0.00	−1.85	1.37	−0.06	−2.04	1.51	−0.06
|error| (MPa)	0.00	1.05	0.21	0.00	1.04	0.21	0.00	1.85	0.38	0.00	2.04	0.40
Slope	0.32	0.55	0.42	0.21	0.47	0.35	−0.07	0.38	0.08	−0.12	0.12	−0.02

better the estimation technique is. In this study the estimation error and the absolute value of the estimation error are used. These indices can be calculated at all measuring points for all depths.[2] Table 2 presents the minimum, maximum and mean of these estimation error-values for the ordinary kriging approach and for the classical statistical approach where the estimate equals the arithmetic mean of the five nearest points. As expected, for an unbiased estimation technique, the average estimation error is for both sites and for both techniques around zero. For the site at Kruibeke the minimum, maximum and mean value of the absolute value of the estimation error are almost equal for both interpolation techniques. At Sint-Anna all three are smaller for ordinary kriging, which shows that the geostatistical approach is beneficial. A scatterplot of estimated versus true values provides additional evidence on how well an estimation method has performed (Isaaks and Srivastava, 1989). In Fig. 4 one can see such graphs for a specific depth at both sites. The best possible estimates would always match the true values and would therefore plot on the 45° line on the scatterplot. In actual practice there are always errors in the estimates, and scatterplots of estimated versus true values always appear as a cloud of points (Isaaks and Srivastava, 1989). Therefore the linear fit through the data is constructed. The closer this line to the 45° line the better the estimation technique. Table 2 gives the minimum, maximum and mean values of those slope values at both sites. All of them are closer to 1 (which is the slope value for the 45° line) for ordinary kriging in comparison to the arithmetic mean. Therefore one can conclude that ordinary kriging is better. The differences between both approaches are more pronounced for the site at Sint-Anna. This is probably, partially, due to the geometry of the dataset. In Kruibeke all the points lay approximately on one line. For geostatistics it is better that the point to be estimated is fully surrounded by known data points. Therefore the geometry of the data is better for the site at Sint-Anna.

[2] The term 'all' depths means the various depths from the top of the Boom clay till 7 m below the top with lags of 0.2 m.

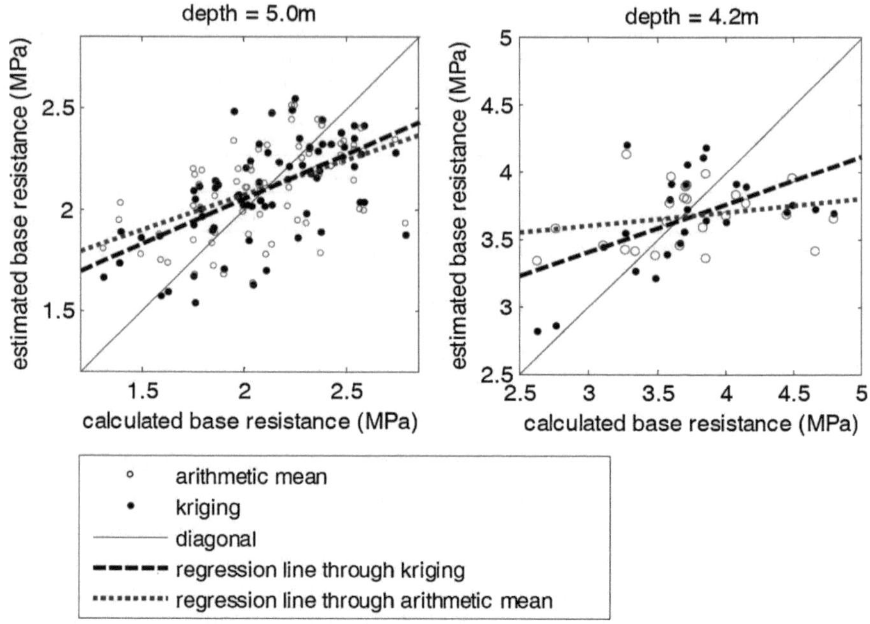

Fig. 4 Estimated unit pile base resistance versus calculated (De Beer) pile base resistance based on the cone resistance at a depth of respectively 5.0 and 4.2 m at Kruibeke (*left*) and Sint-Anna (*right*): comparison of the arithmetic mean and the kriging results

4 Characteristic Resistance of an Axially Loaded Pile

4.1 Design Procedure According to Eurocode 7

In Eurocode 7 (2004), which contains some rules for geotechnical design, the interpolation problem is explicitly mentioned. In Eurocode 7 one can find a design procedure to calculate the characteristic bearing capacity of an axially loaded pile based on cone penetration tests. In this paper the resistance values are calculated for a simple round precast concrete pile with a diameter of 0.4 m and without an enlarged base. According to Eurocode 7 the characteristic value should be derived such that the calculated probability of a less accurate value governing the occurrence of the limit state under consideration is not greater than 5%. In this case it means that the probability of the real resistance being smaller than the calculated characteristic value has to be at maximum 5%. The characteristic value should consider the variability of the compressive resistance of the piles over the site, the number of tests and the stiffness of the structure and its ability to transfer loads from weak to strong spots. In Eurocode 7 the characteristic value of the pile compressive resistance $R_{c,k}$ is obtained according to the following equation:

$$R_{c,k} = min\left(\frac{(R_{c,cal})_{mean}}{\xi_3}, \frac{(R_{c,cal})_{min}}{\xi_4}\right) \tag{1}$$

where ξ_3 en ξ_4 are correlation factors that depend on the number of tested profiles N and $R_{c,cal}$ is the calibrated resistance that is calculated based on the CPT measurements. The correlation factors are determined for an 'average' subsoil. Some additional information about the design procedure and the determination of all the factors can be found in Bauduin (2001, 2002).

4.2 Characteristic Value Based on Geostatistics

In this part of the study the characteristic values determined based on Equation (1) are compared to the characteristic values calculated by geostatistical techniques. The profile of characteristic values on a specific location is considered to be unknown. For both approaches the characteristic profile is calculated based on the five nearest calibrated profiles. In the geostatistical procedure these characteristic values are determined in such a way that the confidence corresponds to 95%. This is possible by means of the kriging variance that is calculated for every geostatistical estimate.

The geostatistical estimates could be done in the same way as the unit pile base resistances are estimated in the previous paragraph. But when looking at Fig. 5, which shows all the calibrated profiles at Kruibeke and Sint-Anna, one can see that all of them have a similar outline. For this reason it is decided to apply a different procedure to determine the characteristic values. Instead of determining a semi-variogram at every depth and making the estimates for all these depths separately, the entire profiles are estimated at once. This is done by an empirical eigenfunction analysis as proposed by Coerts (1996). Based on all the profiles at the site one can determine eigenfunctions. Every profile can be approximated as a linear combination of these eigenfunctions. Because of the fact that the profiles are very

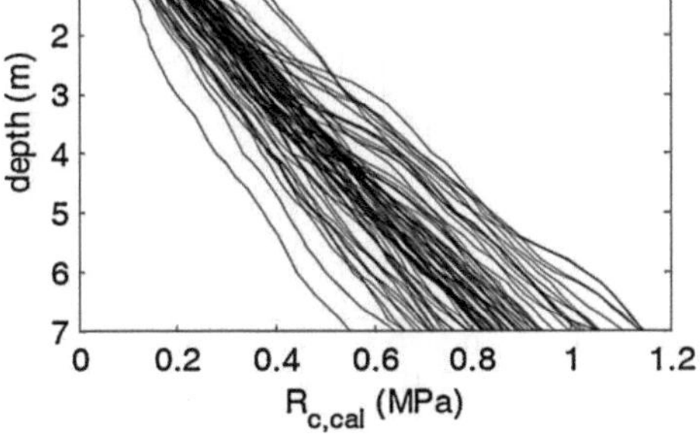

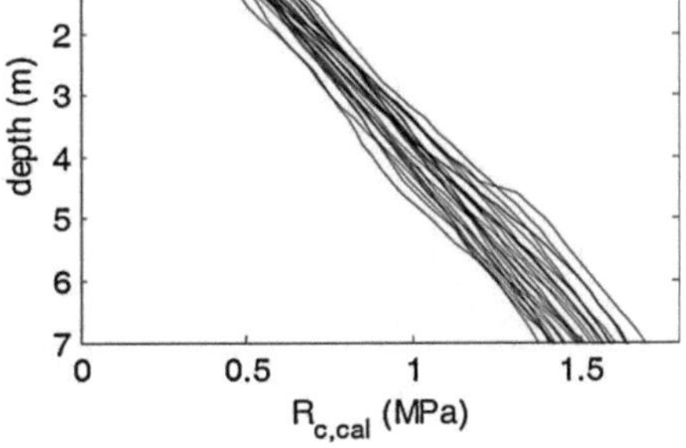

Fig. 5 All profiles of $R_{c,cal}$ from top of the Boom clay to 7 m below at Kruibeke (*left*) and at Sint-Anna (*right*)

much alike, only the first two eigenfunctions are used in this study (Coerts, 1996; Govaerts, 2004).

$$q'_j = \sum_{i=1}^{t} s_{ji} eef_i \qquad (2)$$

where q'_j = jth profile after the eigenfunction analyses, eef_i = ith eigenfunction.

As a result of this eigenfunction analysis, two s-factors are known for every profile or position. A geostatistical study of these s-factors is done. All semivariograms are modeled with a spherical model. In Kruibeke the range, nugget and sill of s_1 and s_2 are respectively 350 m, 0.3 MPa^2, 2.90 MPa^2 and 650 m, 0.3 MPa^2, 0.53 MPa^2. As expected the nugget effect of the s-factor belonging to the first eigenfunction is lower (10%) than the nugget effect of the second s-factor (55%). In Sint-Anna the range of both s-factors is 70 m and there is no nugget effect. Again the ranges at Sint-Anna are smaller than those in Kruibeke and at both sites the ranges are similar to the ranges in the previous part. Those semivariograms are used to krige the s-factors. Based on these estimates and the eigenfunctions the whole profile can be calculated, using Equation (2). To determine the characteristic values the estimated s-factor is reduced (or increased for a negative s-factor) by 1.65 times the square root of the kriging variance. This last is only allowed if the estimation errors are approximately normally distributed around zero. This is checked for both s-factors at both sites. It seems that the errors are symmetrically distributed with slightly larger tails than a normal distribution with the same mean and variance. According to Journel and Huijbregts (1978) this is true for most mining applications. For all four s-factors the interval $[m_E - 1.65\sigma_E, +\infty]$ does indeed contain approximately 95% of the observed errors (respectively 96% and 96% in Kruibeke and 92% and 96% in Sint-Anna). In Fig. 6 the histograms of the difference between the characteristic values, obtained by the geostatistical procedure, and the calibrated values, obtained by the CPT-measurements, are given. For both sites some of the characteristic values are larger than the calibrated values. But this is only true for less

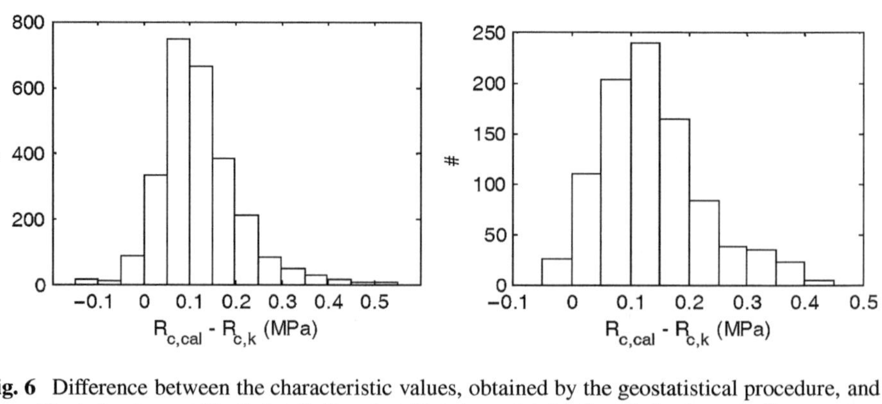

Fig. 6 Difference between the characteristic values, obtained by the geostatistical procedure, and the calibrated values at Kruibeke (*left*) and at Sint-Anna (*right*) for all depths and all measuring points

than 5% of all points (4% in Kruibeke and 3% in Sint-Anna). Thus, as expected, the geostatistically determined characteristic values satisfy the definition described in Eurocode 7. These characteristic values can be compared to the characteristic values which are calculated based on Equation (1). In Kruibeke the geostatistical $R_{c,k}$-values are for 55% of all the values (i.e. 73 locations at various depths) larger than the $R_{c,k}$-values calculated by Equation (1). In Sint-Anna this is so for 99% of the values. The differences between both methods of approach are partially due to the fact that the correlation factors in Eurocode 7 are determined based on the idea that the coefficient of variation of the calibrated values is around 12%. In Kruibeke this variation coefficient is in the first meter below the top of the Boom clay at average 20%. But deeper it decreases to 10% at a depth of 7 m below the top. Therefore the coefficient of variation is smaller than 12% from 2 to 7 m below the top. In Sint-Anna the coefficient of variation decreases from 10% at the top to 5% at 7 m below the top of the Boom clay. Thus in Sint-Anna the coefficient of variation is even lower than in Kruibeke and certainly lower than 12%. Therefore the differences between the two approaches are also more pronounced at this site.

Thus in this case geostatistics is a good alternative method to make a cautious, but not too conservative, estimation of the characteristic resistance of an axially loaded pile.

5 Conclusions

From the study one can conclude that it is indeed possible to estimate the bearing capacity accurately in a point if sufficient measurements are carried out in the immediacy of that point. By taking into account the spatial variation the estimation error is, on average, smaller using ordinary kriging instead of taking the arithmetic mean. Geostatistics also results in information about the estimation error. This information can be used in a probabilistic design.

Geostatistics can also be a good method to make an estimation of the characteristic resistance of an axially loaded pile. In this way the spatial variability is explicitly taken into account and cautious, not too conservative, characteristic values can be determined.

In future research the effect of the geometry of the dataset has to be studied in more detail. In this study both datasets do have a total different geometry (a line at Kruibeke and an irregular grid at Sint-Anna). A similar analysis can also be done for other values of the numbers of data points used in the estimation. Other soils such as the Quaternary and Tertiary sands will be analyzed too. These probably have a different spatial structure. This is going to result in a different relation between the geostatistical calculated characteristic resistances and those calculated using the correlation factors, based on a coefficient of variation of 12%.

Geostatistics takes implicitly into account (a) the locations of the cone penetration tests, (b) the position of the point to be estimated in relation to the data points and (c) the spatial variation of the data. This study shows that this has indeed advantages for practical applications. The final question to be answered is if it is indeed possible to determine other (i.e. better) correlation factors to determine the characteristic resistance of an axially loaded pile (possibly depending on the subsoil and the geometry of the dataset) or if it will remain necessary to conduct the whole, time-consuming, geostatistical analysis.

References

Bauduin C (2001) Design procedure according to Eurocode 7 and analysis of the test results. In: Holeyman A (ed) Screw piles: installation and design in stiff clay. Balkema, Rotterdam pp. 275–303
Bauduin C (2002) Design of axially loaded compression piles according to eurocode 7. In: Conference on piling and deep foundations. ENPC, Nice (F), vol. 3
Brouwer JJ (2007) In-situ soil testing. Lankelma, East Sussex
Coerts A (1996) Analysis of static cone penetration test data for subsurface modelling: a methodology. Koninklijk Nederlands Aardrijkskundig Genootschap/Faculteit Ruimtelijke Wetenschappen Universiteit Utrecht
De Beer E (1971) Méthode de déduction de la capacité portante d'un pieu a partir des résultats des essais de pénétration. Ann Trav Publics Belg 4:191–268
European Committee for Standardization (2004) Eurocode 7: geotechnical design, Part 1: general rules. BSI
Govaerts A (2004) Geostatistical correlation of geotechnical data (Master Thesis, Geotechnical and Mining Engineering). Master's thesis, K.U. Leuven, Leuven
Govaerts A (2006) Geostatistical interpolation of cpt-data in flanders. In: Proceedings of 17th European young geotechnical engineers' conference; Zagreb, Croatia, July 2006
Isaaks EH, Srivastava RM (1989) An introduction to applied geostatistics. Oxford University Press
Jones AL, Kramer SL, Arduino P (2003) Estimation of uncertainty in geotechnical properties for performance-based earthquake engineering. Pacific Earthquake Engineering Research Center
Journel AG, Huijbregts ChJ (1978) Mining geostatistics. Academic Press San Diego, CA
Lunne T, Powell JJM, Robertson PK (1997) Cone penetration testing in geotechnical practice. Routledge, New York
Schittekat J (2001) Engineering geology of the boom clay in the antwerp area. In: Holeyman A (ed) Screw piles: installation and design in stiff clay Balkema, Lisse, pp 11–18
Western AW, Blöschl G (1999) On the spatial scaling of soil moisture. J Hydrol, 217(3–4):203–224

An Examination of Transformation Techniques to Investigate and Interpret Multivariate Geochemical Data Analysis: Tellus Case Study

Jennifer McKinley and Oy Leuangthong

Abstract This research aims to use the multivariate geochemical dataset, generated by the Tellus project, to investigate the appropriate use of transformation methods to maintain the integrity of geochemical data and inherent constrained behaviour in multivariate relationships. The widely used normal score transform is compared with the use of a stepwise conditional transform technique. The Tellus Project, managed by GSNI and funded by the Department of Enterprise Trade and Development and the EU's Building Sustainable Prosperity Fund, involves the most comprehensive geological mapping project ever undertaken in Northern Ireland. Previous study has demonstrated spatial variability in the Tellus data but geostatistical analysis and interpretation of the datasets requires use of an appropriate methodology that reproduces the inherently complex multivariate relations. Previous investigation of the Tellus geochemical data has included use of Gaussian-based techniques. However, earth science variables are rarely Gaussian, hence transformation of data is integral to the approach. The multivariate geochemical dataset generated by the Tellus project provides an opportunity to investigate the appropriate use of transformation methods, as required for Gaussian-based geostatistical analysis. In particular, the stepwise conditional transform is investigated and developed for the geochemical datasets obtained as part of the Tellus project. The transform is applied to four variables in a bivariate nested fashion due to the limited availability of data. Simulation of these transformed variables is then carried out, along with a corresponding back transformation to original units. Results show that the stepwise transform is successful in reproducing both univariate statistics and the complex bivariate relations exhibited by the data. Greater fidelity to multivariate relationships will improve uncertainty models, which are required for consequent geological, environmental and economic inferences.

J. McKinley (✉)
School of Geography, Archaeology and Palaeoecology, Queen's University, Belfast, BT7 1NN, UK
e-mail: j.mckinley@qub.ac.uk

Oy Leuangthong
Centre for Computational Geostatistics (CCG), Department of Civil and Environmental Engineering, University of Alberta, Alberta, Canada
e-mail: oy.leuangthong@ualberta.ca

P.M. Atkinson and C.D. Lloyd (eds.), *geoENV VII – Geostatistics for Environmental Applications*, Quantitative Geology and Geostatistics 16,
DOI 10.1007/978-90-481-2322-3_21,

1 Introduction

The Tellus Project, managed by the Geological Survey of Northern Ireland (GSNI) and funded by the Department of Enterprise Trade and Development and the EU's Building Sustainable Prosperity Fund, involves the most comprehensive geological mapping project ever undertaken in Northern Ireland. The project comprised the collection of both multi-source airborne geophysics and a ground based geochemical survey of soil and streams. The Tellus geochemical survey involved the collection of soil, stream-sediment and stream water samples in rural and urban areas. Rural soil samples were collected at approximately 2 km^2 intervals. Each soil sample comprises a composite of five augers collected from a depth interval of 5–20 cm. The sampling strategy involved the collection of auger samples from each corner of a square with 20 m sides as well as at the centre. Site location is recorded at the central auger hole (Fig. 1).

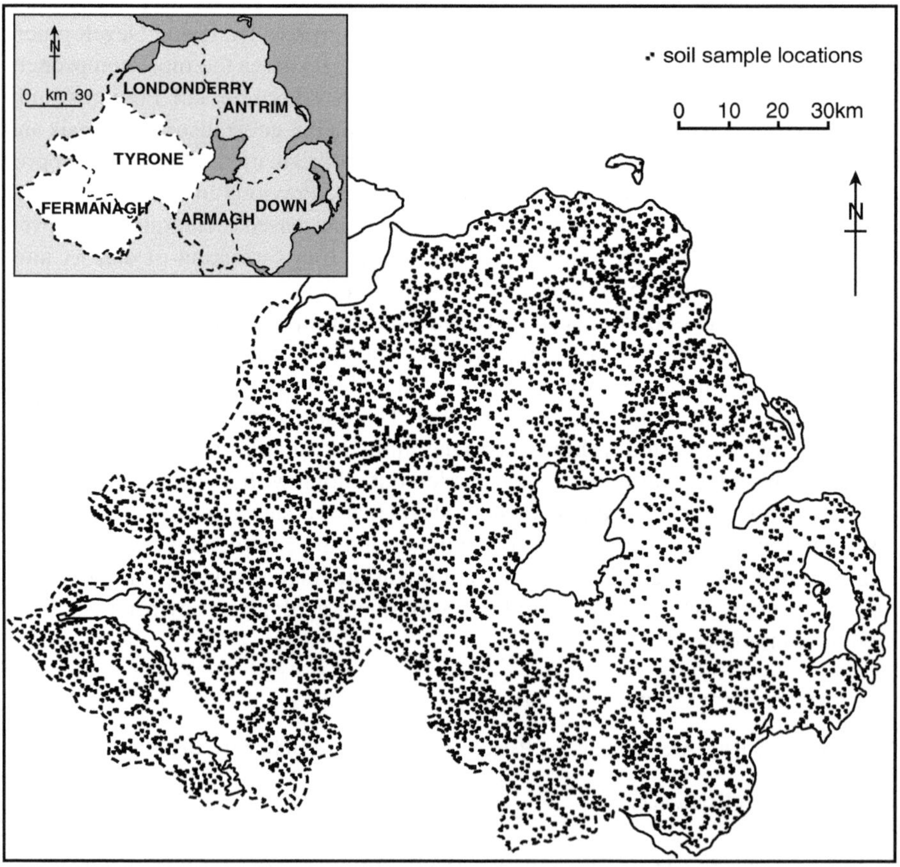

Fig. 1 Location map, counties of Northern Ireland and sample locations for Tellus geochemical survey sampling scheme

1.1 Geological Background

Northern Ireland, despite occupying a limited proportion of the land area represents an almost unparalleled diversity of geology (Mitchell, 2004). The range of rocks presented includes examples of all geological systems up to and including the Palaeogene (comprising basalt lavas and lacustrine sedimentary rocks formed between c. 55 and 62 million years ago). The last 100,000 years of the Northern Ireland's history involves the advance of ice sheets and their meltwaters resulting in a cover of alluvium and peat deposits over at least 80% of bedrock. The economic significance of the Tellus project and the opportunity the multivariate geochemical data offers to decipher and investigate the geological underlay relates to the history of hydrocarbon exploration and mineral prospecting in Northern Ireland. However if geological, environmental and economic inferences are to be made then the integrity of the geochemical data is paramount and manipulation of the soil geochemistry data must honour any inherent geochemical constraints. This study uses the geochemical data to examine the use of transformation methods to maintain the integrity of the data and inherent constrained behaviour in the multivariate relationships. The aim of the research is to enable greater accuracy in the interpretation of the nature of the geochemical variability and consequently any geological, environmental and economic inferences.

2 Previous Research

Previous work (Rawlins et al., 2007; McKinley et al., 2006, 2009 in prep) involved the use of Tellus geochemical data to investigate methods of integrating the geochemical data with the airborne geophysical data to maximise information collected from the ground geochemical survey. The aim of the research was to enable greater interpretation of geological, environmental and economic aspects of Northern Ireland. A Gaussian-based Bayesian updating approach was used by McKinley et al. (2006, 2009 in prep) as a means to improve the resolution of the widely sampled soil geochemistry data by integrating the more closely sampled airborne geophysical data. The advantage of the approach is that multiple variables of different types and different sources (in this case radiometric and soil geochemistry) can be simultaneously integrated and applied to mapping the geochemical variables of economic interest.

2.1 Rationale for the Present Study

The Bayesian updating approach (Deutsch and Zanon, 2004; Ren et al., 2005) used by McKinley et al. (2009 in prep) to improve the resolution of the soil geochemistry data, is a Gaussian-based technique. Hence transformation of data is an integral stage of the approach (normal score transformation was used in this case). Geological data rarely conform to Gaussian behaviour (Leuangthong and Deutsch, 2003), likewise multivariate distributions rarely exhibit Gaussian characteristics such as homoscedasticity and linearity. Common non-Gaussian behaviour for geochemical data is heteroscedasticity, non-linearity and mineralogical constraint. However, Gaussian techniques are often used to represent models of continuous variables. Common practice in geostatistical analysis of multiple-related variables is to transform each variable to a univariate Gaussian distribution one at a time. This ensures each variable is univariate but the multivariate distributions (involving two or more variables at a time) are not explicitly transformed to be multivariate Gaussian and hence does not address the case when the multivariate Gaussian assumption is violated. An alternative transformation technique must be considered.

2.2 The Tellus Geochemical Data

The multivariate geochemical dataset collected by the Tellus project comprise multiple variables that are dependent on each other. This provides an opportunity to investigate the appropriate use of transformation methods such as normal score transform and the stepwise conditional transform (Leuangthong, 2003; Leuangthong and Deutsch, 2003). The transforms need to be implemented with the central aim to maintain the integrity of the geochemical data and honour the inherently constrained behaviour between multiple variables. These relationships often show complex features such as nonlinear relations and/or stoichiometric constraints. This is especially relevant for geochemical data collected as part of the Tellus project and any subsequent geological, environmental and economic inferences. With this in mind, the Clogher Valley area comprising Co. Fermanagh and the southern part of Co. Tyrone, was taken due to the inferred relationship between basement faulting and base metals and renewed interest in mineral prospecting in the area. The Clogher Valley dataset comprised 589 points for seven variables of interest; Cu ppm, Ni ppm, Zn ppm, K2O%, Pb ppm, Co ppm and Cr ppm.

3 Stepwise Conditional Transform (SCT) of Tellus Data

Before any multivariate conditional simulation with dependent variables, we need to understand the univariate distribution of each variable, and any second and higher order relations between the variables. Figure 2 shows the matrix of crossplots illustrating the bivariate relations between the seven variables. Figure 3 shows the same relations as Fig. 2, with the exception that the variables are now normal score transformed. Following a univariate normal score transform, the complex features (e.g. heteroscedasticity, constraints, non-linearity) that are apparent in the original

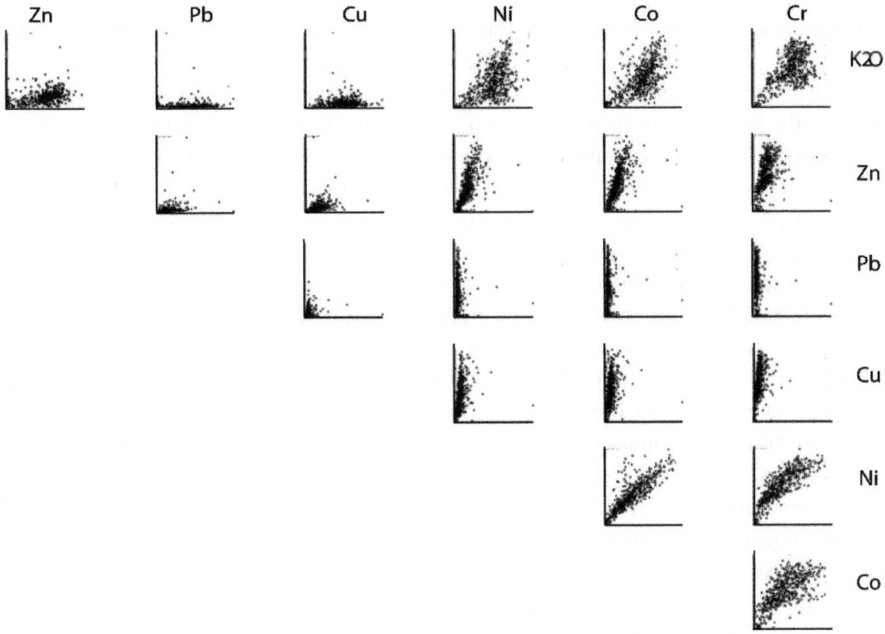

Fig. 2 Crossplot matrix of original variables: Cu, Ni, Zn, K_2O, Pb, Co and Cr

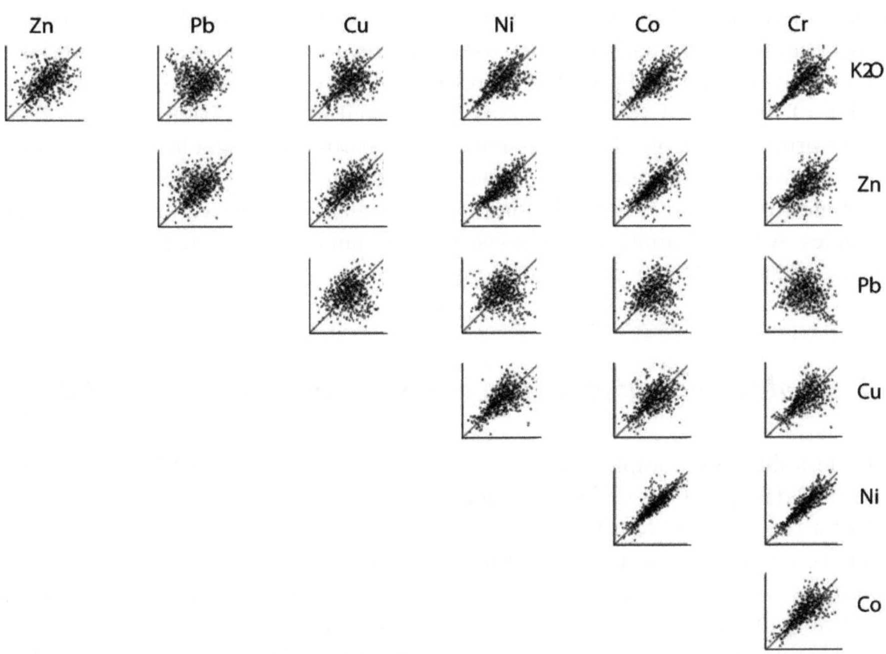

Fig. 3 Crossplot matrix of Normal Score transformed values for Cu, Ni, Zn, K_2O, Pb, Co and Cr

variable crossplots are visibly transferred into Gaussian units; the presence of these relations after Gaussian transform indicates that they may be challenging to reproduce in a conventional Gaussian simulation framework. Normal score transform can usually be effective in mitigating heteroscedastic features but in several of the cross plots it is observed that this clearly not the case. The bivariate distributions are clearly not bivariate Gaussian. This is most evident for crossplots involving K20%, Cr, Cu and Ni. For this reason, an alternative transform is considered for these four variables.

For complex multivariate relations, a number of transformation approaches can be considered. Principal components or factor analysis could be used to generate uncorrelated variables; however, a lack of correlation does not ensure independence. A log ratio transform (Aitchison, 1981, 1999) is another alternative, but is primarily aimed at accounting for compositional data that exhibit constrained behavior. Minimum/maximum autocorrelation factors (Switzer and Green, 1984; Vargas-Guzmán and Dimitrakopoulos, 2003) is yet another technique that is available and is an extension of PCA to a lag $\mathbf{h} \neq 0$ scatterplot. There are many other multivariate transformation approaches available for different purposes; however, in most cases and for those identified here, a transformation to Gaussianity is still required and there is no assurance that even bivariate Gaussianity can be achieved.

The stepwise conditional transform (SCT) was introduced by Rosenblatt (1952) and is described in detail by Leuangthong and Deutsch (2003). The technique applies a quantile transformation technique of observed univariate conditional distributions to standard Gaussian distributions. For the univariate case the SCT technique is identical to the normal score transform. In a bivariate situation, the first variable (Z_1) is transformed using normal scores to yield (Y_1). The normal score transformation of the second variable (Z_2) is conditional to the probability class of the first or primary variable (Y_1). In essence, Z_2 is partitioned into classes conditional to Y_1. A normal score transform is then undertaken for each class of Z_2. For the k-variate case the *kth* variable is conditionally transformed based on the (k-1) first variables. All multivariate distributions are Gaussian in shape at distance lag $\mathbf{h} = 0$. The covariance at $\mathbf{h} > 0$ may not be zero.

3.1 Implementation of SCT in a Nested Fashion for Tellus Data

With less than 600 samples available in the entire dataset for the Clogher Valley area (589 data points), stepwise transformation of four variables will yield poor results for the third and fourth transformed variable due to paucity of information to infer the conditional distributions. Since the transform requires successive conditioning as we increase the number of variables, this effectively means that we are sub setting the data into finer and finer classes, leaving fewer data within each class. For example, in the case of two variables, if we had 100 Cu data points and established that we would subdivide into ten classes, this would mean we had ten Cu data within each class. To define a conditional distribution based on ten data is at

Table 1 Summary of SCT transform orders and corresponding bivariate statistics

SCT order	Primary	Secondary	ρ_{os}	ρ_{ns}	ρ_{sct}
1	K_2O	Cr	0.652	0.543	0.003
2	K_2O	Cu	0.223	0.357	0.000
3	Cr	Ni	0.858	0.879	0.050

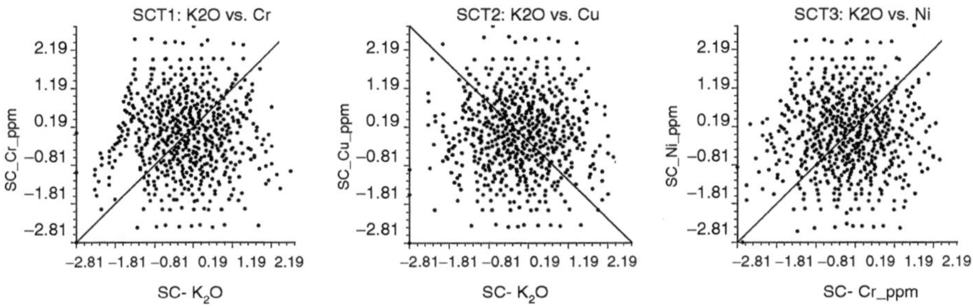

Fig. 4 SCT order 1 (*left*), 2 (*middle*) and 3 (*right*) using ten classes

the very limit of what would be considered reliable. As a result, a nested transform order (Leuangthong et al., 2006) up to two variables is considered (see Table 1). Figure 4 shows the crossplots corresponding to each transform order. Modelling is focussed on the three transformed variables. Note that for the first variable $K_20\%$, SCT K_2O is the same as the normal score (NS) K_2O.

The choice of the primary variable and the ordering of transformation are based on correlation coefficients (Table 1).

For the forward transformation, SCT is run three times with the order shown in Table 1. Following transformation, the variograms corresponding to the four transformed variables are calculated and modelled (see Fig. 5).

The variograms of K_2O and Cr are calculated from the first order transforms, the variogram for Cu from the second order transform and for Ni based on the third order transform. The cross variograms are also calculated to verify that correlation at $\mathbf{h} > 0$ remains relatively small to allow independent modelling to be carried out. Using the SCT values and variograms (based on the orders described in the previous step and Table 1), sequential Gaussian simulation (SGS) is then performed for each of the four SCT variables. Similar to the forward transform, the back transform must also be performed in a stepwise conditional manner.

Using the order outlined in Table 1, the following steps are followed: (a) K_2O and Cr are back transformed first using transform table from order 1; (b) Cu is back transformed based on back transform values of K_20 values given from (a) and the transform table from order 2; and finally Ni is back transformed based on the back transformed values of Cr (from (a)) and transform table from order 3. This way, we avoid multiple values of K20 (although they would have been the same given the order established) and multiple values of Cr, which would be a more critical issue. In total, 100 realizations were generated. For the purposes of comparison, the E-type estimate of the simulations and an arbitrarily chosen realization are plotted

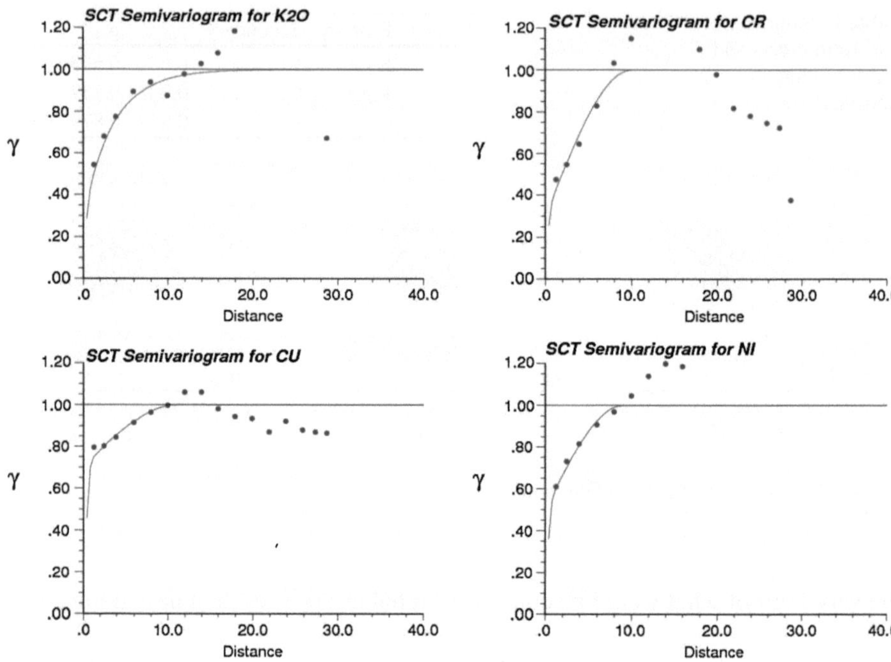

Fig. 5 Variograms corresponding to four SCT variables. A slight trend is apparent in Ni but this was not explicitly modelled in the results shown

in Fig. 6. As expected, the E-type estimate yields a smooth map that is similar to a kriged result while the simulated realization is clearly more variable. Regions of high and low concentrations are easily identifiable in both cases. Moreover the relationship between zones of higher elemental concentration and fault orientation is evident. Following back transformation each model was checked for data reproduction, histogram reproduction, variogram reproduction and multivariate distribution reproduction. Figure 7 shows the reproduction of the bivariate relations (as seen in Fig. 2) following simulation using SCT.

3.2 Data Related Issues with SCT

There are three important issues that need to be addressed with the use of the SCT technique: (1) cross variance for $\mathbf{h} > 0$, (2) the effect of ordering on covariance models, and (3) inference of multivariate distributions with sparse data. (1) There is no guarantee that there is independence beyond $\mathbf{h} = 0$ (i.e. at $\mathbf{h} > 0$). The cross variogram of the transformed variables is checked to ensure that if there is correlation beyond $\mathbf{h} = 0$, that this correlation is relatively negligible (i.e. $\rho(\mathbf{h} > 0) <= 0.2$). If this is not the case then some form of cosimulation may have to be considered.

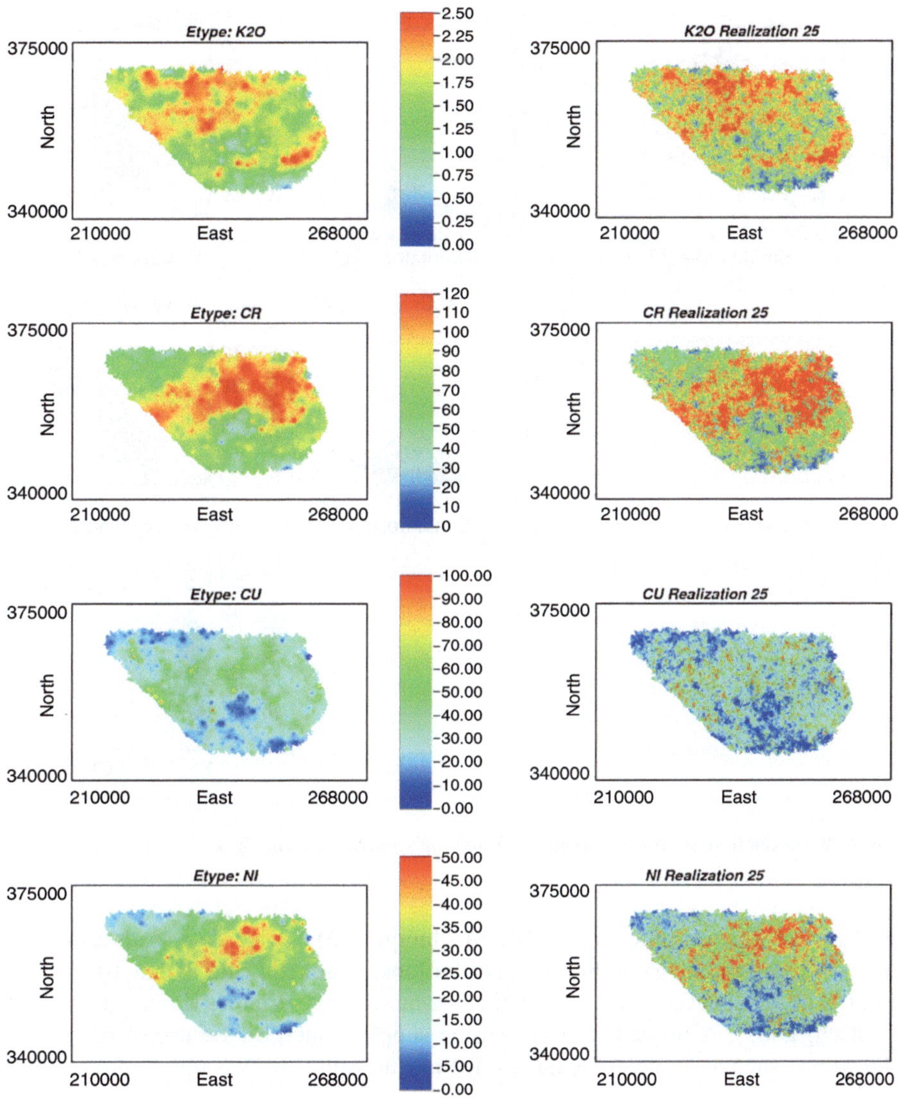

Fig. 6 Comparison between E-type estimate (*left*) and simulated SGS realization (*right*)

However, experience with many data sets (e.g. Leuangthong, 2003; Leuangthong and Deutsch, 2003, 2004; Leuangthong et al., 2006) has shown that this has generally not been required. (2) Leuangthong and Deutsch (2003) found that the effect of transformation ordering was observable in the departure of the variogram of the transformed variable from the original variable. The mismatch can be minimized by choice of the most continuous variable as the primary for the SCT. In the Tellus Clogher Valley data, $K_20\%$ forms the most continuous variable and is

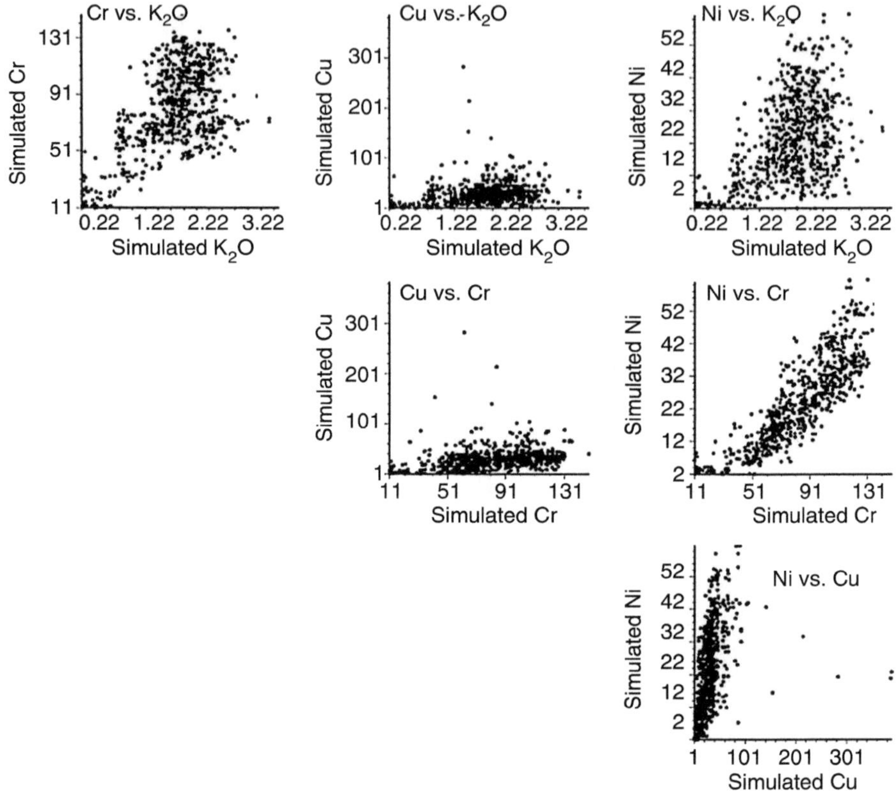

Fig. 7 Reproduction of crossplot features following simulation using SCT

used as the first variable for the SCT technique. (3) Sparse data leads to erratic and non-representative conditional distributions. A general rule is that 10^N to 20^N number of data is acceptable where N is the number of variables (Leuangthong and Deutsch, 2003). A limited data set can be supplemented by the use of smoothing algorithms such as kernel densities to 'fill-in' the gaps; this was not required in this study.

SCT was implemented in a nested fashion for the Tellus Clogher Valley data. In this case a data set totalling less than 600 samples would have yielded poor results for the third and fourth transformed variable due to limited information to infer the conditional distributions. There is an implicit assumption in the implementation of SCT that all data variables are available at all data locations. Therefore the greatest limitation to SCT is non-isotropic sampling. One solution is to transform and simulate the first variable at all locations (Leuangthong and Deutsch, 2003). However, there remains no unique transformed value for the secondary data at all locations of non-isotopic sampling. This was not an issue in the current research.

4 Conclusions

Geostatistical analysis and interpretation of the Tellus geochemical datasets requires use of an appropriate methodology that reproduces the complex multivariate relations that are inherent to the data. SCT is investigated and developed for the geochemical datasets. The transform is shown to reproduce the heteroscedastic, non-linearity and constrained behaviours evident in the data. A nested transform order is considered given the relatively few samples that are available for a multivariate study. These findings are of interest in particular because of the previously recorded relationship between base metals and basin faulting in the Clogher Valley area (Mitchell, 2004; McKinley et al., 2006). The value of the research is reduced uncertainty in modelling of soil geochemistry data, and honouring of inherent geochemical constraints. This will enable more meaningful interpretation of the nature of the geochemical variability in data and consequent geological, environmental and economic inferences. Future work will involve comparison with other transformation methods such the use of the log ratio transform to deal with a greater number of multiple variable and consideration of the constant sum constraint given the geochemical nature of the data.

References

Aitchison J (1981) A new approach to null correlations of proportions. Math Geol 13(2):175–189

Aitchison J (1999) Logratios and natural laws in compositional data analysis. Math Geol 31(5):563–580

Deutsch CV, Zanon S (2004) Direct prediction of reservoir performance with Bayesian updating under a multivariate Gaussian model. Paper presented at the Petroleum Society's 5th Canadian International Petroleum Conference (55th Annual Technical Meeting), Calgary, Alberta, Canada, June 8–10, 2004

Leuangthong O (2003) Stepwise Conditional Transform for Multivariate Geostatistcal Simulation, Ph.D. Thesis, University of Alberta, Edmonton

Leuangthong O and Deutsch CV (2003) Stepwise conditional transformation for simulation of multiple variables. Math Geol 35(2):155–172

Leuangthong O and Deutsch CV (2004) Transformation of residuals to avoid artefacts in geostatistical modelling with a trend. Math. Geol., 36(3), 287–305.

Leuangthong O, Hodson T, Rolley P, Deutsch CV (2006) "Multivariate Simulation of Red Dog Mine, Alaska, USA", CIM Bulletin

McKinley JM, Ruffell A, Lloyd C, van Dam C, Smyth D (2006). Use of geostatistics in the integration of multi-source geophysical and geochemical data in mapping the geological underlay of Northern Ireland. Proceedings of International Congress of the IAMG 2006 Liège, Belgium September 3–8, 2006. (eds) Picard E, Dassargues A and Havenith HB, IAMG

McKinley JM, Ruffell A, Deutsch CV, Neufield C, Young ME (2009 in prep) Use of geostatistics in the integration of multi-source geophysical and geochemical data generated by the Tellus Project, Northern Ireland

Mitchell WI (2004) The Geology of Northern Ireland – Our Natural Foundation, Geological Survey of Northern Ireland, Belfast

Rawlins BG, Lark RM, Webster R (2007) Understanding airborne radiometric survey signals across part of eastern England. Earth Surf Process Landforms 32:1503–1515

Ren W, Leuangthong O, Deutsch CV (2005) Global resource uncertainty using a spatial/multivariate decomposition approach. Paper presented at the Petroleum Society's 6th Canadian International Petroleum Conference (56th Annual Technical Meeting), Calgary, Alberta, Canada, June 7–9
Rosenblatt M (1952) remarks on a multivariate transformation. Ann Math Stat 23(3):470–472
Switzer P, Green AA (1984) Min/Max autocorrelation factors for multivariate spatial imaging. Technical Report No. 6, Department of Statistics, Stanford University, Stanford, CA, 14pp
Vargas-Guzmán JA, Dimitrakopoulos R (2003) Computational properties of min/max autocorrelation factors. Comput Geosci 29:715–723

Shelling in the First World War Increased the Soil Heavy Metal Concentration

Meklit Tariku, Marc Van Meirvenne, and Filip Tack

Abstract A geostatistical analysis of metal concentration data of 2,786 topsoil (0–0.5 m) samples in West-Flanders, Belgium (area approx. 3,100 km^2) revealed a significant increase in the copper (Cu) content over an area of approx. 25 by 25 km around the city of Ypres. On average, the increase in the topsoil within of this area was 6 mg Cu/kg soil which represents several thousand of tons of Cu. Conventional sources of heavy metals, such as metallurgical industry or agricultural could be excluded. The area of Cu enrichment corresponded to the war zone around Ypres of the First World War. Between 1914 and 1918, millions of Cu and lead (Pb) containing shells were fired during several intense battles. Different correlations between several heavy metals were found inside the front zone compared to the rest of the province. Therefore it was concluded that World War I activities were most likely responsible for the overall increased concentrations of Cu, and other heavy metals like Pb, in the topsoil around Ypres. This study illustrates a generally overlooked source of environmental enrichment of heavy metals: historical warfare.

1 Introduction

One of the aims of geochemical surveys is to characterize geochemical background values and to identify areas where concentrations are elevated (Meklit et al., 2008). Therefore, the Flemish Government published official threshold values for background concentrations of a number of heavy metals in soils (Vlaamse Gemeenschap, 1996). Because Flanders, occupying roughly the northern half of Belgium, is mainly covered by quaternary sediments deposited by wind or water, the use of

M. Tariku (✉) and M. Van Meirvenne
Department of Soil Management, Faculty of Bioscience Engineering, Ghent University, Coupure 653, 9000 Gent, Belgium
e-mail: marc.vanmeirvenne@ugent.be

F. Tack
Department of Applied Analytical and Physical Chemistry, Ghent University, Coupure 653, 9000 Gent, Belgium

P.M. Atkinson and C.D. Lloyd (eds.), *geoENV VII – Geostatistics for Environmental Applications*, Quantitative Geology and Geostatistics 16,
DOI 10.1007/978-90-481-2322-3_22,

one reference threshold seemed justified. Although natural causes for increased concentrations can occur in soils weathered in situ, in Flanders concentrations above the official threshold are considered to be man-induced contamination. Industrial activities are the most documented source of elevated concentrations of heavy metals in soils, especially over larger areas (Van Meirvenne and Goovaerts, 2001; Papritz et al., 2005). Other human activities like the application of sludge (Alloway and Jackson, 1991) or animal manure (Payne et al., 1988) can also contribute. However, one generally overlooked source of potential increase of metal concentrations in the fine-earth of soils (i.e. with a particle diameter <2 mm) are historical war activities.

2 Material and Methods

2.1 Study Area

Our study area comprised the entire province of West-Flanders (Fig. 1a), covering an area of 3,144 km^2. The largest part of the province is covered by Pleistocene wind-blown sediments with a sandy-silty texture, besides polders and dunes found along the coastline.

After the initial attack of the First World War (WW I) in August 1914, the western front stabilized along a narrow band running from the North Sea coast at Nieuwpoort into northern France and then eastwards up to the Swiss border in October 1914. This frontline remained largely static during the next 4 years, despite massive attacks by both opponents. One of the locations of intense conflict was the salient of Ypres in the south of West-Flanders. Figure 1b shows the boundaries of the war zone which was delineated by the Belgian Government as "totally destroyed land" after the war (Belgian Law of 15/11/1919; Dendooven, 2006). The city of Ypres itself was never taken by the German army, but it was completely destroyed by artillery fire (see http://www.greatwar.be for an overview of the successive battles). The area between the coast and approximately halfway to Ypres remained rather narrow because this region was kept inundated between 1914 and 1918 and no major attacks were launched in this part during most of the war.

2.2 Data

The data used in the analysis were obtained mainly from the Public Waste Agency of Flanders (Belgium) (OVAM). According to the standard package of analytical determinations for the fine-earth fraction (<2 mm) of soil samples (OVAM, 1997) the total metal analysis is conducted by subjecting 0.5 g of air dry soil to microwave destruction with 6 ml 37% HCl, 2 ml 65% HNO_3 and 2 ml 40% HF (OVAM, 1992; method CMA/2/II/A.3). Analyses were performed either with ICP-AES (OVAM, 1992; method CMA/2/I/B.1) or by graphite furnace atomic absorption

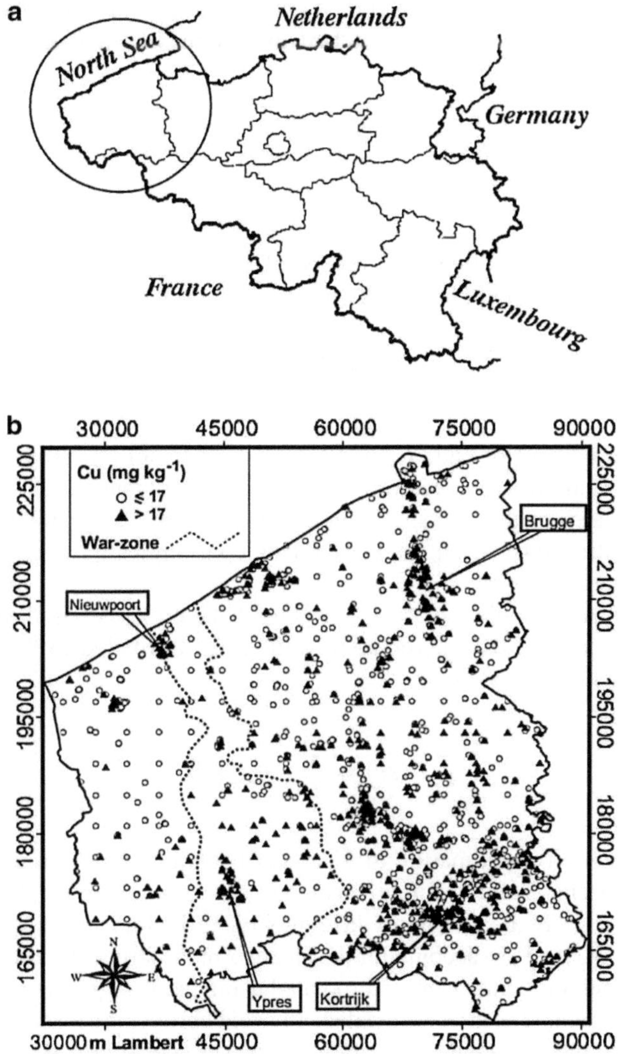

Fig. 1 (**a**) Belgium with identification of W.-Flanders (*circle*); (**b**) W.-Flanders with localization of available topsoil Cu analyses split according to the background threshold value of 17 mg/kg and localisation of the WW I front zone

spectrometry (OVAM, 1992; method CMA/2/I/B.2). Additional data were available from a study conducted in 1998 which aimed at assessing baseline trace element concentrations in Flanders (Tack et al., 2005). This resulted in a data set of 2,786 Cu determinations in topsoil samples (0–0.5 m) taken inside W.-Flanders (Fig. 1b). In those situations where subsamples were provided, a weighted pooled value was calculated to ensure a uniform data support over the 0 to 0.5 m soil depth. Due to missing data, only in a subset of 2,375 locations data were available on several heavy metals (including copper (Cu), lead (Pb) and nickel (Ni)). For Cu the background threshold value is 17 mg/kg for a standard soil (defined as containing 10% of clay and 2% of organic matter).

2.3 Geostatistical Analysis

The variogram $\gamma(\mathbf{h})$ represents the degree of auto-similarity of a variable Z, observed at a number of point locations $z(\mathbf{x}_\alpha)(\alpha = 1, \ldots n)$, in respect to a separation vector $\mathbf{h}$:

$$\gamma(\mathbf{h}) = \frac{1}{2N(\mathbf{h})} \sum_{\alpha=1}^{N(\mathbf{h})} \{z(\mathbf{x}_\alpha + \mathbf{h}) - z(\mathbf{x}_\alpha)\}^2 \tag{1}$$

with $N(\mathbf{h})$ the number of pairs $\mathbf{h}$ apart. The nugget to sill ratio (NSR) reflects the proportion of random errors plus variability at scales less than the shortest sampling distance compared to the overall variance. Because $\gamma(\mathbf{h})$ is sensitive to outliers, it is common to transform strongly skewed data logarithmically, $y(\mathbf{x}) = \ln(z(\mathbf{x}))$, to stabilize the variogram.

Geostatistical interpolation at any unvisited location $\mathbf{x}_0$ is based on solving the kriging system to find the interpolation weights λ_α attributed to the observations within the neighbourhood of $\mathbf{x}_0$:

$$z^*(\mathbf{x}_0) = \sum_{\alpha=1}^{n(\mathbf{x}_0)} \lambda_\alpha z(\mathbf{x}_\alpha) \tag{2}$$

with $z^*(\mathbf{x}_0)$ being the estimation of Z at $\mathbf{x}_0$. The variogram is required to solve the kriging system. Since different variogram structures were identified in different parts of the study area, ordinary kriging with variogram stratification was used (Boucneau et al., 1998) to account for the changes in the structure of the spatial variance. In this method points were interpolated using the variogram of the stratum in which the interpolated point $\mathbf{x}_0$ is located. To avoid abrupt unrealistic discontinuities at the border of the strata the observation points were not stratified.

The observations were logarithmically transformed. So, the estimations $y^*(\mathbf{x}_0)$ had to be back-transformed to original units. Webster and Oliver (2001, p. 180) proposed:

$$z^*(\mathbf{x}_0) = \exp\{y^*(\mathbf{x}_0) + s^2_{\mathrm{OK}_Y}(\mathbf{x}_0)/2 - \psi\} \tag{3}$$

with $s^2_{\mathrm{OK}_Y}(\mathbf{x}_0)$ the ordinary kriging variance of $y^*(\mathbf{x}_0)$ and ψ a Lagrange parameter which is required to include the condition that the sum of the weights equals one into the kriging system. Since Eq. (3) depends on the magnitude of the kriging variance, artificial patterns could be introduced in areas where this parameter varies strongly. Moreover, when data are strongly skewed, the expected value, representing the mean of a distribution, might be less appropriate as a measure of central tendency. Therefore we used a more robust parameter, the median $me_Z(\mathbf{x}_0)$ which was obtained by the simple anti-log operation, on the condition that the distribution was lognormal (Pebesma and Kwaadsteniet, 1997):

$$me_Z(\mathbf{x}_0) = \exp\{y^*(\mathbf{x}_0)\}. \tag{4}$$

The linear correlation between different variables was evaluated by the Pearson correlation coefficient r. But since this parameter is sensitive to outliers the data were first logarithmically transformed. As an alternative the non-parametric rank correlation coefficient r_R was also calculated (Goovaerts, 1997, p. 21).

The spatial correlation between variables was also investigated by an analysis of the coregionalization. This method has been used to detect multivariate spatial correlations between different heavy metals to determine their common sources (Xu and Tao, 2004). In the case of two variables Z_u and Z_v, the method involved fitting a linear model of coregionalization, LMC, to the two autovariograms, $\gamma_u(\mathbf{h})$ and $\gamma_v(\mathbf{h})$, and their cross-variogram $\gamma_{uv}(\mathbf{h})$. The cross-variogram was computed as:

$$\gamma_{uv}(\mathbf{h}) = \frac{1}{2N(\mathbf{h})} \sum_{\alpha=1}^{N(\mathbf{h})} \{z_u(\mathbf{x}_\alpha) - z_u(\mathbf{x}_\alpha + \mathbf{h})\} \{z_v(\mathbf{x}_\alpha) - z_v(\mathbf{x}_\alpha + \mathbf{h})\} \tag{5}$$

with $z_u(\mathbf{x}_\alpha)$ and $z_u(\mathbf{x}_\alpha + \mathbf{h})$ the measured values of Z_u, and $z_v(\mathbf{x}_\alpha)$ and $z_v(\mathbf{x}_\alpha + \mathbf{h})$ for Z_v at $\mathbf{x}_\alpha$ and $x_\alpha + \mathbf{h}$, respectively.

3 Results

3.1 Exploratory Data Analysis

Since we had to rely on samples taken in the frame of soil pollution investigations, a bias to over-sample polluted areas could be expected. A cell-declustering algorithm (Goovaerts, 1997) was used to remove the effects of a preferential sampling in areas with elevated Cu concentrations. The declustering was conducted using cells of 2,400 m, reducing the mean copper concentration from 37.5 to 24.8 mg/kg. Figure 2 shows the histogram of the declustered 2,786 Cu data. The values ranged between 0.2 to 3,600 mg/kg with a median of 12.6 mg/kg, which is about half the mean value. Obviously the distribution is strongly positively skewed with 32.6% of the data exceeding the background threshold of 17 mg/kg and only 1.2% exceeding the Flemish sanitation threshold for agricultural land, which is 200 mg/kg.

Although about one third of the observed Cu data exceeded the background threshold, regional differences occurred, as can be observed on Fig. 1b. To investigate this in more depth, the Cu data were split according to the background threshold. Most samples with Cu $>$ 17 mg/kg were located in the eastern half of the province, especially around larger cities with industrial activities. At some of these locations extremely high Cu contents were measured, but they were typically

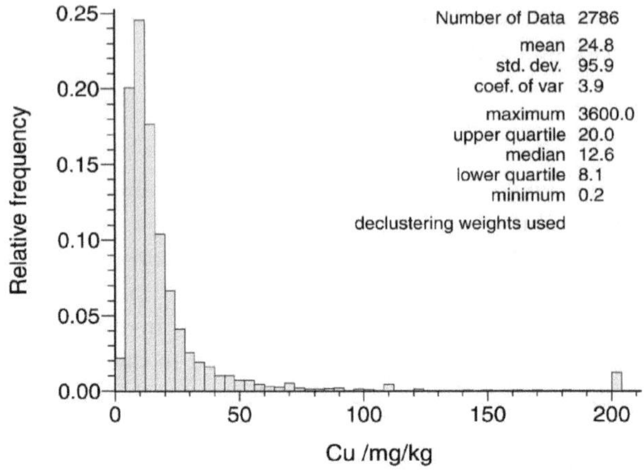

Fig. 2 Histogram and some descriptive statistics of the Cu data in topsoil of W.-Flanders, using declustering weights (last bar groups data exceeding 200 mg/kg)

surrounded by measurement points with Cu concentrations below 17 mg/kg. On the contrary, west of the axis Ypres-Nieuwpoort only a few locations with Cu concentrations exceeded the background threshold. In the area around Ypres, the pattern is different: almost all samples exceeded 17 mg/kg. Therefore we decided to split the province in two zones: one representing the war zone around Ypres, excluding the narrow band towards the sea since in this band war activities were much more limited, and the second being the remaining part of the province. The selected area within the war zone contained 199 data with a median of 18.0 mg Cu/kg, whereas the remaining 2,587 samples had a (declustered) median value of 12.0 mg/kg.

3.2 Mapping the Cu Content

The experimental variograms of the logarithmically transformed Cu observations inside and outside the war zone were obtained through Eq. (1). A spherical model was fitted to them (Fig. 3) which showed clear differences between both zones. Therefore, ordinary lognormal kriging with variogram stratification was used to produce estimations at 500 m intervals. These were back-transformed to estimate the median Cu content according to Eq. (4). Figure 4 shows the result (a similar map produced with block kriging was presented by Van Meirvenne et al., 2008).

The estimates of the median Cu contents were generally below the threshold of 17 mg/kg over most of the province. Locally relatively small patches, usually occurring around a few pixels with strongly increased values reaching occasionally >60 mg Cu/kg, could be observed. Mostly, these patches could be associated with industrial activities around bigger cities or near the harbours. Careful checking

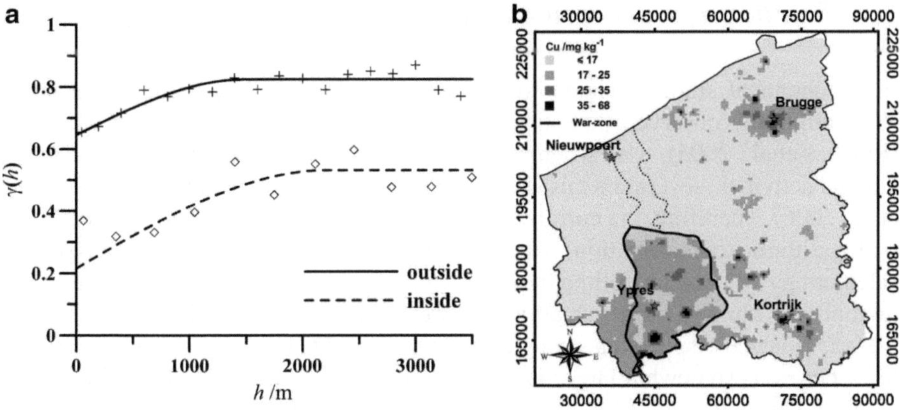

Fig. 3 (**a**) Experimental (*points*) and theoretical (*curves*) variograms of the ln(Cu) data located inside or outside the war zone around Ypres; (**b**) estimations of the median topsoil Cu content obtained with lognormal ordinary kriging with variogram stratification

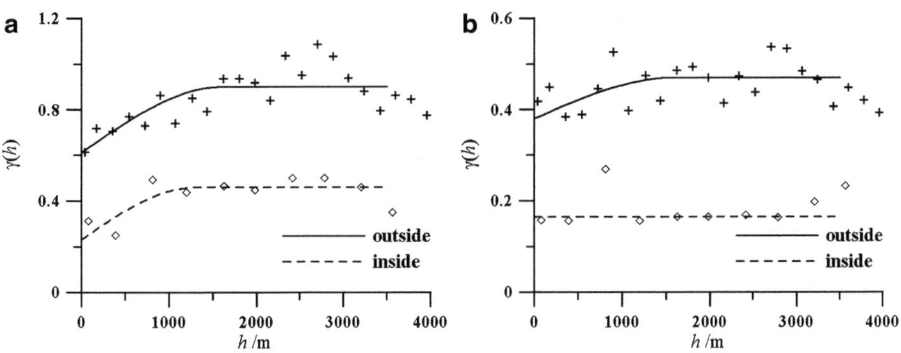

Fig. 4 (**a**) Experimental cross-variograms of lnCu-lnPb and (**b**) lnCu-lnNi for outside and inside the war-zone

confirmed that the patches south of Ypres were indeed related to isolated industrial activities. However, over almost the entire delineated war zone elevated $me_{Cu}(\mathbf{x}_0)$ values were predicted, mainly situated in the range 17–25 mg/kg. Generally this area coincides more or less with the boundaries of the war zone, except in the west where the elevated Cu concentrations extended beyond its limits. This extension could be related to the German Spring Offensive in 1918 during which large parts to the south and southwest of Ypres were captured.

3.3 Relationship Between Cu and Other Heavy Metals

In geochemical studies related to the concentrations of heavy metals in soils, it is commonly observed that strong correlations exist between several elements (Rawlins et al., 2003). Also when the source of heavy metals could be related to industrial activities, several metals show elevated concentrations jointly (e.g. Rawlins et al., 2006). Therefore, the correlation between the heavy metals Cu, Pb and Ni (after logarithmic transformation) was investigated inside and outside the war zone by calculating r, r_R and modelling their coregionalization. Table 1 gives the results, using only the samples on which all metals were determined. Both inside and outside the war zone Cu displayed a strong correlation with Pb. For both zones, r is around 0.74, but r_R is somewhat larger inside than outside the war zone. The relationship between Cu or Pb and Ni is much weaker, especially inside the war zone. These results indicate that: (i) the correlation between Cu and Pb is strong both inside and outside the war zone, but slightly stronger inside; (ii) in both zones the link between Cu or Pb and Ni is weaker than between Cu and Pb; (iii) the relationship between Cu or Pb and Ni is stronger outside the war zone.

The LMC was fitted automatically with the LCMFIT2 computer program (Pardo-Iguzquiza and Dowd, 2002) using a spherical model. The results for Cu-Pb and Cu-Ni are presented in Fig. 4 and Table 2 lists the coefficients for the auto- and the cross-variograms.

The coregionalization analysis showed that Cu, Pb and their cross-variograms displayed a strong structured variability inside the war-zone as confirmed by the lower nugget-to-sill ratio (NSR). Although slightly weaker, outside the war-zone also the correlation between Cu and Pb is spatially structured. Similar results were obtained for the auto-variogram of Zn and its cross-variogram with Cu. With Ni however, although the auto-variogram displayed comparable structures inside as well as outside the war-zone, a structured spatial correlation with Cu was found only outside the war-zone. Inside the war-zone the cross-variogram displayed a pure nugget effect indicating the absence of any spatial similarity in the spatial distribution between Cu and Ni.

Table 1 Pearson (r) and Spearman (r_R – in italic) correlation coefficients between the logarithms of three heavy metals; below diagonal: inside the war zone around Ypres ($n = 160$ – not all metals were determined on all Cu samples), above diagonal: outside the war zone ($n = 2,215$)

	ln Cu	ln Pb	ln Ni
ln Cu		0.742	0.585
		(0.746)	*(0.512)*
ln Pb	0.741		0.518
	(0.791)		*(0.416)*
ln Ni	0.419	0.383	
	(0.424)	*(0.413)*	

Table 2 Coefficients of the auto- variograms of Cu, Pb and Ni and the cross-variograms of Cu × Pb and Cu × Ni both for the inside and outside the war-zone, NSR = nugget-to-sill ratio

		lnCu	lnPb	lnNi	lnCu × lnPb	lnCu × lnNi
	Nugget	0.67	0.82	0.40	0.61	0.38
Outside war-zone	Sill	0.93	1.40	0.51	0.90	0.47
	NSR (%)	73	59	78	68	80
	Nugget	0.25	0.44	0.22	0.23	0.165
Inside war-zone	Sill	0.43	0.82	0.29	0.46	0.165
	NSR (%)	58	36	76	50	100

In general the result of both the correlation and the coregionalization analysis suggested that there could be a difference in the processes causing the enrichment of Cu and Pb inside the war zone compared to the area outside.

4 Discussion

It is unlikely that industrial activities caused a more or less homogenous increase of about 6 mg Cu/kg soil over such a large area. Rawlins et al. (2006) estimated the Pb deposited around a smelter which operated for 53 years to be 2,500 t, which is of the same order as the Cu content found in the topsoil inside the war zone around Ypres. No industrial activities that could have produced a deposition of this magnitude has existed inside or nearby the war zone. An alternative source of Cu could be the application of animal manure (mainly pig slurry) by farmers, as Cu is used as an amendment to pig fodder. De Smet et al. (1996) mapped the phosphate saturation of the soils of this province, and Van Meirvenne et al. (1996) analysed the increase in soil organic carbon content over a time span of some 40 years. Both investigations found strong links between phosphates or organic matter in the soil and the intensive pig breeding due to the use of pig slurry. However, the pattern was not similar to the Cu increase around Ypres. Moreover, pig manure could not explain the correlation between Cu and Pb inside the study area. As textural differences are limited within the war zone, natural pedo-geochemical processes seem to be unlikely as well to have caused such variations in heavy metals.

The possibility that WW I activities could cause soil pollution over a large area seems to have largely missed attention in the environmental literature. Only a few research reports dealt with soil pollution due to WW I activities. Bausinger and Preuss (2005) investigated one site near Ypres which had been used to destroy left-over ammunition after the war and found, besides elevated concentrations of Cu and Pb, also increased amounts of arsenic (As) which was used in chemical warfare to produce nerve gasses. Pirc and Budkovič (1996) reported that Cu and Pb, among other elements, were more or less anomalously high in soils along the Italian-Slovenian WW I front in western Slovenia. Although these studies confirmed the conclusion that war activities could result in elevated concentrations of heavy metals in soil they focussed on only a few locations. No spatial analysis was conducted to evaluate the extent of the environmental impact.

It is estimated that during WW I over 1.45 billion shells were fired by the combined German, French and UK armies. Attacks were usually initiated by massive artillery firing. For example, at the start of the Third Battle of Ypres the British forces fired over four million shells during the 15 days preceding the first infantry attack on July 31, 1917 (Keegan, 2000, p. 361). Every shell contained considerable amounts of Cu, and to a lesser extend Pb and Zn. The body of the shell was made out of iron or steel, but the top fuse, the rotating band and some internal parts were made out of brass, an alloy of about 70% Cu and 30% Zn. The rotating band had to be softer and therefore it was made out of almost pure Cu (containing about 10% Zn) (http://www.madehow.com/Volume-7/Shrapnel-Shell.html on 22/6/2007). In a typical shell the fuse and the rotating band represented about 1 kg (pers. comm. Lt. A. Loncke), so it contained about 0.75 kg of Cu. Upon explosion all parts were deformed or fragmented and spread out although it was a common observation that the brass parts of shells fragmented only partly during explosion. After the war, people searched for these deformed brass pieces since these could be sold. Also, a large scale cleaning up and an overall reconstruction of the area took place.

To evaluate the possibility of a military source of Cu in the soils around Ypres, a simplified mass balance calculation was conducted. It was assumed that Cu was largely immobile and immune to corrosion as soils of this area are not very acid. The increase of the average median Cu concentration inside the war zone around Ypres compared to the rest of the province was 6 mg/kg. With the war zone covering approximately an area of 25 by 25 km, and assuming a soil depth of 0.5 m and an average soil density of 1.5 g/cm then this increase corresponds to 2,813 t of Cu. This 2,813 t of Cu corresponds to 3,750,600 shells if it is assumed that on average a shell contained 0.75 kg Cu. In reality the number of fired shells must have been much larger for the following reasons:

1. A considerable proportion of shells did not explode. Some are still found today. Karg (2005) estimated this proportion to be 10–15%.
2. The Cu concentrations discussed in this paper only refers to the fine-earth fraction of the soil. Particles >2 mm, like pieces of brass, were not included.
3. Only the top 50 cm of the soil profile was considered. As mentioned before, after the war important earth mixing activities took place including deep-soiling.

The exact number of shells fired in the war zone around Ypres during WW I is unknown, but it must have been at least several 10 millions. It will be clear that this order of shells is able to produce a significant increase on a regional scale of the topsoil concentration of Cu and related elements.

5 Conclusions

The geostatistical analysis of the concentrations of heavy metals in the fine-earth fraction of the topsoil of W.-Flanders provided clear indications of a relatively small but significant regional enrichment inside the war zone around Ypres. This enrichment was in the order of 6 mg Cu/kg soil, amounting to approximately 2,800 t of Cu. It could only have been produced by the millions of shells fired during WW I. Other sources, like agricultural amendments or metallurgical industrial activities, were identified as unlikely.

Therefore, we conclude that the WW I activities were most probably responsible for the overall increased concentration of Cu and Pb in the topsoil around Ypres.

Acknowledgment OVAM is gratefully acknowledged for providing the heavy metal dataset of Flanders which was partly used in this study. We also thank Prof. E. Van Ranst for allowing the use of additional metal concentration data in this study. Major B. Vanclooster and Lt A. Loncke are thanked for the details about ammunition used in the first World War.

References

Alloway BJ, Jackson AP (1991) The behaviour of heavy metals in sewage sludge-amended soils. Sci Total Environ 100:151–176

Bausinger T, Preuss J (2005) Environmental remnants of the First World War: soil contamination of a burning ground for arsenical ammunition. Bull Environ Contam Toxicol 74:1045–1052

Boucneau G, Van Meirvenne M, Thas O, Hofman G (1998) Integrating properties of soil map delineations into ordinary kriging. Eur J Soil Sci 49:213–229

Dendooven D (2006) De wederopbouw. In: Chielens P, et al. (eds) De Laatste Getuige. Lannoo, Tielt (in Dutch), pp 97–110

De Smet J, Hofman G, Vanderdeelen J, Van Meirvenne M, Baert L (1996) Phosphate enrichment in the sandy loam soils of West-Flanders, Belgium. Fertilizer Res 43:209–215

Goovaerts P (1997) Geostatistics for natural resources evaluation. Oxford University Press, New York

Karg F (2005) Consideration of toxic metabolites from explosives & chemical warfare agents on polluted military and armament sites for heath risk assessments. In: Uhlmann O, Annokkée G and Arendt F (eds) Consoil 2005 Proceedings, pp. 710–720. Forschungszentrum Karlsruhe

Keegan J (2000) The First World War. Vintage, New York

Meklit T, Van Meirvenne M, Tack F, Verstraete S, Gommeren E, Sevens E (2008) Zinc baseline level and its relationship with soil texture in Flanders, Belgium. In: Soares A, Pereira MJ, Dimitrakopoulos R (eds) geoENV VI – geostatistics for environmental applications. Springer, pp 373–383

OVAM (1992) Compendium voor Monsterneming en Analyse ter uitvoering van het Afvalstoffendecreet en het bodemsaneringsdecreet, Openbare Afvalstoffenmaatschappij voor het Vlaamse Gewest, Mechelen (in Dutch)

OVAM (1997) Oriënterend bodemonderzoek, standaardprocedure. Openbare Afvalstoffenmaatschappij voor het Vlaamse Gewest, Mechelen (in Dutch)

Papritz A, Herzig C, Borer F, Bono R (2005) Modelling the spatial distribution of copper in the soils around a metal smelter in the northwestern Switzerland. In: Renard Ph, Demougeot-Renard H, Froidevaux R (eds) Geostatistics for environmental applications. Springer, Berlin, pp 343–354

Pardo-Iguzquiza E, Dowd PA (2002) FACTOR2D: a computer program for factorial cokriging. Comput Geosci 28:857–875

Payne GG, Martens DC, Kornegay ET, Lindemann MD (1988) Availability and form of copper in three soils following eight annual applications of copper-enriched swine manure. J Environ Qual 17:740–746

Pebesma EJ, de Kwaadsteniet JW (1997) Mapping groundwater quality in the Netherlands. J Hydrol 200:364–386

Pirc S, Budkovič T (1996) Remains of World War I geochemical pollution in the landscape. In: Richardson M (ed) Envionmental xenobiotics. Taylor & Francis, London, pp 375–418

Rawlins BG, Webster R, Lister TR (2003) The influence of parent material on topsoil geochemistry in eastern England. Earth Surf Process Landforms 28:1389–1409

Rawlins BG, Lark RM, Webster R, O'Donnell KE (2006) The use of soil survey data to determine the magnitude and extent of historic metal deposition related to atmospheric smelter emissions across Humberside, UK. Environ Pollut 143:416–426

Tack FMG, Vanhaesebroeck T, Verloo MG, Van Rompaey K, Van Ranst E (2005) Mercury baseline levels in Flemish soils (Belgium). Environ Pollut 134:173–179

Van Meirvenne M, Goovaerts P (2001) Evaluating the probability of exceeding a site specific soil cadmium contamination threshold. Geoderma 102:75–100

Van Meirvenne M, Pannier J, Hofman G, Louwagie G (1996) Regional characterisation of the long-term change in soil organic carbon under intensive agriculture. Soil Use Manag 12:86–94

Van Meirvenne M, Meklit T, Verstraete S, De Boever M, Tack F (2008) Could shelling in the first World War have increased copper concentrations in the soil around Ypres? Eur J Soil Sci 59:372–379

Vlaamse Gemeenschap (1996) Besluit van de Vlaamse regering houdende vaststelling van het Vlaams reglement betreffende de bodemsanering. Belgisch Staatsblad dd. 27.03.1996, pp 7018–7058 (in Dutch)

Webster R, Oliver MA (2001) Geostatistics for environmental scientists. Wiley, Chichester

Xu S, Tao S (2004) Coregionalization analysis of heavy metals in the surface soil of Inner Mongolia. Sci Total Environ 320:73–87

A Geostatistical Analysis of Rubber Tree Growth Characteristics and Soil Physical Attributes

Sidney Rosa Vieira, Luiza Honora Pierre, Célia Regina Grego, Glécio Machado Siqueira, and Jorge Dafonte Dafonte

Abstract The cultivation of rubber trees [*Hevea brasiliensis* (Willd. ex Adr. To Juss.) Müell. Arg.] plays an important role in Brazilian forestry production. However, the relationship between tree production and soil physical attributes is poorly understood. Geostatistical tools such as spatial variability modeling assist the study of the relationships between plant and soil attributes. The objective of this paper is to determine the spatial variability of rubber tree growth characteristics and its relationship to soil–water physical properties (soil mechanical resistance to penetration and field saturated hydraulic conductivity of soil). The experiment was located at Campinas, State of Sao Paulo, Brazil, at a experimental station of the Instituto Agronômico, in a 10 ha area with rubber trees planted in 1992. Samples were taken at 232 points in a 20 × 20 m grid. Average diameter at 1.30 m height and tree height were calculated from average measurements of four trees. The soil physical attributes studied were soil resistance to penetration at 0.40 m depth and field saturated soil conductivity at two depths (0–0.10 m and 0.10–0.20 m). All tree and soil parameters showed moderate to weak spatial dependence among samples. The linear correlation between the attributes of rubber trees and soil was weak. The cross-semivariograms used to evaluate cross-spatial correlations revealed that most of the studied properties did not follow a similar cross-spatial pattern. Spatial variability maps show that areas with higher field saturated hydraulic conductivity of soil have lower soil mechanical resistance to penetration. The field saturated

S.R. Vieira (✉) and L.H. Pierre
Instituto Agronômico (IAC), Av. Barão de Itapura, 1481 CP28, CEP 13020-902, Campinas, SP, Brazil
e-mail: sidney@iac.sp.gov.br

C.R. Grego
EMBRAPA-CNPM, Av. Dr. Júlio Soares de Arruda, 803, Parque São Quirino, 13088-300, Campinas, SP, Brazil
e-mail: crgrego@cnpm.embrapa.br

G.M. Siqueira and J.D. Dafonte
Universidad de Santiago de Compostela (USC), Escuela Politécnica Superior, 27002, 13088-300, Lugo, Spain
e-mail: glecio.machado@rai.usc.es; jorge.dafonte@usc.es

P.M. Atkinson and C.D. Lloyd (eds.), *geoENV VII – Geostatistics for Environmental Applications*, Quantitative Geology and Geostatistics 16,
DOI 10.1007/978-90-481-2322-3_23,

hydraulic conductivity of soil in the 0–10 cm layer showed strong linear and spatial correlation with the diameter of rubber trees, as confirmed by the spatial variability maps of both attributes.

1 Introduction

Rubber trees [*Hevea brasiliensis* (Willd. ex Adr. To Juss.) Müell. Arg.] are native of the Amazon region of Brazil and are currently the main source of natural rubber in the world. Besides the extraction of latex, a number of aspects are contributing to boosting rubber tree production in Brazil, these include factors related to reforestation for soil and water protection (Santos and Mothé, 2007), or to the carbon fixation process, which reduces greenhouse gases (SBS, 2006).

As reported by Vetorazzi and Ferraz (2000), the use of precision forestry improves geospatial data collection and analysis, allowing interventions in forests with sufficient accuracy and precision.

A number of authors (Warrick and Nielsen, 1980; Vieira et al., 1981, 1983; McBratney and Webster, 1986; Rehfeldt et al., 1992; Cambardella et al., 1994; Vetorazzi and Ferraz, 2000; Vieira, 2000; Carvalho et al., 2002; Souza et al., 2004; Siqueira et al., 2008) have shown that the variability of soil properties is spatially dependent, i.e. the difference between the values of a particular property within a certain area can be expressed as a function of the spacing between the sampled points. According to Vieira et al. (1981), the analysis of samples that do not consider the variances calculated and the spacing between samples does not provide a complete description of the variability of a property.

Geostatistics provides information about the spatial structure of the variables and predicts the unknown values and the values of correlated variables. Consequently, geostatistics should be used to determine the spatial dependence of soil properties and the attributes related to the growth and crop production. Thus, data that would be difficult to analyze statistically because of the soil spatial variability can be analyzed more easily using geostatistical tools.

The objective of this study is to evaluate the spatial variability and correlations between tree growth characteristics of *Hevea brasiliensis* and soil physical attributes (soil penetration resistance and field saturated hydraulic conductivity).

2 Materials and Methods

The experimental field was located in a 10 ha area at the *Centro Experimental Central* experiment station, belonging to the Instituto Agronômico (IAC), Campinas, São Paulo, Brazil, at 22°53′S latitude, 47°04′W longitude, with an average elevation of 600 m and a slope of 6.5%.

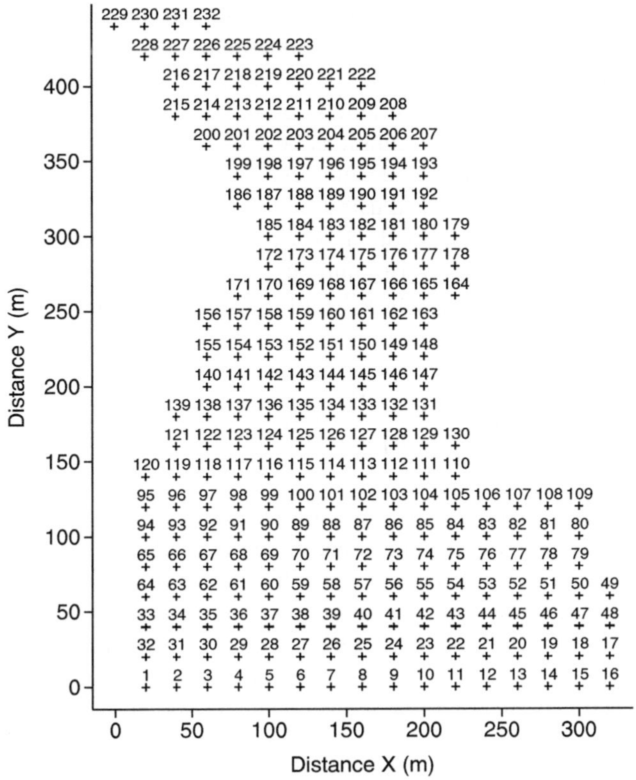

Fig. 1 Grid sampling of tree growth characteristics and soil physical attributes

The soil of the area was classified as a Rhodic Eutrudox (USDA, 1996) and the climate type was Cwa climate according to the Köppen climate classification.

In 1992, 928 trees of *Hevea brasiliensis* were planted using the following clones: IAC 35, PR 261, GT 1, PS 235, RRIM 70, IAN 835, PR 255, RRIM 600 and CR 330. Spacing between trees was 3×7 m.

As shown in Fig. 1, 232 sampling points were located in a 20×20 m grid. Each section contained four trees, named A, B, C and D, which were arranged around the sampling point. The tree growth characteristics evaluated were tree diameter (Diameter, cm) and tree height (Height, m). Diameter at breast height was measured at 1.3 m height. Tree height was determined using a hipsometer. The physical attributes measured were soil mechanical resistance to penetration (RP, MPa) and field saturated hydraulic conductivity of soil (Cond, m/day). Soil resistance to penetration was measured at 0.40 m depth using the STOLF-PLANALSULCAR penetrometer of impact (Stolf et al., 1983), and then calculated for every 0.10 m depth interval (RP_{0-10}, RP_{10-20}, RP_{20-30} and RP_{30-40}). The field saturated hydraulic conductivity of soil was measured using an IAC constant head well permeameter model (Vieira, 1998), while a two constant head well permeameter (3 and 5 cm) was used at two depth intervals: 0.0–0.1 m ($Cond_{0-10}$) and 0.1–0.2 m ($Cond_{10-20}$).

The descriptive statistical parameters (mean, variance, standard deviation, coefficient of variation, minimum value, maximum value, skewness and kurtosis) were obtained in order to verify existence of a central tendency and dispersion of the data using the Stat program (Vieira et al., 1983). Pearson's correlation was used because it often reveals correlations between pairs of variables and helps in the selection of variables for the cokriging estimation (Vieira, 2000).

Spatial variability was analyzed using semivariograms obtained from the Avario software as described in Vieira et al. (1983), and through this the parameters of the models fitted to individual semivariograms and cross-semivariograms were obtained (Vieira, 2000). The semivariogram, $\gamma(\mathbf{h})$, of n spatial observations $z(\mathbf{x}_i), i = 1, n$, can be calculated from Eq. (1):

$$\gamma(\mathbf{h}) = \frac{1}{2N(\mathbf{h})} \sum_{i=1}^{N(\mathbf{h})} [Z(\mathbf{x}_i) - Z(\mathbf{x}_i + \mathbf{h})]^2 \tag{1}$$

where $N(\mathbf{h})$ is the number of pairs of measured values $Z(\mathbf{x}_i)$, $Z(\mathbf{x}_i + \mathbf{h})$, separated by a vector $\mathbf{h}$, which is the distance determined from $Z(\mathbf{x}_i)$ and $Z(\mathbf{x}_i + \mathbf{h})$ coordinates. Calculation of Eq. (1) generates $\gamma(\mathbf{h})$ values corresponding to $\mathbf{h}$ distances for the construction of the semivariogram. According to Vieira (2000), it is expected that measurements located near each other are more similar than measurements separated by great distances, i.e., where $\gamma(\mathbf{h})$ increases with $\mathbf{h}$ until a maximum value is reached at which $\gamma(\mathbf{h})$ stabilizes, at a level that corresponds to the limit distance of spatial dependence, which is the range. Measurements located at distances greater than the range show a random distribution and are therefore independent of each other; beyond such distance, classical statistics can be applied.

Cross-semivariogram analysis was used to determine the spatial cross-correlation between tree growth characteristics of *Hevea brasiliensis* and physical soil attributes (soil penetration resistance and field saturated hydraulic conductivity) (Eq. 2).

$$\gamma(\mathrm{h}) = \frac{1}{2N(\mathbf{h})} \sum_{i=1}^{N(\mathbf{h})} [Z_1(\mathbf{x}_{1i} + \mathbf{h}) - Z_1(\mathbf{x}_{1i})][Z_2(\mathbf{x}_{2j} + \mathbf{h}) - Z_2(\mathbf{x}_{2j})] \tag{2}$$

where $N(\mathbf{h})$ is the number of pairs of measured values Z_1 and Z_2, separated by a vector $\mathbf{h}$.

The spherical model was chosen for fitting to the experimental semivariograms, which allowed for the visualization of the nature of the spatial variation of the variable. The criteria and procedures for fitting the semivariogram models were made according to Vieira et al. (1983). Based on the model used to fit the data, the following semivariogram parameters were defined: (a) nugget effect (C_0), which is the γ value when $\mathbf{h} = 0$; (b) range of the spatial dependence (a), which is the distance beyond which $\gamma(\mathbf{h})$ remains approximately constant, after increasing as h increases; (c) threshold ($C_0 + C_1$), which is the $\gamma(\mathbf{h})$ value beyond the range approaching the data variance, if it exists.

where $N(\mathbf{h})$ is the number of pairs of observations separated by a distance $\mathbf{h}$. Spatial dependence ratio (*SDR*) was calculated using Eq. (3).

$$SDR = \left(\frac{C_0}{C_0 + C_1}\right) * 100 \tag{3}$$

According to Cambardella et al. (1994), SDR represents spatial randomness and can be used to classify spatial dependence as strong if SDR $< 25\%$, moderate if SDR is between 26% and 75% and weak is SDR $> 75\%$.

For the tree growth characteristics and physical properties measured, the semi-variance was shown to be dependent on distance. The variables were interpolated without bias and with minimum variance using the ordinary kriging method with Krige software as described in Vieira et al. (1983) in order to properly build contour maps using Surfer software (Golden Software, 1999).

3 Results and Discussion

Table 1 shows the descriptive statistics for the tree growth characteristics and physical parameters measured. The tree growth characteristics of *Hevea brasiliensis* show low values of the coefficient of variation (CV), while physical attributes show higher CV values, particularly field saturated hydraulic conductivity data, which is in agreement with the classification by Warrick and Nielsen (1980). The CV values for field saturated hydraulic conductivity and soil mechanical resistance to penetration coincide with the values reported by Vieira (1998) and Souza et al. (2004), respectively. The values of the coefficients of skewness and kurtosis suggest that all data show a lognormal frequency distribution, insofar as these parameters are distanced from 0 to 3, according to Carvalho et al. (2002).

The values obtained for Pearson's correlation (Table 2) between *Hevea brasiliensis* tree growth characteristics (Diameter and Height) and field saturated hydraulic

Table 1 Statistical parameters, tree growth characteristics and soil physical attributes (SD = standard deviation; CV = coefficient of variation; Min = minimum; Max = maximum; Skew = skewness coefficient; Kurt = kurtosis coefficient)

Attributes	Mean	SD	CV	Min	Max	Skew	Kurt
Diameter (cm)	19.79	2.39	12.12	9.01	26.52	−0.40	1.80
Height (m)	10.92	1.92	17.61	6.40	16.3	0.46	−0.13
$Cond_{0-10}$ (m/dia)	97.66	68.32	69.96	7.52	330.8	0.97	0.55
$Cond_{10-20}$ (m/dia)	101.7	79.20	77.86	3.75	390.9	1.29	1.39
RP_{0-10} (MPa)	1.183	0.37	31.71	0.59	2.84	1.262	2.08
RP_{10-20} (MPa)	1.988	0.97	49.10	0.89	6.27	1.762	4.00
RP_{20-30} (MPa)	3.764	2.14	57.05	1.11	11.55	1.488	2.12
RP_{30-40} (MPa)	4.933	2.10	42.74	1.46	12.92	1.123	1.783

Table 2 Correlation coefficients for the parameters measured

	Diameter	Height	$Cond_{0-10}$	$Cond_{10-20}$	RP_{0-10}	RP_{10-20}	RP_{20-30}	RP_{30-40}
Diameter	1							
Height	0.528	1						
$Cond_{0-10}$	0.420	0.480	1					
$Cond_{10-20}$	0.413	0.506	0.974	1				
RP_{0-10}	0.157	0.049	0.028	0.032	1			
RP_{10-20}	0.206	0.082	0.061	0.014	0.493	1		
RP_{20-30}	0.178	0.285	0.196	0.187	0.320	0.667	1	
RP_{30-40}	0.132	0.155	0.026	0.033	0.130	0.401	0.613	1

conductivity are intermediate at both depths studied ($Cond_{0-10}$ and $Cond_{10-20}$). Moreover, the values of correlation between rubber tree growth characteristics and soil resistance to penetration at different depths (RP_{0-10}, RP_{10-20}, RP_{20-30} and RP_{30-40}) are very weak. Likewise, the correlation coefficient among soil physical attributes is very weak. It must be emphasized that the intermediate values obtained for the correlation between the parameters of rubber tree and field saturated hydraulic conductivity may represent a greater production of natural rubber insofar as higher field saturated hydraulic conductivity in the upper layers of soil results in higher soil water content at deeper layers, which increases rubber tree productivity. Santos (1982) suggested that the highest production of rubber is obtained in trees with large diameters. Thus, the presence of areas with high values of field saturated hydraulic conductivity in the study area supports the further development and increased productivity of rubber trees.

The spherical model was fitted to the semivariogram for all attributes; this confirms that this model is the most suitable for soil and plant data, as reported by McBratney and Webster (1986), Cambardella et al. (1994), Souza et al. (2004) and Siqueira et al. (2008) (Table 3 and Fig. 2). However, $Cond_{10-20}$ and RP_{0-10} showed a pure nugget effect, or perhaps the spacing used was not sufficient to detect the spatial variability of these attributes.

Field saturated hydraulic conductivity at 10 cm depth ($Cond_{0-10}$) showed the highest nugget effect value (C_0). According to Vieira (2000), the nugget effect accounts for the discontinuity between samples or the variability not detected during sampling. This fact is mainly due to the great variability of the data, as shown by the standard deviation (68.32) and coefficient of variation (69.96%). The other attributes showed low nugget effect values.

The values obtained for tree growth parameters (diameter and height) were the lowest values of range of the spatial dependence (a), with 56.74 and 91.78 m respectively. The physical attributes pertaining to soil water showed larger range values, between 150 and 234.71 m. Cambardella et al. (1994) described the spatial dependence ratio (SDR) for the attributes involved in this study as moderate to low.

Vieira (2000) suggested that when two variables are correlated spatially, their cross-semivariogram must reach its sill near the value of covariance. Accordingly, it appears that the height and diameter of rubber trees are close to its present value

Table 3 Fitted semivariogram models for the parameters measured

Attributes	Model	C_0	C_1	a	SDR
Diameter (cm)	Spherical	4.25	1.50	56.74	26.11
Height (m)	Spherical	1.07	2.38	91.98	69.02
$Cond_{0-10}$ (m/dia)	Spherical	4,300	1,700	150.00	28.33
$Cond_{10-20}$ (m/dia)		Pure nugget effect			
RP_{0-10} (MPa)		Pure nugget effect			
RP_{10-20} (MPa)	Spherical	0.73	0.33	230.47	31.08
RP_{20-30} (MPa)	Spherical	2.81	2.56	234.71	47.70
RP_{30-40} (MPa)	Spherical	2.60	2.00	150.00	43.48

C_0: nugget effect; C_1: structural variance; a: range; SDR: spatial dependence ratio.

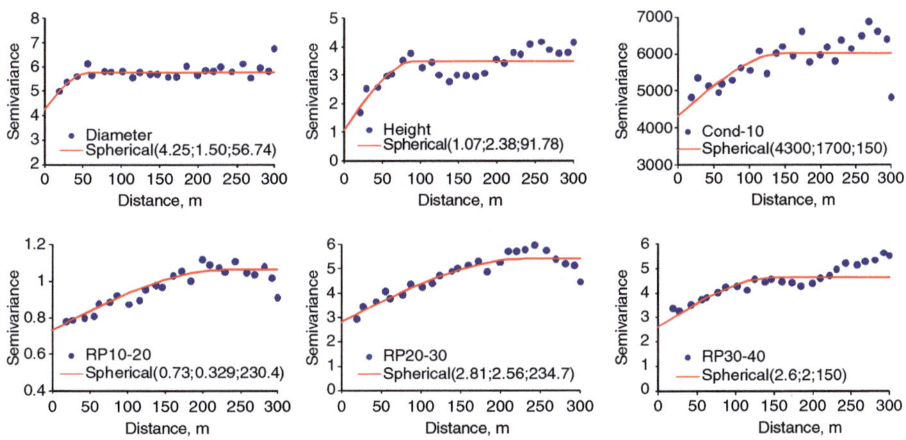

Fig. 2 Experimental and fitted model semivariograms of the studied variables

of covariance (Fig. 3). The cross-semivariogram between rubber tree growth parameters (diameter and height) and field saturated hydraulic conductivity of soil for the 0–10 cm layer shows drift. The cross-semivariogram between Diameter and $Cond_{0-10}$ shows inverse spatial correlation, while the cross-semivariogram between tree diameter and soil resistance to penetration (RP_{10-20}, RP_{20-30} and RP_{30-40}) shows spatial cross-correlation.

The maps of tree diameter and height (Fig. 4) are not similar, and show different spatial patterns. The distribution of contour lines for the field saturated hydraulic conductivity of soil at the 0–10 cm layer shows wide variation. Vieira et al. (1981) and Rehfeldt et al. (1992) found high spatial variability of the saturated hydraulic conductivity of soil for a floodplain, and attributed such variability to soil heterogeneity factors.

The maps of soil mechanical resistance to penetration confirm the increase in average values with depth, as shown in Table 1. The maps of soil penetration resistance look similar. Likewise, the cross-semivariogram (Fig. 3) reveals a similar spatial pattern for soil mechanical resistance to penetration in the different layers.

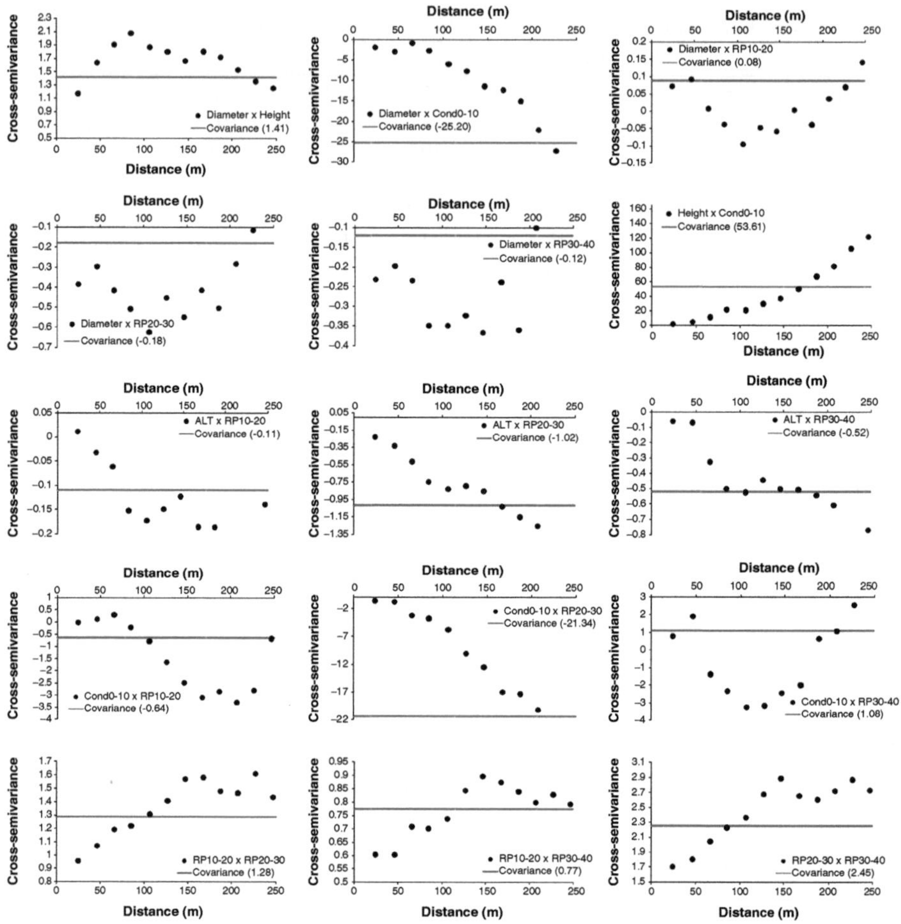

Fig. 3 Cross-semivariograms for the studied variables

The cross-semivariogram of the soil physical attributes reveals a clear cross-spatial structure for $Cond_{0-10} \times RP_{10-20}$ and $Cond_{0-10} \times RP_{30-40}$ (Fig. 3). However, there is a strong inverse cross-spatial correlation between the $Cond_{0-10} \times RP_{20-30}$.

According to Abrams et al. (1992) and Mesquita et al. (2006), the plant water status stands out among the many factors that influence the production of natural rubber. The plant water status results from the interaction of other factors (evaporative demand of the atmosphere, soil water content, density of planting, cultivation system and physiological processes). In this respect, Devakumar et al. (1988) suggested that the dry periods induce physiological changes in rubber trees, which results in lower productivity. Thus, areas with higher field saturated hydraulic conductivity and lower soil resistance to penetration favor the development of plants with greater heights and diameters, which increases production of natural rubber, as reported by Santos (1982).

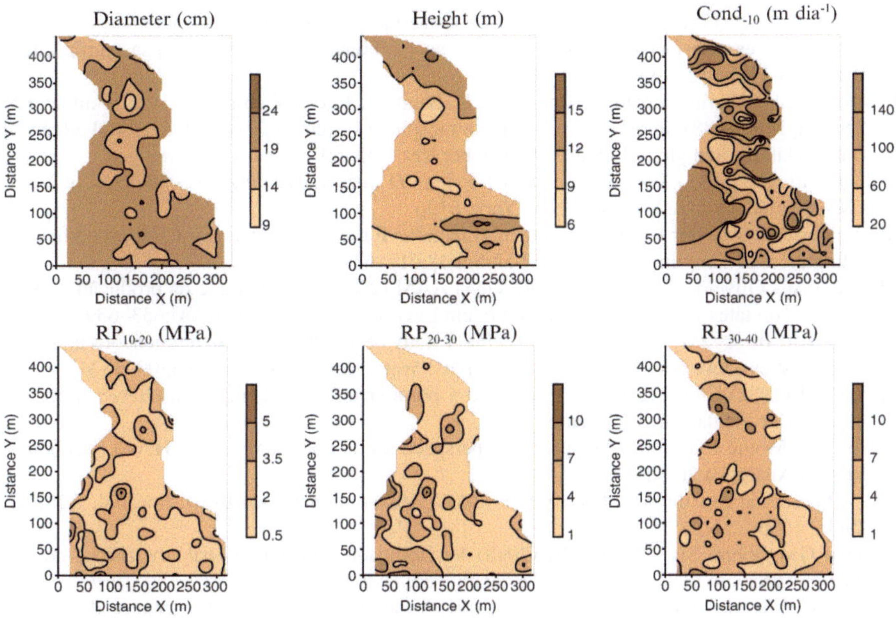

Fig. 4 Maps of the estimated values obtained using ordinary kriging

4 Conclusions

The soil-water parameters studied showed higher CV values than the tree growth parameters of *Hevea brasiliensis*. The spatial dependence ratio was weak to moderate for all the studied properties. There was a cross-spatial pattern between tree diameter and field saturated hydraulic conductivity. According to this pattern, the areas with high field saturated hydraulic conductivity show the highest values of diameter and height, such that this soil physical property (field saturated hydraulic conductivity) can be used as an indicator of rubber tree production.

Acknowledgements The authors are grateful to the *Ministerio de Asuntos Exteriores y de Cooperación* (MAEC-AECID) from Spain for the granting of scholarships for PhD studies. This work has been funded by *Ministerio de Educación y Ciencia*, within the framework of research project CGL2005-08219-C02-022, co-funded by *Xunta de Galicia*, within the framework of research project PGIDIT06PXIC291062PN and by the European Regional Development Fund (ERDF).

References

Abrams MD, Kloeppel BD, Kubiske ME (1992) Ecophysiological and morphological responses to shade and drought in two contrasting ecotype of Pinnus serotina. Tree Physiol 10(4):343–355

Cambardella CA, Mooman TB, Novak JM, Parkin TB, Karlem DL, Turvo RF, Konopa AE (1994) Field scale variability of soil properties in central Iowa soil. Soil Sci Amer J 47:1501–1511

Carvalho JRP, Silveira PM, Vieira S (2002) Geoestatística na determinação da variabilidade espacial de características químicas do solo sob diferentes preparos. Pesquisa Agropecuária Brasileira 37(8):1151–1159

Devakumar AS, Gururaja Rao G, Rajagopal R, Sanjeeva Rao P, George MJ, Vijayakumar KR, Sethuraj MR (1988) Studies on soil–plant–atmosphere system in *Hevea*: II. Seasonal effects on water relations and yield. Indian J Nat Rubber Res 1:45–60

Golden Software (1999) *Surfer 7.0. contouring and 3D surface mapping for scientist's engineers.* User's Guide. Golden Software, New York, 619p

McBratney AB, Webster R (1986) Choosing functions for semi-variograms of soil properties and fitting them to sampling estimates. J Soil Sci 37:617–639

Mesquita AC, Oliveira LEM, Cairo PAR, Viana AAM (2006) Sazonalidade da produção e características do látex de clones de seringueira em Lavras, MG. Bragantia 65(4):633–639

Rehfeldt KR, Boggs JM, Gelhar LW (1992) Field study of dispersion in a heterogeneous aquifer 3. Geostatistics analysis of hydraulic conductivity. Water Resour Res 28(12):3309–3324

Santos PM (1982) Efeito da interação enxerto x porta-enxerto em seringueira (*Hevea spp.*) MS. Thesis, Universidade de São Paulo

Santos GR, Mothé CG (2007) Prospecção e perspectivas da borracha natural, Hevea brasiliensis. Revista Analytica 26:32–41

SBS – Sociedade Brasileira de Silvicultura (2006) Fatos e números do Brasil florestal. SBS, São Paulo, 106p

Siqueira GM, Vieira SR, Ceddia MB (2008) Variabilidade espacial de atributos físicos do solo determinados pro métodos diversos. Bragantia 67(1):203–211

Souza ZM, Marques Júnior J, Pereira GT (2004) Variabilidade espacial de atributos do solo em diferentes formas do relevo sob cultivo de cana-de-açúcar. Revista Brasileira de Ciência do Solo 28:937–944

Stolf R, Fernandes J, Furlani Neto VL (1983) Penetrômetro de impacto IAA/PLANALSUCARS tolf: recomendação para seu uso. STAB 1(3):18–23

USDA – United States Department of Agriculture (1996) Keys to soil taxonomy, 7th edn. USDA, Washington, 644p

Vetorazzi CA, Ferraz SFB (2000) Silvicultura de precisão: uma nova perspectiva para o gerenciamento de atividades florestais. In: Borém A, Queiroz DM (eds) Agricultura de Precisão. Viçosa, pp 65–75

Vieira SR (1998) Permeâmetro: novo aliado na avaliação de manejo do solo. O Agronômico 47–50:125

Vieira SR (2000) Geoestatística em estudos de variabilidade espacial do solo. In: Novais RF, Alvarez VH, Schaefer GR (eds) Tópicos em Ciência do solo. Sociedade Brasileira de Ciência do Solo Viçosa, vol 1, pp 1–54

Vieira SR, Nielsen DR, Biggar JW (1981) Spatial variability of field-measured infiltration rate. Soil Sci Soc Amer J 45:1040–1048

Vieira SR, Hatfield JL, Nielsen DR, Biggar JW (1983) Geostatistical theory and application to variability of some agronomical properties. Hilgardia 51(3):1–75

Warrick AW, Nielsen DR (1980) Spatial variability of soil physical properties in the field. In: Hillel D (ed) Applications of soil physics. Academic Press, New York

Investigating the Potential of Area-to-Area and Area-to-Point Kriging for Defining Management Zones for Precision Farming of Cranberries

Ruth Kerry, Daniel Giménez, Peter Oudemans, and Pierre Goovaerts

Abstract Cranberries are harvested by flooding the field and agitating vines so the fruit, which float can be skimmed from the surface and loaded into barrels. This harvesting method makes application of standard precision farming practices difficult. This paper investigates the potential of combining Area-to-Area (AtoA) and Area-to-Point (AtoP) kriging of yield totals from individual fields with remotely sensed data for defining within-field management zones.

1 Introduction

Cranberry (*Vaccinium macrocarpon* Ait.) is a high value intensively managed perennial crop that grows on wetlands. Given strict federal guidelines that prohibit the expansion of cranberry acreage on wetlands, increasing profitability of cranberry production is most likely to be achieved by precision management (Pozdnyakova et al., 2002, 2005). Perennial crops like cranberry seem ideal for precision management as they often develop patterns of yield variability that are relatively stable in time in response to spatial variation in disease and soil properties. Over the lifespan of the cranberry plant, patterns in external factors can result in genotypic heterogeneity (Novy et al., 1996). Various approaches have been used to classify fields into

R. Kerry (✉) and D. Giménez
Department of Geography, Brigham Young University, Provo, UT, USA
and
CRSSA, Rutgers, The State University of NJ, 14 College Farm Road, New Brunswick, NJ, USA
e-mail: ruth_kerry@byu.edu; gimenez@envsci.rutgers.edu

P. Oudemans
Department of Plant Biology & Pathology, Rutgers, The State University of NJ, 59 Dudley Road, New Brunswick, NJ, USA
e-mail: oudemans@AESOP.Rutgers.edu

P. Goovaerts
Biomedware, Inc., 516 North State Street, Ann Arbor, MI, USA
e-mail: goovaerts@terraseer.com

P.M. Atkinson and C.D. Lloyd (eds.), *geoENV VII – Geostatistics for Environmental Applications*, Quantitative Geology and Geostatistics 16,
DOI 10.1007/978-90-481-2322-3_24,

management zones for combinable crops (Lark and Stafford, 1996; Grenzdörffer and Gebbers, 2001; Khosla et al., 2008). These usually recognize that the temporal variability in yield can be greater than the spatial variability, so temporal variability should not be ignored.

Most precision farming studies begin by characterizing the within-field variation in yield, then look at potential causes for that variation, such as soil type and prevalence of disease (Johnston et al., 1998). Yield mapping for combinable crops using weight or volume sensors associated with a differential global positioning system (DGPS) on the combine (Auernhammer et al., 1993; Blackmore and Moore, 1999) is now quite commonplace. In contrast, the cranberry crop, presents a particular challenge to within-field yield characterization because it is harvested by flooding the cranberry bog and agitating the vines to loosen the fruit. The fruit, which float, are then collected from the surface and packaged into barrels, giving one total yield value per field.

To characterize within-field variability of cranberry yield, remotely sensed images have been used (Hughes et al., 1998; Oudemans et al., 2002). In an intensive study of one field Pozdnyakova et al. (2002) showed that remotely sensed imagery indicated patterns of within-field variation in cranberry yield, infiltration rate and vine density. Intense ground surveys are not, however, a practical or economic way forward for characterizing cranberry yield within fields. Pozdnyakova et al. (2005) conducted a spatial analysis of cranberry yield at three scales but noted that differences in sampling support, which were not explicitly taken into account, affected the yield distribution statistics more than the spatial scale of the measurements.

To address these problems, yield values for cranberry fields that were not sampled in a given year can be estimated using Area-to-Area (AtoA) kriging which incorporates the size and shape of the fields in variogram deconvolution and kriging. Area-to-Point kriging (AtoP) (Kyriakidis, 2004) uses irregularly sized and shaped areal data to make predictions to a point support, creating surfaces that depict smooth trends in cranberry yield and, thus, inform on within-field variation. Estimates from AtoP kriging are coherent in that the average of yield values from all points within an areal unit returns the original value for that areal unit (Kyriakidis, 2004). Here, we present a preliminary study of AtoA and AtoP kriged surfaces of cranberry yield (1991–2004) from about 700 cranberry beds in a region of southern NJ. Surfaces from years with similar weather were combined with a vegetation index to classify the region into management zones at a regional and at a field scale.

2 Methods

2.1 Yield Data

Yield data for over 700 cranberry fields in the Chatsworth area of southern NJ were obtained for a 14 year period. The average field area is 16,830 m^2 but size and shape of fields varies considerably. There is one yield value per field expressed in Mg/ha, but several of these values were missing for any given year, particularly in 1991 and 1992. There are several different cultivars of cranberry grown in the area, but the three main varieties are Ben Lear (BL), Stevens (ST) and Early Black (EB). For the former two cultivars a yield of about 34 Mg/ha is considered good whereas for EB, the most common cultivar, 26 Mg/ha is reasonable. With good management, however, yields can reach as high as 55 Mg/ha (Pozdnyakova et al., 2002).

Information on the crops from individual fields was provided by a growers cooperative (Ocean Spray, Lakeville-Middleboro, MA). Yield values for each year were standardized to zero mean and unit variance so that direct comparisons between years could be made using a single scale (Figs. 3 and 4).

Within-field yield variations were estimated as part of an intensive survey of one cranberry bed, the Nadine bed, on an unaligned survey grid with an interval of 20 m. Berries were counted at each location using a 30.5 × 30.5 cm^2 frame (two replications per location) and approximated into Mg/ha for subsequent analysis (Pozdnyakova et al., 2002).

2.2 *Weather Data*

Weather information was obtained for the Indian Mills, NJ station which is about 16 km from the main cranberry growing area of Chatsworth, NJ. Monthly 30 year normals (1971–2000) and actual monthly, minimum, mean and maximum temperatures and precipitation totals for the period between 1991 and 2004 were obtained from http://climate.rutgers.edu/stateclim_v1/monthlydata/index.html. Similar statistics were derived for the cranberry growing season that lasts from April to September.

2.3 *Aerial Imagery*

Geocorrected colour-infra red images from July 2002 and 2004 with ground pixel sizes of 7 and 4 m were obtained from the Ikonos (Space Imaging, Inc.) and Quickbird (Digital Globe, Inc.) satellites, respectively. The Enhanced Vegetation Index (EVI) was calculated for each pixel and an average EVI value for all pixels within a given field was calculated and assigned to the coordinates of the field centroid. The EVI is useful for detecting differences in the canopy structure including leaf area index. The EVI was developed to increase the sensitivity of the vegetation signal over the normalized difference vegetation index (NDVI) in high biomass regions and reduce atmospheric influences (see Huete et al., 2002 for details of the EVI). The images were sub-sampled (e.g. every fifth pixel was extracted) and kriged to a 20 m grid to improve computational manageability and smooth some small scale variability that was not of interest.

2.4 Geostatistical Methods

Following exploratory data analysis, variograms of yield (areal) data were calculated using 25 lags of 100 m. The corresponding point-support variogram models were then inferred using an iterative deconvolution procedure that seeks the point-support model that, once regularized, is the closest to the model fitted to the areal data (Goovaerts, 2008). The model was used to estimate the yield and the associated kriging variance for the unit X using K neighbouring field data:

$$\hat{z}(X) = \sum_{i=1}^{K} \lambda_i z(v_i) \quad \sigma^2(X) = C(0) - \sum_{i=1}^{K} \lambda_i \overline{C}(v_i, X) - \mu(X) \tag{1}$$

where the unit X represents either an area (i.e. field) v_α (AtoA kriging) or a point u_s within that area (AtoP kriging). The kriging weights and the Lagrange parameter $\mu(X)$ are computed by solving the following system of equations:

$$\begin{aligned} &\sum_{j=1}^{K} \lambda_j \overline{C}(v_i, v_j) + \mu(X) = \overline{C}(v_i, X) \quad \mathrm{i} = 1, ..., K \\ &\sum_{j=1}^{K} \lambda_j = 1. \end{aligned} \tag{2}$$

The area-to-area covariances $\overline{C}(v_i, v_j)$ and area-to-point covariances $\overline{C}(v_i, X = \mathbf{u}_s)$ are approximated as the average of the point-support covariance $C(\mathrm{h})$ computed between any two locations discretizing the areas v_i and v_j, or v_i and $\mathbf{u}_\mathrm{s}$. By construction, aggregating the AtoP kriging estimates within each area yields the AtoA kriged map, as long as the same K areal data are used for both types of kriging.

2.5 Statistical Methods

Principal components analysis (PCA) was conducted in GenStat (Payne, 2006) using the average mean, minimum and maximum growing season temperature, growing season precipitation total, mean yield and average yield values of the cultivars Ben Lear, Stevens and Early Black for each year. Groupings of years were interpreted in relation to weather data for the 14 year period. GenStat was also used for non-hierarchical classification of the AtoA and AtoP kriged yield along with aggregated and kriged EVI data, respectively. First, observations are ordered and assigned to a user-specified number of groups of equal proportion. An iterative procedure is then used to transfer observations between-groups until the between-groups sum of squares can no longer be increased. This criterion amounts to minimizing the trace of the pooled within-class dispersion matrix (Payne, 2006).

Non-hierarchical classifications with 1–10 groups were performed and the best number of groups for classification was determined from marked deviations in the trend line of plots of the between-groups sum of squares criterion value for different numbers of classes.

3 Results and Discussion

3.1 Weather Data

The first two PCs accounted for 67% of variation in weather and by cultivar yield data. Figure 1a shows that precipitation is plotted near to the origin of the principal component plot suggesting it has little influence on yield. Also the variation in precipitation totals over the study period (Fig. 2b) does not explain any of the groupings of years in Fig. 1b. This result is expected since the cranberry crop is irrigated. The main factor for the grouping of years identified in Fig. 1b is mean temperature (Fig. 2a). Three years had temperatures higher than normal (1998, 2002, 2004) and for 2 years they were lower than normal (1992, 1997). There were two groups of years with average temperatures. Those with higher yield (2003, 1999, 2001) had low temperatures in the previous winter (Fig. 1, ii) suggesting that lower temperatures may have restricted the incidence of pests on yield. Those with lower yield

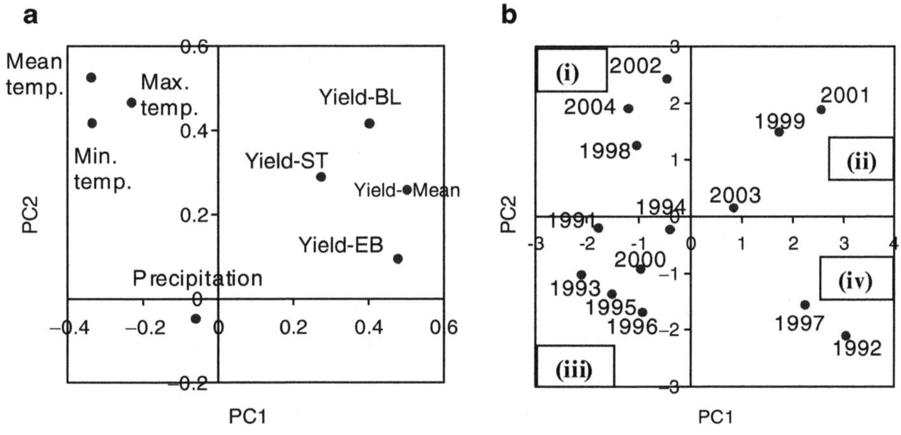

Fig. 1 Scatterplots for principal components 1 and 2 of (**a**) Latent vector loadings for weather variables, mean yield and yield of Ben Lear (BL), Stevens (ST) and Early Black (EB) cultivars and (**b**) principal component scores for different years with: (i) higher than average growing season temperatures, (ii) average growing season temperatures after a colder than average winter, (iii) average growing season temperatures after a warmer than average winter and (iv) lower than average growing season temperatures

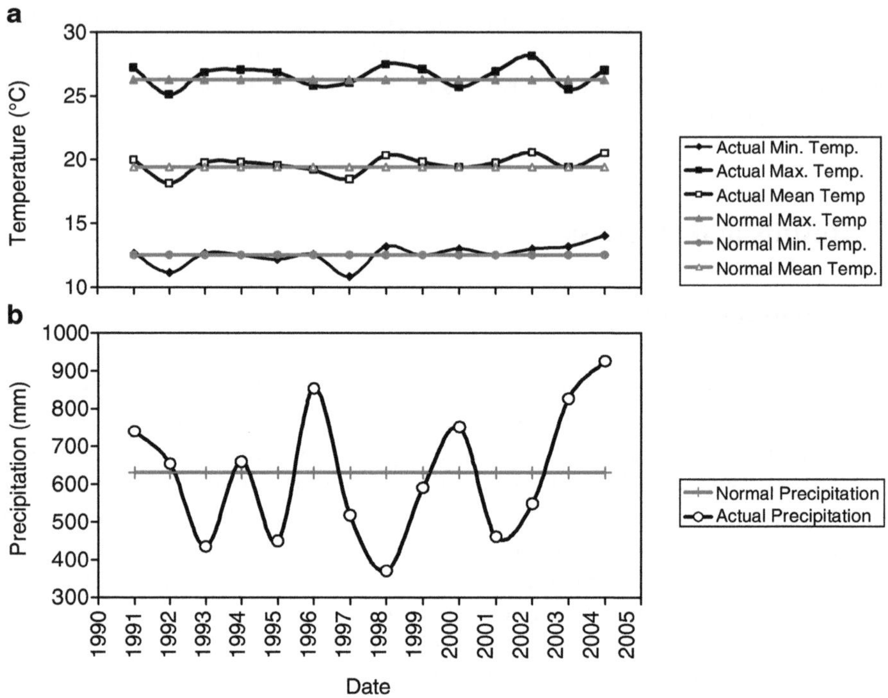

Fig. 2 Thirty year normal (1971–2000) and actual growing season (**a**) average minimum, mean and maximum temperatures and (**b**) precipitation totals for Indian Mills, NJ

(1991, 1993, 1994, 1995, 1996, 2000) had average/warm winter temperatures prior to the growing season (Fig. 1, iii). Years with growing season temperatures higher and lower than normal are used in subsequent analysis.

3.2 *Yield Data*

The standardized AtoA kriged yield data for various years are shown in Fig. 3. The maps show only the 308 fields at the centre of the growing area in Chatsworth, NJ. There are more similarities in the patterns of standardized yield between years with similar weather conditions than years with different weather conditions; however, there are some areas that are consistently high yielding in all years. These tend to be beds with a large perimeter:area ratio or those that have drainage ditches that run through the centre of the field. Pozdnyakova (2001, unpublished) noted that the perimeter:area ratio of the beds influences yield because each bog is surrounded by drainage ditches so those with a larger perimeter:area ratio have better control over drainage. Indeed, Kruskall Wallis H tests for 1991–1993 based on four groups of perimeter:area showed differences in yield at levels of significance of 0.108, 0.008

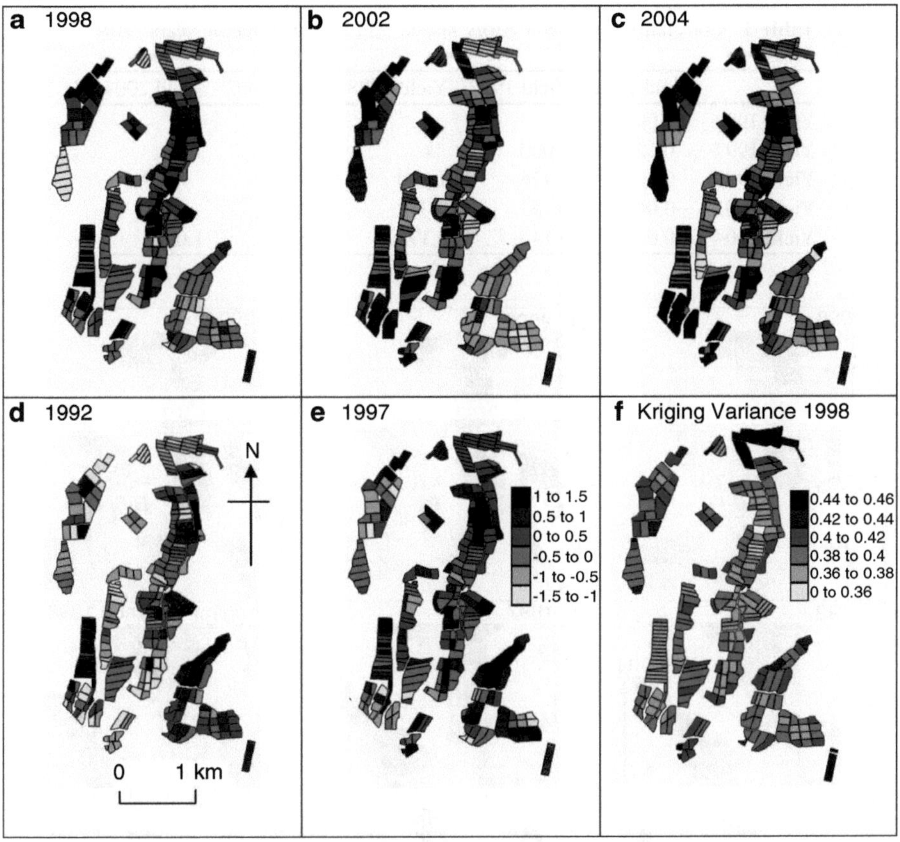

Fig. 3 Maps of AtoA kriged standardized cranberry yield for warmer than average growing seasons (**a–c**) and colder than average growing seasons (**d–e**). Classes are expressed as standard deviations of yield. Map of AtoA kriging variance of 1998 standardized yield (**f**)

and 0.016, respectively. In contrast Fig. 3f shows that the variance associated with the predictions for each bed is not related to bed geometry, but rather its location in relation to other beds. Beds near the centre of the study area have smaller kriging variances than those near the edges. A brief comparison between AtoA kriged yield and yield kriged using a centroid based approach which does not take into account differences in the size and shape of beds showed that the differences in kriged standardized yield were negligible and of the magnitude of 0.005 for the vast majority of fields. The long thin beds in the far north east of the study area, however, were the most prone to larger over- and under-estimations by the centroid-based approach and this could have an effect on the classification of these fields.

Some groups of fields that are high yielding in warm years switch to being low yielding in cold years and *vice versa*. Groups of fields that behave similarly can also be identified and these could be potentially managed in a uniform manner. The correlation coefficients between yields from the different years in this central growing

Table 1 Correlations between AtoA kriged yield from different years (308 fields)

	Yield 1992	Yield 1997	Yield 1998	Yield 2002	Yield 2004
Yield 1992	1.000				
Yield 1997	0.128	1.000			
Yield 1998	0.108	0.476	1.000		
Yield 2002	−0.061	0.153	0.319	1.000	
Yield 2004	0.034	0.143	0.172	0.571	1.000

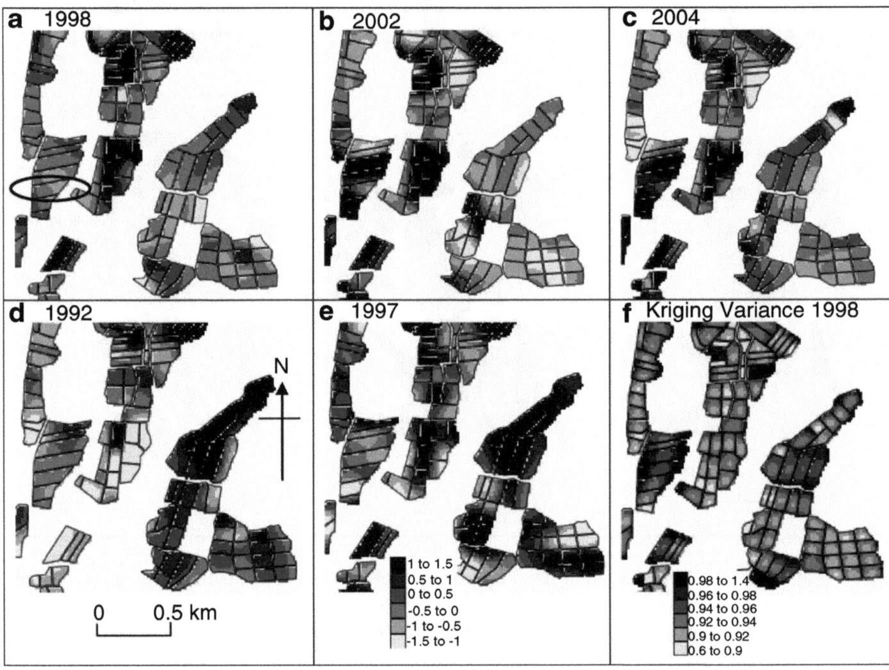

Fig. 4 Area to Point (AtoP) kriged maps of cranberry yield for warmer than average growing seasons (**a–c**) and colder than average growing seasons (**d–e**). Classes are expressed as standard deviations of yield. Map of AtoP kriging variance of 1998 standardized yield (**f**)

area (Table 1) confirm the visual results from Fig. 3: yields recorded in years with similar weather tend to be more strongly correlated, whilst the correlation for those with different weather conditions is generally weaker and sometimes negative. The stronger than expected correlation between 1997 and 1998 yields given the different weather conditions can probably be attributed to the vegetative growth in 1 year that contributes to higher yield in the following year.

The standardized AtoP kriged yield data for various years are shown in Fig. 4 for a small area so that within-field details can be observed. There are more similarities in the patterns of standardized yield between years with similar weather conditions than years with different weather conditions. Some fields exhibit within-field

Table 2 Correlations between AtoP kriged yield from different years (20,179 points)

	Yield 1992	Yield 1997	Yield 1998	Yield 2002	Yield 2004
Yield 1992	1.000				
Yield 1997	0.065	1.000			
Yield 1998	−0.083	0.441	1.000		
Yield 2002	−0.092	0.061	0.287	1.000	
Yield 2004	0.071	0.128	0.185	0.587	1.000

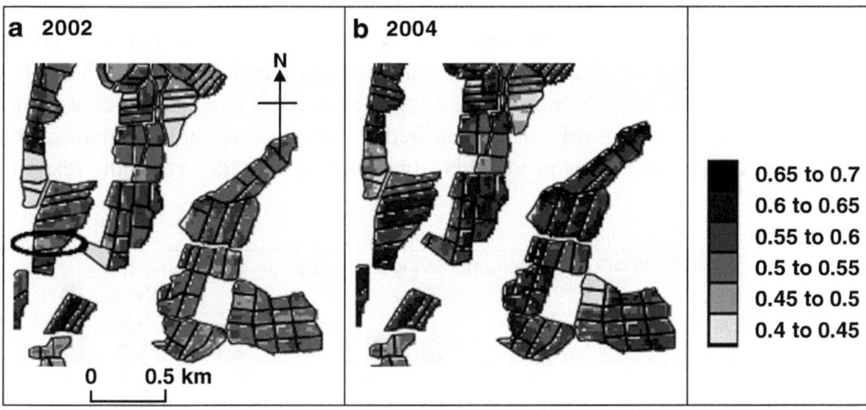

Fig. 5 Maps of kriged EVI for (**a**) July 2002 and (**b**) July 2004

variability, whereas others do not and some patterns of within-field variability are evident in more than 1 year whereas others are not. The correlation coefficients computed from AtoP kriging results (Table 2) lead to the same conclusions as the results obtained in Table 1 for AtoA kriging.

As bed centroids were used as the coordinates for each yield value, the kriging variance (Fig. 4f) within beds tends to be least at the centre and increases towards the edge. The kriging variances are generally large in bigger beds and where the beds are long and thin, the kriging variance shows distinct anisotropy.

3.3 Enhanced Vegetation Index (EVI) Data

Figure 5a and b shows maps of kriged EVI for 2002 and 2004. The patterns of variability for the 2 years have some features in common and there is quite a bit of variation in EVI within fields. If this is a true reflection of plant health, such small and disjointed zones might be difficult to manage. The correlations between aggregated EVI and yield in 2002 (2004) were 0.174 (0.167) and 0.184 (0.255) for AtoA and AtoP kriging, respectively. These weak correlations probably result from the large amounts of small scale variation in the EVI data. Correlation between EVI 2002 and 2004 was 0.542 and 0.803 for aggregated and kriged EVI, respectively. This suggests that there is more correlation in the fine detail than the broad patterns, and that these within-field variations seem to be consistent in time.

3.4 Classification Analysis

Figure 6 shows the results of non-hierarchical classification using AtoA kriged yield and aggregated EVI from various years. According to plots of the between-groups sum of squares criterion, classifications with both two and five classes might be appropriate in warm years, which suggests the existence of two scales of variation in yield in such years. Classification identified seven zones when information from colder than average years was used. This suggests that in colder years the response

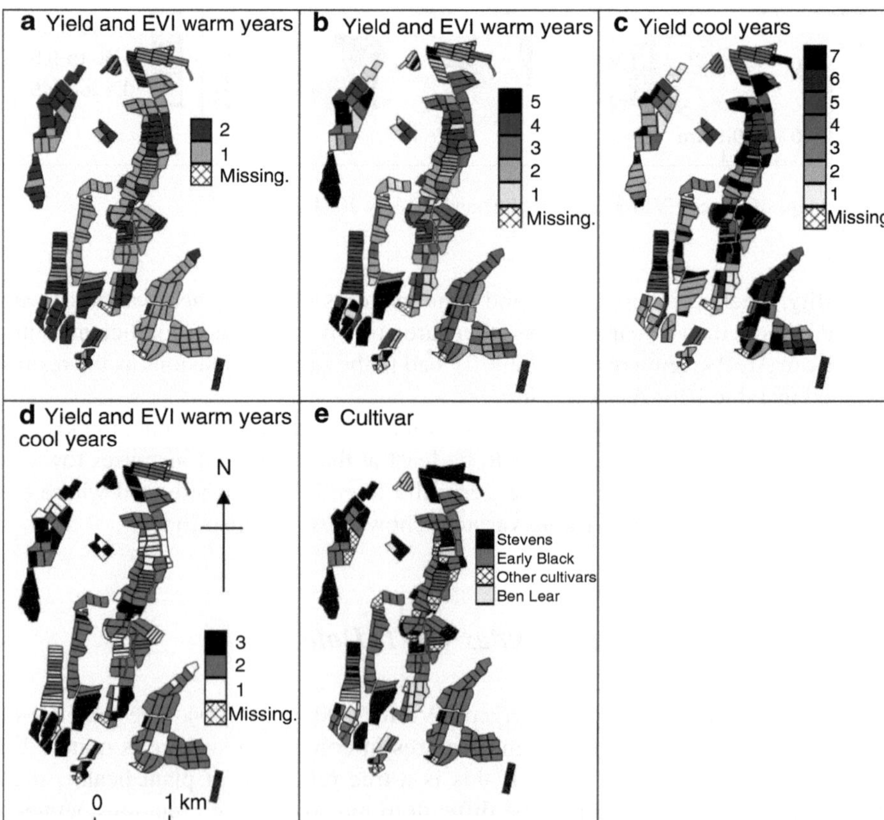

Fig. 6 Classifications of AtoA yield and aggregated EVI from different years (**a–d**) and map of cultivars grown (**e**)

of the crop is far more varied and less predictable. The differences between warm and cold years could be observed in the similarities of variogram parameters (not shown) for years with a given type of weather. Both types of year exhibited two scales of variation, but the range for each structure was smaller for cooler years and the variance associated with the first structure greater too, whilst in warmer years, the second structure accounted for more of the variation. Nevertheless, there are some similarities in the extent of some zones for both cold and warm years which help identify areas with consistent yields. When information from cold and warm years was included in classification, three main classes were identified (Fig. 6d), and these broadly matched the distribution of the three main cultivars (Fig. 6e). Although there are exceptions to this pattern, it suggests that when weather is ignored, the main influence on yield at this scale is not location, but cultivar and the main management differences should be between cultivars. The differences between Fig. 6d and e could, however, be used to identify fields that tend to be higher or lower yielding than their counterparts with the same cultivar and should therefore be managed differently.

Figure 7 shows the results of non-hierarchical classification using AtoP kriged yield and kriged EVI from various years. Due to computational difficulties, these data had to be aggregated from a 20 m to a 40 m grid. According to the plots of

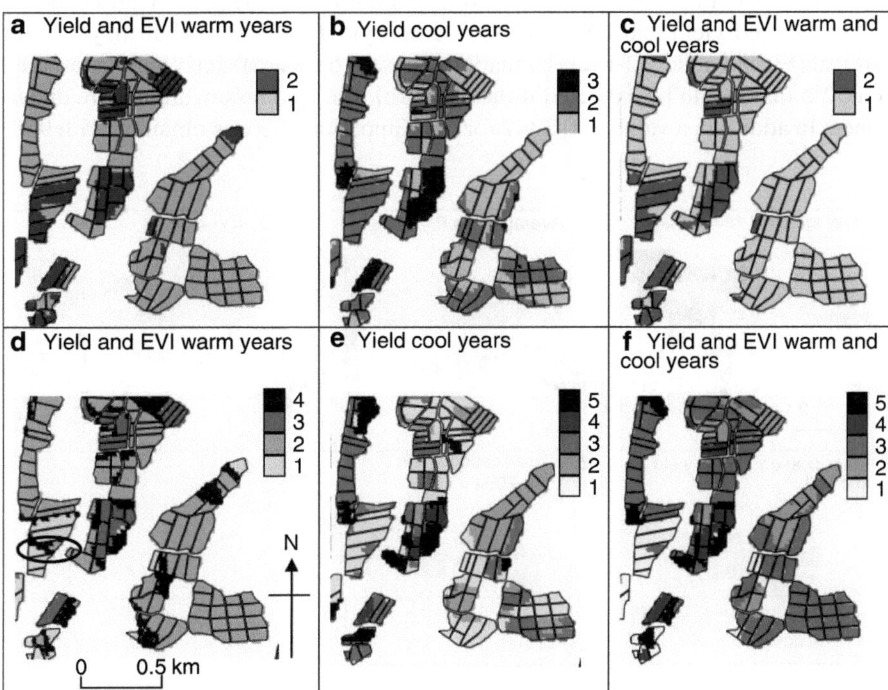

Fig. 7 Classifications of AtoP kriged yield and kriged EVI from warm years (**a** and **d**), cool years (**b** and **e**) and cool and warm years together (**c** and **f**)

the between-groups sum of squares criterion, two different numbers of classes were optimal for each group of years suggesting that there are two scales of variation evident in the data that produce two or three classes and four or five classes. Classification identified two and four, three and five and two and five zones for warm, cold and both types of year, respectively (Fig. 7). More than one class is identified for more fields in cold years than warm years and even when four and five classes are used for warm years and both types of year (Fig. 7d and f), the within-field variability is confined to certain fields while other fields show no within-field variation. There are several fields which exhibit no within-field variability in classification for each year and some fields which consistently exhibit within-field variability. An example of the latter is the Nadine bed which is circled in Fig. 7d and shows within-field variation in all maps except Fig. 7b.

Within-field yield information from the Nadine bed based on pre-harvest berry counts were used as an initial assessment of the AtoP classification. Figure 8a shows standardized average yield based on berry counts from 1999–2004. The patterns were similar to this average in most years, but were less well defined in 2003 and 2004. Figure 8b shows average AtoP kriged yield and Fig. 8c shows observed EVI values from 7 m pixels for the field in 2002. Two non-hierarchical classifications with two classes were computed, one based on the berry count yield data from 1999–2004 and the other based on AtoP kriged yield from warm and cold years and EVI 2002 (7 m pixels). There are some similarities in the classifications based on AtoP kriged yield and EVI and berry count yield 1999–2004 but they are not identical (Fig. 8d and e). The classifications showed 68% areal agreement compared with 52% that would be expected if the similarities in the classification were due to chance. In addition, a value of 0.3342 for the Kappa statistic was obtained with 95%

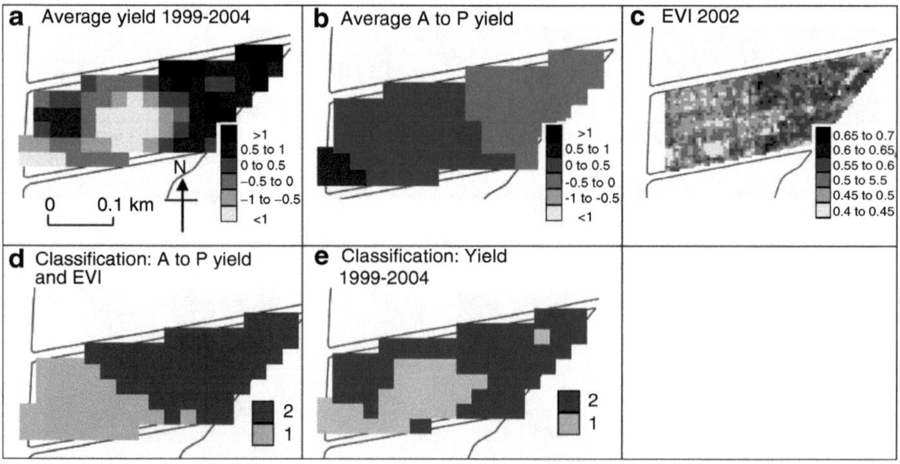

Fig. 8 Maps of average yield for the Nadine bed based on (**a**) berry counts from 1999–2004, (**b**) AtoP kriged yield from cold and warm years; (**c**) observed EVI 2002, and two class classifications based on (**d**) AtoP kriged yield and EVI 2002, and (**e**) berry counts 1999–2004

confidence intervals of 0.1259 and 0.5425. Landis and Koch (1977) state that a value of 0.3342 shows fair agreement between the two classifications suggesting that the AtoP kriged surfaces combined with EVI data can give a reasonable indication of possible within-field management zones for cranberries.

4 Conclusions

This paper shows the potential for AtoA and AtoP kriging to identify management zones in cranberry crops at two scales. The first scale involves groups of fields being treated similarly and the second involves differential management within fields. Combining AtoP kriged yield data with vegetation indices has advantages over using remotely sensed data on its own as the broad underlying patterns of yield variation are brought out without them being masked by management features and small scale variability that could not be managed. The analysis suggests the existence of groups of fields that behave similarly in certain types of growing season and others that have marked within-field variability over a number of years and could benefit from precision management. The reasons for these field heterogeneities need to be investigated as well as more detailed validation of the AtoP kriged yield and EVI patterns with yield variability data based on berry counts for more fields.

This paper compared years with different weather conditions to gain insight into spatial patterns of yield for different kinds of growing season. Variograms identified more short range variation in colder than average growing seasons, leading to a greater number of classes. This suggests that the crop would be more difficult to manage in such years. Future analysis will involve integrating information from all years in the 14 year period to discriminate between zones with consistently high and low yielding and those that have variable yield in time. Mapping within-field yield to a finer spatial resolution will be investigated using about 100 fields, rather than all fields in the whole region. This will suppress the need to sub-sample EVI data to be computationally manageable and AtoP kriging can be conducted for finer interpolation grids. Unwanted management effects in EVI data will be filtered using factorial kriging and additional variables, such as the area:perimeter ratio and age of cranberry bed, will be incorporated as secondary information in the interpolation procedure.

Acknowledgements We would like to thank Ocean Spray Cranberries Inc. for providing yield data, Larisa Pozdnyakova (Golovko) of RiceTec, Alvin, TX for collecting and pre-processing much of the data used in this paper and Dan A. Sims, Ball State University for calculating the EVI values. Funding was provided as part of USDA-IFAFS grant # 2001-52103-11310.

References

Auernhammer H, Demmel M, Muhr T, Rottmeier J, Wild K (1993) Yield measurements on combine harvesters. Amer Soc Agric Eng Paper No. 93-1506

Blackmore S, Moore M (1999) Remedial correction of yield map data. Prec Agric 1:53–66

Goovaerts P (2008) Kriging and semivariogram deconvolution in the presence of irregular geographical units. Math Geosci 40:101–128

Grenzdörffer G, Gebbers RIB (2001). Seven years of yield mapping – analysis and possibilities of multi-year yield mapping data. In: Grenier G, Blackmore S (eds) ECPA 2001. Agro-Montpellier, Montpellier, France, pp 31–36

Huete A, Didan K, Miura T, Rodriguez EP, Gao X, Ferreira LG (2002) Overview of the radiometric and biophysical performance of the MODIS vegetation indices. Remote Sens Environ 83:195–213

Hughes MG, Oudemans PV, Davenport JR, Ayres K, Airola TM, Lee A (1998) Evaluating commercial cranberry beds for variability and yield using remote sensing techniques. In: Robert PC, Rust RH, Larson WE (eds) Precision agriculture. ASA, CSSA, SSSA. Madison, WI, pp 1493–1500

Khosla R, Inman D, Westfall DG, Reich RM, Frasier M, Mzuku M, Koch AB, Hornung A (2008) A synthesis of multi-disciplinary research in precision agriculture: site-specific management zones in the semi-arid western Great Plains of the USA. Prec Agric 9:85–100

Johnston AE, Barraclough PB, Poulton PR, Dawson CJ (1998) Assessment of some spatially variable soil factors limiting crop yields. The International Fertiliser Society, York, United Kingdom

Kyriakidis P (2004) A geostatistical framework for area-to-point spatial interpolation. Geo Anal 36:259–289

Landis JR, Koch GG (1977) The measurement of observer agreement for categorical data. Biometrics 33:159–174

Lark RM, Stafford JV (1996) Consistency and change in spatial variability of crop yield over successive seasons: Methods of data analysis. In: Robert PC, Rust RH, Larson WE (eds) Precision agriculture. ASA, CSSA, SSSA. Madison, WI, pp 141–150

Novy RG, Vorsa N, Patten K (1996) Identifying genotypic heterogeneity in 'McFarlin' cranberry: a randomly-amplifies polymorphic DNA (RAPD) and phenotypic analysis. J Amer Soc Horticult Sci 121:210–215

Oudemans PV, Pozdnyakova L, Hughes MG, Rahman F (2002) GIS and remote sensing for detecting yield loss in cranberry culture. J Nematol 34:207–212

Payne RW (ed) (2006) The Guide to GenStat Release 9 – Part 2: Statistics. (VSN International, Hemel Hempstead, UK)

Pozdnyakova L, Oudemans PV, Hughes MG, Giménez D (2002) Estimation of spatial and spectral properties of phytophthora root rot and its effects on cranberry yield. Comput Electronics Agric 37:57–70

Pozdnyakova L, Giménez D, Oudemans PV (2005) Spatial analysis of cranberry yield at three scales. Agronomy J 97:29–57

Estimating the Local Small Support Semivariogram for Use in Super-Resolution Mapping

Peter Michael Atkinson and Chockalingam Jeganathan

Abstract Three methods were introduced for estimating the local semivariogram for use in procedures such as super-resolution pattern prediction. The first is simply to use a training image to estimate the global semivariogram required. The second method employs a deconvolution–convolution procedure to estimate the local semivariogram. The estimated semivariogram represents proportions and so a further step is required to convert the proportions semivariogram to represent a binary field. The third method is an integration of the first two methods obtained by weighted linear combination across the lags of the semivariograms. The results are evaluated using the known target local semivariogram. The integrated method provides some advantages. The discussion points to problems and potential future improvements on the method.

1 Introduction

Super-resolution mapping techniques have been proposed for remote sensing classification (e.g., Tatem et al., 2001a, b; Atkinson, 2005). The basic principle of such super-resolution techniques is as follows. A land cover proportions image is assumed as input to the algorithm. This proportions image is assumed to have been produced through some prior area proportions prediction technique applied to a multispectral or hyperspectral remotely sensed image at a given coarse spatial resolution **V**. Area proportions techniques, sometimes referred to as soft classifiers, are common and include mixture models (Adams et al., 1985), artificial neural networks, and fuzzy c-means classifiers (Foody, 1996; Atkinson et al., 1997). Super-resolution mapping algorithms transform the proportions image into a hard land cover classification at a finer spatial resolution **v** than that of the original image.

There are many different approaches to achieve the image downscaling operation implicit in super-resolution mapping. Spatial optimization techniques have included

P.M. Atkinson (✉) and C. Jeganathan
School of Geography, University of Southampton, Highfield, Southampton, United Kingdom SO17 1BJ
e-mail: P.M.Atkinson@soton.ac.uk

P.M. Atkinson and C.D. Lloyd (eds.), *geoENV VII – Geostatistics for Environmental Applications*, Quantitative Geology and Geostatistics 16,
DOI 10.1007/978-90-481-2322-3_25,

approaches based on the Hopfield neural network (HNN) (Tatem et al., 2001a, b, 2002, 2003; Nguyen et al., 2006), genetic algorithms (Mertens et al., 2003) and those based on simple pixel-swapping techniques (Atkinson, 2005; Thornton et al., 2006). Alternative techniques have included regression-based approaches such as the standard feed-forward back-propagation artificial neural network (Mertens et al., 2004), an extension of linear mixture modelling (Zhan et al., 2002; Verhoeye and De Wulf, 2002), geostatistics (Boucher and Kyriakidis, 2006) and a Bayesian approach (Kasetkasem et al., 2005). Some approaches involve spatial optimization where the objective function is defined based on a target that is required to be known *a priori* and, thus, such approaches can be considered to involve some learning or training (e.g., Tatem et al., 2002; Atkinson, 2004).

Here, two objectives are distinguished: (i) mapping objects that are larger than the size of the image pixels (the high-resolution or H-resolution case; Woodcock and Strahler, 1987) and (ii) mapping objects that are smaller than the image pixels (the low-resolution or L-resolution case). While in the H-resolution case it is possible to provide solutions that do not require any learning or training, some learning is more likely to be required in the L-resolution case. This paper is concerned with the L-resolution case, for which spatial optimization based on training is appropriate. The question that is addressed is how best to provide the training.

The particular case that provides the context for the research reported here is the paper by Tatem et al. (2002) in which the HNN was developed as a pattern prediction technique specifically for the L-resolution case. The method is described briefly here. The Energy function of the HNN was, in simple terms,

$$E = k_1 G + k_2 C \tag{1}$$

Where E is the energy, G is the goal and C is a constraint, and k_1 and k_2 are weights where $k_1 = (1 - k_2)$. In the standard version of the algorithm designed for the H-resolution case, the goal was to maximise the spatial correlation between neighbouring (sub-)pixels and the constraint was to reduce the error between the predicted proportions obtained by upscaling the predicted fine spatial resolution binary field and the original proportions used as input to the algorithm. In the formulation designed for pattern prediction in the L-resolution case the Goal G was replaced by a semivariogram goal:

$$E = \sum_{i=1}^{n(h)} k_i V_i + k_C C \tag{2}$$

Where the V_i are the semivariogram goals for each distance band $i = 1, 2, \ldots n(h)$, assuming isotropy and k_C is now the weight associated with the same proportions constraint. In this method, the V_i act as training data. They must be obtained prior to application of the method. Possible sources of information with which to estimate the required V_i include (i) prior expert knowledge and (ii) cartographic maps, but most likely (iii) a training image classified to the required classes and at the desired spatial resolution for a different area that is deemed representative of the target area.

There are several problems with the use of training data as follows:

(i) The training area may not be representative of the target area for which downscaling is required.
(ii) The global semivariogram model may not be representative of the local situation.

This paper explores new methods for localising the training image semivariogram using the available coarse spatial resolution proportions imagery for use in super-resolution pattern prediction. Although super-resolution mapping provides the context and a potential requirement for a local semivariogram, this paper deals only with prediction of the local semivariogram.

2 Methods

2.1 Background: The Semivariogram and Regularization

The semivariogram, the central tool of geostatistics, is one of several functions that may be used to characterise the scales of spatial variation present in remotely sensed imagery (Journel and Huijbregts, 1978; Curran and Atkinson, 1998). It is assumed that readers are familiar with the basic concepts. The experimental semivariogram, $\hat{\gamma}(\mathbf{h})$, can be estimated from $p(\mathbf{h})$ paired observations, $z(\mathbf{x}_\alpha)$, $z(\mathbf{x}_\alpha + \mathbf{h})$, $\alpha = 1, 2, \ldots p(\mathbf{h})$ using:

$$\hat{\gamma}(\mathbf{h}) = \frac{1}{2p(\mathbf{h})} \sum_{\alpha=1}^{p(h)} \{z(\mathbf{x}_\alpha) - z(\mathbf{x}_\alpha + \mathbf{h})\}^2 \tag{3}$$

To use the semivariogram in most geostatistical procedures it is necessary to fit a mathematical model to the empirical values.

The semivariogram is dependent on the support at which it is observed (Atkinson, 1993, 1995). It is possible to define a model of the regularizing or convolving effect of the support on the semivariogram (Clark, 1977; Journel and Huijbregts, 1978; Jupp et al., 1988, 1989). One model is defined by Journel and Huijbregts (1978) as:

$$\gamma_v(\mathbf{h}) = \bar{\gamma}(v, v_\mathbf{h}) - \bar{\gamma}(v, v) \tag{4}$$

where $\bar{\gamma}(v, v_\mathbf{h})$ represents the integral punctual semivariance between two pixels of size v whose centroids are separated by $\mathbf{h}$ and $\bar{\gamma}(v, v)$ represents the average punctual semivariance within a pixel of size v (i.e., the within-block variance). This model can be implemented in practice through numerical approximation.

2.2 *Methods for Estimating the Local Target Semivariogram*

The goal is to obtain a locally conditioned semivariogram that can be used, for example, in the objective function of super-resolution mapping for pattern prediction. Below, we detail three different approaches for the above task.

- Method 1: It is assumed that a training image and, thereby, training semivariogram is available at the desired fine spatial resolution. The global semivariogram estimated and modelled for this image is used to represent the desired local target semivariogram. This is the current practice in super-resolution mapping.
- Method 2: The training image is assumed to be unavailable and instead the local semivariogram at the fine spatial resolution is estimated through a deconvolution-convolution procedure based on the available coarse spatial resolution input image.
- Method 3: A hybrid approach is explored which combines linearly the outputs of methods 1 and 2.

An assessment of the accuracy of the three methods is presented based on the local semivariogram observed for a known target image at the desired fine spatial resolution.

2.2.1 Method 1

Method 1 is straightforward as it involves nothing more than estimating the experimental semivariogram $\hat{\gamma}^{T}_{vI}(\mathbf{h})$ of the available training image T of binary (0,1) values (represented by I) at fine spatial resolution v as described above in Section 2.1. The values of semivariance at the defined set of lags j provide the information that is required in Eq. (2). This procedure was used in Tatem et al. (2002). In this paper, this objective was modified slightly. The fine spatial resolution binary training image semivariogram $\hat{\gamma}^{T}_{vI}(\mathbf{h})$ is replaced by the proportions training image semivariogram $\hat{\gamma}^{T}_{v}(\mathbf{h})$ in order to facilitate comparison with method 2 below.

A problem, as outlined in the introduction, is that $\hat{\gamma}^{T}_{vI}(\mathbf{h})$ (or $\hat{\gamma}^{T}_{v}(\mathbf{h})$) may not be representative of the *target* image locally. This method serves as a useful benchmark for the method proposed below.

2.2.2 Method 2

In method 2, it is assumed that a training image defined at the desired fine spatial resolution is *not* available. In the absence of prior information, the only option is to attempt to estimate the desired local semivariogram from the available coarse spatial resolution data. The proposed local deconvolution-convolution method is as follows:

1. Guess a candidate model for the punctual semivariogram globally $\gamma.(\mathbf{h},\theta)$. In this research, the global semivariogram $\hat{\gamma}^{O}_{V}(\mathbf{h})$ of the available coarse spatial

resolution proportions image $\hat{S}_V^O(\mathbf{x})$ was estimated and a model $\gamma_V^O(\mathbf{h}, \theta)$ fitted, where θ represents the fitted model parameters. The starting value of the punctual model range was the same as for $\gamma_V^O(\mathbf{h}, \theta)$, while the starting punctual model sill was double that for $\gamma_V^O(\mathbf{h}, \theta)$.
2. Use the regularization equation (4) to convolve the punctual semivariogram $\gamma_.(\mathbf{h}, \theta)$ to the coarse spatial resolution V, thus, estimating $\tilde{\gamma}_V(\mathbf{h})$ where $\sim$ denotes the induced model.
3. Compare $\tilde{\gamma}_V(\mathbf{h})$ with the observed local semivariogram $\hat{\gamma}_V^O(\mathbf{h}, W(\mathbf{x}))$ and use the error to adjust the model parameters defining $\gamma_.(\mathbf{h}, \theta)$ such as to estimate $\gamma_.(\mathbf{h}, W(\mathbf{x}), \theta)$, the local punctual semivariogram.
4. Convolve $\gamma_.(\mathbf{h}, W(\mathbf{x}), \theta)$ to estimate the local semivariogram $\hat{\gamma}_v(\mathbf{h}, W(\mathbf{x}))$ defined for the desired fine spatial resolution.

The actual required set of semivariograms is not $\hat{\gamma}_v(\mathbf{h}, W(\mathbf{x}))$, but $\hat{\gamma}_{vI}(\mathbf{h}, W(\mathbf{x}))$ (i.e., the set of semivariograms defined for a binary, or classified, field). Thus, it is necessary to add to the above sequence a procedure to obtain the binary semivariograms $\hat{\gamma}_{vI}(\mathbf{h}, W(\mathbf{x}))$, as follows:

1. For every local window W, use $\hat{\gamma}_v(\mathbf{h}, W(\mathbf{x}))$ to simulate a random field $\hat{S}_v(\mathbf{x})$ of large, but arbitrary dimensions with values constrained to lie within the bounds (0,1).
2. Transform the proportions image to a binary field $\hat{S}_{vI}(\mathbf{x})$ by applying an indicator transform $I_k(\cdot)$, where k can be any value, but here is set to 0.5.
3. Estimate the desired target experimental semivariogram $\hat{\gamma}_{vI}(\mathbf{h}, W(\mathbf{x}))$ defined locally.

Alternative procedures would be:

1. Convert the available proportions image at the coarse spatial resolution to a binary field by applying an indicator threshold (e.g., 0.5), and applying method 2 as above. A problem is that this approach involves a loss of information. A further problem is that the indicator transform to a binary field implies a change to the punctual scale, complicating the scaling process.
2. Apply method 2 to the available proportions image as above, but then transform the sill and range according to some empirical or mathematical relation.

In this paper, this second sequence is omitted; it is assumed that this or an equivalent procedure can be followed. Thus, the focus of attention is on estimating the local semivariogram of proportions $\hat{\gamma}_v(\mathbf{h}, W(\mathbf{x}))$ defined on the desired fine spatial resolution locally. It is for this reason that $\hat{\gamma}_v^T(\mathbf{h})$ was estimated in place of $\hat{\gamma}_{vI}^T(\mathbf{h})$ in method 1 above.

2.2.3 Method 3

It was considered that the outputs of method 1 (well estimated global semivariogram of training image, but potentially not representative of the target area, especially locally) and method 2 (local semivariogram of the target area, but potentially poorly estimated) bring different benefits. It was, therefore, of interest to explore their combination. If the uncertainty associated with each prediction were known, then the two predictions could be integrated in a Bayesian sense using the inverse of the prediction variances as weights. Without such information on uncertainty, a weighted linear combination of the semivariance values on a lag-by-lag basis was investigated for a range of weights as follows:

$$\hat{\gamma}(\mathbf{h}) = \lambda_1 \hat{\gamma}_v(\mathbf{h}, W(\mathbf{x})) + \lambda_2 \hat{\gamma}_v^T(\mathbf{h}) \quad (5)$$

Where, λ_1 varies in (0,1) and $\lambda_1 + \lambda_2 = 1$.

3 Data

A fine spatial resolution (nominally 5 m) image of (nominally land cover) proportions of dimension 139 by 94 pixels was simulated. Only two classes were simulated in this example. This provides the target image $\hat{S}_v(\mathbf{x})$ (Fig. 1). The upper left quarter of this target image was taken to represent the training image $\hat{S}_v^T(\mathbf{x})$. This represents the case in remote sensing where the training image is usually available at a finer spatial resolution, but for a smaller area than, or different area to, the coarse spatial resolution input image of proportions.

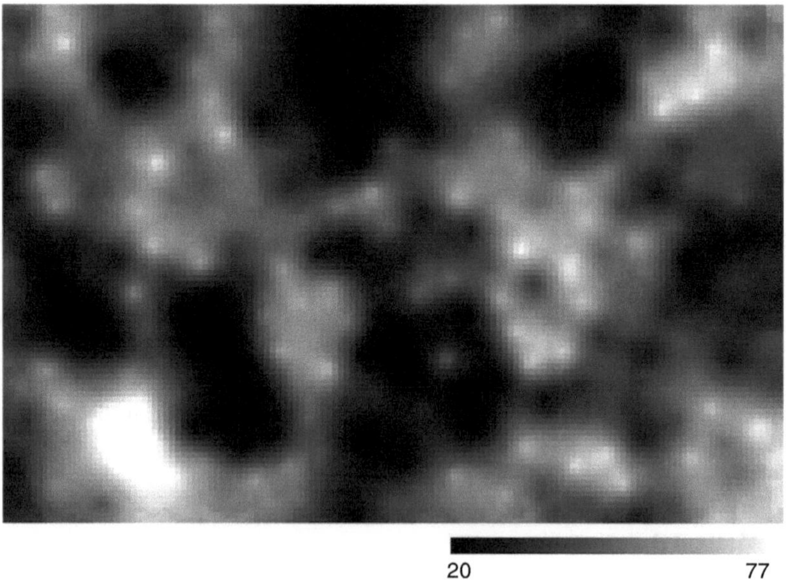

Fig. 1 Fine spatial resolution (5 m) proportions image $\hat{S}_v(\mathbf{x})$

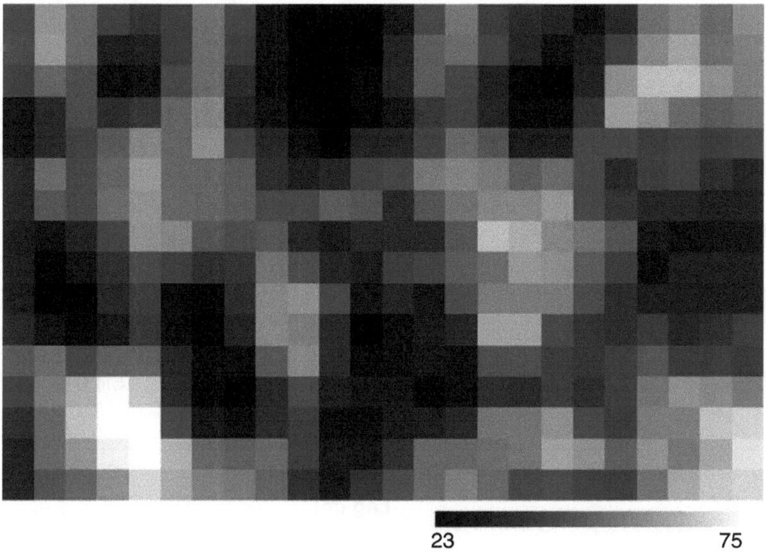

Fig. 2 Coarse spatial resolution (30 m) proportions image $\hat{S}_v(\mathbf{x})$

The fine spatial resolution simulated image $\hat{S}_v(\mathbf{x})$ obtained above was degraded (i.e., convolved with a square wave filter equal to the coarse pixel size, nominally 30 m) to simulate an observed coarse spatial resolution image $\hat{S}_V(\mathbf{x})$ of land cover proportions (Fig. 2).

4 Analysis

The global experimental semivariogram $\hat{\gamma}_V(\mathbf{h})$ was estimated for the coarse spatial resolution (30 m) observed image of proportions $\hat{S}_V(\mathbf{x})$ and a spherical model fitted using weighted least squares (Fig. 3). Although not available for analysis in the practical case, the global experimental semivariogram $\hat{\gamma}_v(\mathbf{h})$ was also estimated for the fine spatial resolution (5 m) observed image of proportions $\hat{S}_v(\mathbf{x})$ and a spherical model fitted (Fig. 4). This semivariogram provides a reference for future comparison.

4.1 Method 1

The global experimental semivariogram $\hat{\gamma}_v^T(\mathbf{h})$ was estimated for the fine spatial resolution (5 m) training image of proportions $\hat{S}_v^T(\mathbf{x})$ and a spherical model fitted (Fig. 5). This $\hat{\gamma}_v^T(\mathbf{h})$ estimates directly the desired local semivariogram of proportions $\hat{\gamma}_v(\mathbf{h}, W(\mathbf{x}))$.

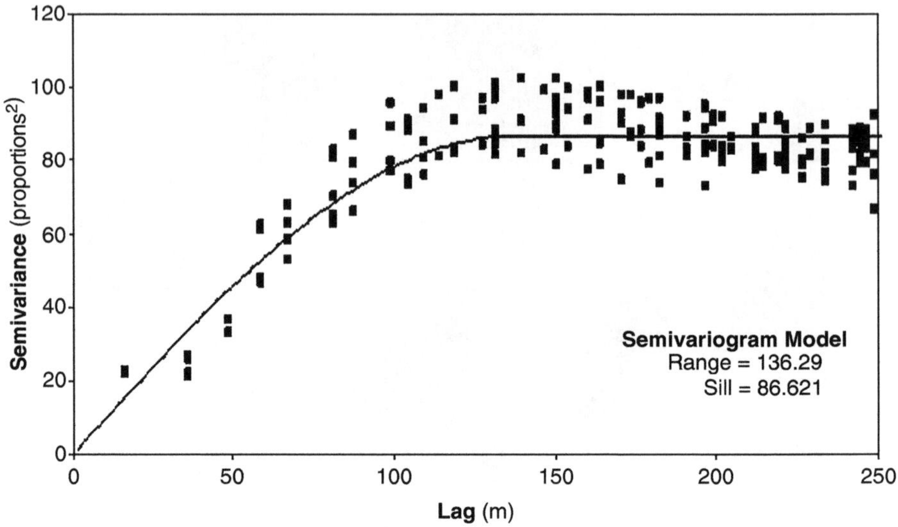

Fig. 3 Spherical model fitted to semivariogram for coarse spatial resolution (30 m) proportions image

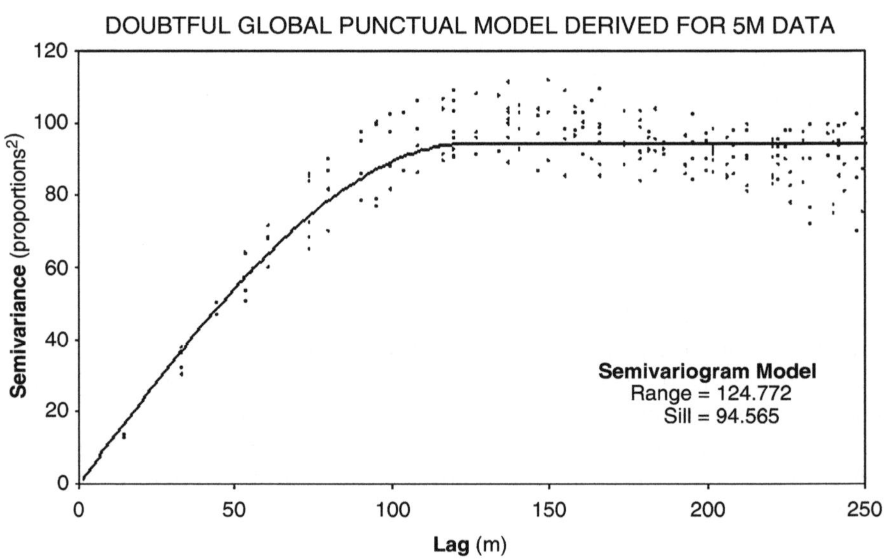

Fig. 4 Spherical model fitted to semivariogram for fine spatial resolution (5 m) target proportions image

Since the local target image $\hat{S}_v(\mathbf{x})$ is available it was possible to evaluate the accuracy of $\hat{\gamma}_v^T(\mathbf{h})$ as an estimate of $\gamma_v(\mathbf{h}, W(\mathbf{x}))$. The residual sum of squares is shown as a function of lag in Fig. 6.

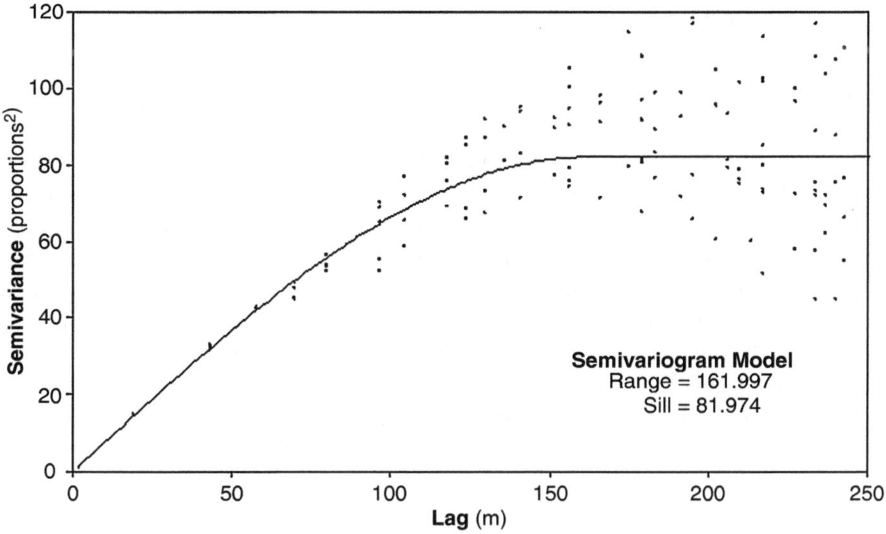

Fig. 5 Spherical model fitted to semivariogram for fine spatial resolution (5 m) training proportions image

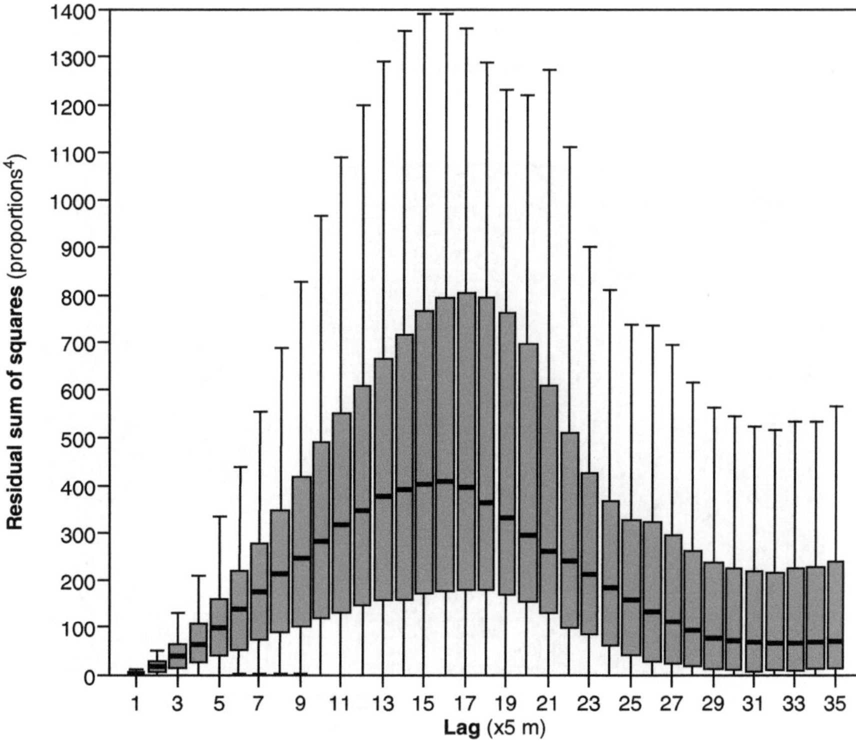

Fig. 6 Lag-wise variation in the residual sum of squares for method 1

4.2 Method 2

Method 2 was followed to estimate the local semivariogram $\gamma_v(\mathbf{h}, W(\mathbf{x}))$ at discrete lags at the desired fine spatial resolution of 5 m. Local variation in the parameters of the fitted punctual semivariogram $\gamma_{\bullet}(\mathbf{h}, W(\mathbf{x}), \theta)$ are displayed in Figs. 7 and 8. Figure 7 shows the local variation in the range of the punctual model and Fig. 8 shows the local variation in the sill of the punctual model obtained using Method 2.

The range varies in a spatially structured way, with values between 94 and 373 m. The largest values of the range correspond to large proportions (hot spots) in Figs. 1 and 2. The sill varies from 78 to 147 semivariance units (in this case, the squared proportion). Again, the largest estimates of the sill parameter correspond to the hot spots in Figs. 1 and 2.

Since the local target image $\hat{S}_v(\mathbf{x})$ is available it was possible to evaluate the accuracy of $\tilde{\gamma}_v(\mathbf{h}, W(\mathbf{x}))$ as an estimate of $\gamma_v(\mathbf{h}, W(\mathbf{x}))$. The residual sum of squares is shown as a function of lag in Fig. 9.

4.3 Method 3

The estimates made using methods 1 and 2 were combined using a linear weighting. Linear weights ranging between 0 (for method 1) and 1 (for method 2) were tested

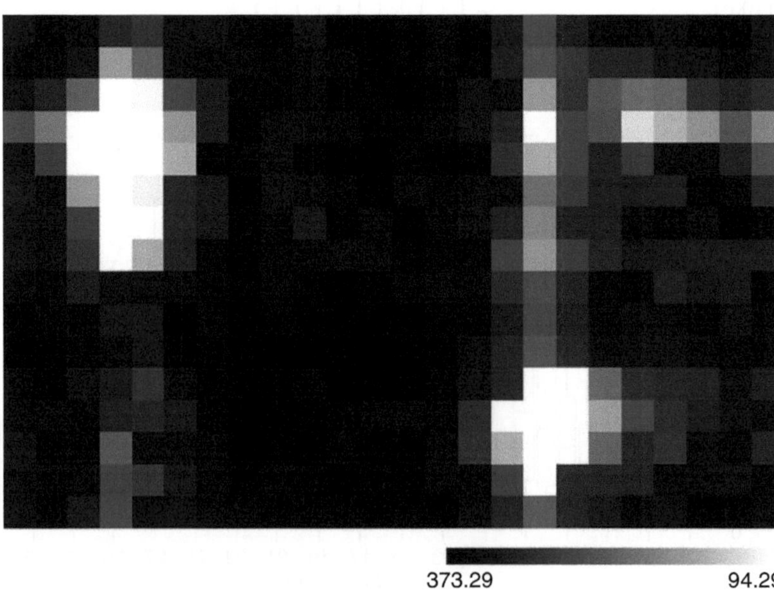

Fig. 7 Local variation in the range of the punctual model for method 2

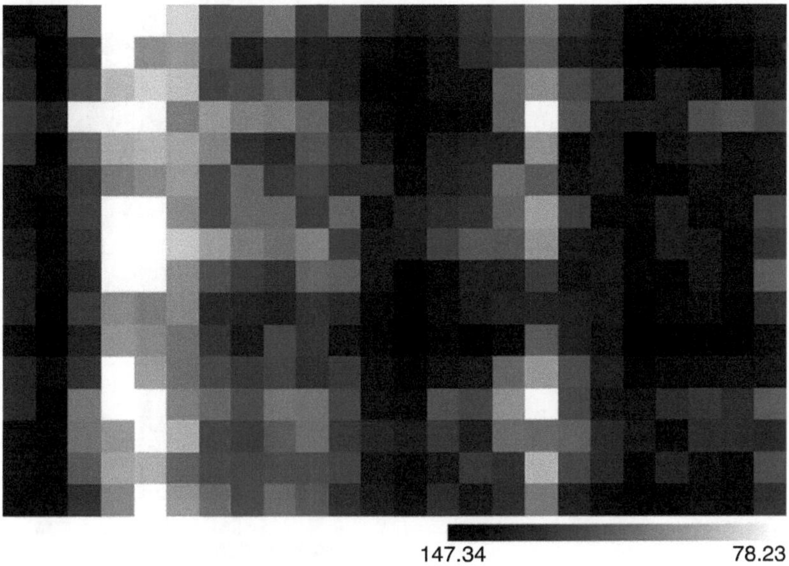

Fig. 8 Local variation in the sill of the punctual model for method 2

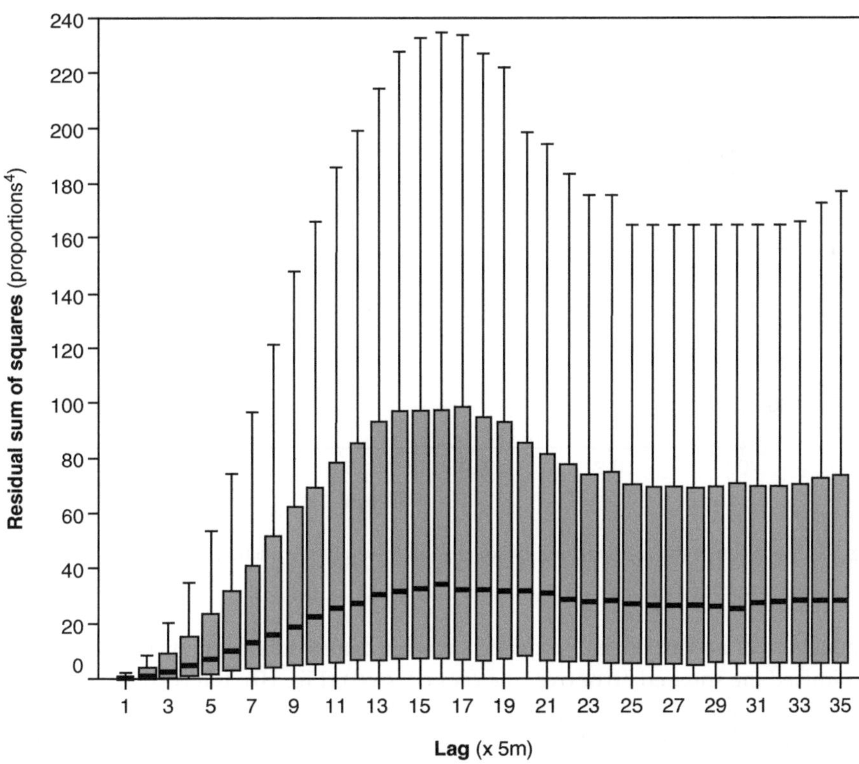

Fig. 9 Lag-wise variation in the residual sum of squares for method 2

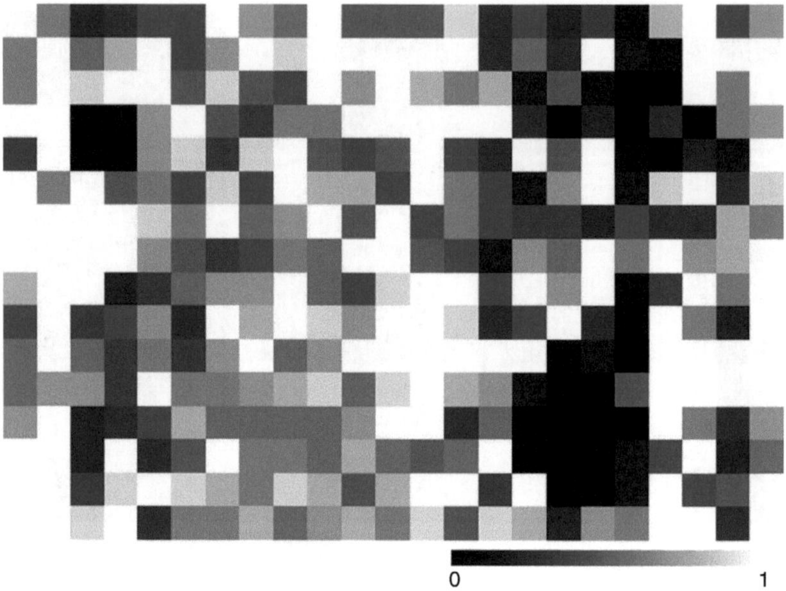

Fig. 10 Variation in the optimised local weight for method 3

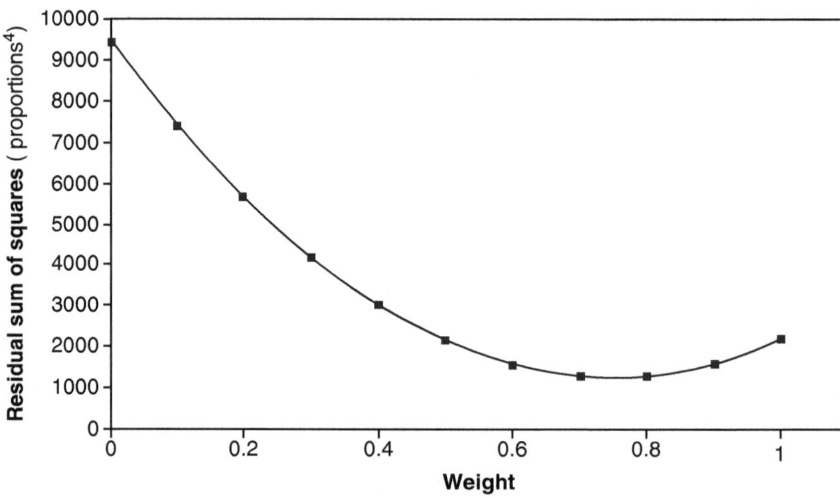

Fig. 11 Variation in the residual sum of squares with weight for method 3

for each pixel and an image produced showing the optimal weight (i.e., the one that minimised the squared error over the range tested) (Fig. 10). Interestingly, method 2 is more accurate on more occasions than method 1.

A plot was also produced showing the sum of squared errors against weight when a single weight was chosen across all pixels (Fig. 11). This represents the real case

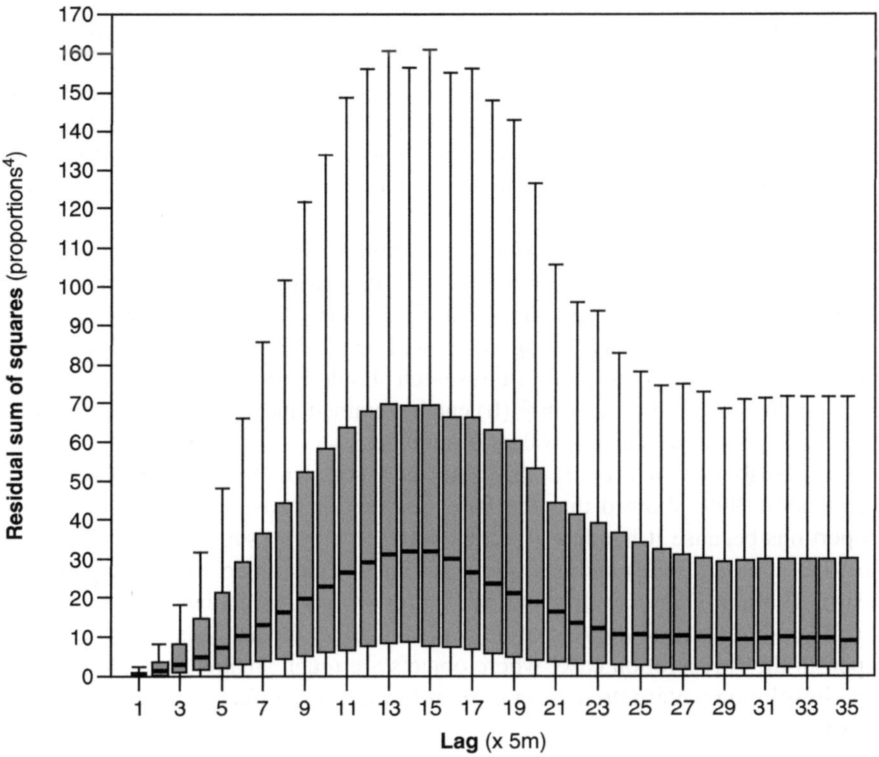

Fig. 12 Lag-wise variation in the residual sum of squares for method 3

where the target image of proportions is not available and the investigator must select a single weight for the method as a whole. The plot shows that on average, a weight of 0.75 (in favour of the local deconvolution-convolution method) minimises the sum of squared errors.

Based on Fig. 11, a weight of 0.75 was selected and method 3 applied to the whole image. A plot of the sum of squared errors against lag was produced for method 3 as for methods 1 and 2 (Fig. 12). The error is on average smaller for method 3 than for either method alone.

5 Discussion

The rather simplistic analysis presented above omits some important considerations. These are discussed here. The simulated image was small in size. The small image was adequate to develop and demonstrate the methods only. It would make sense to expand the simulated image extent or proceed to a real remotely sensed image in future analysis.

The simulated image $\hat{S}_v(\mathbf{x})$ resulted in a semivariogram that was adequately fitted using a spherical model with a large range. This does not adequately represent the L-resolution case in remote sensing for which super-resolution pattern prediction may be desirable. It more adequately represents the H-resolution case. Therefore, the results of this paper must be seen as demonstrating a potential method only. Further testing is required on a simulated field or real imagery in which micro-scale structures are present.

The lack of micro-scale variation is unfortunate because it is such micro-scale variation that may be captured in the global semivariogram of proportions $\hat{\gamma}_v(\mathbf{h})$, but omitted from the local semivariogram $\hat{\gamma}_V(\mathbf{h}, W(\mathbf{x}))$ at the coarse spatial resolution. Thus, the benefit of a training image will be most apparent where micro-scale variation exists. The focus of future research should be to concentrate on how the micro-scale structure revealed in $\hat{\gamma}_v(\mathbf{h})$ can be retained and injected into the estimate of $\gamma_v(\mathbf{h}, W(\mathbf{x}))$ through an equivalent of method 3.

The distribution of proportions values is a concern. Specifically, Collins and Woodcock (1999) showed that the Beta distribution provides a useful model for proportions because it handles all cases between a point support (all values are 0 or 1, i.e., a binary field) and a single pixel image (the single value is equal to the mean). The non-Gaussian distribution of proportions, particularly where the support is small relative to the size of objects may affect the analysis described in this paper. In the L-resolution case, where objects are much smaller than the pixel and super-resolution pattern prediction is appropriate, these effects are likely to be less relevant.

It is interesting to note that the punctual sill variance can be estimated from the coarse spatial resolution proportions image $\hat{S}_V(\mathbf{x})$. Similarly, the relation between the range and spatial resolution is known (see Section 4.3.2). Therefore, it is possible that the deconvolution-convolution procedure can be simplified based on empirical relations.

The method demonstrated in this paper is one of two basic approaches for combining a training image semivariogram and a local deconvolved-convolved estimate of the local semivariogram. In method 3, the combination is undertaken at the fine spatial resolution based on deconvolving and convolving from $\hat{\gamma}_v(\mathbf{h}, W(\mathbf{x}))$. However, a simpler and potentially preferable alternative is to deconvolve and convolve the training image semivariogram $\hat{\gamma}_v^T(\mathbf{h})$ to the coarse spatial resolution and make a comparison with $\hat{\gamma}_V(\mathbf{h}, W(\mathbf{x}))$ at the coarse spatial resolution. This alternative method has the benefit that the deconvolution–convolution process needs to be undertaken only once. Once the punctual model has been tuned to take into account the local information it can be convolved to the fine spatial resolution to estimate $\gamma_v(\mathbf{h}, W(\mathbf{x}))$.

The results of this paper are preliminary and the methods need to be evaluated on a wider range of data, both simulated and real, and tested across a wider range of parameters. A required next step is also to apply the local semivariogram estimates within a super-resolution mapping pattern prediction algorithm to evaluate the benefit of the local estimates for super-resolution mapping.

6 Conclusion

This paper has introduced a new method for combining a global training image semivariogram $\hat{\gamma}_v^T(\mathbf{h})$ that is well estimated but not necessarily representative of the target, and not representative locally within the target, with an estimate of the local semivariogram of proportions $\hat{\gamma}_v(\mathbf{h}, W(\mathbf{x}))$ that was obtained by a deconvolution-convolution procedure based on a locally available coarse spatial resolution image of proportions. The method works sufficiently well to merit further investigation along the lines suggested in the discussion section above.

Acknowledgments The authors would like to thank the University of Southampton for providing funding to support this research.

References

Adams JB, Smith MO, Johnson PE (1985) Spectral mixture modelling: a new analysis of rock and soil types at the Viking Lander 1 site. J. Geophys. Res 91:8098–8112

Atkinson PM (1993) The effect of spatial resolution on the experimental variogram of airborne MSS imagery. Int J Remote Sens 14:1005–1011

Atkinson PM (1995) Regularizing variograms of airborne MSS imagery. Can J Remote Sens 21:225–233

Atkinson PM (2004) Super-resolution land cover classification using the two-point histogram, GeoENV IV: Geostatistics for Environmental Applications, pp 15–28

Atkinson PM (2005) Super-resolution target mapping from soft classified remotely sensed imagery. Photogrammetric Eng Remote Sens 71:839–846

Atkinson PM, Cutler MEJ, Lewis HG (1997) Mapping sub-pixel proportional land cover with AVHRR imagery. Int J Remote Sens 18:917–935

Boucher A, Kyriakidis PC (2006) Super-resolution land cover mapping with indicator geostatistics. Remote Sens Environ 104:264–282

Clark I (1977) Regularization of a semi-variogram. Comput Geosci 3:341–346

Collins JB, Woodcock CE (1999) Modelling the distribution of cover fraction of a geophysical field. In: Atkinson, PM, Tate NJ (eds) Advances in Remote Sensing and GIS Analysis. Wiley, Chichester, pp 119–133

Curran PJ, Atkinson PM (1998) Geostatistics and remote sensing, Prog Phys Geo 22:61–78

Foody GM (1996) Approaches for the production and evaluation of fuzzy land cover classifications from remotely-sensed data. Int J Remote Sens 17:1317–1340

Journel AG, Huijbregts CJ (1978) Mining geostatistics. Academic Press, London

Jupp DLB, Strahler AH, Woodcock CE (1988) Autocorrelation and regularization in digital images. I. Basic theory. IEEE Trans Geosci Remote Sens 26:463–473

Jupp DLB, Strahler AH, Woodcock CE (1989) Autocorrelation and regularization in digital images. II. Simple image models. IEEE Trans Geosci Remote Sens 27:247–258

Kasetkasem T, Arora MK, Varshney PK (2005) Super-resolution land cover mapping using a Markov random field based approach. Remote Sens Environ 96:302–314

Mertens KC, Verbeke LPC, Ducheyne EI, De Wulf RR (2003) Using genetic algorithms in sub-pixel mapping. Int J Geo Inf Sci 24:4241–4247

Mertens KC, Verbeke LPC, Westra T, De Wulf RR (2004) Sub-pixel mapping and sub-pixel sharpening using neural network predicted wavelet coefficients. Remote Sens Environ 91:225–236

Nguyen MQ, Atkinson PM, Lewis HG (2006) Super-resolution mapping using a Hopfield neural network with fused images. IEEE Trans Geosci Remote Sens 44:736–749

Tatem AJ, Lewis HG, Atkinson PM, Nixon MS (2001a) Super-resolution target identification from remotely sensed images using a Hopfield neural network. IEEE Trans Geosci Remote Sens 39:781–796

Tatem AJ, Lewis HG, Atkinson PM, Nixon MS (2001b) Multiple-class land-cover mapping at the sub-pixel scale using a Hopfield neural network. Int J Appl Earth Obs Geoinf 3:184–190

Tatem AJ, Lewis HG, Atkinson PM, Nixon MS (2002) Super-resolution land cover pattern prediction using a Hopfield neural network. Remote Sens Environ 79:1–14

Tatem AJ, Lewis HG, Atkinson PM, Nixon MS (2003) Increasing the spatial resolution of Landsat TM imagery for land cover mapping in agricultural areas. Int J Geo Inf Sci 17:647–672

Thornton MW, Atkinson PM, Holland DA (2006) Super-resolution mapping of rural land cover features from fine spatial resolution satellite sensor imagery. Int J Remote Sens 27:473–491

Verhoeye J, De Wulf R (2002) Land cover mapping at sub-pixel scales using linear optimization techniques. Remote Sens Environ 79:96–104

Woodcock CE, Strahler AH (1987) The factor of scale in remote sensing. Remote Sens Environ 21:311–322

Zhan Q, Molenaar M, Lucieer A (2002) Pixel unmixing at the sub-pixel scale based on land cover class probabilities: application to urban areas. In: Foody GM, Atkinson PM (eds) Uncertainty in remote sensing and GIS. Wiley, Chichester, pp 59–76

Modeling Spatial Uncertainty for Locally Uncertain Data

Elena Savelyeva, Sergey Utkin, Sergey Kazakov, and Vasyliy Demyanov

Abstract The work discusses methods dealing with "soft" input data, where local uncertainty is represented by a variance. Modifications of ordinary kriging and sequential direct stochastic simulations based on such data are applied to a real hydrogeological case study and a synthetic environmental contamination study. The modification performed on direct simulation approach does not require any data transformation assumptions. The method is compared with Bayesian Maximum Entropy (BME) based stochastic simulations, which provide an alternative way of integrating "soft" information.

1 Introduction

Uncertainty manifests at all stages of data handling starting with data acquisition and propagating through fitting model parameters process and the actual modeling. Nowadays results of data analysis (modeling and forecasting) are accompanied by an estimate of the model uncertainty and assessment of uncertainty is considered as an important part of the data analysis process. However, many conventional methods assume input data as exact, which is clearly questionable. Usually, raw data carry internal uncertainty caused by equipment errors, calibration and/or different kinds of methodological assumptions and expert's judgment. For example, dealing with analysis of marine living resources acoustic and trawl surveys' measurements are calibrated to the actual variables of interest (model input raw data) using some heuristic constants (for example, trap effective radius) (Sokolov, 2006). Thus, the stochastic nature represented by additional information is not taken into account (Barange et al., 1996 etc.). Monitoring measurements are also recalculated usually averaging over some period or surface, but there are other possible expert

E. Savelyeva (✉), S. Utkin, and S. Kazakov
Nuclear Safety Institute Russian Academy of Sciences, B.Tulskaya 52, 113191, Moscow, Russia
e-mail: esav@ibrae.ac.ru

V. Demyanov
Institute of Petroleum Engineering, Heriot-Watt University, Edinburgh, UK
e-mail: vasily.demyanov@pet.hw.ac.uk

P.M. Atkinson and C.D. Lloyd (eds.), *geoENV VII – Geostatistics for Environmental Applications*, Quantitative Geology and Geostatistics 16,
DOI 10.1007/978-90-481-2322-3_26,

recalculations. The same situation is expected in other branches dealing with measurements. Thus, all "raw" data are generally what is referred to as "soft" data. Really exact "hard" data do not exist.

The time has come for development of approaches dealing with real raw measurements, not cleaned and processed in an ad hoc way, smearing possible uncertainties. At the current stage we consider uncertain data as data with associated local uncertainty (uncertainty for each datum). Such an approach can be referred as "soft" geostatistics.

Parkin et al. (2005) discuss methods that allow estimation of local probabilistic features based on "soft" data only (multiple measurements reproduced into local probability density functions). However, local uncertainty of the model is not always enough for data description. For instance, levels of uncertainty for fish total biomass estimate require modeling of spatial uncertainty (Savelieva et al., 2007) – by means of stochastic simulations approach. Thus, our goal is to reproduce true spatial variability conditioned to given uncertain "soft" data by means of "soft" stochastic simulations.

In the present work we consider two types of "soft" stochastic simulation approaches: the Bayesian Maximum Entropy (BME) based stochastic simulations and the generalization of sequential direct stochastic simulations to the case of uncertain data. Bayesian Maximum Entropy (BME) is a general tool rigorously incorporating the different kinds of available knowledge including "soft" site specific data (Christakos, 2000; Christakos et al., 2002). The result of BME local estimation is a posterior probability density function, which allows extension of BME to produce sequential stochastic simulations. In general, the BME methodology allows consideration of the whole estimation grid (spatial pattern) simultaneously without the sequential principle which sometimes underestimates the variability when dealing with local probability distribution functions, but in the current work we use only the sequential approach.

The other approach we consider originates from classical geostatistics – to make geostatistical fans rejoice – modified sequential direct stochastic simulations. Kriging technology allows incorporation of measurement errors considering them as local variances. Kriging with measurement errors is a simple modification of ordinary kriging. Direct simulations is a stochastic kriging based algorithm to describe spatial uncertainty (Soares, 2001). Unlike other geostatistical simulation algorithms it does not require any data transformation and assumptions about the distribution. Direct simulations based on kriging estimates in the original data scale open the possibility to replace ordinary kriging with the kriging accounting for measurement errors. This approach is simpler than the BME based method and requires less computational time.

Development of these methods was initially motivated by analysis and mapping of the commercial marine living resources (Savelieva et al., 2007). In this work they are applied to hydrogeologic and environmental contamination data to assess the feasibility of the approach.

2 Theoretical Background

2.1 Uncertainty Description

Uncertainty of data requires detailed formalization before incorporation into any kind of analysis. Description of data uncertainty can be crudely divided into the following main groups: level of confidence, range interval, variance and probability density function (pdf). Every type of uncertainty description is associated with measurement methodology.

Measurement creditability is an expert's decision on the quality of measurement. It depends on subjective features, such as, the measurement session procedure, quality of a sample and others. Creditability is usually given as a categorical value from a specified range. Level of creditability can be used in model parameter fitting (for example, in machine learning approaches) – greater attention is paid to more reliable data. In this work we will not consider such types of uncertainty.

Uncertainty of data also originates from the measurement device error. Usually devices are calibrated so as to provide a variance (σ^2) as a guide to the measurement uncertainty. Such variances depend only on the device and are constant over the measuring area. Such variances allow estimation of data confidence levels. Under the assumption of local Gaussian distribution the 95% confidence level is given by a 4σ interval. Avoiding any distributional assumption the 95% confidence level increases to $6\sigma(2\sigma$ as a penalty for unknown distribution). This result, based on application of Visochansky-Pitunin equation, is discussed in Chiles and Delfiner (1999). Thus, device error can be considered as a variance or formalized as interval of possible data values.

Many sampling methodologies use repeated measurements performed in close proximity (much smaller than the distances between sample locations) – for example, in soil sciences. Usually, these repeated measurements are subject to preliminary expert analysis (using a specially prescribed methodology depending on the science domain) with a unique value as a result. This value is considered as a measurement in the following analysis. Sometimes it is accompanied with a creditability level as discussed above.

In reality, repeated measurements provide the widest possibility for data uncertainty description. Considering them as realizations of a random process with unknown probability distribution function one can treat their uncertainty as an interval (after estimation of most probable value and variance or confidence intervals) or as a model of the probability density function (non-parametric – a histogram based on cut-offs or parametric – model with fitted parameters). Examples of such descriptions (after Parkin et al., 2005 and Savelieva et al., 2005) are presented in Figs. 1 and 2. The dataset was devoted to ^{137}Cs soil contamination due to the Chernobyl accident (26/04/1986). Repeated measurements were taken close to each other and recalculated to the date of the accident. Figure 1 presents local pdfs based on a small number of values (below 15), allowing estimation of the minimum,

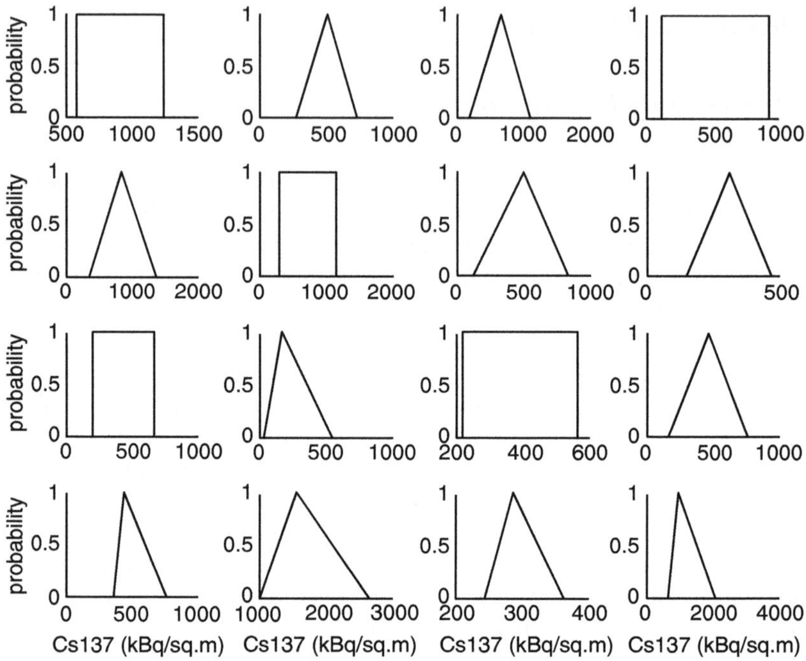

Fig. 1 Examples of "soft" pdfs based on minimum, maximum and median

maximum and a median considered in this case as the most probable value. Figure 2 presents examples of raw histograms and the fitted models.

2.2 Kriging with Measurement Errors

Ordinary kriging is a well-known geostatistical estimator described in the literature, e.g. by Goovaerts (1997), Chiles and Delfiner (1999), etc. Below we briefly outline the modification of ordinary kriging to take into account measurement errors.

Following conventional geostatistics, let us suppose that there is a random field $Z(x)$ represented by a set of given values Z_i measured at locations x_i with measurement errors ε_i:$Z(x_i) = Z_i \pm \varepsilon_i$. Several assumptions concerning the measurement errors are made: the errors are uncorrelated ($S_{ij} = E\{\varepsilon_i \varepsilon_j\} = 0$), the errors are not correlated with the value ($E\{Z_i \varepsilon_i\} = 0$) and the error variances ($E\{\varepsilon_i \varepsilon_i\} = \sigma_i^2$) are known. Kriging is a linear estimator, for an unmeasured location x_0 ($Z^*(x_0)$) given by:

$$Z^*(x_0) = \sum_{i=1}^{N(x_0)} \lambda_i(x_0) Z(x_i), \tag{1}$$

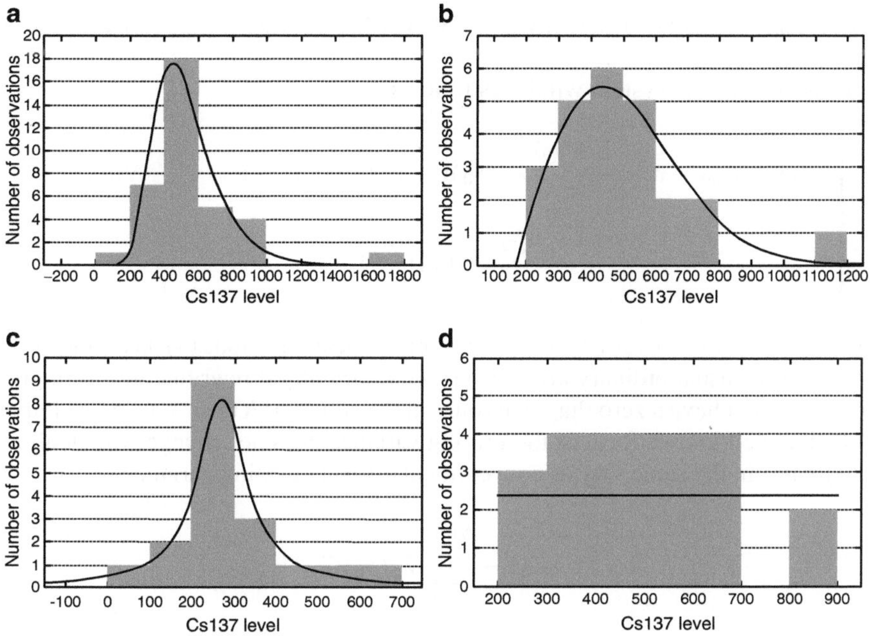

Fig. 2 Examples of raw histograms and fitted models of the pdf (scaled to be compared with the histogram): (**a**) extreme distribution; (**b**) Weibul distribution; (**c**) Cauchy distribution; (**d**) uniform distribution

where $N(x_0)$ is the number of samples from the neighbourhood of x_0 taken into account for the estimation, and $\lambda_i(x_0)$ are the kriging weights. The neighbourhood depends on a user-defined search rule.

As for all members of the kriging family, the set of weights (λ_i) is determined by minimizing the estimation variance under the unbiasedness constraint:

$$\min_{\lambda}\left(Var\left(Z*(x_0;\lambda)-Z(x_0)\right)-2\mu\left(\sum_{i=1}^{N(x_0)}\lambda_i(x_0)-1\right)\right) \tag{2}$$

where μ is the Lagrangian multiplier. Let us consider the estimation variance

$$Var\left(Z^*(x_0;\lambda)-Z(x_0)\right)=E\left(\sum_{i=1}^{N(x_0)}\lambda_i(x_0)Z(x_i)-Z(x_0)\right)^2=$$
$$\sum_{i=1}^{N(x_0)}\lambda_i^2(x_0)\left(\sigma_i^2+C_{00}\right)+2\sum_{i=1}^{N(x_0)}\sum_{\substack{j=1\\j\neq i}}^{N(x_0)}\lambda_i(x_0)\lambda_j(x_0)C_{ij}$$
$$-2\sum_{i=1}^{N(x_0)}\lambda_i(x_0)C_{i0}+C_{00} \tag{3}$$

Where $C_{00} = Var(ZZ)\forall Z(x)$, $C_{ij} = Var(Z_iZ_j) = Var(Z(x_i)Z(x_j))$ and $C_{i0} = Var(Z(x_i)Z(x_0))$.

Thus, the kriging system will be as follows:

$$\begin{cases} \lambda_i(x_0)(C_{00} + \sigma_i^2) + \sum\limits_{\substack{j=1 \\ j \neq i}}^{N(x_0)} \lambda_j(x_0)C_{ij} = C_{i0} + \mu, \quad i = 1 \ldots N(x_0) \\ \sum\limits_{i=1}^{N(x_0)} \lambda_i(x_0) = 1 \end{cases}, \quad (4)$$

where σ_i^2 are the error variances. The system of linear equations (4) has some difference from the usual ordinary kriging system. Especially if rewritten to a variogram form it will not have a zero diagonal (only in the equation $N(x_0) + 1$). Most importantly, the well-known form of the kriging variance does not change, which allows us to treat it in the same way as conventional ordinary kriging variance:

$$\sigma_K{}^2 = C_{00} - \sum_{i=1}^{N(x_0)} \lambda_i(x_0)C_{i0} + \mu. \quad (5)$$

2.3 Direct Sequential Simulations with Measurement Errors

The direct sequential simulations (DSS) approach (Soares, 2001) is based on the sequential principle for constructing spatial uncertainty (global multivariable probability distribution function) and kriging technology for estimation of the local cumulative distribution function (cdf). Direct sequential simulations appears to be well adapted to highly skewed data (Savelieva et al., 2007).

Each realization (pattern) is characterized by two main features: a global cdf of the random field $Z(x)$ and its spatial correlation structure. Kriging is responsible for reproduction of spatial correlation of a pattern. To reproduce a global cdf a set of specially organized classes are used. Each class is provided by a class-specific pdf to draw a current value. A selected class depends on the value of the kriging estimate.

One of the possible ways to construct a system of classes is to use a normalization transformation function ϕ (presented as a table) linking data quintile values with normal distribution quintile values. The local cdf at location x_0 while spatial pattern modeling uses kriging and a transformation function ϕ. The kriging estimate transformed to the normal distribution ($y(x_0) = \phi(z^*(x_0))$) indicates the location moment of the local cdf. The kriging variance ($\sigma_K{}^2(x_0)$) depends mostly on the sill value of the spatial correlation model of the underlying process, and its transformation is a normalization on C_{00}. Thus, the kriging variance ($\sigma_K{}^2(x_0)/C_{00}$) gives a space moment of the local cdf. In the normalized space, the local cdf is considered to be Gaussian – $N(y(x_0), \sigma_K{}^2(x_0)/C_{00})$. A value for the current realization is drawn using this local cdf (y') and back-transformed to initial data space $z'(x_0) = \phi^{-1}(y')$.

The main advantage of the direct sequential simulation approach is that no transformation of the data is required. It allows the modification of this approach to take measurement errors (presented by error variances) into account. Ordinary (or simple) kriging applied in the original version is replaced by kriging with measurement errors, as described in Section 1.2. The current realization value along with the kriging variance (treated equivalent to a measurement error variance) is stored in the database for the following application within the sequential simulation framework.

Other traditional geostatistical sequential stochastic simulation approaches such as Gaussian or indicator sequential simulations require additional assumptions on the pdf of measurement errors. In the Gaussian context, measurement errors need to be considered as normally distributed. The indicator approach does not require any specific type of distribution, but "soft" indicator transformation requires knowledge of the local pdf.

2.4 Several Remarks on BME

A detailed description of computational and theoretical aspects of the Bayesian Maximum Entropy (BME) theory and practical recommendations concerning its application can be found in Christakos (2000, 2002). Here we briefly outline the basic features of the BME method that are relevant to the present work.

The spatial distribution of a physical variable is routinely represented by means of a spatial random field (SRF) $X(\boldsymbol{s})$, where the vector $\boldsymbol{s}$ denotes spatial location. The BME mapping framework integrates various physical knowledge bases, such as the general knowledge base $\mathcal{G}$ (physical laws, empirical relations, statistical moments of any order, scientific theories etc.) and the site-specific knowledge base $\mathcal{S}$ (real measurements, uncertain observations, secondary information etc.) to construct the posterior pdf of $X(\boldsymbol{s})$ at any mapping point $\boldsymbol{s}_k$. It is performed in several stages: first the structural (or prior) pdf model, f_G, of SRF $X(\boldsymbol{s})$ at all mapping points $\boldsymbol{s}_{map} = (\boldsymbol{s}_{soft}, \boldsymbol{s}_k)$ is derived from the general knowledge $\mathcal{G}$. After that, the prior pdf is conditioned (by means of an operational Bayesian conditionalization rule) to the site-specific knowledge $\mathcal{S}$ leading to the posterior pdf f_K:

$$f_K(\chi_k) = A^{-1} \int d\chi_{soft} f_S(\chi_{soft}) f_G(\chi_{map}), \tag{6}$$

where A is a normalization parameter. Clearly this posterior pdf is not limited by any specific distribution, giving a realistic stochastic description of a physical variable across space.

In the framework of sequential stochastic simulations the prior pdf f_G is derived once, while the conditioning stage is performed consequently for mapping points $\boldsymbol{s}_k$, taking them one by one. The posterior pdf, $f_K \boldsymbol{s}_k)$, allows the drawing of a stochastic value as a realization and characteristics analogous to soft description of site-specific information to be added to S as new points $\mathbf{s}_{soft}$.

3 Data Description

The methods described above are applied to synthetic and real data: groundwater level monitoring in the vicinity of a nuclear waste storage facility and air quality monitoring around a nuclear power plant (NPP) in a normal situation and in a situation of an accidental release of radioactivity.

Groundwater levels are measured at 52 spatially distributed monitoring wells (Nuzhny et al., 2007). The total number of measurements is 26,999, covering the period from April 1970 to January 2006. The time is not treated as an additional coordinate, so each well is presented by a set of measurements providing uncertainty. The local difference between maximum and minimum values ranges from 1.7 up to 14.6 m. The mean and variance per well were considered as a measurement value and a measurement variance. Local variability varies in space.

The synthetic data used in the current analysis were used in the Spatial Interpolation Comparison (SIC) 2004 contest (Dubois and Galmarini, 2005). The participants were provided with the prior information representing the spatial behavior of the monitoring data (200 sample locations) in ten different days. These data can be considered as the repeated measurements, which describe the local uncertainties (for example, local variances). Two new data sets were distributed during the competition for interpolation at 808 validation locations by a method tuned on prior data. One of the data sets described an ordinary monitoring situation around a NPP, the other one was obtained as a model of an accidental release of radioactivity. In our study we used these data sets as measurements together with the uncertainty obtained on the basis of the prior data. Presence of the validation data allows us to check the reliability of the methods' performance.

4 Results and Discussion

Figures 3 and 4 present comparative results of ordinary kriging and kriging with measurement errors applied to the groundwater levels. It can be observed that the estimates of kriging with measurement errors are smoother (Fig. 4). The same conclusion can be reached from the comparison of the minimum and the maximum values. Kriging with measurement errors gives higher minimum (219.4 vs. 217.8 m) and lower maximum (251.7 vs. 251.94 m), while the median estimates for the both methods present the same value – 237.47 m. Kriging variances in accordance with Eq. (5) are very similar to each other but they are not identical due to the difference in the coefficients obtained through solving different systems of linear equations.

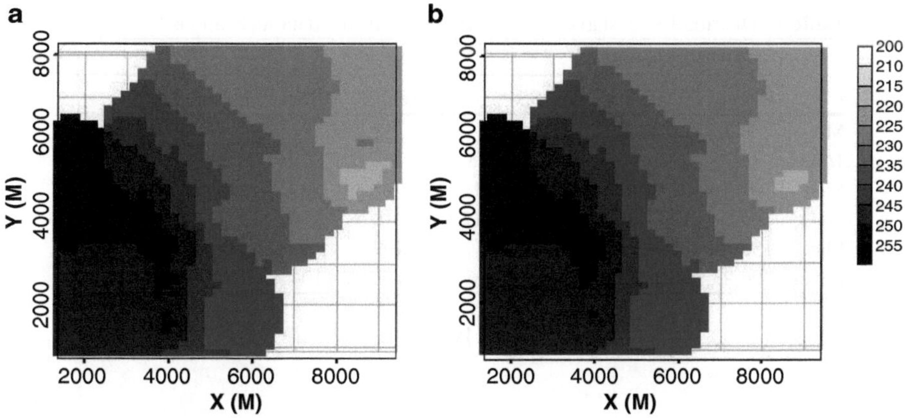

Fig. 3 Kriging estimate: (**a**) ordinary kriging; (**b**) kriging with measurement error

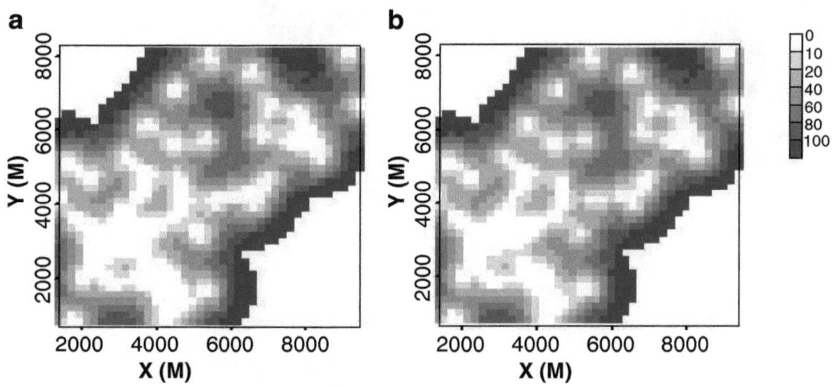

Fig. 4 Kriging variance: (**a**) ordinary kriging; (**b**) kriging with measurement error

Validation of kriging with measurement errors on the SIC'2004 data showed better results compared with ordinary kriging. Thus: Pearson correlation coefficient for the ordinary data set increased to 0.78 from 0.73, and for data with release up to 0.69 from 0.56.

Results of stochastic simulations performed for SIC'2004 data are presented in Table 1, as a distribution of statistical characteristics over 50 DSS and 20 BME realizations for each data set. DSS realizations appeared to be more variable, which is indicated by larger range of values. The simulations were performed on a regular grid, as statistics of the real values are collected on a more dense validation set. BME-based simulations present higher variability in statistical features.

Figures 5a–c and 6a–b present examples of stochastic realizations by direct sequential simulations (Fig. 5) and BME-based sequential simulations (Fig. 6). DS realizations look rather strange, but averaging over 50 realizations (Fig. 5d) indicates some correspondence with kriging estimates (Fig. 3).

Table 1 Distribution in statistical characteristics of SIC data realizations

–	Ordinary situation			Accident case		
–	Real	BME	DSS	Real	BME	DSS
Min	58.2	57–58.2	57–57.2	58.2	57–57.3	57–57.4
1/4Q	82.2	82.2–86.7	82.3–85.8	82.2	82.1–86.12	82.3–85.5
Med	97.5	97.9–101.1	97.5–101	97.6	97.8–109.8	97.5–101.4
3/4Q	109.5	109–112.5	108.5–112.8	110.5	110.2–113.7	109.6–114
Max	153	151–170.3	153–168	1,499	1,300–1,501	1,499–1,500
Mean	96.23	96.42–99.45	95.8–98.6	108.99	104.38–108.2	104.3–107.4

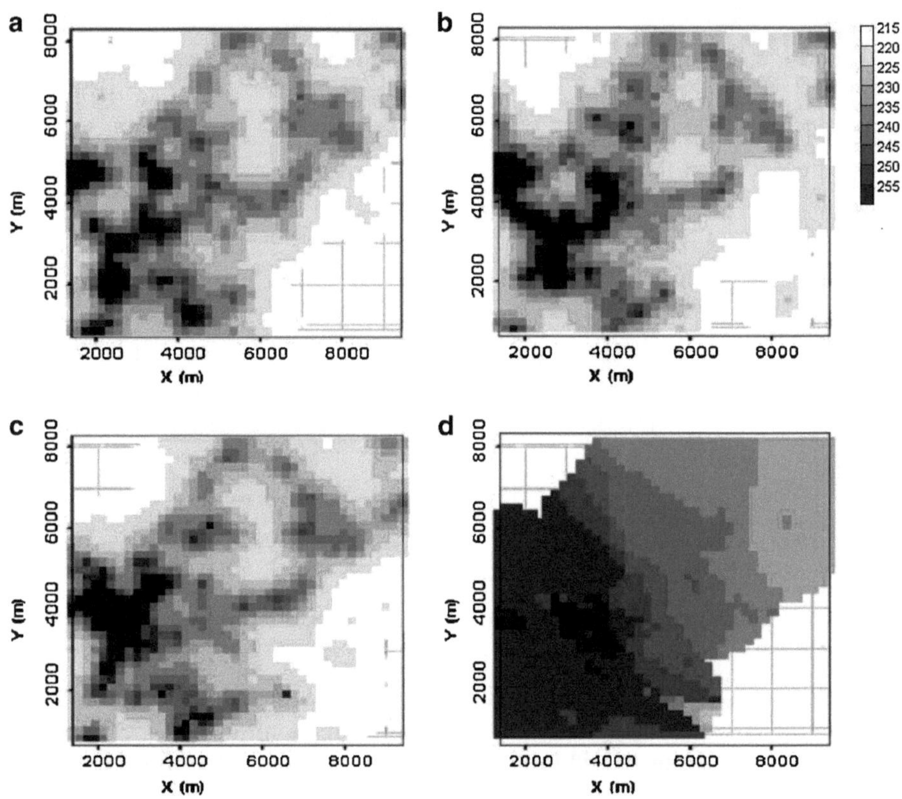

Fig. 5 Examples of direct sequential simulations with measurement error (**a–c**) and average over 50 realizations (**d**)

5 Conclusions

The main conclusion of this work is that there are different ways in which pure "soft" data can be used for analysis and modeling of spatial uncertainty. BME-base stochastic simulations – a BME extension – has demonstrated advantages over kriging-based methods with accounting for the measurement error. The presented

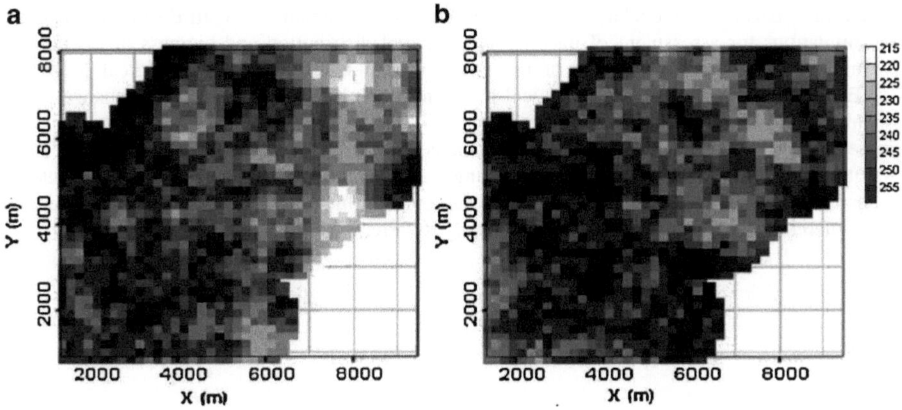

Fig. 6 Examples of BME based stochastic simulations

case studies showed the feasibility of the approach. However, more research is needed to make more general conclusions about the proposed method. The method can be used in the case studies where the raw data feature very high local uncertainty. Further research will be devoted to incorporation in the methodology of different types of local uncertainty descriptions.

Acknowledgments The work was partly supported by Russian fund for fundamental researches (RFFI) 07-08-00257.

References

Barange M, Hampton I, Soule M (1996) Empirical determination of in situ target strengths of three loosely aggregated pelagic fish species. ICES J Marine Sci 53:225–232

Chiles J-P, Delfiner P (1999) Geostatistics. Modeling spatial uncertainty. Wiley, New York

Christakos G (2000) Modern spatiotemporal geostatistics. Oxford University Press, New York

Christakos G, Bogaert P, Serre M (2002) Temporal GIS. Springer, New York

Goovaerts P (1997) Geostatistics for natural resources evaluation. Oxford University Press, New York

Dubois G, Galmarini S (2005) Introduction to the spatial interpolation comparison (SIC) 2004 exercise and presentation of the data sets. Appl GIS 2:9-01–9-10

Nuzhny A, Savelieva E, Jastrebkov A (2007) Statistical analysis for extracting features on the groundwater level dynamics. In: Zhao P, Agterberg F, Cheng Q (eds) Proceedings of IAMG2007, Geomathematics and GIS analysis of resources, Environment and Hazards, Beijing, pp 723–726

Parkin R, Savelieva E, Serre M (2005) "Soft" geostatistical analysis of radioactive soil contamination. In: Renard P, Demougeot-Renard H, Froidevaux R (eds) Geostatistics for environmental applications. Springer, Berlin, pp 331–342

Savelieva E, Demyanov V, Kanevski M, Serre M, Christakos G (2005) BME_based uncertainty assessment of the chernobyl fallout. Geoderma 128:312–324

Savelieva E, Bizikov V, Goncharov S, Popov S, Mazzola S, Bonanno A, Patti B (2007) Stochastic simulations for assessment of uncertainty of spatial distribution and biomass of marine living resources. In: Proceedings of the 6th European Conference on Ecological Modeling (ECEM'07), Challenges for ecological modeling in a changing world: Global changes, sustainability and ecosystem management, pp 457–458. Trieste

Soares A (2001) Direct sequential simulation and cosimulation. Math Geol 33:911–926

Sokolov VI (2006) Stock assessment of red king crab (Paralithodes camtschaticus) in the Russian part of the Barents Sea basing on the trap survey data. In: Abstracts of the VIIth All-Russian Conference on Commercial Invertebrates. VNIRO Press, Murmansk, pp 129–132

Spatial Interpolation Using Copula-Based Geostatistical Models

Hannes Kazianka and Jürgen Pilz

Abstract It is common practice in geostatistics to use the variogram to describe the spatial dependence structure of the underlying random field. However, the variogram is sensitive to outlying observations and strongly influenced by the marginal distribution of the random field. As an alternative to spatial modeling using the variogram we consider describing the spatial correlation by means of copula functions. We present three methods for performing spatial interpolation using copulas. By exploiting the relationship between bivariate copulas and indicator covariances, the first method performs indicator kriging and disjunctive kriging. As a second method we propose a simple kriging of the rank-transformed data. The third method is a plug-in Bayes predictor, where the predictive distribution is calculated using the conditional copula given the observed data and the model parameters. We show that the latter approach generalizes the frequently applied trans-Gaussian kriging. Finally, we report on the results obtained for the so-called Joker data set from the spatial interpolation comparison SIC2004.

1 Introduction

Copulas describe the dependence between random variables independently of their marginal distributions. They are commonly used in financial and actuarial statistics, however, they are just beginning to become popular in geostatistics. Spatial dependence is traditionally described using the variogram which is strongly influenced by the univariate distribution of the random field. Extreme outlying observations adversely affect the empirical and theoretical variogram estimates. Moreover, spatial modeling often relies on the Gaussian assumption which is hardly fulfilled for environmental processes. To circumvent these disadvantages Bardossy (2006) proposed the use of copulas to describe the spatial variability. In the following we adopt this

H. Kazianka (✉) and J. Pilz
Institute of Statistics, University of Klagenfurt, Universitätsstraße 65-67,
9020 Klagenfurt, Austria
e-mail: hannes.kazianka@uni-klu.ac.at; juergen.pilz@uni-klu.ac.at

P.M. Atkinson and C.D. Lloyd (eds.), *geoENV VII – Geostatistics for Environmental Applications*, Quantitative Geology and Geostatistics 16,
DOI 10.1007/978-90-481-2322-3_27, © Springer Science+Business Media B.V. 2010

methodology and use it not only for spatial modeling of dependence structures but also for spatial interpolation. We present three methods for estimating the values of the random field at unknown locations. The first method we suggest is indicator and disjunctive kriging. The second method is rank-order kriging, originally proposed by Journel and Deutsch (1996), where we calculate the covariance function through its relationship to the Spearman rank correlation. The third method is a plug-in Bayes predictor and can be used if all multivariate distributions of the random field are modeled using the copula.

The paper is organized as follows. Section 2 reviews the basic properties of copulas, while Section 3 briefly describes the spatial copula methodology. In Section 4 the spatial interpolation techniques using copulas are presented and in Section 5 they are used to analyze the Joker data set from the spatial interpolation comparison SIC2004 (Dubois, 2005). Section 6 is devoted to conclusions.

2 Copulas

The word "copula" was first used by Sklar (1959) to describe distribution functions on the n-dimensional unit cube, $\boldsymbol{I}^n$, that link multivariate distributions to their one-dimensional margins. To be precise, an n-dimensional copula is an n-dimensional real function $C : \boldsymbol{I}^n \to \boldsymbol{I}$ which satisfies the following properties:

1. For every $\boldsymbol{u} \in \boldsymbol{I}^n$

$$C(\boldsymbol{u}) = 0 \text{ if at least one coordinate of } \boldsymbol{u} \text{ equals } 0,$$
$$C(\boldsymbol{u}) = u_k \text{ if all coordinates of } \boldsymbol{u} \text{ are 1 except } u_k.$$

2. For every $\boldsymbol{a}, \boldsymbol{b} \in \boldsymbol{I}^n$ with $\boldsymbol{a} \leq \boldsymbol{b}$ the n-th order difference of C on $[\boldsymbol{a}, \boldsymbol{b}]$

$$V_C([\boldsymbol{a}, \boldsymbol{b}]) = \Delta_{a_n}^{b_n} \Delta_{a_{n-1}}^{b_{n-1}} \dots \Delta_{a_1}^{b_1} C(\boldsymbol{u}) \geq 0.$$

In the above expression a first order difference is defined as $\Delta_{a_k}^{b_k} C(\boldsymbol{u}) = C(u_1, \dots, u_{k-1}, b_k, u_{k+1}, \dots, u_n) - C(u_1, \dots, u_{k-1}, a_k, u_{k+1}, \dots, u_n)$.

From this definition it is clear that a copula is a distribution function on the n-dimensional unit cube with uniformly distributed margins. The most important theoretical result about copulas was proved by Sklar (1959) and expresses the ability of copulas to describe the dependence between random variables without information about their marginal distributions: If H denotes an n-dimensional distribution function with margins $F_1, \dots, F_n$, then there exists an n-dimensional copula C such that for all $\boldsymbol{x} \in \overline{\mathbb{R}}^n$,

$$H(x_1, \dots, x_n) = C(F_1(x_1), \dots, F_n(x_n)). \tag{1}$$

If $F_1, \dots F_n$ are all continuous, then C is unique. Conversely, if C is an n-dimensional copula and $F_1, \dots F_n$ are distribution functions, then the function

H is an n-dimensional distribution function with margins $F_1, \ldots F_n$. Moreover, if $F_1^{-1}, \ldots, F_n^{-1}$ are the inverse distribution functions of $F_1, \ldots, F_n$, we get

$$C(u_1, \ldots, u_n) = H\left(F_1^{-1}(u_1), \ldots, F_n^{-1}(u_n)\right). \tag{2}$$

If C is an absolutely continuous copula, its density can be written as

$$c(u_1, \ldots, u_n) = \frac{\partial^n C(u_1, \ldots, u_n)}{\partial u_1 \ldots \partial u_n} = \frac{h\left(F_1^{-1}(u_1), \ldots, F_n^{-1}(u_n)\right)}{\prod_{i=1}^{n} f_i\left(F_i^{-1}(u_i)\right)}, \tag{3}$$

where h denotes the density of H and the f_i denote the densities of F_i. One of the advantages of working with copulas is that they are invariant under strictly increasing transformations of the random variables. Therefore, typical data transformation methods, such as taking the logarithm or performing a Box–Cox transformation, have no impact on the copula. Bivariate copulas are directly linked to the scale free measure of association known as Spearman's rho. The Spearman rank correlation between two random variables X_1 and X_2 with copula C can be calculated as

$$\rho_{X_1,X_2} = 12 \int\int_{I^2} u_1 u_2 dC(u_1, u_2) - 3 = 12 \int\int_{I^2} C(u_1, u_2)\, du_1 du_2 - 3. \tag{4}$$

For a thorough introduction to copulas the reader is referred to Nelsen (2006).

3 Spatial Modeling Using Copulas

Although copulas are widely used for describing the dependence between random variables, for example in financial statistics, there are only a few papers about incorporating copulas into the geostatistical framework so far. In the following assume that we have a second-order stationary random field $\{Z(\boldsymbol{x}) \mid \boldsymbol{x} \in \boldsymbol{S}\}$, where $\boldsymbol{S} \subseteq \mathbb{R}^2$ is the area of interest.

3.1 Describing the Random Field Using Copulas

Bardossy (2006) presented a method for spatial modeling using copulas that generalizes the concept of the variogram. Let F_Z denote the univariate distribution of the random process which is the same for each location $\boldsymbol{x}$ due to stationarity. All multivariate distributions of the random field are described using multivariate copulas with the help of Sklar's theorem (see Eq. 1). For example, the relation between two locations separated by the vector $\boldsymbol{h}$ is characterized by the bivariate distribution

$$P(Z(\boldsymbol{x}) \le z_1, Z(\boldsymbol{x}+\boldsymbol{h}) \le z_2) = C_{\boldsymbol{h}}(F_Z(z_1), F_Z(z_2)), \tag{5}$$

whose dependence structure is described by the copula $C_{\boldsymbol{h}}$. The copula becomes a function of the separating vector $\boldsymbol{h}$ (or the separating distance $h := \|\boldsymbol{h}\|$ if the random field is isotropic) and does not depend on the location $\boldsymbol{x}$. Hence, the spatial copula describes the dependence over the whole range of quantiles and not only the mean dependence as the variogram does. Every spatial copula is symmetric by definition. This means that $C_{\boldsymbol{h}}(u_1, \ldots, u_n) = C_{\boldsymbol{h}}\left(u_{\pi(1)}, \ldots, u_{\pi(n)}\right)$ for an arbitrary permutation π and $n \geq 2$. Moreover, we want to add the following two restrictions: $\|\boldsymbol{h}\| \to \infty$ implies $C_{\boldsymbol{h}}(\boldsymbol{u}) \to \prod_{i=1}^{n} u_i$ and $\|\boldsymbol{h}\| \to 0$ implies $C_{\boldsymbol{h}}(\boldsymbol{u}) \to \min_i u_i$. These restrictions ensure that far distant observations have almost no dependence and observations that are very close to each other have a strong dependence.

The special case of a Gaussian random field, where the copula can be written as $C(u_1, \ldots, u_n) = \boldsymbol{\Phi}_{\boldsymbol{0},\boldsymbol{\Lambda}}\left(\Phi^{-1}(u_1), \ldots, \Phi^{-1}(u_n)\right)$ with $\operatorname{diag}(\boldsymbol{\Lambda}) = (1, \ldots, 1)^T$ and the marginal distribution is $F_Z = \Phi_{m,\sigma^2}$, is included in this model. Here, $\boldsymbol{\Phi}_{\boldsymbol{\mu},\boldsymbol{\Sigma}}$ denotes the distribution function of the multivariate Gaussian distribution with mean vector $\boldsymbol{\mu}$ and covariance matrix $\boldsymbol{\Sigma}$. The Gaussian copula becomes a function of $\boldsymbol{h}$ by assuming that its correlation function follows one of the well-known parametric geostatistical models, e.g. the Matern model. However, the Gaussian copula, as well as the Student-t copula, does not only express a symmetric but also a radially symmetric dependence, $C(u_1, u_2) = u_1 + u_2 - 1 + C(1 - u_1, 1 - u_2)$. This means that high and low values of the distribution have equal dependence properties. To allow for more flexibility Bardossy (2006) introduced a non-Gaussian copula family which is constructed from a multivariate non-central χ^2-distribution. Squaring the entries of a Gaussian random vector $\boldsymbol{Y} \sim \mathcal{N}(\boldsymbol{m}, \boldsymbol{\Sigma})$, where $\boldsymbol{m} = (m, \ldots, m)$ and $\boldsymbol{\Sigma}$ denote the mean vector and the correlation matrix respectively, leads to a multivariate distribution with margins having a non-central χ^2-distribution with 1 degree of freedom and non-centrality parameter $\lambda = m^2$. The distribution function D and density d can be calculated as

$$D(z_1, \ldots, z_n) = \sum_{i=0}^{2^n-1} (-1)^{\sum_{j=1}^{n} i_j} \boldsymbol{\Phi}_{\boldsymbol{m},\boldsymbol{\Sigma}}(\boldsymbol{\varepsilon}_i),$$

$$d(z_1, \ldots, z_n) = \frac{\sum_{i=0}^{2^n-1} \boldsymbol{\phi}_{\boldsymbol{m},\boldsymbol{\Sigma}}(\boldsymbol{\varepsilon}_i)}{2^n \sqrt{\prod_{i=1}^{n} z_i}},$$

where $i_j \in \{0, 1\}$, $i = \sum_{j=1}^{n} i_j 2^{j-1}$, $\boldsymbol{\varepsilon}_i = \left((-1)^{i_1}\sqrt{z_1}, \ldots, (-1)^{i_n}\sqrt{z_n}\right)$ and $\boldsymbol{\phi}_{\boldsymbol{\mu},\boldsymbol{\Sigma}}$ denotes the Gaussian density function. Using Eqs. 2–3 the copula and its density can be evaluated.

3.2 Parameter Estimation

In the spatial copula model we have mainly three types of parameters. We have parameters $\boldsymbol{\theta}$ defining the correlation structure, copula parameters $\boldsymbol{\lambda}$ and parameters

η for the family of marginal distributions F_Z. Inference for all the parameters can be based on the maximum likelihood approach. If the copula density can be evaluated for all $n \geq 2$ dimensions, maximization of the likelihood is not difficult. However, as is the case for the non-central χ^2-copula, it may occur that calculation of the copula density is infeasible for higher dimensions. Here we proceed to perform maximum likelihood estimation only with the bivariate copula densities. Under the assumption that different pairs of observations are treated as independent we have to maximize

$$l\left(\boldsymbol{\Theta}; Z\left(\boldsymbol{x}\right)\right) = \prod_{\substack{i,j \in \{1,\ldots,n\} \\ i \neq j}} c_{\theta,\lambda}\left(F_\eta\left(Z\left(\boldsymbol{x}_i\right)\right), F_\eta\left(Z\left(\boldsymbol{x}_j\right)\right)\right) \prod_{k \in \{i,j\}} f_\eta\left(Z\left(\boldsymbol{x}_k\right)\right),$$

where $\boldsymbol{\Theta} = (\boldsymbol{\theta}, \boldsymbol{\lambda}, \boldsymbol{\eta})$ is the parameter vector, $c_{\theta,\lambda}$ is the copula density, F_η is the marginal distribution as a function of $\boldsymbol{\eta}$ and f_η denotes its density. The procedure works well as long as there is no intention to estimate anisotropy.

An advantage of working with copulas that are constructed from elliptical distributions is that the correlation matrix explicitly appears in their analytical expression. If we parameterize the correlation matrix using a geostatistical covariance model, we no longer need to estimate a sill since it is equal to 1. The reason for this is that the overall variance of the random field is a property of the marginal distribution and the copula describes the dependence structure without information about the margins.

3.3 Goodness-of-Fit Testing for Spatial Copulas

For selecting a spatial copula model that suits the given data we have to perform a goodness-of-fit test. We use a blanket test recently presented and validated by Genest and Remillard (2008) and apply it to the different lag classes $h_1, \ldots, h_r$. Although the test is designed for n-dimensional copulas we recommend working only with bivariate copulas for simplicity. The test is based on a parametric bootstrapping procedure and makes use of the Kolmogorov-Smirnov statistic, T_n, or the Cramer-von Mises statistic, S_n:

$$S_n = \int_{[0,1]^2} \mathbb{C}_n\left(\boldsymbol{u}\right)^2 dC_n\left(\boldsymbol{u}\right) \quad \text{and} \quad T_n = \sup_{\boldsymbol{u} \in [0,1]^2} \left|\mathbb{C}_n\left(\boldsymbol{u}\right)\right|,$$

where $\mathbb{C}_n = \sqrt{n}\left(C_n - C_\theta\right)$, C_n is the empirical copula calculated using the n data points and C_θ is the estimation under the null hypothesis. The steps of the algorithm are as follows:

1. For each of the lags $h_1, \ldots, h_r$ compute the empirical copula $C_n^{h_1}, \ldots, C_n^{h_r}$.
2. Estimate the theoretical copula, for example using the maximum likelihood approach. Denote the estimated parameters by $\boldsymbol{\theta}$. For every lag class there is a corresponding theoretical copula $C_\theta^{h_1}, \ldots, C_\theta^{h_r}$.

3. Calculate the Cramer-von Mises or the Kolmogorov-Smirnov statistic for every lag class, $T_n^{h_1}, \ldots, T_n^{h_r}$ or $S_n^{h_1}, \ldots, S_n^{h_r}$.
4. For a large integer N, repeat the following steps for every $k \in \{1, \ldots, N\}$
   (a) Simulate a random field whose copula is exactly the estimated theoretical copula from Step 2.
   (b) Compute the empirical copula for every lag class, $C_{n,k}^{h_1}, \ldots, C_{n,k}^{h_r}$.
   (c) Estimate the theoretical copula of the simulated field and denote the estimated parameters by $\boldsymbol{\theta}_k$. For every lag class there is a corresponding theoretical copula, $C_{\boldsymbol{\theta}_k}^{h_1}, \ldots, C_{\boldsymbol{\theta}_k}^{h_r}$.
   (d) Evaluate the test statistics $T_{n,k}^{h_1}, \ldots, T_{n,k}^{h_r}$ or $S_{n,k}^{h_1}, \ldots, S_{n,k}^{h_r}$.
5. An approximate p-value for every lag class $h_1, \ldots, h_r$ is given by

$$p_{h_j} = \frac{1}{N}\sum_{k=1}^{N} I\left(S_{n,k}^{h_j} > S_n^{h_j}\right) \quad \text{or} \quad p_{h_j} = \frac{1}{N}\sum_{k=1}^{N} I\left(T_{n,k}^{h_j} > T_n^{h_j}\right),$$

   where $I(.)$ is an indicator function and $j = 1, \ldots, r$.

In the case where the spatial copula is constructed from a multivariate distribution, simulation of a random field with a predefined copula means simply simulating from the multivariate distribution.

4 Spatial Interpolation Using Copulas

After having modeled the spatial data we are interested in predicting the values of the random field at unknown locations. In the following we propose three different methods for performing spatial interpolation using copulas.

4.1 Indicator Kriging and Disjunctive Kriging

Indicator kriging is used to estimate the conditional distribution of the random field given the data. This is done by cokriging of indicator variables $I(Z(x_i) \leq z_j)$, where $i = 1, \ldots, n$ and the z_j are certain thresholds, e.g. quantiles. Simple calculation shows that bivariate copulas are related to indicator covariances and cross-covariances via

$$\begin{aligned} \gamma_{z_j}(h) &= C_h\left(F_Z(z_j), F_Z(z_j)\right) - F_Z(z_j)^2, \\ \gamma_{z_j,z_k}(h) &= C_h\left(F_Z(z_j), F_Z(z_k)\right) - F_Z(z_j)\, F_Z(z_k). \end{aligned} \tag{6}$$

Plugging these relationships in the cokriging procedure, we arrive at an indicator kriging that is based on the spatial copula model. The fact that only bivariate copulas are needed makes it possible to use one of the numerous flexible copulas that do not have multivariate extensions or that have too few parameters for using them in a multivariate approach. The Gumbel-Hougaard extreme value copulas would be an example.

Indicators are not only used in indicator kriging but also in non-linear geostatistics. If the random field is discretized and takes only a finite number of values, say 1 to m, every function of $Z(\boldsymbol{x})$ can be written as

$$f(Z(\boldsymbol{x})) = f_1 I(Z(\boldsymbol{x}) \leq 1) + \cdots + f_m I(Z(\boldsymbol{x}) \leq m).$$

The disjunctive kriging estimator is calculated by cokriging of the indicators

$$[f(Z(\boldsymbol{x}))]^{DK} = f_1 [I(Z(\boldsymbol{x}) \leq 1)]^{CK} + \cdots + f_m [I(Z(\boldsymbol{x}) \leq m)]^{CK}.$$

Again, the relationships from Eq. 6 are used in the cokriging system. Rivoirard (1994) argued that "in the same way that kriging is based on the variogram, so disjunctive kriging is based on the bivariate distributions". In our case the bivariate distribution of the random field is defined in terms of the bivariate copula and so is disjunctive kriging.

4.2 Rank-Order Kriging

Assume we have an isotropic random field with known univariate distribution F_Z and the bivariate distributions can be described by the copula C_h. Furthermore, we have a realization $\{z(\boldsymbol{x}_i) \mid \boldsymbol{x}_i \in \boldsymbol{S}\}$ of the random field and we want to predict the values at $\boldsymbol{x}_{n+1}, \ldots, \boldsymbol{x}_{n+m}$. Applying Eq. 4 we can calculate the Spearman rank correlation curve ρ as a function of h which is exactly the correlation function for the rank-transformed variable $V(\boldsymbol{x}) = F_Z(Z(\boldsymbol{x}))$. Since $V(\boldsymbol{x})$ is a uniform distribution on [0, 1], $\frac{\rho}{12}$ gives the corresponding covariance function. Journel and Deutsch (1996) proposed to apply a simple kriging of ranks, where the linear predictor at the unknown locations $\boldsymbol{x}_j$ is given by

$$v^*(\boldsymbol{x}_j) = \sum_{i=1}^{n} \lambda_i v(\boldsymbol{x}_i) + \frac{1}{2}\left(1 - \sum_{i=1}^{n} \lambda_i\right),$$

where $j = n+1, \ldots, n+m$. Since back-transforming $v^*(\boldsymbol{x}_j)$ using F_Z^{-1} would lead to a biased estimate for $Z(\boldsymbol{x}_j)$, a bias correction is introduced

$$z^*(\boldsymbol{x}_j) = F_Z^{-1}(v^*(\boldsymbol{x}_j)) + \lambda(\boldsymbol{x}_j)\left[F_Z^{-1}(L(v^*(\boldsymbol{x}_j))) - F_Z^{-1}(v^*(\boldsymbol{x}_j))\right], \quad (7)$$

where $L(\cdot)$ is the distribution function of all kriged values $v^*(\boldsymbol{x}_j)$ and $\lambda(\boldsymbol{x}_0) = \left(\frac{\sigma_K^2(\boldsymbol{x}_j)}{\sigma_{max}^2}\right)^{\omega}$ with σ_K^2 being the kriging variance, σ_{max}^2 being the maximal kriging variance of all estimations and $\omega > 0$ a correction level parameter.

Although this method reproduces the original distribution of the data and z^* is an unbiased estimate, the covariance structures of $V(\boldsymbol{x})$ and $Z(\boldsymbol{x})$ are not reproduced. Another disadvantage of rank-order kriging is that there is no guarantee for the estimated ranks to be in the interval [0, 1]. To ensure that all estimates are between 0 and 1 it is sufficient to force all kriging weights to be non-negative, however, this is accompanied by a loss in accuracy. Moreover, z^* has no minimum kriging variance, only v^* has that property.

To partially overcome these drawbacks a direct sequential simulation of the ranks at the kriging locations $\boldsymbol{x}_j$, $j=n+1,\ldots,n+m$, is proposed. The simulated ranks are drawn from a uniform distribution with mean equal to the kriging predictor and variance equal to the kriging variance, $\left[v^*(\boldsymbol{x}_j) - \sqrt{3}\sigma_K(\boldsymbol{x}_j), v^*(\boldsymbol{x}_j) + \sqrt{3}\sigma_K(\boldsymbol{x}_j)\right]$. At each step, the kriging system consists of the original data and the previously sampled data. It may occur that the endpoints of the uniform distribution are outside the [0, 1] interval leading to simulated ranks outside [0, 1]. In this case they have to be set to 0 or 1, depending on whether they are <0 or >1. After simulation the bias correction described in Eq. 7 is applied to the estimated ranks. For a large number N the sequential simulation is repeated N times and the resulting predictors are back-transformed using F_Z^{-1} and averaged. This procedure yields estimates that are exact at known data locations, unbiased, follow the univariate distribution F_Z and reproduce the covariance of the random field. Sequential simulation is a time-consuming method for large data sets. Hence, we adapt a method proposed by Saito and Goovaerts (2000) who used it in the case of a normal-score transformation. Again, the simple kriging predictor, $v^*(\boldsymbol{x}_j)$, and the simple kriging variance, $\sigma_K^2(\boldsymbol{x}_j)$, are calculated. The conditional distribution of $V(\boldsymbol{x}_j)$ given the data is modeled as a uniform distribution with mean equal to $v^*(\boldsymbol{x}_j)$ and variance equal to $\sigma_K^2(\boldsymbol{x}_j)$. If the endpoints a and b of the uniform distribution are outside the [0, 1] interval, they are reset to 0 and 1, respectively, and the density of the local distribution changes to

$$d(x) = \begin{cases} \min\left\{\frac{1}{2(v^*(\boldsymbol{x}_j)-a)}, \frac{1}{2v^*(\boldsymbol{x}_j)}\right\}, & \text{if } x \in \left[\max\{0,a\}, v^*(\boldsymbol{x}_j)\right], \\ \max\left\{\frac{1}{2(b-v^*(\boldsymbol{x}_j))}, \frac{1}{2(1-v^*(\boldsymbol{x}_j))}\right\}, & \text{if } x \in \left[v^*(\boldsymbol{x}_j), \min\{b,1\}\right]. \end{cases}$$

The 100 percentiles, $v_p(\boldsymbol{x}_j)$, of this local distribution are calculated, where $p = \frac{k}{100} - \frac{0.5}{100}$ and $k = 1,\ldots,100$. After back-transformation, $z_p(\boldsymbol{x}_j) = F_Z^{-1}(v_p(\boldsymbol{x}_j))$, the $z_p(\boldsymbol{x}_j)$ are unbiased estimators for the quantiles of the local distribution of $Z(\boldsymbol{x})$. Their average is an unbiased estimator for the mean, hence, the kriging estimate is defined as

$$\hat{Z}\left(\boldsymbol{x}_j\right) = \frac{1}{100}\sum_{k=1}^{100} z_p\left(\boldsymbol{x}_j\right) \quad \text{with } p = \frac{k}{100} - \frac{0.5}{100}.$$

4.3 Plug-In Bayes Estimation

The copula enters the rank order kriging procedure only through the Spearman rank correlation. Furthermore, both rank order kriging and disjunctive kriging only use bivariate copulas. On the one hand these facts may be useful since flexible bivariate copula families can be applied, but on the other hand these methods do not fully exploit the spatial copula model presented in Section 3.1. When we go the Bayesian way, we can take account of the uncertainty of parameter estimation. Moreover, there is a predictive distribution for every rank-transformed variable $V\left(\boldsymbol{x}_0\right)$ at an unknown location $\boldsymbol{x}_0$,

$$p\left(v\left(\boldsymbol{x}_0\right) \mid \mathcal{D}\right) = \int p\left(v\left(\boldsymbol{x}_0\right) \mid \boldsymbol{\Theta}, \mathcal{D}\right) p\left(\boldsymbol{\Theta} \mid \mathcal{D}\right) d\boldsymbol{\Theta},$$

where $\mathcal{D} = \{z\left(\boldsymbol{x}_1\right), \ldots, z\left(\boldsymbol{x}_n\right)\}$ denotes the set of all n known data values. When we (falsely) assume that the maximum likelihood estimates, $\hat{\boldsymbol{\Theta}}$, of all the parameters are the true values, we get that $p\left(v\left(\boldsymbol{x}_0\right) \mid \mathcal{D}\right) = p\left(v\left(\boldsymbol{x}_0\right) \mid \hat{\boldsymbol{\Theta}}, \mathcal{D}\right)$. In the spatial copula model this is exactly the density of the conditional copula of $V\left(\boldsymbol{x}_0\right)$ given the rank-transformed data and the estimated parameters

$$p\left(v\left(\boldsymbol{x}_0\right) \mid \hat{\boldsymbol{\Theta}}, \mathcal{D}\right) = c_h\left(v\left(\boldsymbol{x}_0\right) \mid \hat{\boldsymbol{\Theta}}, \mathcal{D}\right) = \frac{c_h\left(v\left(\boldsymbol{x}_0\right), v\left(\boldsymbol{x}_1\right), \ldots, v\left(\boldsymbol{x}_n\right) \mid \hat{\boldsymbol{\Theta}}\right)}{c_h\left(v\left(\boldsymbol{x}_1\right), \ldots, v\left(\boldsymbol{x}_n\right) \mid \hat{\boldsymbol{\Theta}}\right)},$$

where $v\left(\boldsymbol{x}_i\right) = F_Z\left(z\left(\boldsymbol{x}_i\right)\right)$ and $i = 1, \ldots, n$. If the copula is constructed from a multivariate distribution with conditional density d and marginal distribution F with density f, Eq. 3 tells us that the conditional copula can be written as

$$c_h\left(v\left(\boldsymbol{x}_0\right) \mid \hat{\boldsymbol{\Theta}}, \mathcal{D}\right) = \frac{d\left(F^{-1}\left(v\left(\boldsymbol{x}_0\right)\right) \mid \hat{\boldsymbol{\Theta}}, \mathcal{D}\right)}{f\left(F^{-1}\left(v\left(\boldsymbol{x}_0\right)\right)\right)}.$$

In the case of a Gaussian copula $F = \Phi$, $f = \phi$ and $d = \phi_{\mu,\sigma^2}$ is a Gaussian density with mean $\mu = \boldsymbol{\Sigma}_{12}\boldsymbol{\Sigma}_{22}^{-1}\boldsymbol{a}$ and variance $\sigma^2 = 1 - \boldsymbol{\Sigma}_{12}\boldsymbol{\Sigma}_{22}^{-1}\boldsymbol{\Sigma}_{21}$, where $\boldsymbol{a} = \left(\Phi^{-1}\left(v\left(\boldsymbol{x}_1\right)\right), \ldots, \Phi^{-1}\left(v\left(\boldsymbol{x}_n\right)\right)\right)^T$, $\boldsymbol{\Sigma}_{22}$ is the correlation matrix of the known locations and $\boldsymbol{\Sigma}_{12} = \boldsymbol{\Sigma}_{21}^T$ is the vector of correlations between the known locations and the location where prediction should take place.

Since the predictive density of $V\left(\boldsymbol{x}_0\right)$ is defined on the unit interval, we avoid estimated ranks outside [0, 1]. Furthermore, the predictive density of $Z\left(\boldsymbol{x}_0\right)$ can be calculated by just using a Jacobian transformation. To get back from the ranks to the original scale the transformation is F_Z^{-1}. The corresponding Jacobian determinant is exactly the density f_Z. Hence,

$$p\left(z(\boldsymbol{x}_0) \mid \hat{\boldsymbol{\Theta}}, \mathcal{D}\right) = c_h\left(F_Z(z(\boldsymbol{x}_0)) \mid \hat{\boldsymbol{\Theta}}, \mathcal{D}\right) f_Z(z(\boldsymbol{x}_0)). \tag{8}$$

The Bayes estimator for $Z(\boldsymbol{x}_0)$ under the quadratic loss is the mean of the predictive distribution, $\hat{Z}(\boldsymbol{x}_0) = E\left(Z(\boldsymbol{x}_0) \mid \hat{\boldsymbol{\Theta}}, \mathcal{D}\right)$. Of course it can be evaluated using Eq. 8, but with the help of an integral transformation we can also derive an estimator similar to the one by Saito and Goovaerts (2000) which we have already used for rank order kriging:

$$\begin{aligned}\hat{Z}(\boldsymbol{x}_0) &= \int_{-\infty}^{\infty} z(\boldsymbol{x}_0)\, c_h\left(F_Z(z(\boldsymbol{x}_0)) \mid \hat{\boldsymbol{\Theta}}, \mathcal{D}\right) f_Z(z(\boldsymbol{x}_0))\, dz(\boldsymbol{x}_0)\\ &= \int_0^1 F_Z^{-1}(v(\boldsymbol{x}_0))\, c_h\left(v(\boldsymbol{x}_0) \mid \hat{\boldsymbol{\Theta}}, \mathcal{D}\right) dv(\boldsymbol{x}_0).\end{aligned}$$

Analogously, the prediction variance $\hat{\sigma}^2(\boldsymbol{x}_0)$ can be calculated as

$$\hat{\sigma}^2(\boldsymbol{x}_0) = \int_0^1 \left(F_Z^{-1}(v(\boldsymbol{x}_0)) - \hat{Z}(\boldsymbol{x}_0)\right)^2 c_h\left(v(\boldsymbol{x}_0) \mid \hat{\boldsymbol{\Theta}}, \mathcal{D}\right) dv(\boldsymbol{x}_0).$$

Similarly to copula kriging the frequently applied trans-Gaussian kriging (Diggle and Ribeiro, 2007) also works with a marginal transformation of the random field. The aim of trans-Gaussian kriging is to deal with non-Gaussian random fields by assuming that the transformed random field, $Y(\boldsymbol{x}) = g(Z(\boldsymbol{x}))$, is Gaussian and g is a suitable transformation that has to be determined. In most applications the transformation g is chosen from the Box–Cox family of transformations. In the following we show that there is a direct relationship between the trans-Gaussian kriging model and the spatial copula model.

Theorem 1. *The trans-Gaussian kriging model using an almost surely strictly monotone transformation is equivalent to the Gaussian spatial copula model.*

Proof. Assume we have a trans-Gaussian random field with an almost surely strictly monotone transformation g. Hence, $Y(\boldsymbol{x}) = g(Z(\boldsymbol{x})) \sim \mathcal{N}(\mu, \sigma^2)$. From the invariance theorem mentioned in Section 2 we get that the copula corresponding to the multivariate distributions of $Z(\boldsymbol{x})$ must be the Gaussian copula corresponding to $Y(\boldsymbol{x})$. Using $Z(\boldsymbol{x}) = g^{-1}(Y(\boldsymbol{x}))$ we obtain the univariate marginal distribution of $Z(\boldsymbol{x})$ as

$$F_Z(z) = \int_{-\infty}^{z} \phi_{\mu,\sigma^2}(g(t)) \left|g'(t)\right| dt,$$

and the Gaussian spatial copula model is fully determined. If we assume that the random field follows the Gaussian spatial copula model with known F_Z, then $g(z) := \Phi^{-1}(F_Z(z))$ is a suitable transformation.

Since we can also use any other copula different from the Gaussian copula in our approach, we observe that the spatial copula model is a generalization of the trans-Gaussian model. Even if we want to stay within the Gaussian framework, it is more convenient to use the copula methodology because it is easier to specify the univariate distribution of the random field than to determine a suitable transformation function. Especially when we work with multimodal or extreme value data this fact becomes obvious.

For certain copula families not all data values can be used to build the predictive distribution. For the non-central χ^2-copula mentioned in Section 3.1 this happens because one would need to evaluate 2^n terms for the calculation of the conditional copula. In these cases we propose a local prediction.

5 Application: SIC2004 Joker Data

In this section we test our methodology by means of the Joker data set, which was investigated in detail during the spatial interpolation comparison SIC2004 (Dubois 2005). This extreme value data set simulates an accidental release of radioactivity using a dispersion process. The 200 training points have a mean of 108.99, a standard deviation of 121.96 and a skewness of 9.92. Figure 1a displays the training data as gray dots and the 808 test data as gray circles. The two extreme observations (1,070.4 and 1,499) are indicated by the black dots.

At first we fit a quadratic trend surface model to the data. The residuals follow a generalized extreme value distribution. Using the goodness-of-fit test described

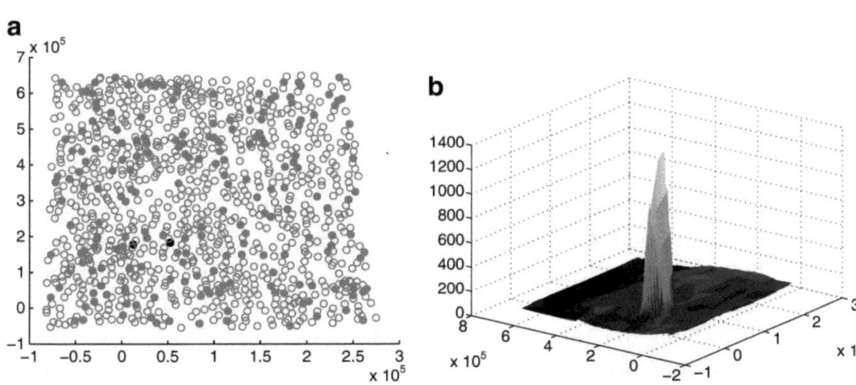

Fig. 1 The locations of the Joker training (*dots*) and test data (*circles*) are displayed in (**a**). A surface plot of the Joker data is shown in (**b**)

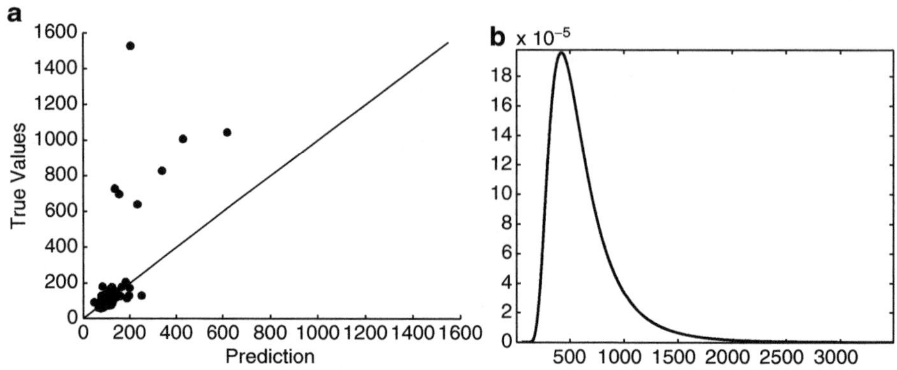

Fig. 2 The predicted values of the test set are plotted against the true values in (**a**). The predictive density at a hotspot is displayed in (**b**)

in Section 3.3 we find out that it is sufficient to work with the Gaussian spatial copula. The correlation matrix of the Gaussian distribution is parameterized (cf. Section 3.1) by a mixture of a Gaussian and an exponential correlation model. Geometric anisotropy is considered by a 2×2 transformation matrix, where one entry is fixed to avoid interference with the ranges of the correlation models. All parameters, including the nugget, two ranges, one mixing parameter, three parameters for the generalized extreme value distribution and three anisotropy parameters, are estimated using the maximum likelihood approach. Note that there is no need to estimate a sill. Prediction is performed using the plug-in Bayes approach. The predicted values are plotted against the true values in Fig. 2a. The predictive density at a hotspot is visualized in Fig. 2b and it shows that values around 1,500 are still contained in the body of the distribution. The results for the test data are: RMSE $= 65.87$, MAE $= 16.22$, ME $= -2.58$ and Pearson-r $= 0.71$. Compared to the results of more than 30 participants of the SIC2004 this would be the third smallest RMSE, the second smallest MAE and the third largest Pearson correlation.

6 Conclusion

Copulas can be used to describe spatial dependence and in this way generalize the concept of the variogram. Moreover, the spatial copula model can be used to perform spatial interpolation. It generalizes the trans-Gaussian kriging method and is therefore a flexible tool for working with non-Gaussian, multimodal and extreme value data. Specifying the univariate distribution of the random field is easier than finding a suitable transformation for trans-Gaussian kriging, which makes the spatial copula model attractive even if the Gaussian copula is used. Another advantage is that the sill need not to be estimated and, hence, the model contains one parameter less. Results on the SIC2004 Joker data demonstrate that the presented approach could be applied for emergency monitoring and estimating exceedance probabilities for certain emergency thresholds in environmental monitoring systems.

Acknowledgement This work was partially funded by the European Commission, under the Sixth Framework Programme, by the Contract N. 033811 with DG INFSO, Action Line IST- 2005-2.5.12 ICT for Environmental Risk Management. The views expressed herein are those of the authors and are not necessarily those of the European Commission.

References

Bardossy A (2006) Copula-based geostatistical models for groundwater quality parameters. Water Resour Res 42(W11416). doi:10.1029/2005WR004754
Diggle P, Ribeiro P (2007) Model-based geostatistics. Springer, New York
Dubois G (2005) Automatic mapping algorithms for routine and emergency monitoring data. EC Joint Research Centre, Belgium
Genest C, Remillard B (2008) Validity of the parametric bootstrap for goodness-of-fit testing in semiparametric models. Ann I H Poincare B 44:1096–1127
Journel A, Deutsch C (1996) Rank order geostatistics. In: Baafi E, Schofield N (eds) Geostatistics Wollongong '96. Kluwer, Dordrecht, pp 174–187
Nelsen R (2006) An introduction to copulas. Springer, New York
Rivoirard J (1994) Introduction to disjunctive kriging and non-linear geostatistics. Oxford University Press, Oxford
Saito H, Goovaerts P (2000) Geostatistical interpolation of positively skewed and censored data in a dioxin-contaminated site. Environ Sci Technol 34:4228–4235
Sklar A (1959) Fonctions de repartition a *n* dimensions et leurs marges. Publ Inst Stat Univ Paris 8:229–231

Exchanging Uncertainty: Interoperable Geostatistics?

Matthew Williams, Dan Cornford, Lucy Bastin, and Ben Ingram

Abstract This paper discusses a solution to providing interoperable, automatic geostatistical processing through the use of Web services, developed in the INTAMAP project (INTeroperability and Automated MAPping). The project builds upon Open Geospatial Consortium standards for describing observations, typically used within sensor webs, and employs Geography Markup Language (GML) to describe the spatial aspect of the problem domain. Thus, the interpolation service is extremely flexible, being able to support a range of observation types, and can cope with issues such as change of support and differing error characteristics of sensors (by utilising descriptions of the observation process provided by SensorML).

XML is accepted as the *de facto* standard for describing Web services, due to its expressive capabilities which allow automatic discovery and consumption by 'naïve' users. Any XML schema employed must, therefore, be capable of describing every aspect of a service and its processes. However, no schema currently exists that can define the complex uncertainties and modelling choices that are often present within geostatistical analysis. We show a solution to this problem, developing a family of XML schemata to enable the description of a full range of uncertainty types. These types will range from simple statistics, such as the kriging mean and variances, through to a range of probability distributions and non-parametric models, such as realisations from a conditional simulation. By employing these schemata within a Web Processing Service (WPS) we show a prototype moving towards a truly interoperable geostatistical software architecture.

1 Introduction

Uncertainty in geographic information is ubiquitous, be it from measurement error, observation operator error or modelling error. It is how we process and propagate this uncertainty that is of importance, especially when high-risk decisions are to be

M. Williams (✉), D. Cornford, L. Bastin, and B. Ingram
Knowledge Engineering Group, School of Engineering and Applied Science, Aston University, Birmingham B4 7ET, UK
e-mail: williamw@aston.ac.uk

P.M. Atkinson and C.D. Lloyd (eds.), *geoENV VII – Geostatistics for Environmental Applications*, Quantitative Geology and Geostatistics 16,
DOI 10.1007/978-90-481-2322-3_28,

made based on such information (Atkinson, 1999; Couclelis, 2003; Heuvelink and Goodchild, 1998). In the field of geostatistics, uncertainty from multiple sources is encountered routinely. Consider, for example, a user in the field collecting soil samples. Inputting the data onto a small footprint machine (e.g. Personal Digital Assistant) the user is able to store the data for lab processing, or alternatively, to submit the data for processing to a Web service. A typical process in this scenario might use the available data to predict where the user should next sample optimally, or to provide an estimate of the soil properties at an unsampled location. Errors in the original measurements, stemming from systematic sensor effects and random fluctuations, will combine with errors in the models used to process and interpolate the data, to produce significant levels of uncertainty (which must be explicitly estimated and quantified) in the final predictions. Traditionally, the soil data in this example would be processed from start to finish within a single software package to produce, for example, an interpolated map of heavy metal concentration, with estimation uncertainty represented as variance at each predicted location. The uncertainty in prediction might also be crystallised as exceedance probabilities, showing the likelihood that a critical threshold is exceeded at any location, or as sets of realised samples from the predicted distribution. While traditional geostatistical applications recognise and model the uncertainty at the end of the analysis, a conceptual model for describing and communicating uncertainties is of less importance, since the data are not usually shared with other applications. Uncertainty at the intermediate stages of analysis is, therefore, rarely explicitly characterised. However, if different processing steps (e.g. outlier detection, data harmonisation, parameter estimation, interpolation) are delegated to separate Web services, it becomes necessary for each service to receive an understandable summary of the uncertainty inherent in the sample data, and introduced by the intervening processing steps. Currently, there is a trend in software engineering to move away from tightly coupled legacy systems and towards loosely coupled, interoperable, services (Erl, 2005) based on XML. A conceptual design which allows the communication of uncertain results is of foremost importance in the development of such an interoperable geostatistical application. This paper introduces a conceptual model of uncertainty and examples of how one might encode uncertainty in XML, motivated by examples arising within the INTAMAP project.

2 XML, Web Services and SOAs

Interoperability is defined as "the ability of two or more systems or components to exchange information and to use the information that has been exchanged" (IEE, 18 Jan 1991). This section provides an overview of several technologies and concepts that provide the foundations of an 'interoperable' application.

XML (Yergeau et al., 2006) is a structured language that allows metadata to be integrated with content, thus, adding a layer of intelligence to information (Erl, 2004). XML is implemented by defining a set of *elements* and *attributes* that are unique to a particular context, or domain. A collection of such elements and attributes is often referred to as a *vocabulary*. Vocabularies can be defined formally using a *schema definition language*, typically 'XML Schema' language (Fallside and Walmsley, 2004), but is not a requirement of XML. The descriptive nature and extensibility of XML are two key ingredients that contribute towards it being a suitable language for interoperability.

The concept of a 'service' in software engineering is not a new term and typically refers to an independent building block within a larger application environment, or distributed system (Erl, 2004). A Web service is an implementation of a service that uses XML to describe the operations available including the data inputs and outputs. There are other types of Web service (RESTful) which do not rely so heavily on XML. However, we do not discuss these in this paper and from hereon the term *Web service* refers specifically to an XML Web service.

Communication of data to and from a Web service is encoded as XML and transported via an Internet protocol (this is usually HTTP). Adhering to these requirements provides an interoperable framework that allows software applications, written in different languages and on different platforms, to communicate seamlessly. A collection of these services, 'loosely coupled', forms the basis of a design philosophy called Service Oriented Architecture.

The term Service Oriented Architecture has many definitions, perhaps one or the more concise is defined in Josuttis (2007) as:

> SOA is an architectural paradigm for dealing with business processes distributed over a large landscape of existing and new heterogeneous systems that are under the control of different owners.

A SOA is usually realised as a collection of Web services, that may be governed by different owners, communicating with one another to form a processing chain. In context, this could be a risk management or decision support chain.

3 The INTAMAP Project

Introducing interoperability into the field of geostatistics, INTAMAP seeks to provide a fully automated interpolation service implementing a Web Processing Service interface (Schut, 2007). A WPS is a restriction on a normal Web service, governed by the Open Geospatial Consortium, that is suited to processing of geospatial data. Simply, a Web Processing Service can be thought of as a function that can be called over the Web. Within INTAMAP we are also developing a range of novel automatic mapping algorithms including Bayesian trans-Gaussian kriging, fast anisotropy detection, data harmonisation for heterogeneous networks and fast approximate techniques that can deal with multiple sensor and error characteristics (Ingram et al., 2008). Key to all the methods we employ is a description of the uncertainties on the inputs and outputs of the interpolation process. Currently no such XML vocabulary, or *schema*, exists to allow the description of uncertainty, hence our

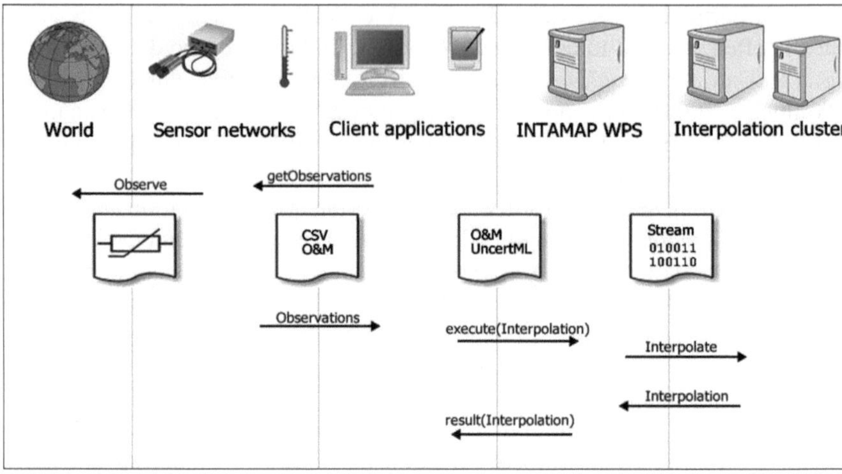

Fig. 1 Example workflow for an interpolation request in the INTAMAP project. All geostatistical processes are carried out using R and C++ on a separate server to the WPS. A client may obtain observations from multiple sensor systems before submitting them to INTAMAP for processing. Clients may also be services; chaining of services in this way underpins the foundation of Service Oriented Architectures

development of UncertML. The inputs to the INTAMAP Web service are XML files describing the observations, with UncertML being used to characterise the observation errors (see Section 4). The results produced by INTAMAP contain inherent, and additional, uncertainty introduced by the interpolation process which must be communicated for the results to be of any subsequent utility. Figure 1 provides an overview of a typical workflow, integrating the INTAMAP service with existing client applications. The rest of this paper discusses a solution to the problem, UncertML, and investigates the integration into INTAMAP, providing 'interoperable geostatistics'.

4 Describing Uncertainty in XML

In this section, we discuss the design of an XML language for describing uncertainty, UncertML, depicted using the Unified Modeling Language (UML). The UML diagrams used within this paper are static structure diagrams, whose notation is clearly defined in Section 4.4 of Portele (2007). Examples in XML are given, where necessary, to illustrate how it may be used.

4.1 Conceptual Model and Examples

The core design of UncertML is split into three distinct sections; summary statistics, distributions and realisations (Fig. 2). Aggregate types for statistics, distributions

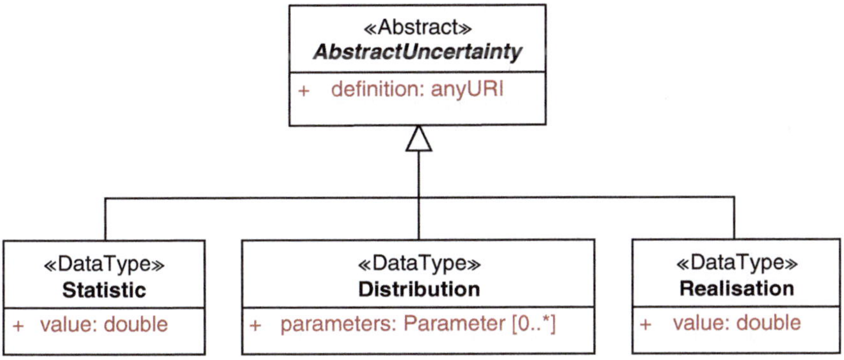

Fig. 2 Conceptual overview of UncertML. Three main types extend the abstract uncertainty type; 'Statistic', 'Distribution' and 'Realisation'. Other types are also available and discussed in more detail later

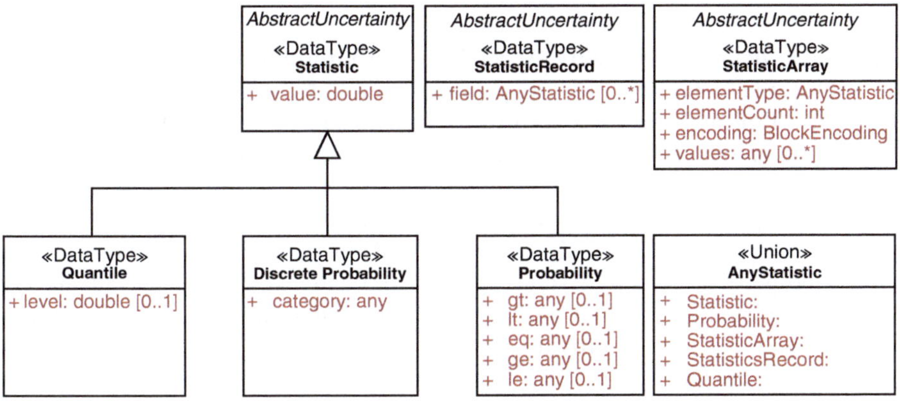

Fig. 3 UncertML model for summary statistics

and realisations also exist where deemed necessary. It is important to note that UncertML does not provide a framework for describing phenomena or their units of measure, nor does it provide any geospatial attributes. Removing this level of detail allows UncertML to be integrated into a diverse range of domains.

Throughout UncertML we follow a weak-typed design pattern that offers improved extensibility at the cost of strict validation. Weak-typing works by providing generic types with generic properties, in contrast to a strongly-typed design with concrete types and well-defined properties (Figure 4). Figure 3 introduces the hierarchy for summary statistics; the base type is a generic 'Statistic' that can be used for most summary statistics such as mean, median, variance or standard deviation. The value of a statistic is given through the 'value' property that holds a single real number. All types within UncertML extend the 'AbstractUncertainty' type and therefore

```
<Statistic definition="Mean">         <Mean>
  <value>34.5</value>                   <value>34.5</value>
</Statistic>                          </Mean>
```

Fig. 4 Comparison of a weak-typed (*left*) and strong-typed (*right*) representation of a mean value. Weak-typing is more generic and provides greater extensibility, however, strong-typing provides easier validation

```
<Statistic definition="Mean">
  <value>26.5</value>
</Statistic>

<Probability definition="Probability" gt="23.4" lt="33.4">
  <value>0.34</value>
</Probability>
```

Fig. 5 Two XML instances, the first represents a mean value while the latter shows the probability that a value falls between 23.4 and 33.4

inherit the ‘definition’ property. Accepting any Uniform Resource Identifier (URI) as a value, the ‘definition’ property provides a level of semantics to the weak-typed elements. Typically the URIs resolve to a dictionary entry describing the uncertainty type of interest. However, other methods of description may be used such as ontologies.

Certain summary statistics require additional information than the generic ‘Statistic’ type provides. A ‘Quantile’ is used for describing quantiles where a ‘level’ property, accepting a value between 0.0 and 1.0, defines the quantile of interest. Probabilities offered through either the ‘DiscreteProbability’ or ‘Probability’ types. The former provides a ‘category’ property which may contain any information, and the latter offers a range of properties including ‘equal to’, ‘greater than’ and ‘less than’; a combination of which may be used. Figure 5 demonstrates how these statistics can be encoded in UncertML. It should be noted that probabilities differ from other summary statistics in that their ‘value’ property contains a *probability* (0.0–1.0) rather than an actual *value* (with units of measure etc.).

It is often the case that one would wish to describe a collection of individual statistics to provide a summary of a particular variable. A ‘StatisticsRecord’ is used for this exact purpose and groups different statistics into a unified structure (Figure 6). When dealing with multiple instances of the same statistic it is more appropriate to use the ‘StatisticArray’ type. The flexibility of UncertML allows any combination of records and arrays to be created including arrays of records and records of arrays. All aggregate types within UncertML utilise the Sensor Web Enablement (SWE) common encoding schema (Botts and Robin, 2007) to provide an extensive list of options for encoding the data, including most MIME types.

A ‘Distribution’ type in UncertML follows a similar pattern to a ‘StatisticsRecord’ (Fig. 7) due to it containing a collection of parameters. An example encoding of a distribution is shown in Fig. 8 and consists of a reference to a

```
<StatisticRecord>
  <field>
    <Statistic definition="Mean">
      <value>34.5</value>
    </Statistic>
  </field>
  <field>
    <Statistic definition="Standard_Deviation>
      <value>12.4</value>
    </Statistic>
  </field>
</StatisticRecord>
```

Fig. 6 A collection of individual statistics can be grouped into a 'StatisticRecord' to provide a summary of a variable. This example shows a mean and standard deviation

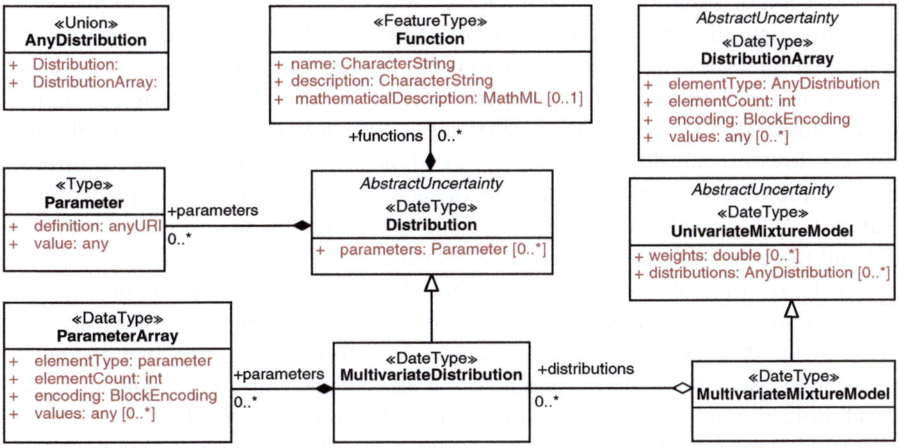

Fig. 7 UncertML model for distributions and other related types. The base 'Distribution' type is similar to the 'StatisticsRecord' discussed earlier, however, the addition of 'functions' provides a mechanism for describing a cumulative distribution function

```
<Distribution definition="Gaussian_Distribution">
  <parameters>
    <Parameter definition="Mean">
      <value>23.4</value>
    </Parameter>
    <Parameter definition="Variance">
      <value>56.7</value>
    </Parameter>
  </parameters>
</Distribution>
```

Fig. 8 A typical distribution encoded in UncertML. Reference to a dictionary entry is made through the 'definition' property which provides a complete description of a distribution, including its cumulative distribution function dictionary through the 'definition' property as well the distribution parameters and their values. When a link to such a definition is not available a 'Distribution' can be extended to include a set of functions inline, encoded in MathML (Carlisle et al., 2003), such as a cumulative distribution function, probability density function or other arbitrary functions that may be performed on a distribution. Such flexibility allows users to work with distributions about which they have no prior knowledge.

There are many instances where a single distribution is not sufficient or where it is desirable to work with multivariate distributions, interpolation being one such example. The 'DistributionArray' type takes the form of an 'array of records' mentioned earlier and allows multiple instances of a particular distribution to be encoded efficiently using the SWE encoding schema. The 'MultivariateDistribution' type shown in Fig. 7 is an extension of the base 'Distribution' type, differentiated by the inclusion of a number of 'ParameterArray' properties. This is due to the nature of multivariate distributions having more than a single value for each parameter. UncertML provides two mixture model types that may be used for encoding a collection of distributions, each of which describe a variable by different amounts. Conceptually the 'UnivariateMixtureModel' is similar to the standard 'DistributionArray'. However, an additional property yields an array of values between 0.0–1.0 to indicate the relative fraction, or weight, of each distribution, the total of which must sum to 1. A 'MultivariateMixtureModel' is a restriction on the univariate model that only allows a collection of multivariate distributions.

The final strand of UncertML is concerned with realisations, or samples, seen in Fig. 9. A single realisation is encoded using the 'Realisation' type which is identical to a 'Statistic'. However, we feel it necessary to make a conceptual distinction between the two. Typically one would not wish to work with single realisations, instead preferring to encode large arrays; UncertML provides the 'RealisationArray' type as a solution. A 'RealisationArray' utilises the SWE encoding block to provide an efficient means of encoding vast quantities of data. A small example can be seen in Fig. 10.

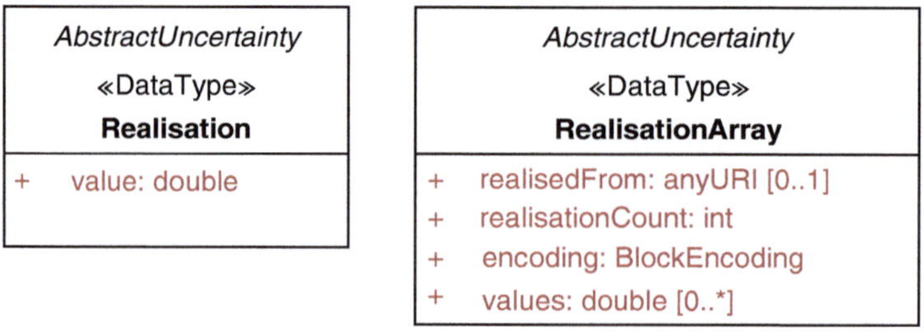

Fig. 9 A single realisation may be encoded using the 'Realisation' type, however, a typical user would wish to encode multiple realisations for which scenario a 'RealisationArray' is provided

```
<RealisationArray realisedFrom="Gaussian_Distribution">
  <realisationCount>5</realisationCount>
  <swe:encoding>
      <swe:TextBlock tokenSeparator="," tupleSeparator=" "
      decimalSeparator="." />
  </swe:encoding>
  <values>
    53.2,58.4,51.3,42.9,60.02
  </values>
</RealisationArray>
```

Fig. 10 Realisations (or samples) can be encoded using the 'RealisationArray' type. If the distribution from which these samples were realised is known then the 'realisedFrom' property may be used. A 'tokenSeparator' is used to identify individual values within a tuple and a 'tupleSeparator' is used to separate tuples

5 Integrating UncertML into the INTAMAP Project

The Observations & Measurements schema (Cox, 2007) provides an extensive model for describing the act of observing. Accompanying the 'result' property, this model may include properties for documenting the observation time ('samplingTime'), the feature or location ('featureOfInterest'), the property being measured ('observedProperty') and the procedure or instrument used to generate the result ('procedure'). Typically the 'procedure' property will contain a sensor model encoded in SensorML (Botts and Robin, 2007) which can describe the error characteristics of a sensor (e.g. bias).

Within INTAMAP a request for interpolation is made by sending a collection of observations, encoded in the O&M schema, to the Web Processing Service interface. UncertML is used within the 'result' of an observation (Fig. 11) to describe the uncertainty inherent in observed values. Utilising both the error characteristics of a sensor and the observation uncertainty allows use of the arbitrary likelihood estimation techniques mentioned briefly in Section 3.

Due to UncertML types not encoding phenomena or geospatial attributes it is envisaged that a three layered architecture, seen in Fig. 12, will be employed, where each layer adds an extra level of detail. It should be stressed that this chain is not a part of UncertML, nor is it mandatory that UncertML be implemented in this way, it is simply an abstract notion of how one may wish to use UncertML when dealing with geographic data.

Depending on user preferences made in the request, the result of an interpolation can take several forms. The bulk of the data will be encoded in any one of the uncertainty types within UncertML and additional information may be added by separate schemata. A typical result may consist of a regular grid, possibly defined in GML (Portele, 2007), of some variable defined by a series of Gaussian distributions encoded in UncertML. Figure 1 in Section 3 displays the lineage of an interpolation request in INTAMAP.

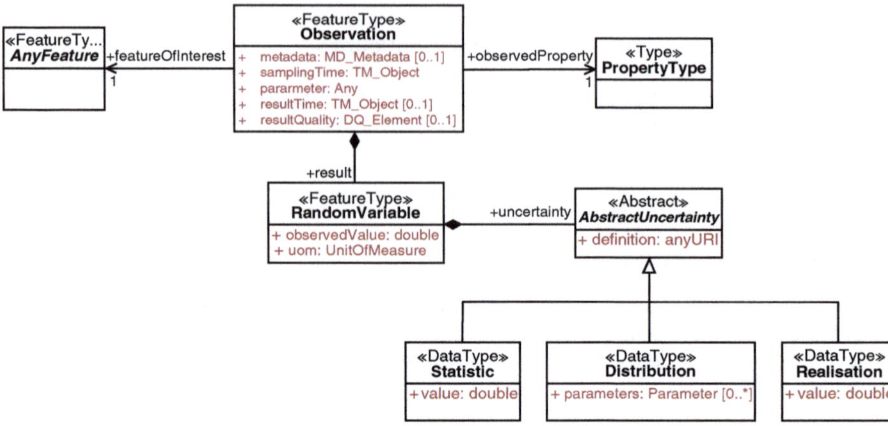

Fig. 11 An observation model within the O&M schema. The result can be of any type, in this instance it is a 'RandomVariable' which uses any uncertainty type from the UncertML schema to encode the value

Fig. 12 Three layered implementation of UncertML. At it's simplest, UncertML only encodes the values of an uncertainty type. A 'RandomVariable' type adds a link to a phenomena and its units of measure and a 'GeospatialRandomVariable' adds further detail with an attached geometry. These random variable types are not included in UncertML and represent only one possible implementation

6 Conclusion

Embracing the ongoing evolution in software engineering to adopt a loosely coupled, interoperable, framework will make geostatistical methods available to a larger array of users. With the development of UncertML, as part of the INTAMAP project, a large step has been taken towards achieving this goal. The European Radiological Data Exchange Platform (EURDEP) provides a case study for the INTAMAP project and demonstrates a clear need for real-time interpolation across a Service Oriented Architecture.

However, for truly interoperable geostatistics, several areas require greater attention. A conceptual model for supporting the use of UncertML within geostatistical models will see the inclusion of variograms, covariance functions and other random functions. Other extensions to the UncertML model will include the addition of fuzzy memberships.

Currently, we are undergoing discussions with the Open Geospatial Consortium with the view of making the UncertML specification an official, governed, standard. This *may* be included as part of the OWS-6 request for quotation.

A working interpolation service will be available for testing online shortly. More information and latest developments can be found at the INTAMAP website (http://www.intamap.org).

Acknowledgements This work is funded by the European Commission, under the Sixth Framework Programme, by Contract 033811 with DG INFSO, action Line IST-2005-2.5.12 ICT for Environmental Risk Management.

References

IEEE Standard Computer Dictionary (18 Jan 1991) compilation of IEEE standard computer glossaries. IEEE Std 610

Atkinson PM (1999) Geographical information science: geostatistics and uncertainty. Prog Phys Geogr 23:134–142

Botts M, Robin A (2007) OpenGIS sensor model language (SensorML) implementation specification. OpenGIS standard 07-000, Open Geospatial Consortium Inc, July 2007. http://www.opengeospatial.org/standards/sensorml

Carlisle D, Miner R, Ion P, Poppelier N (2003) Mathematical markup language (MathML) version 2.0 (2nd edn). W3C recommendation, W3C, October 2003. http://www.w3.org/TR/2003/REC-MathML2-20031021/

Couclelis H (2003) The certainty of uncertainty: GIS and the limits of geographic knowledge. Trans GIS 7(2):165–175. doi: 10.1111/1467-9671.00138

Cox S (2007) Observations and measurements – Part 1 - observation schema. OpenGIS standard 07-022r1, Open Geospatial Consortium Inc, December 2007. URL http://www.opengeospatial.org/standards/om. http://www.opengeospatial.org/standards/om

Gregoire Dubois and Stefano Galmarini. Introduction to the spatial interpolation comparison (SIC) 2004 exercise and presentation of the datasets. *Applied GIS*, 1:1–11, 2005.

Erl T (2004) Service-oriented architecture: a field guide to integrating XML and web services. Prentice Hall PTR, Upper Saddle River, NJ. ISBN 0131428985

Erl T (2005) Service-oriented architecture: concepts, technology, and design. Prentice Hall PTR, Upper Saddle River, NJ ISBN 0131858580

Fallside DC, Walmsley P (2004) XML schema part 0: primer, 2nd edn. W3C recommendation, W3C, October 2004. http://www.w3.org/TR/2004/REC-xmlschema-0-20041028/

Heuvelink GBM, Goodchild MF (1998) Error propagation in environmental modelling with GIS. Taylor & Francis, London

Ingram B, Cornford D, Csató L (2008) Robust automatic mapping algorithms in a network monitoring scenario. 7th international conference on geostatistics for environmental applications (geoENV 2008), this volume

Josuttis NM (2007) SOA in practice: the art of distributed system design (Theory in practice). O'Reilly Media. ISBN 0596529554

Portele C (2007) OpenGIS Geography Markup Language (GML) encoding standard. OpenGIS standard 07-036, Open Geospatial Consortium Inc, August 2007. URL http://www.opengeospatial.org/standards/gml. http://www.opengeospatial.org/standards/gml

Schut P (2007) OpenGIS web processing service 1.0.0. OpenGIS standard 05-007r7, Open Geospatial Consortium Inc, June 2007. URL http://www.opengeospatial.org/standards/wps. http://www.opengeospatial.org/standards/wps

Yergeau F, Maler E, Bray T, Paoli J, Sperberg-McQueen CM (2006) Extensible markup language (XML) 1.0 (4nd edn). W3C recommendation, W3C, August 2006. http://www.w3.org/TR/2006/REC-xml-20060816

Hierarchical Bayesian Model for Gaussian, Poisson and Ordinal Random Fields

Pierrette Chagneau, Frédéric Mortier, Nicolas Picard, and Jean-Noël Bacro

Abstract As most georeferenced data sets are multivariate and concern variables of different kinds, spatial mapping methods must be able to deal with such data. The main difficulties are the prediction of non Gaussian variables and the modelling of the dependence between processes. The aim of this paper is to propose a new approach that permits simultaneous modelling of Gaussian, count and ordinal spatial processes. We consider a hierarchical model implemented within a Bayesian framework. The method used for Gaussian and count variables is based on the generalized linear mixed models. Ordinal variable is taken into account through a generalization of the ordinal probit model. We use a moving average approach to model the spatial dependence between the processes. The proposed model is applied to pedological data collected in French Guiana.

1 Introduction

Soil maps are more and more used as input in environmental and ecological studies. In fact, soil characterization could explain landscapes and vegetation stand. So modelling spatial distribution of soil properties has been a challenge for ecologists. In such geological studies, there are few available data as they are expensive to collect. Moreover, data are often of different nature. Element concentration, granularity and coloration are usually measured for soil characterization. With increased collection of such multivariate geostatistical data, there arises the need for spatial mapping methods to handle related data of different nature. This raises two difficulties: the prediction of multivariate discrete random fields and the modelling of the dependence between continuous and discrete variables.

P. Chagneau (✉), F. Mortier, and N. Picard
CIRAD, UR 37, Campus international de Baillarguet, 34 398 Montpellier Cedex 5, France
e-mail: pierrette.chagneau@cirad.fr; frederic.mortier@cirad.fr; nicolas.picard@cirad.fr

J.-N. Bacro
I3M, Université de Montpellier 2, Place Eugène Bataillon, 34 095 Montpellier Cedex 5, France
e-mail: bacro@math.univ-montp2.fr

P.M. Atkinson and C.D. Lloyd (eds.), *geoENV VII – Geostatistics for Environmental Applications*, Quantitative Geology and Geostatistics 16,
DOI 10.1007/978-90-481-2322-3_29,

In the univariate case, the prediction of continuous spatial processes has been widely studied (Cressie, 1991; Wackernagel, 2003). On the contrary, few models were developed for discrete random fields. The most commonly used methods for binary and ordinal variables is the disjunctive kriging (Journel and Huijbregts, 1978; Chilès and Delfiner, 1999). The main drawback of this method is that it requires the determination of bivariate distributions to model the dependence, which can lead to heavy computational costs. The univariate modeling of discrete random fields has received increasing attention in years. New models have been defined, particularly to deal with count variables. Diggle et al. (1998) proposed to embed linear kriging methodology within the framework of the generalized linear mixed model where the random effect is modelled by a Gaussian spatial process. Wolpert and Ickstadt (1998) proposed to model count data with a Poisson distribution with random intensity. They modelled the random intensity using Gamma process. Such models are now often described in the hierarchical Bayesian framework (Banerjee et al., 2004). More recently, the BME approach first introduced by Christakos (1990, 1998) to predict continuous variables has been extended to predict categorical variables (Bogaert, 2002; D'Or and Bogaert, 2004).

The prediction of multivariate spatial processes has been widely studied in the last few decades (Cressie, 1991; Wackernagel, 2003). Cokriging methods are the most popular. They are efficient but they request some restricting assumptions of normality. Modelling of the dependence between variables in this model are based on full covariance structure model. The choice of the covariance framework (intrinsic covariance model [Wackernagel, 2003], coregionalization model [Grzebyk and Wackernagel, 1994; Banerjee et al., 2004]) leads to more or less flexible models. Ver Hoef and Barry (1998) defined a new family of flexible variograms using moving average functions. In the literature, few methods have been developed for ordinal random fields. In general, disjunctive cokriging is used. As for the disjunctive kriging, the modelling of the dependence between variables needs to know all bivariate distributions. In practice, the determination of the bivariate distributions can be tedious and requests to use isofactorial models. Finally, there are few methods allowed to deal simultaneously with continuous and discrete variables. Recently, we proposed an approach which enables simultaneously modelling Gaussian, Poisson and ordinal spatial processes (Chagneau et al., 2008). Our model is based on a hierarchical Bayesian framework. The method used for Gaussian and count variables is the same as Diggle's one based on the generalized linear mixed model. Unlike his approach, our model can take into account ordinal random fields through a generalization of the multivariate ordinal probit models to the spatial case (Chaubert et al., 2008). Ver Hoef and Barry (1998)'s approach is used to model the dependence between the related spatial Gaussian latent processes. These dependent processes are built by convolving white noise processes with a moving average function.

The aim of this paper is thus to apply this method to predict soil properties from pedologic data collected in French Guiana. In Section 2, we present the data and we describe the spatial hierarchical model. Results are given in Section 3. Finally, in Section 4, we draw some conclusions and give some perspectives for future work.

2 Materials and Methods

2.1 Data

Data were collected in the Paracou experimental forest in French Guiana (5°15′N, 52°55′W; 0–50 m elevation), 15 km inland from the coast (Gourlet-Fleury et al., 2004). The climate is the humid tropical type with a mean annual precipitation of around 2,980 mm. The study site is characterized by a patchwork of hills (100–300 m in diameter and 20–50 m in height) separated by humid valleys (Epron et al., 2006). Part of the site is permanently waterlogged.

Soils are mostly acrisol (FAO-ISRIC-ISSS, 1998) developed over a Precambrian metamorphic formation. The soil is characterized by schists and sandstones and locally crossed by veins of pegmatite, aplite and quartz. Soil properties were measured in 12 permanent sample plots of the experimental site. We were only interested in four permanent plots located at the south of the experimental site. They were located at some distance from one another and elevation and slope are known on these plots. Each plot measured 250 × 250 m. Around 70 points were recorded in each plot. These points were randomly chosen. A 1.2 m core of soil was extracted in each location for characterization. Soil texture, soil colour, and the presence of stones or coloured spots were used to classify the soils. Manual perception of clay content and silt dryness was used to distinguish soils exhibiting vertical drainage from soils exhibiting superficial lateral drainage. Six levels of drainage were distinguished to classify varying degrees of hydromorphism. Further details concerning the drainage characteristics can be found in Sabatier et al. (1997).

2.2 Spatial Hierarchical Model

2.2.1 Model

The spatial hierarchical model we proposed is specifically designed to take into account variables of different kinds. The model can be defined for any number L of response variables but, for sake of simplicity, we restrict ourselves to $L = 3$ variables of different kinds: a Gaussian variable, a Poisson variable and an ordinal variable. Clearly the definition we give below for $L = 3$ can readily be extend to any number of variables. Before describing the model, let us first introduce some notations.

Let $(\mathbf{s}_1, \ldots, \mathbf{s}_N)$ be the sampled locations. Let $Y_1(\mathbf{s}_i)$ (resp. $Y_2(\mathbf{s}_i)$, $Y_3(\mathbf{s}_i)$) be a Gaussian variable (resp. a Poisson variable, an ordinal variable with J modalities) at location $\mathbf{s}_i$. Let $\mathbf{Y}_k(\mathbf{s}) = (Y_k(\mathbf{s}_1), \ldots, Y_k(\mathbf{s}_N))$, $k = 1, 2, 3$ be the vector of the variable Y_k observed at all locations. Let $\mathbf{Y}(\mathbf{s}) = (\mathbf{Y}_1(\mathbf{s}), \mathbf{Y}_2(\mathbf{s}), \mathbf{Y}_3(\mathbf{s}))$ be the vector of all variables observed at all locations.

The spatial model is based on a hierarchical framework like Wolpert and Ickstadt's one (Wolpert and Ickstadt, 1998). This approach accommodates complexity in high-dimension models by decomposing a model into a series of simpler conditional levels. Each random variable $Y_k(\mathbf{s}_i)$ depends on a latent variable $\beta_k(\mathbf{s}_i)$. Conditionally to $\beta_k(\mathbf{s}_i)$ and $\beta_m(\mathbf{s}_j)$, the variables $Y_k(\mathbf{s}_i)$ and $Y_m(\mathbf{s}_j)$ are independent. The $\beta_k(.)$ processes are dependent. For Gaussian and Poisson variables, we follow the generalized linear mixed model proposed by Diggle et al. (1998):

$$Y_1(\mathbf{s}_i)|\mu_1, \beta_1(\mathbf{s}_i), \nu_1 \sim \mathcal{N}(\mu_1 + \beta_1(\mathbf{s}_i), \nu_1^2), \tag{1}$$

$$Y_2(\mathbf{s}_i)|\mu_2, \beta_2(\mathbf{s}_i) \sim \mathcal{P}(\exp(\mu_2 + \beta_2(\mathbf{s}_i))). \tag{2}$$

$\mathcal{N}(m, \sigma^2)$ denotes the normal distribution with mean m and variance σ^2 and $\mathcal{P}(\lambda)$ the Poisson distribution with parameter λ. μ_1 and μ_2 are the trends of $Y_1(\mathbf{s})$ and $Y_2(\mathbf{s})$ respectively. ν_1^2 corresponds to the nugget effect of the variogram of the Gaussian variable Y_1, that's why it is constant in space.

Unlike Diggle's approach, the present model can take into account ordinal spatial processes through a generalization of the multivariate ordinal probit model to the spatial case (Ashford and Swoden, 1970; Chib and Greenberg, 1998). The principle consists in introducing and truncating an underlying Gaussian random field in the same way as in the truncated Gaussian simulation technique (Matheron et al., 1987):

$$\mathbb{P}(Y_3(\mathbf{s}_i) = j|Z_3(\mathbf{s}_i), \alpha_3, \beta_3(\mathbf{s}_i), \mu_3) = \mathbb{P}(Z_3(\mathbf{s}_i) \in]\alpha_{3;j-1}, \alpha_{3;j}]|\beta_3(\mathbf{s}_i), \mu_3),$$

$$Z_3(\mathbf{s}_i)|\beta_3(\mathbf{s}_i), \mu_3 \sim \mathcal{N}(\mu_3 + \beta_3(\mathbf{s}_i), 1). \tag{3}$$

$\boldsymbol{\alpha}_3 = (\alpha_{3;0}, \alpha_{3;1}, \ldots, \alpha_{3;J})$ denotes the vector of thresholds related to the Gaussian variable Z_3. By convention, $\alpha_{3;0} = -\infty$ and $\alpha_{3;J} = +\infty$. μ_3 corresponds to the trend of the variable Z_3. In the same way, we can deal with nominal variables by generalizing the multinomial probit model (Daganzo, 1979; Natarajan et al., 2000) to the spatial case. Expressions (1), (2) and (3) make the first level of the hierarchical model.

The spatial dependence between the processes $Y_k(.)$ is carried by the latent Gaussian processes $\beta_k(.)$, $k = 1, 2, 3$. The processes are built according to the moving average construction proposed by Ver Hoef and Barry (1998), that is to say by convolving a moving average function with a mixture of white noise processes.

Let f_k, $k = 1, 2, 3$ be a moving average function defined on $\mathbb{R}^2$. $\boldsymbol{\theta}_k$ denotes the vector of parameters of f_k. Let T_k, $k = 1, 2, 3$ be a linear combination of white noise processes:

$$T_k(\mathbf{x}|\rho_k, \boldsymbol{\Delta}_k) = \sqrt{1 - \rho_k^2}\, W_k(\mathbf{x}) + \rho_k W_0(\mathbf{x} - \boldsymbol{\Delta}_k)$$

where $W_k(.)$, $k = 0, 1, 2, 3$ is a white noise process,ρ_k, $k = 1, 2, 3$ belongs to the interval $[-1, 1]$ and $\boldsymbol{\Delta}_k = (\Delta_{k,x}, \Delta_{k,y}) \in \mathbb{R}^2$. The process $W_0(.)$ induces a dependence between the T_k processes since

$$\text{Cor}\left[\int_{\mathbb{R}^2} T_k(\mathbf{x}+\boldsymbol{\Delta}_k)d\mathbf{x}, \int_{\mathbb{R}^2} T_m(\mathbf{x}+\boldsymbol{\Delta}_m)d\mathbf{x}\right] = \rho_k\rho_m \equiv \rho_{km},\ k \neq m.$$

ρ_{km} can be seen as the cross correlation between the white noise processes T_k and T_m. $\boldsymbol{\Delta}_k$ allows a spatial "shift" in the correlation between the two processes T_k and T_m. If $\boldsymbol{\Delta}_k = \boldsymbol{\Delta}_m = (0,0)$ then we have bivariate correlation at each location $\mathbf{x}$ with independence among locations $\mathbf{x}$ and $\mathbf{t}$ when $\mathbf{x} \neq \mathbf{t}$ (Ver Hoef and Barry, 1998). The variable $\beta_k(\mathbf{s}_i)$ is defined by:

$$\beta_k(\mathbf{s}_i) = \int_{\mathbb{R}^2} f_k(\mathbf{x}-\mathbf{s}_i|\boldsymbol{\theta}_k)T_k(\mathbf{x}|\rho_k, \boldsymbol{\Delta}_k)d\mathbf{x}.$$

So the conditional distribution of $\boldsymbol{\beta}(\mathbf{s}) = (\boldsymbol{\beta}_1(\mathbf{s}), \boldsymbol{\beta}_2(\mathbf{s}), \boldsymbol{\beta}_3(\mathbf{s}))$ is a $3N$-dimensional Gaussian distribution with zero mean and covariance matrix $\mathbf{C}$:

$$\boldsymbol{\beta}(\mathbf{s})|\boldsymbol{\theta}_1, \boldsymbol{\theta}_2, \boldsymbol{\theta}_3, \boldsymbol{\rho}, \boldsymbol{\Delta} \sim \mathcal{N}_{3N}(\mathbf{0}, \mathbf{C})$$

where $\boldsymbol{\rho} = (\rho_1, \rho_2, \rho_3)$ and $\boldsymbol{\Delta} = (\boldsymbol{\Delta}_1, \boldsymbol{\Delta}_2, \boldsymbol{\Delta}_3)$. This makes the second level of the hierarchy. One advantage of this construction is that the expression of the covariance matrix $\mathbf{C}$ is known:

$$C_{kk}(\mathbf{h}) = \text{Cov}[\boldsymbol{\beta}_k(\mathbf{s}), \boldsymbol{\beta}_k(\mathbf{s}+\mathbf{h})] = \int_{\mathbb{R}^2} f_k(\mathbf{x}) f_k(\mathbf{x}-\mathbf{h})d\mathbf{x}, \tag{4}$$

$$C_{km}(\mathbf{h}) = \text{Cov}[\boldsymbol{\beta}_k(\mathbf{s}), \boldsymbol{\beta}_m(\mathbf{s}+\mathbf{h})] = \rho_k\rho_m \int_{\mathbb{R}^2} f_k(\mathbf{x}) f_m(\mathbf{x}-\mathbf{h}+\boldsymbol{\Delta}_m - \boldsymbol{\Delta}_k)d\mathbf{x}. \tag{5}$$

$\boldsymbol{\rho}$ and $\boldsymbol{\Delta}$ express the strength and the shift-asymmetry of cross spatial dependence for cross-covariances.(Ver Hoef et al., 2004). Depending on the choice of the moving average functions, the calculation of the integral is either explicit or complex. In the latter case, each element of the matrix can be seen as an autocorrelation in signal theory and can be calculated with the Fast Fourier Transform (Ver Hoef et al., 2004).

The third level of the hierarchical model consists in giving the *prior* distributions on the parameters. The *prior* on μ_1, μ_2, μ_3 is an uniform distribution. For ν_1^2, we chose to use an inverse gamma conjugate *prior* specification $\nu_1^2 \sim IG(a,b)$ where a and b are fixed. We assign an independent uniform *prior* to each spatial dependence parameter $\boldsymbol{\theta}_i, i = 1,2,3$, $\boldsymbol{\rho} = (\rho_1, \rho_2, \rho_3)$ and $\boldsymbol{\Delta} = (\boldsymbol{\Delta}_1, \boldsymbol{\Delta}_2, \boldsymbol{\Delta}_3)$. The *prior* distribution of the thresholds $\boldsymbol{\alpha}_3$ is the order distribution of $J-2$ uniform random variables.

2.2.2 Model Implementation

While the classical approach by maximum likelihood is difficult, the use of conditional independence and the introduction of the latent Gaussian variable Z_3 in the ordinal case allow the evaluation of the *posterior* distribution of the parameters. Using the *prior* distributions, the joint distribution is given by:

$$\begin{aligned}
&\pi(\mu_1, \mu_2, \mu_3, \boldsymbol{\beta}(\mathbf{s}), \mathbf{Z}_3(\mathbf{s}), \nu_1, \boldsymbol{\alpha}, \boldsymbol{\theta}_1, \boldsymbol{\theta}_2, \boldsymbol{\theta}_3, \boldsymbol{\rho}, \boldsymbol{\Delta}|\mathbf{Y}(\mathbf{s}))\\
&\quad\propto \exp\left\{-\frac{1}{2\nu_1^2}{}^{\mathrm{T}}(\mathbf{Y}_1(\mathbf{s}) - \mu_1\mathbf{1} - \boldsymbol{\beta}_1(\mathbf{s}))(\mathbf{Y}_1(\mathbf{s}) - \mu_1\mathbf{1} - \boldsymbol{\beta}_1(\mathbf{s}))\right\}\\
&\quad\times \prod_{i=1}^{N}\left[\frac{[\exp(\mu_2 + \beta_2(s_i))]^{Y_2(s_i)}\exp\{\exp(\mu_2 + \beta_2(s_i))\}}{Y_2(s_i)!}\right]\\
&\quad\times \prod_{i=1}^{N}\left[\exp\left\{-\frac{1}{2}(Z_3(s_i) - \mu_3 - \beta_3(s_i))^2\right\} \mathbb{1}(Z_3(s_i) \in]\alpha_{3;Y_3(s_i)-1}; \alpha_{3;Y_3(s_i)}])\right]\\
&\quad\times \exp\left\{-\frac{1}{2}{}^{\mathrm{T}}\beta(\mathbf{s})C^{-1}\beta(\mathbf{s})\right\}\pi(\nu_1^2)
\end{aligned}$$

where $\mathbb{1}$ denotes the indicator function and $\mathbf{1}$ a vector of length N with all terms equal to 1.

The marginal *posterior* distributions for each of these parameters can be obtained through the implementation of a Markov chain Monte Carlo (MCMC) simulation scheme. Parameters μ_1, μ_3, $\boldsymbol{\beta}_1(\mathbf{s})$, $\boldsymbol{\beta}_3(\mathbf{s})$, ν_1, $\boldsymbol{\alpha}_3$ are drawn iteratively from their full conditional distributions:

$$\mu_1|\ldots \sim \mathcal{N}\left(\frac{1}{N}\sum_{i=1}^{N}(Y_1(\mathbf{s}_i) - \beta_1(\mathbf{s}_i)), \frac{\nu_1^2}{N}\right),$$

$$\mu_3|\ldots \sim \mathcal{N}\left(\frac{1}{N}\sum_{i=1}^{N}(Z_3(\mathbf{s}_i) - \beta_3(\mathbf{s}_i)), \frac{1}{N}\right),$$

$$\nu_1^2|\ldots \sim IG\left(a + \frac{N}{2}, b + \frac{\sum_{i=1}^{N}(Y_1(\mathbf{s}_i) - \mu_1 - \beta_1(\mathbf{s}_i))^2}{N}\right),$$

$$\begin{aligned}
&\alpha_{3;j}|\ldots \sim\\
&\quad\mathcal{U}[\max(\max(Z_3(\mathbf{s}_i)|Y_3(\mathbf{s}_i) = j), \alpha_{3;j-1}); \min(\min(Z_3(\mathbf{s}_i)|Y_3(\mathbf{s}_i) = j + 1),\\
&\qquad\alpha_{3;j+1})],
\end{aligned}$$

$$Z_3(\mathbf{s}_i)|Y_3(\mathbf{s}_i), \beta_3(\mathbf{s}_i), \mu_3 \sim \mathcal{N}(\mu_3 + \beta_3(\mathbf{s}_i)) \text{ truncated on } [\alpha_{3,Y_3(\mathbf{s}_i)-1}; \alpha_{3,Y_3(\mathbf{s}_i)}],$$

$$\boldsymbol{\beta}_1(\mathbf{s})|\ldots \sim \mathcal{N}_N(\mathbf{m}_1^*, \mathbf{V}_1^*) \text{ with } \begin{cases} \mathbf{V}_1^* = \left(\mathbf{V}_1^{-1} + \frac{1}{\nu_1^2}\mathbf{I}\right)^{-1} \\ \mathbf{m}_1^* = \mathbf{V}_1^*\left(\mathbf{V}_1^{-1}\mathbf{m}_1 + \frac{1}{\nu_1^2}(\mathbf{Y}_1(\mathbf{s}) - \mu_1\mathbf{1})\right) \end{cases} \text{where}$$

$\mathbf{m}_1$ and $\mathbf{V}_1$ are respectively the conditional expectancy and the covariance matrix of $\boldsymbol{\beta}_1(\mathbf{s})$ given $\boldsymbol{\beta}_2(\mathbf{s})$ and $\boldsymbol{\beta}_3(\mathbf{s})$,

$$\boldsymbol{\beta}_3(\mathbf{s})|\ldots \sim \mathcal{N}_N(\mathbf{m}_3^*, \mathbf{V}_3^*) \text{ with } \begin{cases} \mathbf{V}_3^* = (\mathbf{V}_3^{-1} + \mathbf{I})^{-1} \\ \mathbf{m}_3^* = \mathbf{V}_3^*(\mathbf{V}_3^{-1}\mathbf{m}_3 + (\mathbf{Z}_3(\mathbf{s}) - \mu_3\mathbf{1})) \end{cases} \text{where}$$

$\mathbf{m}_3$ and $\mathbf{V}_3$ are respectively the conditional expectancy and the covariance matrix of $\boldsymbol{\beta}_3(\mathbf{s})$ given $\boldsymbol{\beta}_1(\mathbf{s})$ and $\boldsymbol{\beta}_2(\mathbf{s})$.

The vector $\boldsymbol{\beta}_2(\mathbf{s})$ is updated by an adaptative version of a Metropolis Langevin algorithm (Atchade, 2006). Let $\pi(\boldsymbol{\beta}_2(\mathbf{s}))$ be the target distribution. The proposal distribution is given by:

$$q_h(\boldsymbol{\beta}_2^*(\mathbf{s})|\boldsymbol{\beta}_2(\mathbf{s})) \sim \mathcal{N}_N\left(\boldsymbol{\beta}_2(\mathbf{s}) + \frac{h^2}{2}D(\boldsymbol{\beta}_2(\mathbf{s})), h^2\mathbf{I}\right)$$

where

$$D(\boldsymbol{\beta}_2(\mathbf{s})) = \frac{\delta}{\max(\delta, |\nabla \ln(\pi(\boldsymbol{\beta}_2(\mathbf{s})))|)}\nabla \ln(\pi(\boldsymbol{\beta}_2(\mathbf{s}))).$$

∇ is the gradient operator, $\delta > 0$ is a fixed constant and $h > 0$ is a scale parameter. The proposed value $\boldsymbol{\beta}_2^*(\mathbf{s})$ is accepted with probability

$$\min\left(1, \frac{\pi(\boldsymbol{\beta}_2^*(\mathbf{s}))q_h(\boldsymbol{\beta}_2(\mathbf{s})|\boldsymbol{\beta}_2^*(\mathbf{s}))}{\pi(\boldsymbol{\beta}_2(\mathbf{s}))q_h(\boldsymbol{\beta}_2^*(\mathbf{s})|\boldsymbol{\beta}_2(\mathbf{s}))}\right).$$

The scale parameter h is updated at each iteration of the algorithm in order to obtain a acceptance rate of 0.574.

The parameter μ_2 and the spatial dependence parameters $\boldsymbol{\theta}_i, i = 1, 2, 3$, $\boldsymbol{\rho}$ and $\boldsymbol{\Delta}$ are sampled from a Metropolis step (Hastings, 1970). Each vector $\boldsymbol{\theta}_i$, each term of $\boldsymbol{\rho}$ and each vector $\boldsymbol{\Delta}_k$ is updated separately. The proposal distribution of each parameter is a normal distribution centered on the current value of the parameter. If there are constraints on the parameter, the value is proposed according to a truncated normal distribution.

In the bivariate case, we can notice that the parameters ρ_k and ρ_m are not identifiable; only the product $\rho_{km} = \rho_k\rho_m$ can be identify. To ensure that all parameters are identifiable, the threshold $\alpha_{3,1}$ related to the ordinal variable is fixed to 0 (Cowles, 1996). In the same way, we can let $\boldsymbol{\Delta}_1 = (0, 0)$. All shifts $\boldsymbol{\Delta}_k, k = 2, 3$ are relative to $\boldsymbol{\Delta}_1$ and $\boldsymbol{\Delta}$ is reduced to $(\boldsymbol{\Delta}_2, \boldsymbol{\Delta}_3)$. Initial values of the parameters for the MCMC inference are randomly chosen. But it is better to run the algorithm in the univariate case for each variable and to take the obtained estimations as initial values for the multivariate procedure.

The predictions of the random field at unknown locations are obtained by following the method described by Kern (2000).

3 Results

The model is applied to pedological data described in Section 2.1. We consider a random field made of by two variables: a Gaussian variable, the slope (Y_1) and an ordinal variable, the soil drainage (Y_2). The last one counts six ordered modalities. Some of these modalities are rarely observed, so we gather the observations in four modalities in order to have sufficiently observations by modalities. The modalities are ordered from well drained soils to hydromorphic soils. Three hundred and twenty seven data were available. Two hundred locations were sampled for the estimation. The remaining 127 values were used as validation data set. As data were collected in permanent sample plots, the spatial pattern is aggregated.

The chosen moving average functions had a Gaussian form:

$$f_k(x,y) = \sqrt{\frac{4\sigma_k}{\pi\varphi_k^2}} \exp\left(-\frac{2(x^2+y^2)}{\varphi_k^2}\right) \text{ with } \boldsymbol{\theta}_k = (\sigma_k, \varphi_k).$$

No asymmetry-shift was introduced, so $\boldsymbol{\Delta}_k = (0,0), \forall k$.

To check model estimation, we use some summary statistics for validation. Let $\widehat{Y_1}(\mathbf{s}_i)$ be the predicted value of the Gaussian variable at location $\mathbf{s}_i$ for the ith datum of the validation data set. Let $\widehat{\text{var}}(\widehat{Y_1}(\mathbf{s}_i))$ (resp. $\widehat{\text{sd}}(\widehat{Y_1}(\mathbf{s}_i))$) be the estimated prediction variance (resp. standard deviation) at location $\mathbf{s}_i$. Let n be the number of data in the validation data set. For the Gaussian variable, we compute:

- biais $= \frac{1}{n}\sum_{i=1}^{n}(\widehat{Y_1}(\mathbf{s}_i) - Y_1(\mathbf{s}_i))$
- RMSPE $= \sqrt{\frac{\sum_{i=1}^{n}(\widehat{Y_1}(\mathbf{s}_i) - Y_1(\mathbf{s}_i))^2}{n}}$
- RMEV $= \sqrt{\frac{\sum_{i=1}^{n}\widehat{\text{var}}(\widehat{Y_1}(\mathbf{s}_i))}{n}}$
- 80%PI $= \frac{1}{n}\sum_{i=1}^{n}\mathbb{1}[|\widehat{Y_1}(\mathbf{s}_i) - Y_1(\mathbf{s}_i)| < 1.28\widehat{\text{sd}}(\widehat{Y_1}(\mathbf{s}_i))]$

If the estimated prediction variances are corrected, then RMEV should be closed to RMSPE (Ver Hoef et al., 2004). The prediction interval coverage 80%PI should be about 80%. For the ordinal variable, we give the percentage of well predicted values in the validation data set.

Figure 1 gives the sampling from full conditional distributions.

Table 1 summarizes the estimation of the parameters for the bivariate data set. The estimations are the *posterior* means of the distribution sampled from the MCMC scheme. The standard deviation of the distribution is given in brackets.

The speed convergence is high for parameters related to the Gaussian variable. On the contrary, parameters estimation for the ordinal variable requires more iterations to obtain the convergence of the chains (Fig. 1). In fact, the burn-in duration is

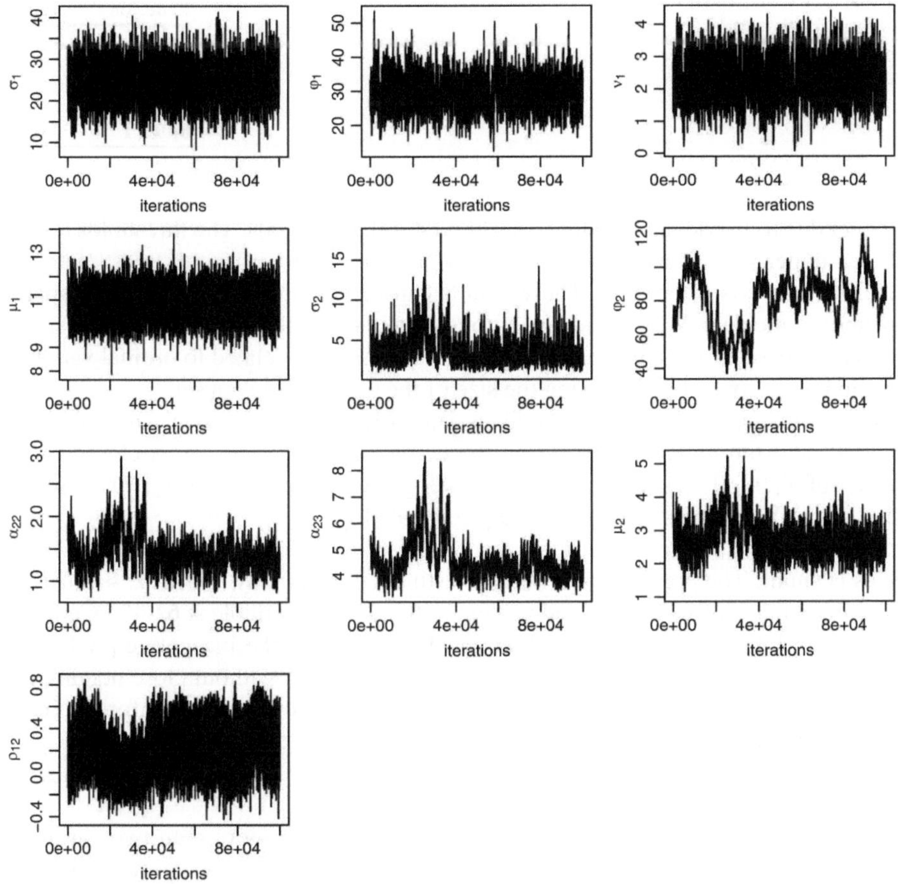

Fig. 1 Sampling from full conditional distributions

Table 1 Estimation of parameters from the data set made up by slope and soil drainage at Paracou

Parameter	Bivariate	Estimates
σ_1	23.70	(4.32)
φ_1	29.89	(4.90)
ν_1	2.18	(0.62)
μ_1	10.89	(0.54)
σ_2	3.08	(1.20)
φ_2	87.05	(10.45)
$\alpha_{2,2}$	1.35	(0.18)
$\alpha_{2,3}$	4.28	(0.39)
μ_2	2.61	(0.38)
ρ_{12}	0.28	(0.19)

Table 2 Cross-validation criteria for the slope predictions

Criterion	Values
Bias	−0.20
RSMPE	4.17
RMEV	4.54
80%PI	0.87

longer due to two levels of latent variables ($\mathbf{Z}_2(\mathbf{s})$ and $\boldsymbol{\beta}_2(\mathbf{s})$). The thresholds related to the underlying Gaussian variable $Z_2(\mathbf{s})$ are particularly difficult to estimate and their variance is often high. These results can be explained by the specific spatial pattern of the data. We have shown through simulations that the accuracy of estimations decreases if the data are aggregated. The estimates related to ordinal variables could be improved by increasing the size of the calibration data set.

The obtained estimates for the Gaussian variable are consistent with the range, the sill and the nugget observed on the empirical variogram. μ_1 is closed to the mean of $Y_1(\mathbf{s})$. The standard deviation is high for the parameter φ_2. The variable $\mathbf{Y}_1(\mathbf{s})$ and $\mathbf{Y}_2(\mathbf{s})$ are slightly positively correlated.

The cross validation criteria for the Gaussian variable are given in Table 2. The bias is small. The prediction variance is estimated accurately (RSMPE $\approx$ RMEV). The percentage of well predicted values for the ordinal variable is 61.4%. Concerning the ordinal variable, it should be noted that most of the inaccurate predictions concerned locations near the boundaries of plots or sites without close neighbours. These locations coincide with locations where the prediction variance for the Gaussian variable is high. Moreover, for the ordinal variable, one of the modalities is under observed. The lack of information about this modalities may explain some mistakes in the predictions.

It is possible to simulate and estimate parameters for more than two variables. But the inference procedure becomes computationally intensive and time consuming because of the size of handled covariance matrix in this case.

The choice of the moving average function can be questioned. The chosen form f_k is particularly pleasant since few parameters are nedeed and integrals in Equations (4) and (5) are easy to evaluate. More flexible functions could be used like disk-based kernel (Kern, 2000) or anisotropic function (Ver Hoef et al., 2004) if the number of parameters is reasonable. They could improve the modelling of the dependence between variables and consequently improve the predictions.

4 Conclusion

The proposed approach permits modelling a spatial multivariate random field made of variables of different nature. A unified methodology (generalized linear model) can be applied for Gaussian, Poisson and ordinal variables through the introduction of Gaussian latent variable in the discrete case. Although the estimation procedure is time consuming, this approach is an interesting alternative to disjunctive cokriging for the prediction of ordinal variables. However, the number of data used for estimation must be sufficient to obtain accurate estimates. The modelling of the dependence between the processes by the moving average approach has the advantage to be very flexible. Anisotropic data can be dealt with if a convenient moving average function is chosen. An extension of the model can be considered for nominal variable. In the same way we have generalized the ordinal probit model to deal with ordinal variable, we can generalize the multinomial probit model to take into account nominal variables.

References

Ashford S, Swoden R (1970) Multivariate probit analysis. Biometrics 26:535–546

Atchade Y (2006) An adaptive version for the Metropolis Adjusted Langevin Algorithm with a truncated drift. Methodol Comput Appl Probab 8(2):235–254

Banerjee S, Carlin B, Gelfand A (2004) Hierarchical modeling and analysis for spatial data. Number 101 in Monogr. Statist. Appl Probab Chapman and Hall/CRC Boca Raton, FL

Bogaert P (2002) Spatial prediction of categorical variables: the Bayesian maximum entropy approach. Stoch Environ Res Risk Assess 16:425–448

Chagneau P, Mortier F, Picard N, Bacro J (2008) Hierarchical Bayesian model for spatial prediction of multivariate non-Gaussian random fields (submitted)

Chaubert F, Mortier F, Saint-André L (2008) Multivariate dynamic model for ordinal outcomes. J Multivariate Anal 99:1717–1732

Chib S, Greenberg E (1998) Analysis of multivariate probit models. Biometrika 85(2):347–361

Chilès J, Delfiner P (1999) Geostatistics. Modeling spatial uncertainty. Wiley Ser. Probab. Stat. Wiley, New York

Christakos G (1990) A bayesian maximum-entropy view to the spatial estimation problem. Math Geol 22(7):763–777

Christakos G (1998) Bayesian Maximum Entropy analysis and mapping: a farewell to kriging estimators? Math Geol 30(4):435–462

Cowles M (1996) Accelerating Monte Carlo Markov chain convergence for cumulative-link generalized linear models. Statist Comput 6:101–111

Cressie N (1991) Statistics for spatial data. Wiley, New York

Daganzo C (1979) Multinomial probit. The theory and its application to demand forecasting. Academic Press, A Subsidiary of Harcourt Brace Jovanovich, New York

Diggle P, Tawn J, Moyeed R (1998) Model-based geostatistics (With discussion). J R Stat Soc Ser C 47(3):299–350

D'Or D, Bogaert P (2004) Spatial prediction of categorical variables with the Bayesian Maximum Entropy approach: the Ooypolder case study. Eur J Soil Sci 55:763–775

Epron D, Bosc A, Bonal D, Freycon V (2006) Spatial variation of soil respiration across a topographic gradient in a tropical rain forest in French Guiana. J Trop Ecol 22:565–574

FAO-ISRIC-ISSS (1998) World reference base for soil resources. In: World Soil Resour Rep 84:109

Gourlet-Fleury S, Guehl J, Laroussinie O (2004) Ecology and management of neotropical rainforest lessons drawn from Paracou, a long-term experimental research site in French Guiana. Elsevier Paris

Grzebyk M, Wackernagel H (1994) Multivariate analysis and spatial/temporal scales: real and complex models. In: Proceedings of the XVIIth international biometric conference, Hamilton, Ontario

Hastings W (1970) Monte Carlo sampling methods using Markov chains and their applications. Biometrika 57:97–109

Journel A, Huijbregts C (1978) Mining geostatistics. Academic Press, London

Kern J (2000) Bayesian process-convolution approaches to specifying spatial dependence structure. Ph.D. thesis, Institute of Statistics and Decision Sciences, Duke University, Durhan, NC

Matheron G, Beucher H, Fouquet C, Galli A, Guerillot D, Ravenne C (1987) Conditional simulation of the geometry of fluvio-deltaic reservoirs. In: Proceedings of the SPE annual technical conference and exhibition, Dallas, Texas, 237–30 September 1987, reservoir engineering

Natarajan R, McCulloch C, Kiefer N (2000) A Monte Carlo EM method for estimating multinomial probit models. Comput Stat Data Anal 34:33–50

Sabatier D, Grimaldi M, Prevost M, Guillaume J, Godron M, Dosso M, Curmi P (1997) The influence of soil cover organization on the floristic and structural heterogeneity of a Guianan rain forest. Plant Ecol 131:81–108

Ver Hoef J, Barry R (1998) Constructing and fitting models for cokriging and multivariable spatial prediction. J Stat Plann Infer 69(2):275–294

Ver Hoef J, Cressie N, Barry R (2004) Flexible spatial models for kriging and cokriging using moving averages and the Fast Fourier Transform (FFT). J Comput Graph Stat 13:265–289

Wackernagel H (2003) Multivariate geostatistics. An introduction with applications, 3rd completely revised edn. Springer, Berlin

Wolpert RL, Ickstadt K (1998) Poisson/gamma random field models for spatial statistics. Biometrika 85(2):251–267

Detection of Optimal Models in Parameter Space with Support Vector Machines

Vasily Demyanov, Alexei Pozdnoukhov, Mike Christie, and Mikhail Kanevski

Abstract The paper proposes an approach aimed at detecting optimal model parameter combinations to achieve the most representative description of uncertainty in the model performance. A classification problem is posed to find the regions of good fitting models according to the values of a cost function. Support Vector Machine (SVM) classification in the parameter space is applied to decide if a forward model simulation is to be computed for a particular generated model. SVM is particularly designed to tackle classification problems in high-dimensional space in a non-parametric and non-linear way. SVM decision boundaries determine the regions that are subject to the largest uncertainty in the cost function classification, and, therefore, provide guidelines for further iterative exploration of the model space. The proposed approach is illustrated by a synthetic example of fluid flow through porous media, which features highly variable response due to the parameter values' combination.

1 Introduction

Mathematical models used for computing predictions of many geo- and environmental systems are traditionally of parametric nature. The model parameters, usually, bear a large degree of uncertainty due to lack of knowledge about the particular phenomenon. Conventionally, the model parameters can be fitted to the available observations using various optimisation techniques, which raises the problem of the confidence of such model fit, e.g. in Demyanov et al., (2006).

Uncertainty of a forecast is based on the probabilistic analysis of prediction model solutions. A probabilistic approach was applied to quantify uncertainty of

V. Demyanov (✉) and M. Christie
Institute of Petroleum Engineering, Heriot-Watt University, Edinburgh, UK
e-mail: vasily.demyanov@pet.hw.ac.uk; Mike.Christie@pet.hw.ac.uk

A. Pozdnoukhov and M. Kanevski
Institute of Geomatics and Analysis of Risk, University of Lausanne, Switzerland
e-mail: alexei.pozdnoukhov@unil.ch; Mikhail.Kanevski@unil.ch

P.M. Atkinson and C.D. Lloyd (eds.), *geoENV VII – Geostatistics for Environmental Applications*, Quantitative Geology and Geostatistics 16,
DOI 10.1007/978-90-481-2322-3_30,

production forecasts in petroleum reservoirs (Subbey et al., 2002). The prediction uncertainty is based on the inference from a set of multiple reservoir models that match well past production observations (historic data). Accuracy of such inferences largely depends on the number of good fitting models with different combinations of the parameter values. Multiple good fitting models found in different regions of the parameter space would provide a more robust uncertainty assessment (Christie et al., 2006).

The search for good fitting models in a high dimensional parameter space is a challenging task. The misfit surface, which reflects the goodness of model fit in the parameter space, usually has a complicated structure with multiple local minima. Conventional gradient optimisation methods can be successfully used to find a single local minimum. Stochastic optimisation methods (e.g. simulated annealing, genetic algorithms [GA], particle swarm) are used to find multiple local minima, which correspond to multiple good fitting models.

Adaptive stochastic sampling algorithms (e.g. GA, Neighbourhood Approximation [NA] algorithm [Sambridge, 1999], etc.) iteratively refine regions of low misfit in the parameter space (see Fig. 1) based on previously evaluated models. Thus, the search for good models is based on some sort of interpolation (or pre-evaluation). In case of NA it is Voronoi interpolation, which implies a discontinuous constant value in the polygon neighbourhood.

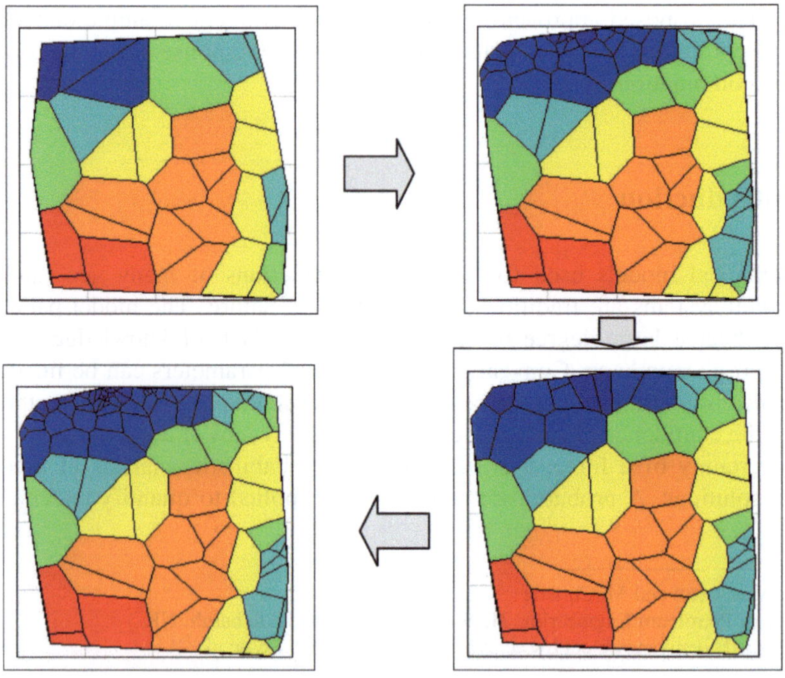

Fig. 1 Adaptive sampling in the search for good fitting models: *blue colour* corresponds to low misfit (each model is displayed by a Voronoi cell following NA sampling)

Every misfit evaluation entails forward simulation of the reservoir model, which is a computationally costly process and entails solving finite-difference flow equations on a fine grid. A guided sampling approach was proposed to increase the computational efficiency of the adaptive sampling by reducing the number of forward reservoir simulations (Demyanov, 2007). The approach proposed using fast misfit artificial neural network interpolation models to compute approximate misfit values instead of the exact ones from forward flow simulation. Artificial neural networks were used as non-linear and non-parametric regression models to evaluate the misfit at chosen locations based on the previously simulated pool of models (Christie et al., 2006).

The principal question in the search for good fitting models is whether to run a flow simulation to evaluate the goodness of model fit. This splits all possible models into two classes – the likely good fitting models for which we need to run the flow simulation and the models for which no flow simulation run is needed as they are unlikely to provide a reasonable match. Thus, we can reformulate the adaptive sampling problem using classification to decide whether to run the flow simulation at any particular location. The parameter space becomes separated by the classifier into the areas where the flow simulation is to run and where it is not. Therefore, instead of solving a misfit interpolation problem we have a classification problem. The classification problem is focused on detection of the regions where evaluation of the misfit is needed via flow simulation to find good history match models. Application of classification algorithms rather than regression algorithms to the guided sampling allows more flexibility and it is less influenced by noise and inaccuracy of the approximate interpolation solution.

There exists a wide selection of statistical and data driven algorithms to solve the classification problem. Machine learning also includes a number of classification tools such as Support Vector Machines, probabilistic neural networks, self-organising maps, k-nearest neighbours, etc. (Haykin, 1999). A comprehensive review of applications of traditional and recent machine learning algorithms to spatial prediction problems is presented in Kanevski et al., 2009.

The high dimensionality of the parameter space is another problem that complicates the search for good fitting models. Even when large numbers of potential models are evaluated the space remains fairly empty and poorly explored. Therefore, the search for multiple low misfit models becomes difficult even for adaptive algorithms; this is true especially in the presence of local minima in the misfit surface. The exploration of high dimensional space is burdened by the curse of dimensionality problem (Hastie et al., 2001). The problem of curse of dimensionality is in exponential increase in volume with adding an extra dimension to the parameter space and is illustrated in Fig. 2 (Kanevski et al., 2009). Suppose 1,000 data samples are drawn uniformly in the unit volume around the origin of the coordinate system. Then, the distance from the origin to the nearest sample increases drastically with dimension. In high dimensional space, the notion of nearest neighbour becomes obsolete – all the samples are equally far or, in other words, are located in the corners of the high dimensional space.

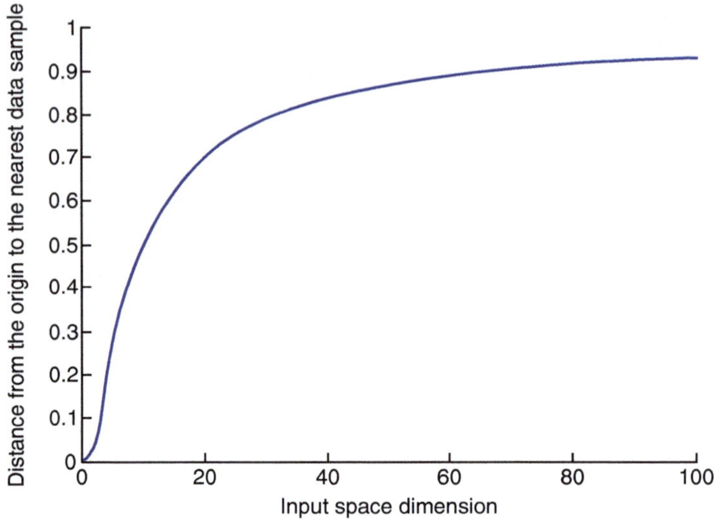

Fig. 2 Dependence of the distance from the origin to the nearest sample from the space dimension

Summarising the complications of uncertainty qualification with multiple generated models we can outline the following:

1. Difficulty in finding good fitting models in high dimensional parameter space due to complex dependencies of the misfit from the different combinations of parameter values.
2. The sampling approach requires thousands of generated models, nevertheless the space remains poorly explored and populated by models.
3. The misfit surface in the parameter space is, usually, not smooth with numerous local variations. Therefore, gradient search methods become trapped in local minima. Misfit interpolation models may not be accurate enough and suffer from the curse of dimensionality because the nearest neighbour notion, which most interpolation algorithms are based on, becomes obsolete in high dimensions.

1.1 Aims

The aim of the paper is to propose a way to improve the search for good fitting models with a robust classification method, which can overcome the curse of the dimensionality problem.

Support Vector Machine (SVM) is a data driven classification method which provides a non-linear and a non-parametric classification in high dimensional input space and effectively handles the curse of dimensionality. As a machine learning algorithm it is able to capture dependencies from the data in the model parameter space whilst training. Based on the captured dependencies SVM is used to classify models in parameter space according to their goodness of fit.

The purpose of using SVM classification is to improve the sampling efficiency by reducing the computational effort spent on forward reservoir simulation for every generated model. Guided sampling based on the classification results would be able to find good fitting models faster with fewer forward simulations computed only for the models from the regions classified as "good fit". SVM classification would improve sampling quality because this data-driven algorithm overcomes the curse of dimensionality problem.

In this paper we propose the methodology of classification for sampling for the good fitting models and illustrate it with a synthetic feasibility example. However, the illustrative case study we used is of low dimension – with just three model parameters. A higher dimensional study, which would tackle the curse of dimensionality problem, will be the subject of future research.

2 Support Vector Machine (SVM) Classification

SVM is a machine learning approach, derived from Statistical Learning Theory, which aims to deal with data of high dimensionality by approaching the nonlinear problems in a robust and non-parametric way. Interested readers can to refer to some of the key introductions to the theory of SVMs and related algorithms (Vapnik, 1995; Scholkopf and Smola, 2002). Here we only mention the main principles of SVMs which will be applied to the sampling of models in a high-dimensional parameter space.

Suppose we deal with the linearly separable data $(x_1, y_1), \dots (x_N, y_N)$, where x are the input features and $y \in \{+1, -1\}$ are the binary labels, corresponding, for example, to the models subject to computer simulation and "not worth" simulating. By linearly separable we mean data that can be discriminated into two classes by a hyperplane. The idea of SVM is to separate this dataset by finding the hyperplane that is, roughly speaking, the farthest apart from the closest training points. The minimal distance between the hyperplane and the training points is called the margin, which is maximized by the SVM algorithm (see Fig. 3).

The maximum margin principle, derived from Statistical Learning Theory, prevents over-fitting in high-dimensional input spaces, and thus leads to good generalization abilities. The decision function used to classify the data is a linear one, as follows:

$$f(x, w) = w \cdot x + b, \tag{1}$$

where the coefficient vector w and the threshold constant b are optimised in order to maximise the margin. This is a quadratic optimization problem with linear

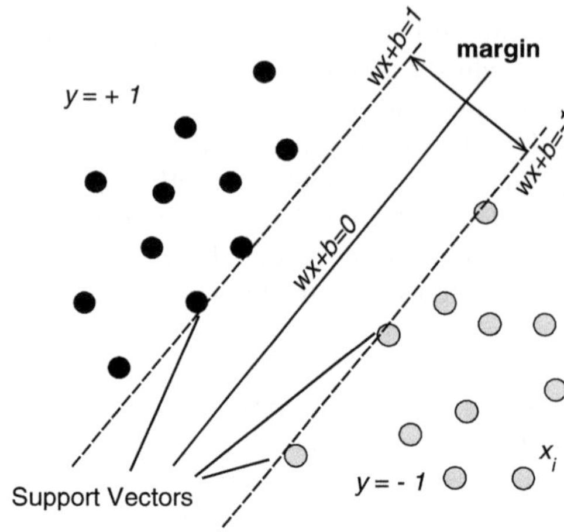

Fig. 3 Margin maximization principle: the basic idea of Support Vector Machine

constraints which has a unique solution. Moreover, w is a linear combination of the training samples y_i, taken with some weights α_i:

$$w = \sum_{i=1}^{N} y_i \alpha_i x_i . \tag{2}$$

The samples with non-zero weights are the only ones which contribute to this maximum margin solution. They are the closest samples to the decision boundary and are called Support Vectors (SVs) (see Fig. 3). SVs are penalized such that $0 < \alpha < C$ to allow for misclassification of training data (taking into account the mislabelled samples or noise).

A so-called kernel trick is used to make this classifier non-linear. A kernel is a symmetric semi-positive definite function $K(x, x')$. According to the Mercer theorem, this implies that it corresponds to a dot product in some space (Reproducing Kernel Hilbert Space, RKHS). Generally, given a (linear) algorithm, which includes data samples in the form of dot products only, one can obtain a (non-linear) kernel version of it by substituting the dot products with kernel functions. This is the case for linear SVM, where the decision function (1) relies on the dot products between samples, as clearly seen by substituting (2) into (1). The final classification model is a kernel expansion:

$$f(x, \alpha) = \sum_{i=1}^{N} \alpha_i K(x, x_i) + b \tag{3}$$

The choice of the kernel function is an open research issue. Using some typical kernels like Gaussian RBF, one takes into account some knowledge like distance-based similarity of the samples. The parameters of the kernel are the hyper-parameters of SVM and have to be tuned using cross-validation or another similar technique.

2.1 Probabilistic Post-processing

A probabilistic interpretation of the outputs of an SVM is often desirable for uncertainty assessment. To introduce a characterization of prediction uncertainty, the values of the decision function (1) or (3) can be transformed into the smooth confidence measure, $0 < p(y = 1|x) < 1$. This is done, for example, through taking a sigmoid transformation of $f(x, \alpha)$ (Platt, 1999):

$$p(y = 1\,|x) = \frac{1}{(1 + \exp(a \cdot f(x) + b))}, \qquad (4)$$

where a and b are constants. These constants are tuned using a maximum likelihood (usually, the negative log-likelihood to simplify the optimisation) on the testing dataset. The value of a is negative, and if b is found to be close to zero, then the default SVM decision threshold $f(x) = 0$ coincides with a confidence threshold level of 0.5.

2.2 Support Vector Exploration of Potential Sampling Locations

Let us stress the importance of the weights α_i in the prediction (3). The following cases are possible:

- If $\alpha_i = 0$, then $y_i f(x_i) \geq 1$, the point is well described by the others.
- If $C > \alpha_i > 0$, then $y_i f(x_i) = 1$, meaning that the point is a Support Vector(SV).
- If $\alpha_i = C$, then $y_i f(x_i) \leq 1$, meaning that it is either noise, untypical or mislabelled point.

Note that removing all other points except the SVs from the training data set and training SVM on the SVs only leads to the same decision boundary. SVs have the determinant meaning for the given classification task. If one, on the other hand, would add more data samples from the correctly classified zones and they appear to be of the correct class, these samples would not change the decision boundary of SVM. These facts give us an opportunity to use the locations and the corresponding weights of SVs as the criteria for the search for the sampling locations to improve the classification model.

The proposed method of sampling optimisation is based on the described properties of Support Vectors (Pozdnoukhov and Kanevski, 2007). Given a new prospective sampling location, one iteratively includes it into the current model with first positive and then negative labels assigned to it. After the re-training of the SVM the model weights α^+ and α^-, have to be analysed. If the new sample obtains zero weight and does not become a SV, it doesn't contribute to the prediction model and is somehow "useless". On the other hand, a sample that becomes a SV is of a particular importance to the task since it defines the decision function.

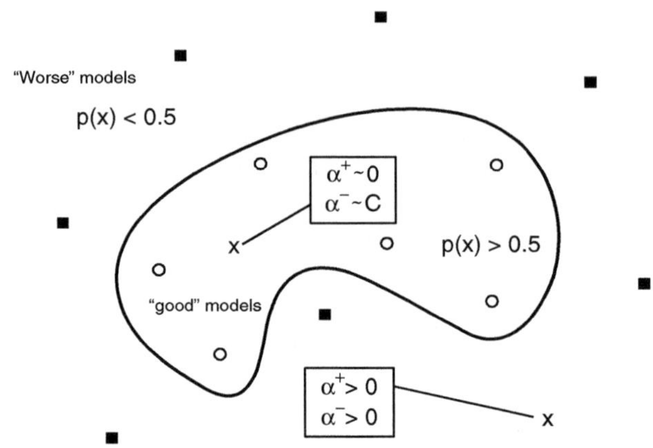

Fig. 4 The sampling point inside the class region of "good" models can be identified by considering the predicted probability and ranked following its potential weight as if the sample would become a Support Vector. Predicted "worse" models can be filtered out. The unexplored regions can be identified as well

In our search for the prospective good models we are interested in the following situations illustrated in Fig. 4. A sample lies inside the region of "good" models if it simultaneously satisfies the following two conditions:

- When labelled positively it obtains zero or some small weight $\alpha^+ \sim 0$.
- And when labelled negatively its weight is bounded with $C(\alpha^- \sim C)$.

Such samples are subject to simulation. Although, it is well described by the available data set and does not provide new information about the parameter space. Those samples with both $\alpha^+ > 0$ and $\alpha^- > 0$ are inside the unexplored regions of the parameter space and can be proposed for consideration.

Another important issue is the ranking by significance of those samples which were found to be interesting. Since here we are only interested in a relative measure in order to provide the ranking, the following value will be used:

$$\eta(x_k) = \begin{cases} 0,\ if\ \alpha_k^+ = 0,\ \alpha_k^- = C;\ or\ \alpha_k^+ = C,\ \alpha_k^- = 0, \\ \frac{\alpha_k^+ + \alpha_k^-}{2C}, \quad otherwise. \end{cases} \tag{5}$$

Since it is time-consuming to explore the multi-dimensional parameter space by retraining an SVM classifier two times for every prospective sampling location, one may carry out this procedure for the samples with sufficient probabilities $p(y = 1|x) > \eta$. The estimation of the latter involves a simple computation of (3) and (4) which is reasonably fast.

3 Classification in Model Parameter Space: IC Fault Model Example

3.1 SVM Application

SVM classification was applied to detect the areas of good fitting models in a synthetic case study with the IC Fault model. This model has been used extensively for uncertainty quantification exercises where different adaptive stochastic sampling models were used to find multiple history matched models (Christie et al., 2006; Erbaş, 2006). The IC Fault model is really simple with just three free parameters, however it exhibits a highly complex misfit surface with multiple local minima, which makes it quite challenging for the uncertainty quantification study.

The goal of the application was to demonstrate SVM capability to classify the model goodness of fit based on a limited number of data, decide on whether it is worth running a flow simulation model for a particular parameter combination in a search good fit and propose new sampling locations for further sampling.

3.2 IC Fault Model

The IC Fault model is an extremely simple three-parameter model set up in Imperial College by J. Carter (Carter et al., 2004) as a test example for automated history matching. It has proved extremely difficult to history match, and one of the conclusions published in Carter et al. (2004) has been that the best history matched model (from $\sim$160, 000 models) is of limited use as a predictor.

The geological model consists of six layers of alternating high and low permeability sands (see Fig. 5). The three high permeability layers have identical properties, and the three low permeability layers have a different set of identical properties. The porosities of the high and low permeability layers are 0.30 and 0.15 respectively. The width of the model is 1,000 ft, with a simple fault in the middle. There is a water injector well at the left-hand edge, and a producer well on the right-hand edge. The simulation model is 100×12 grid blocks, with each geological layer divided into two simulation layers with equal thickness (Tavassoli et al., 2004).

The model has three unknown parameters for history matching: high and low permeability (k_{high} and k_{low}) and the fault throw (h). Our prior model has uniform distribution with ranges: $k_{high} \in [100, 200]$, $k_{low} \in [0, 50]$ and $h \in [0, 60]$.

The IC Fault model study specified a point in the parameter space as the true parameter values: $k_{high} = 131.6$, $k_{low} = 1.3$ and $h = 10.4$. The misfit was defined using a standard least squares model using the discrepancy between the simulated and the observed oil and water production rates for the 3-year history-matching period.

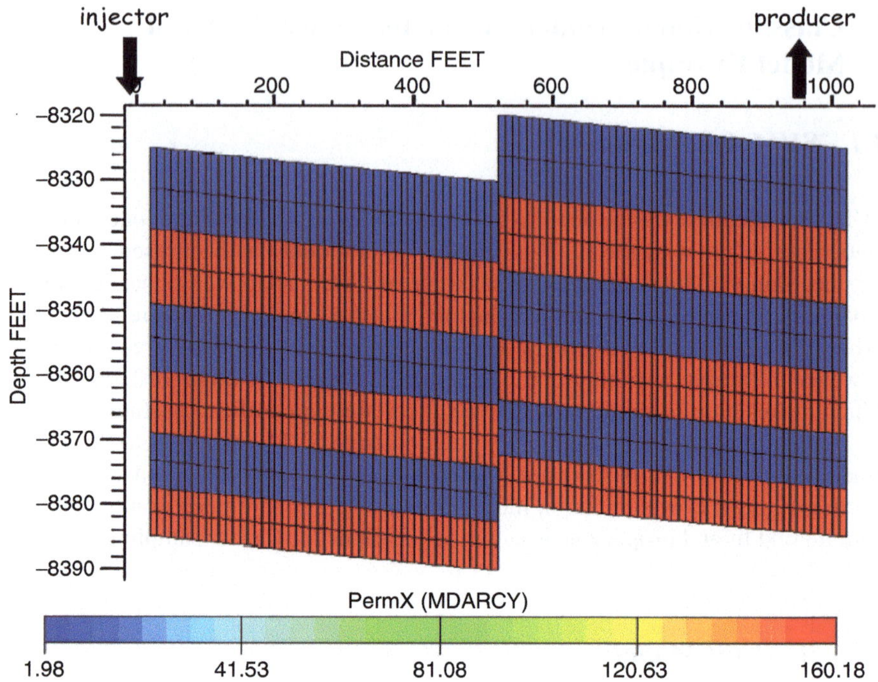

Fig. 5 The IC Fault Model (Tavassoli et al., 2004)

3.3 SVM Classification of Good Fitting Models

A binary SVM classifier was trained with a supervised learning rule, which implies preparation of a training data set. We used initial random sampling to generate a limited number of models with the corresponding misfit values computed based on the forward flow simulations. Adaptive sampling (e.g. NA) was not used to generate the training set. Although a random set of models has a smaller chance to include any models with low misfit, it provides a better coverage of the parameter space and does not concentrate on any particular locality.

Then, the SVM classifier was trained using the 200 initial random samples. The classification results were then validated using the exhaustive data base of 160,000 models generated with uniform Monte Carlo (Tavassoli et al., 2004). The exhaustive data base exhibits a good coverage of the parameter space and depicts the regions of the low misfit models with a high resolution represented by a large number of good fitting models. Validation analysis was aimed at checking the accuracy of the SVM classifier.

The location of the initial 200 models used for training is presented in Fig. 6a in a 2D projection of the 3D parameter space. A threshold at $misfit = 70$ was chosen to split the data into two classes: those data for which to run the flow simulation and the others for which the flow simulation run is not computed. There appeared to be just

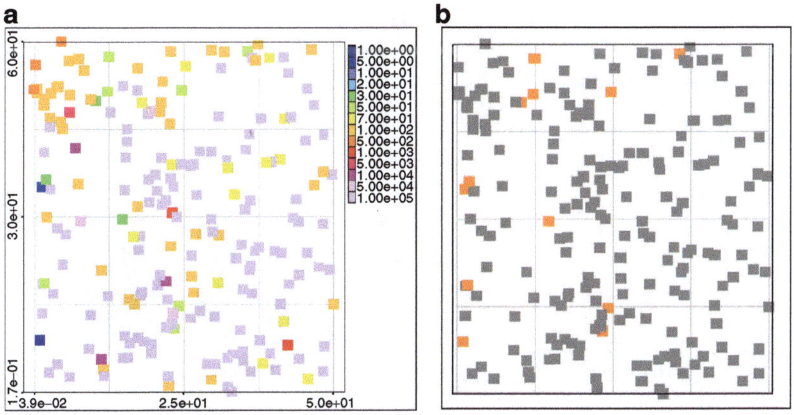

Fig. 6 Misfit for 200 initial models for training (**a**), model classes for 200 training models (**b**)

a handful of models with a misfit below 70 (see Fig. 6b). However, they are located in different regions of the parameter space, which is important for maintaining good generalisation ability of the classifier.

Validation with the exhaustive database of 160,000 models was carried out to check whether SVM classification is accurate; i.e., do the SVM models, which were classified to run the flow simulation for, match the production well? All 160,000 model locations were classified by the SVM to determine which misfit class they belong to – below 70 or above 70. SVM classified 15,000 models out of 160,000 as the models with the misfit <70, for which the flow simulation is to be computed. This is just under 10% of the total number of models. Once the actual misfit of the classified 15,000 models was obtained based on the flow simulation runs, it became possible to analyse how good were the "classified" modes in terms of the history match. Figure 7a shows the cumulative histogram of the actual misfit of the 15,000 models classified to have the misfit below 70. It demonstrates that just under 90% of the models classified for running the simulation have the actual misfit below 70, which confirms that the SVM classifier is accurate.

The next stage entails checking if the SVM classifier has missed any of the good models in the database; i.e. how many good fitting models from the data base have not been classified to run the flow simulation for. Figure 7b shows the lower end of the misfit histogram for all the models from the data base. The bars in the histogram indicate the proportion of the models classified for running the simulation (according to SVM) in each interval of the misfit values. It can be seen that over 75% of the models with the misfit below 20, which can be considered as reasonably low, were classified by SVM for running the simulation (see the first four bars in Fig. 7b). As the model misfit increases the chance of a model to be classified by SVM for running the flow simulation decreases. This is illustrated by the decrease of the bar proportion representing the models classified for flow simulations out of all the models from the database with the misfit above 30 (see Fig. 7b).

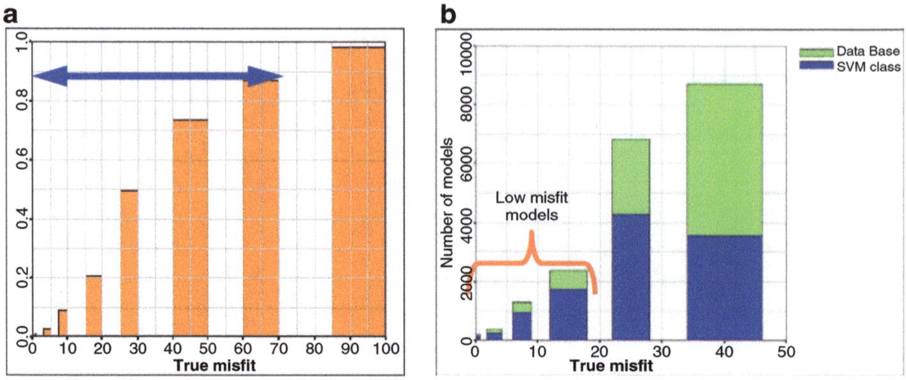

Fig. 7 Cumulative histogram of the true misfit for the modes classified by SVM to have the misfit below 70 (**a**); histogram for the misfit of models with low misfit: in the data base (DB) versus the low misfit models classified by SVM to have the misfit below 70 (**b**)

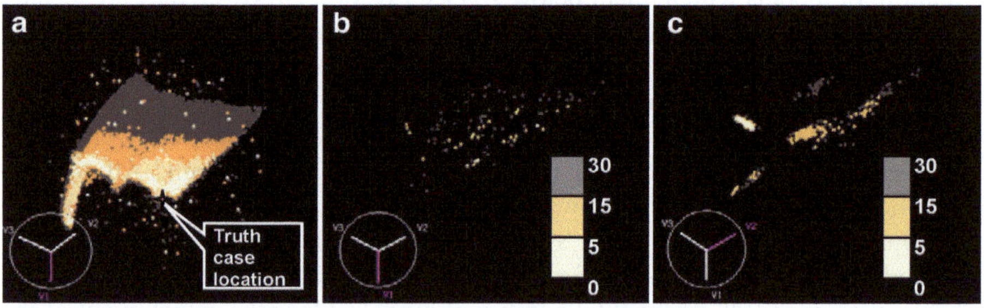

Fig. 8 Good fitting models with the misfit below 30 from the three ensembles of models: exhaustive data set of 160,000 models (**a**), 500 samples based on SVM classification (**b**), conventional NA sampling of 7,000 models (**c**)

Figure 8 shows the locations of the good fitting models with the misfit below 30 in the 3D parameter space. Good fitting models from the exhaustive database of 160,000 models shows how complicated the surface of the good models is (see Fig. 8a). The SVM classifier was able to capture some of the basic structure of the misfit surface with just 500 samples and detect diverse good fitting models in different areas (see Fig. 8b), which provided more robust predictions. NA sampling with 1,700 flow simulations concentrated on good fitting models in a particular cluster away from the truth case model (see Fig. 8c), which resulted in deviation of the forecasts from the truth case solution (see Fig. 9b) (Erbaş, 2006).

The production profiles computed by flow simulation of the models inferred from the generated ensembles are shown in Fig. 9, where the models are run into the forecasting period past the period where the fitted data were available. Inferred models from both SVM classification based ensemble (500 flow simulations from Fig. 8b) demonstrate fair spread of forecasts, which comfortably encompass the truth case

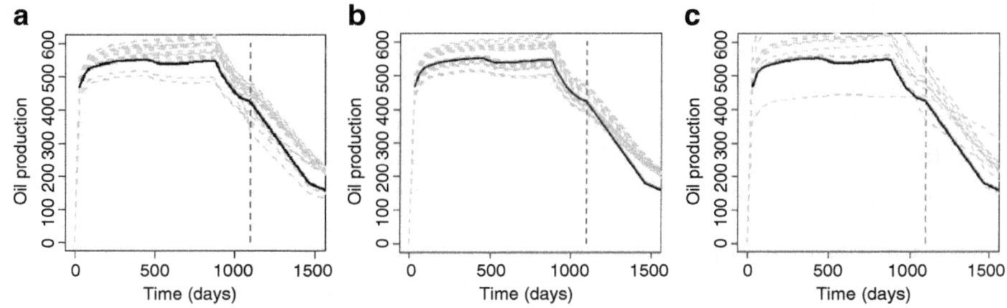

Fig. 9 Good fitting models inferred from the three ensembles of models: ensemble of 500 models based on SVM classification (**a**), conventional NA sampling of 1,700 models (**b**) and long GA sampling run of 2,000 models (**c**). Vertical line shows the start of the forecasting period

(see Fig. 9a). The results were compared with conventional sampling methods – Neighborhood Approximation (NA) and Genetic Algorithms (GA) (Erbaş, 2006). The conventional NA run with a limited number of generated models (1,700 flow simulations) demonstrates reduction of the spread in the predictions for the forecasting period (see Fig. 9b). The inferred ensemble of models generated by GA based on 2,000 flow simulations provide a fair spread in the forecasting period, but some of the models do not fit the history well (see Fig. 9c). For all approaches the truth case model solution lies towards the edge of uncertainty envelope. This can be explained by a peculiar choice of the truth case model, which lies on the edge of the good fitting model region (see Fig. 8a). This suggests a low posterior probability of the truth case model.

4 Conclusions

An SVM classification model was used to detect the models for which to run flow simulations. Thus classification is able to separate the regions in the parameter space where to search for good fitting models. Classification is more robust than regression, which aims at accurate estimation of the actual misfit value given by the interpolation model. The SVM approach is beneficial in high dimensional space relative to other machine learning algorithms, because it provides the best prediction in terms of error without compromising the complexity of the classification model.

The SVM classifier applied to a synthetic model demonstrated accurate results, which captured better the pattern of the low misfit regions in the parameter space than other data driven interpolation algorithms (e.g. a multi-layer perceptron). Models found with SVM classification based on just 200 forward simulations are as good as the ones generated from a traditional adaptive NA sampling, which entailed more forward flow simulations.

We have to admit that the observed performance of SVMs in terms of predicting the low misfit models is not better than equivalent NA techniques reported for this case study by (Erbaş, 2006). Rather, it is in the possibly wider application of SVMs to the exploration of the parameter space and uncertainty characterization that we see considerable potential. This view is justified since SVMs are specifically designed to handle high-dimensional data and extract a sparse set of support vectors from them. Thus, SVM being a data-driven method is useful to describe the previously unexplored regions of the parameter space and find hidden dependencies in it.

Acknowledgements Funding of this work was provided by UK Engineering and Physical Sciences Research Council (GR/T24838/01) and by the industrial sponsors of the Heriot-Watt Uncertainty Project.

The authors acknowledge Swiss National Science Foundation funding "GeoKernels: kernel-based methods for geo and environmental sciences" project N°200021 − 113944.

The authors would like to thank J. Carter of Imperial College for providing the model and the data for the IC Fault case study.

References

Carter JN, Ballester PJ, Tavassoli Z, King PR (2004) Our calibrated model has no predictive value: an example from the petroleum industry. Proceedings of the Fourth International Conference on Sensitivity Analysis

Christie M, Demyanov V, Erbas D (2006) Uncertainty quantification for porous media flows. J Comput Phys 217:143–158

Demyanov V (2007) Neural network guided sampling for uncertainty quantification of production forecasts. Presentation at UFORDS, Scarborough

Demyanov V, Wood SN, Kedwards TJ (2006) Improving ecological impact assessment by statistical data synthesis using process based models. J Royal Stat Soc, Appl Stat (Ser C) 55(1, Part 1):41–62

Erbaş D (2006) Sampling strategies for uncertainty quantification in oil recovery prediction. Thesis for the Degree of Doctor of Philosophy, Heriot-Watt University, August 2006

Hastie T, Tibshirani R, Friedman J (2001) The elements of statistical learning. Springer, New York

Haykin S (1999) Neural networks: a comprehensive foundation. Pearson Higher Education, New Delhi

Kanevski M, Pozdnoukhov A, Timonin V (2009) Machine learning algorithms for geoSpatial data. 369, EPFL Press, Switzerland, p 300

Platt J (1999) Probabilistic outputs for support vector machines and comparison to regularized likelihood methods. In: Smola AJ, Bartlett P, Scholkopf B, Schuurmans (eds) Advances in large margin classifiers. MIT Press, Cambridge, MA

Pozdnoukhov A, Kanevski M (2007). Interactive monitoring network optimization using support vector machines. In Spatial Statistics and GIS conference (Stat-GIS 2007), Klagenfurt

Sambridge M (1999) Geophysical inversion with a neighbourhood algorithm – I. Searching a parameter space. Geophys J Int 138:479–494

Scholkopf B, Smola AJ (2002) Learning with kernels. MIT press, Cambridge, MA

Subbey S, Christie MA, Sambridge M (2002) Uncertainty reduction in reservoir modelling. In: Chen Z, Ewing RE (eds) Fluid flow and transport in porous media: mathematical and numerical treatment. American Mathematical Society Contemporary Mathematics Monograph, Providence, Rhode Island, pp 457–467

Tavassoli Z, Carter JN, King PR (2004) Errors in history matching, SPE Journal, September 2004, pp 352–361

Vapnik V (1995) The nature of statistical learning theory. Springer, New York

Robust Automatic Mapping Algorithms in a Network Monitoring Scenario

Ben Ingram, Dan Cornford, and Lehel Csató

Abstract Automatically generating maps of a measured variable of interest can be problematic. In this work we focus on the monitoring network context where observations are collected and reported by a network of sensors, and are then transformed into interpolated maps for use in decision making. Using traditional geostatistical methods, estimating the covariance structure of data collected in an emergency situation can be difficult. Variogram determination, whether by method-of-moment estimators or by maximum likelihood, is very sensitive to extreme values. Even when a monitoring network is in a routine mode of operation, sensors can sporadically malfunction and report extreme values. If this extreme data destabilises the model, causing the covariance structure of the observed data to be incorrectly estimated, the generated maps will be of little value, and the uncertainty estimates in particular will be misleading.

Marchant and Lark (2007) propose a REML estimator for the covariance, which is shown to work on small data sets with a manual selection of the damping parameter in the robust likelihood. We show how this can be extended to allow treatment of large data sets together with an automated approach to all parameter estimation. The projected process kriging framework of Ingram et al. (2008) is extended to allow the use of robust likelihood functions, including the two component Gaussian and the Huber function. We show how our algorithm is further refined to reduce the computational complexity while at the same time minimising any loss of information.

To show the benefits of this method, we use data collected from radiation monitoring networks across Europe. We compare our results to those obtained from traditional kriging methodologies and include comparisons with Box–Cox

B. Ingram (✉)
Facultad de Ingeniería, Universidad de Talca, Camino Los Niches, Curicó, Chile
e-mail: bri@utalca.cl

D. Cornford
Neural Computing Research Group, Aston University, Aston Street, Birmingham, B4 7ET, United Kingdom
e-mail: d.cornford@aston.ac.uk

L. Csató
Faculty of Mathematics and Informatics, Universitatea BABES-BOLYAI, Str. Mihail Kogalniceanu, Nr. 1 RO-400084 Cluj-Napoca, Romania
e-mail: lehel.csato@cs.ubbcluj.ro

P.M. Atkinson and C.D. Lloyd (eds.), *geoENV VII – Geostatistics for Environmental Applications*, Quantitative Geology and Geostatistics 16,
DOI 10.1007/978-90-481-2322-3_31,

transformations of the data. We discuss the issue of whether to treat or ignore extreme values, making the distinction between the robust methods which ignore outliers and transformation methods which treat them as part of the (transformed) process. Using a case study, based on an extreme radiological events over a large area, we show how radiation data collected from monitoring networks can be analysed automatically and then used to generate reliable maps to inform decision making. We show the limitations of the methods and discuss potential extensions to remedy these.

1 Introduction

Choosing an appropriate overall model is an important part of interpolating and analysing observations collected from sensor networks. The model should be based on assumptions about the underlying process that generated the observations. Practically speaking, it is almost impossible to exactly specify the correct model which introduces difficulties when attempting to estimate parameters within the model. In this paper we consider the concept of robust geostatistics. By applying robust geostatistical methods we aim to limit the effects of observations that do not correspond to our chosen model. Robust models are frequently employed with datasets where outliers are present, as might often be the case in an automatic monitoring scenario.

The idea of robust geostatistics is not new and has been studied in geostatistics for many years (Cressie and Hawkins, 1980). In this paper we avoid parameter estimation techniques using method-of-moments based estimators such as those described by Genton (1998) and instead focus on likelihood based approaches such as those proposed by Marchant and Lark (2007). In this paper, we show how a fast Bayesian projected process kriging framework can be used for robust parameter estimation to generate accurate maps of an area of interest. Using this framework allows the efficient utilisation of most commonly used likelihood functions without having to resort to computationally expensive Markov Chain Monte Carlo (MCMC) sampling techniques as used in other Bayesian methods (Diggle et al., 1998). As a result, we can experiment with a number of robust likelihood models, in an near real-time framework.

In this paper, by applying a variety of non-Gaussian likelihood models that have heavier tails which help to account for outliers, we compare a number of robust methods. Specification of an appropriate robust likelihood model could be specific to the domain to which it is being applied; our results are particularly relevant to environmental monitoring of radioactivity.

2 Gaussian Process

Model based geostatistics makes the assumption that any finite collection of random variables is jointly Gaussian. Here we assume that the data takes the form:

$$(x_i, y_i) : i = 1, \ldots, n, \tag{1}$$

where we denote spatial location by x_i and observations at the location x_i are denoted by y_i. Each observation, y_i, is assumed to be a realisation of a random variable Y_i which is dependant on the value of an unobserved random process $S(x)$ (Diggle et al., 1998).

We assume observations have the following relationship to the underlying process:

$$Y_i = S(x_i) + Z_i, \tag{2}$$

where Z_i is an additive, potentially non-Gaussian, error on the observations that is assumed to be independent for each observation. Equation (2) defines an *arbitrary likelihood function*, $p(Y_i|S(x))$, which we will generally assume has heavy tails to model the outlying observations.

2.1 Gaussian Process Approximations

We adopt a Bayesian framework for our iterative algorithm. Our aim is to infer the posterior distribution of the underlying random process $S(x)$ given the observed data, $Y = \{Y_i\}_{i=1..n}$. This has the standard form:

$$p(S(x)|Y, \theta) = \frac{\left[\prod_i p(Y_i|S(x))\right] p(S(x)|\theta)}{\int \left[\prod_i p(Y_i|S(x))\right] p(S(x)|\theta) dS(x)} \tag{3}$$

where the posterior is the product of the likelihood terms and the Gaussian process prior, divided by a normalising constant, often called the marginal likelihood, $p(Y|\theta)$.

2.2 Parametrisation of Posterior Moments

Since we allow for arbitrary likelihood models, in this case robust likelihood models, an exact solution would require the application of MCMC sampling from this very high dimensional posterior distribution, which will be prohibitively computationally expensive for large datasets in our real-time setting. Our approach is to approximate the true non-Gaussian posterior by the *optimal* Gaussian process posterior that minimises the Kullback–Leibler (KL) divergence measure between the *true* posterior distribution and the approximating posterior distribution. By minimising the KL divergence, we match the first two moments of the two distributions (Csató and Opper, 2002).

To enable the use of the arbitrary likelihoods, Equation (2), we represent the Gaussian process by a parametrisation of the posterior moments. The posterior mean is parametrised as:

$$\mu_{posterior}(x) = \mu_{prior}(x) + \sum_i^m \alpha_i c(x, x_i), \tag{4}$$

where $c(x, x_i)$ is the (*a priori*) covariance function between the point x and the points x_i used in the approximation. We write the covariance between two spatial locations as $c(x, x_i) = cov(x, x_i)$. $\boldsymbol{\alpha} = \{\alpha_i\}_{i=1..n}$ is then the vector of the parameters of the posterior mean of the process. The posterior variance is parametrised as:

$$c_{posterior}(x, x') = c_{prior}(x, x') + \sum_{i,j=1}^{m} c(x, x_i) C_{(i,j)} c(x_j, x') \tag{5}$$

where $\mathbf{C} = \{C_{i,j}\}_{i,j=1..n}$ is a matrix of parameters for the posterior covariance.

Given the above parametrisation of the posterior moments, we now show how these parameters $\boldsymbol{\alpha}$ and $\mathbf{C}$ can be updated in an iterative algorithm. It was shown in Csató and Opper (2002) that the parametrisation can be applied recursively to give an iterative update rule:

$$\mu_{t+1} = \mu_t + q_{t+1} c_t(x, x_{t+1}), \tag{6}$$

$$c_{t+1}(x, x') = c_t(x, x') + r_{t+1} c_t(x, x_{t+1}) c_t(x_{t+1}, x') \tag{7}$$

where t indicates the pseudo-time step in the algorithm or iteration, and x_{t+1} is the spatial location of the new observation being included at iteration $t + 1$. The scalar coefficients q_{t+1} and r_{t+1}, which update the model at each iteration can be computed analytically or numerically. The analytic update equations derived in Csató and Opper (2002) are given by:

$$q_{t+1} = \frac{\partial}{\partial[S(x)]} log\langle p(Y_{t+1}|S(x))\rangle_t, \tag{8}$$

$$r_{t+1} = \frac{\partial^2}{\partial[S(x)]^2} log\langle p(Y_{t+1}|S(x))\rangle_t, \tag{9}$$

where the derivatives are with respect to the mean function at time $t + 1$ and the expectations, denoted $\langle\cdot\rangle_t$, are taken with respect to the posterior Gaussian process at algorithm pseudo-time t. These update equations essentially process the observations one at a time and update the posterior parametrisation by matching the moments of the updated parametrised posterior to the true, potentially non-Gaussian posterior. Further details can be found in Csató and Opper (2002).

3 Robust Likelihood Models

Robust likelihood models facilitate the estimation of the variogram parameters in the case where outlying observations are present in the data. If likelihoods which model a 'robust' error distribution are used within a traditional model based geostatistical approach then sampling from a potentially high dimensional distribution is required and can be very time consuming.

The method we presented earlier in this paper allows for the specification of arbitrary likelihoods without the large computational overhead that comes with existing

MCMC based model based geostatistics. We now present and discuss some robust likelihoods that can be used and compare them to some existing techniques for treating data with outliers.

3.1 Two Component Gaussian

We could assume that the observations come from separate processes: a routine process and an extreme process. One approach that seems intuitive is to introduce two components into the likelihood model, one component to model the routine observations and another component to model the extreme observations or outliers. We need not necessarily restrict ourselves to a two component Gaussian likelihood model, but for the purposes of this paper we employ a mixture of two components. Assuming that the routine observations follow a Gaussian distribution is a common hypothesis although this is often an approximation. However assuming that the extreme or outlier observations follow a Gaussian distribution with a large variance could be debated; empirically we have found it works well, although there is little theoretical justification.

The two component Gaussian mixture is constructed by summing two weighted Gaussian distributions to create the mixture likelihood:

$$p(Y_i|S(x_i)) = \beta \mathcal{N}_a(Y_i|S(x_i)) + (1-\beta)\mathcal{N}_b(Y_i|S(x_i)) \tag{10}$$

where β gives the weight of the mixture, or the fraction of the observations that belong to the routine process $\mathcal{N}_a(Y_i|S(x_i))$. We set the variance or noise σ_a^2, of the routine process to model our assumptions about the error in the observation process. The extreme process is denoted by $\mathcal{N}_b(Y_i|S(x_i))$ and a much larger noise σ_b^2 is defined, which represents our beliefs about the extreme process. Alternative mixtures of likelihoods could be considered, but in this paper we will only look at the case where the likelihood models are summed.

3.2 Laplace

A alternative approach which makes yields a cruder robust likelihood model is to assume that the likelihood function has a Laplace distribution. The Laplace distribution has the probability density function:

$$Laplace(x|\mu, b) = \frac{1}{2b} exp\left(-\frac{|x-\mu|}{b}\right) \tag{11}$$

where μ is the location parameter and b is a scale parameter. The Laplace distribution is also known as the double sided exponential distribution.

3.3 Huber Functions

One approach to determining robust likelihood models was presented by Marchant and Lark (2007). Here the Huber function is used in the likelihood term. The Huber function is given by:

$$\rho(d) = \begin{cases} \frac{1}{2}d^2 & \text{if } |d| \leq c \\ c\,|d| - \frac{1}{2}c^2 & \text{otherwise} \end{cases} \tag{12}$$

where c is a constant determining the robustness of the estimator. In the case $c = \infty$ the model is equivalent to the standard maximum likelihood estimator, with a Gaussian likelihood model. Rather than optimising the parameter c, here we choose a number of values for c and see which gives the best results. Future work will investigate the selection c using alternative methods.

4 Box–Cox Transformations

A standard alternative that is commonly used when a dataset is contaminated with outliers or at least when the dataset is assumed to be non-Gaussian distributed is that of the Box–Cox transformation Box and Cox (1964). The data is transformed to be approximately Gaussian distributed using:

$$y^{(\lambda)} = \begin{cases} \frac{y^{\lambda}-1}{\lambda} & \text{if} \lambda \neq 0 \\ \log(y) & \text{if} \lambda = 0 \end{cases} \tag{13}$$

and thus the effect of outliers can be reduced, but not completely removed. In this paper we try a number of values for λ to identify which is the most appropriate for the given data. We should note that the Box–Cox approach is very different in character to the preceding approaches, since in the previous methods we have assumed that the outliers arise because of a local corruption to observations, whereas in the Box–Cox approach we transform the entire field, albeit in a manner that attempts to maximise the (marginal) Gaussianity of the observations.

5 Covariance Selection

We follow the methodology of Ingram et al. (2005) for determining the covariance function used in the experiments. We use a nested covariance model which has a linear sum of a Gaussian and exponential covariance function components:

$$c_{mix}(u) = \pi\sigma_{gau}^2 exp\left(\frac{u^2}{\phi}\right) + (1-\pi)\sigma_{exp}^2 exp\left(\frac{u}{\phi}\right). \tag{14}$$

We assume that the exponential component models the short range rough process and that the Gaussian component models the smoother properties of the process at longer lag separations, which is consistent with a belief that at short ranges the radioactivity field is dominated by turbulent mixing processes, while at longer range large scale weather, soil and geological differences dominate.

6 Datasets

To demonstrate the various methods discussed previously, we will use a radiation data collected over the German monitoring network. Radiation data for most countries in Europe is available from the EURDEP (EUropean Radiological Data Exchange Platform) website.[1] We use a dataset with a simulated release of radiation into the environment prepared by BfS,[2] which uses the real EURDEP observed background radiation with an added deposition generated from a radiation dispersion model. The simulated release represents some kind of disaster that could potentially take place. The event in this case is not a serious disaster, but rather a small release into the environment over a large area. The release is dispersing in the E–W direction more rapidly than the N–S direction. Anisotropy in the contamination process will present problems for the models. In total there are 1,900 observations. We divide this into two sets, a set for estimating the model parameters (1,200 observations) and a prediction set for cross validation (700 observations).

7 Results

Contour maps have been produced to show the mean predictions and estimates for the kriging variance. These can be seen in Figs. 1–5. The first thing to note is that each seems to capture the features of the simulated contaminant along the lower middle section of the area. Looking at the Gaussian range (Table 1) for the default method shows how this parameter has become extremely large and this effect can be seen as over smoothing the effect of the contamination. The Huber function and Box–Cox transformation model also suffer somewhat from over estimating the Gaussian range parameter in the E–W direction, but to a lesser degree. The Gaussian Mixture and Laplace models further improve, but anisotropy in the estimation is still marked, which is realistic.

The summary statistics show that the Gaussian Mixture has the lowest error (both MAE and RMSE) of all the methods investigated. The predictions are also

[1] http://eurdep.jrc.it/.

[2] German Federal Office for Radiation Protection.

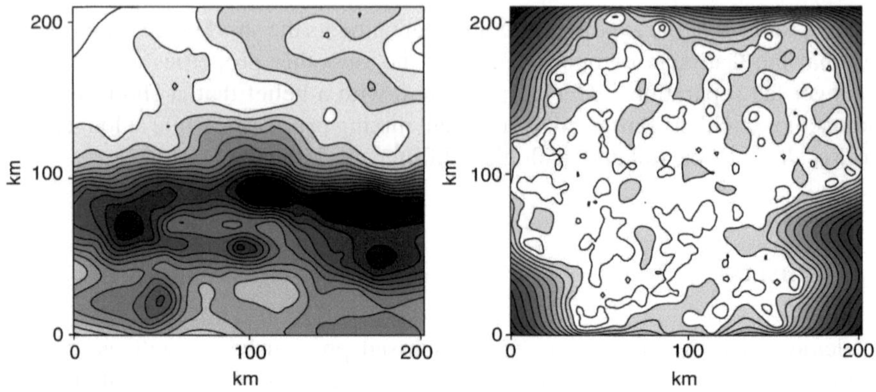

Fig. 1 Contour plot of (*left*) mean predictions and (*right*) variance estimates for default (Gaussian likelihood) model

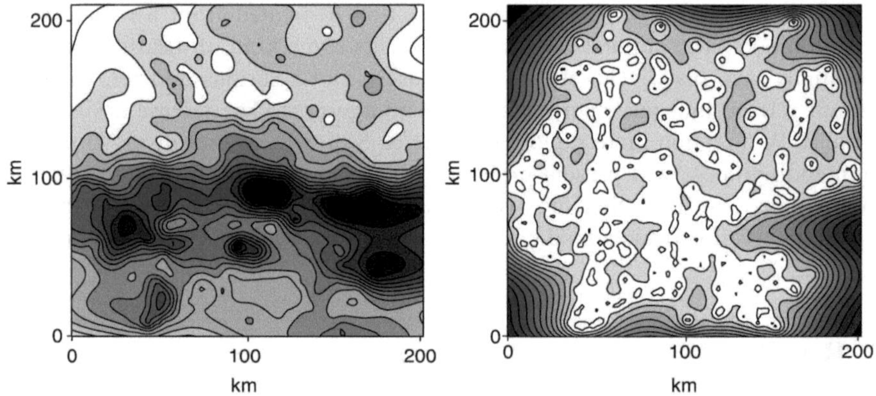

Fig. 2 Contour plot of (*left*) mean predictions and (*right*) variance estimates for mixture likelihood model

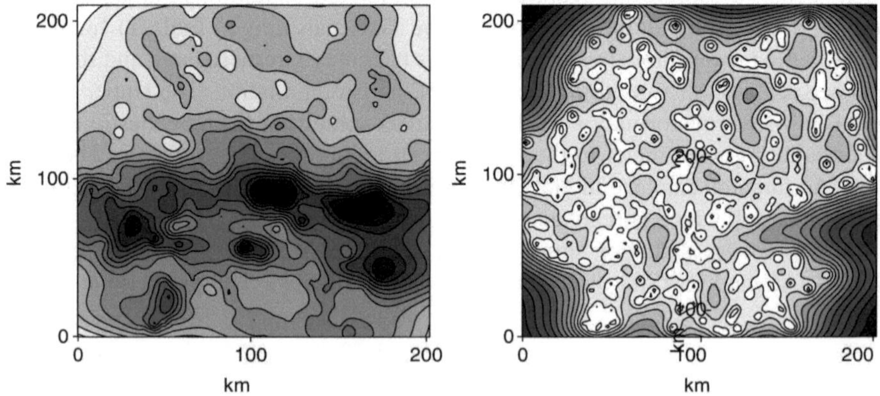

Fig. 3 Contour plot of (*left*) mean predictions and (*right*) variance estimates for Laplace likelihood model

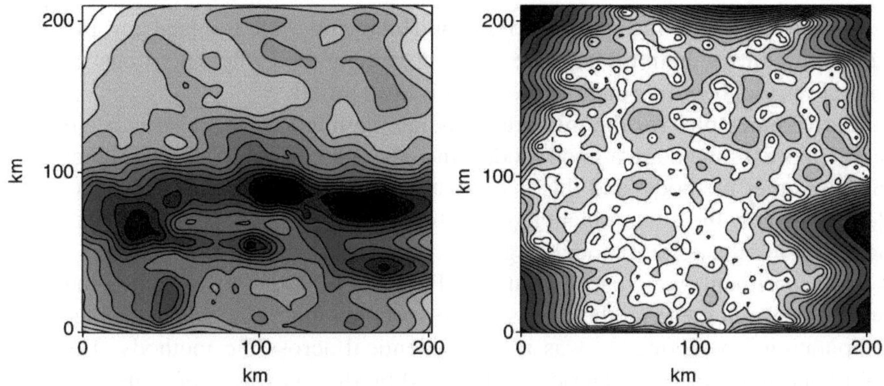

Fig. 4 Contour plot of (*left*) mean predictions and (*right*) variance estimates for Huber likelihood model

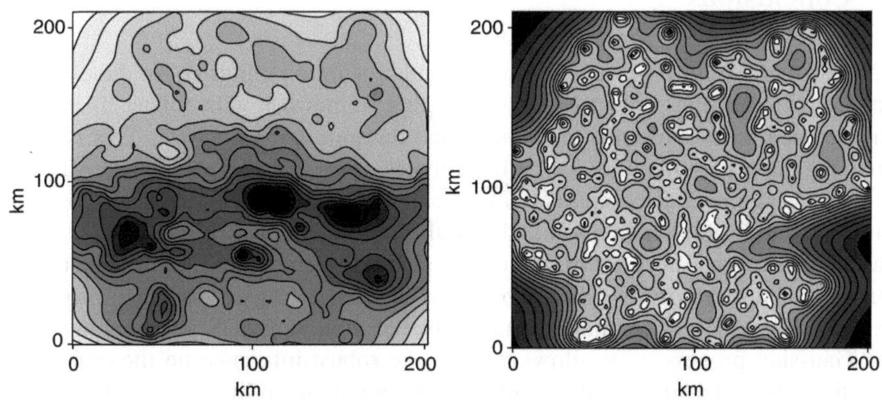

Fig. 5 Contour plot of (*left*) mean predictions and (*right*) variance estimates for Box–Cox transformation based model

Table 1 Covariance parameters for the different methods considered. Sill gives the overall sill, summing both components

Method	Nugget	Sill	Gaussian range	Exp. range	MAE	RMSE	R
Default	0.32	1.35	(530.30, 1.67)	(0.23, 0.12)	0.0010	0.0210	0.83
Gaussian mixture	0.11	0.67	(48.36, 7.60)	(0.09,0.07)	0.0004	0.0131	0.87
Laplace	0.16	0.75	(58.23, 5.42)	(0.13, 0.19)	0.0006	0.0175	0.86
Huber function	0.22	1.01	(148.37,0.09)	(0.43, 0.09)	0.0009	0.0192	0.86
Box–Cox	0.19	0.90	(136.89, 0.60)	(0.82, 0.77)	0.0007	0.0190	0.86

more correlated with the observations. The Huber function, Laplace and Box–Cox transformation methods all improve on the default method where no robust assumptions are made, however the improvement is quite small.

The variance plot for the mixture Gaussian likelihood (Fig. 2) indicates that the parameters estimated are a good model since there are lower kriging variances than with the other methods, and this is consistent with the observed errors. The mean plot shows how the contaminant has a distinct pattern which cannot be observed in the plot using the default model (Fig. 1).

All experiments were carried out on a Pentium 4 2 Ghz PC. Since the main difference between these methods was in the specification of the likelihood term, the computational performance was roughly identical across the methods. The computational time, for parameter estimation and prediction was approximately 2 min per model.

8 Conclusions

In this paper we have presented four methods for treating outliers in datasets. We have shown that the projected process kriging framework with robust likelihoods can be used in the presence of outliers, and on quite large datasets. This is based on using maximum likelihood type II estimates of the parameters in the covariance functions. The overall computational time is under 2 min. This is using an unoptimised Matlab implementation and initial work on a C++ library suggests this can be reduced by an order of magnitude simply by changing the implementation language. Furthermore in other experiments, not shown here, we have processed over 10,000 observations in reasonable time. Employing a Bayesian framework, the Gaussian process prior allows us to make robust inference on the covariance function parameters despite the complex structure in the observations, with possible outliers, which would not be possible with standard method of moments estimators. The Bayesian approach taken here should be called an empirical Bayes (or plug-in) method since maximum a-posteriori estimates of covariance function parameters are used; it would be interesting to assess the impact of sampling from (and then marginalising with respect to) the parameters in the covariance function. This would require far more computationally expensive sampling methods, but would give a clear indication of the role of parameter uncertainty in (posterior) predictive uncertainty.

The radiological dataset that we have used shows that all four 'robust' methods offer an improvement over standard kriging results, in terms of some standard metrics, however the Gaussian mixture likelihood seems to perform slightly better that other methods in this example. An explanation might be that the second Gaussian component of the mixture likelihood seems to better model the contamination process, although we have not rigorously shown this. The contamination process is more than a few outlying observations, but rather a large number ofobservations from a second process. The other models may not be able to capture this 'second process' since they are based on heavy tailed distributions which, conceptually at least, arise as the result of a single process.

There is a difference between the robust likelihood methods and the Box–Cox transformation. The robust likelihoods all assume an underlying latent Gaussian process, with observations that are contaminated by heavy tailed, zero mean, symmetric, noise models; their aim is essentially to represent the underlying process filtering the noise appropriately. In the Box–Cox approach the observations are transformed such that their marginal distribution is approximately Gaussian, by a range of transformations from the identity to the log transform. Thus the robustness arises from the squashing affect of the transformation which reduces the impact of large observations (i.e. deals with the skew of the distribution) – the outliers. The key question to consider in choosing an appropriate method is probably more related to assumptions about the form of the noise on the observations together with assumptions about the distribution of the latent process. Note the Box–Cox transformation can only transform variables such that they are marginally Gaussian, not jointly so. In practice, to confirm ones beliefs, it seems that it will always be necessary to compare a range of methods using validation or cross validation to select the empirically best method, even when strong prior information is available.

There are a number of aspects to the modelling process that were only touched on and require further investigation. The selection of the parameters of the likelihood models, for example, estimating the mixing coefficient and variances for each component in the Gaussian mixture model, could be performed automatically rather than being specified *a-priori*. This would also be possible for the c parameter for the Huber function. This is not trivial however, as there is a conceptual difficulty in partitioning the observation errors without additional knowledge, and would probably require a Bayesian treatment, with the effort being applied to defining appropriate priors. So called Trans-Gaussian Kriging (Pilz et al., 2004) incorporates a method to estimate the Box–Cox transformation parameter, which could be incorporated into future models. It is interesting to speculate whether other approaches, such as indicator kriging or copula based methods might also be employed in circumstances where the underlying process has a skewed or otherwise non-Gaussian distribution, potentially also using robust likelihood models to account for the presence of outliers caused by a heavy tailed noise distribution. Finally, although we have not directly tackled this here, in some cases it might be preferable to remove the outlier prior to processing, for example in cases where the outlier represent failure of the observing system or some other catastrophic error.

Acknowledgements This work is funded by the European Commission, under the Sixth Framework Programme, by the Contract N. 033811 with the DG INFSO, action Line IST-2005-2.5.12 ICT for Environmental Risk Management. The views expressed herein are those of the authors and are not necessarily those of the European Commission.

References

Box GEP, Cox DR (1964) An analysis of transformations. J R Stat Soc 26(2):211–252

Cressie N, Hawkins DM (1980) Robust estimation of the variogram: I. Math Geol 12(2):115–125

Csató L, Opper M (2002) Sparse online Gaussian processes. Neural Comput 14(3):641–669

Diggle PJ, Tawn JA, Moyeed RA (1998) Model-based geostatistics. Appl Stat 47:299–350

Genton MG (1998) Highly robust variogram estimation. Math Geol 30(2):213–221

Ingram B, Csató L, Evans D (2005) Fast spatial interpolation using sparse Gaussian processes. Appl GIS 1(2):15:1–17

Ingram B, Cornford D, Evans D (2008) Fast algorithms for automatic mapping with space-limited covariance functions. Stoch Environ Res Risk Assess 22(5):661–670

Marchant BP, Lark RM (2007) Robust estimation of the variogram by residual maximum likelihood. Geoderma 140(1–2):62–72

Pilz J, Pluch P, Spoeck G (2004) Bayesian Kriging with lognormal data and uncertain variogram parameters. In: Proceedings of the Fifth European Conference on geostatistics for environmental applications. Springer, Berlin

Parallel Geostatistics for Sparse and Dense Datasets

Ben Ingram and Dan Cornford

Abstract Very large spatially-referenced datasets, for example, those derived from satellite-based sensors which sample across the globe or large monitoring networks of individual sensors, are becoming increasingly common and more widely available for use in environmental decision making. In large or dense sensor networks, huge quantities of data can be collected over small time periods. In many applications the generation of maps, or predictions at specific locations, from the data in (near) real-time is crucial. Geostatistical operations such as interpolation are vital in this map-generation process and in emergency situations, the resulting predictions need to be available almost instantly, so that decision makers can make informed decisions and define risk and evacuation zones. It is also helpful when analysing data in less time critical applications, for example when interacting directly with the data for exploratory analysis, that the algorithms are responsive within a reasonable time frame.

Performing geostatistical analysis on such large spatial datasets can present a number of problems, particularly in the case where maximum likelihood. Although the storage requirements only scale linearly with the number of observations in the dataset, the computational complexity in terms of memory and speed, scale quadratically and cubically respectively. Most modern commodity hardware has at least two processor cores if not more. Other mechanisms for allowing parallel computation such as Grid based systems are also becoming increasingly commonly available. However, currently there seems to be little interest in exploiting this extra processing power within the context of geostatistics.

In this paper we review the existing parallel approaches for geostatistics. By recognising that different natural parallelisms exist and can be exploited depending on whether the dataset is sparsely or densely sampled with respect to the range of variation, we introduce two contrasting novel implementations of parallel

B. Ingram (✉)
Facultad de Ingeniería, Universidad de Talca, Camino Los Niches, Curicó, Chile
e-mail: bri@utalca.cl

D. Cornford
Neural Computing Research Group, Aston University, Aston Street, Birmingham, B4 7ET, United Kingdom
e-mail: d.cornford@aston.ac.uk

P.M. Atkinson and C.D. Lloyd (eds.), *geoENV VII – Geostatistics for Environmental Applications*, Quantitative Geology and Geostatistics 16,
DOI 10.1007/978-90-481-2322-3_32,

algorithms based on approximating the data likelihood extending the methods of Vecchia (1988) and Tresp (2000). Using parallel maximum likelihood variogram estimation and parallel prediction algorithms we show that computational time can be significantly reduced. We demonstrate this with both sparsely sampled data and densely sampled data on a variety of architectures ranging from the common dual core processor, found in many modern desktop computers, to large multi-node super computers. To highlight the strengths and weaknesses of the different methods we employ synthetic data sets and go on to show how the methods allow maximum likelihood based inference on the exhaustive Walker Lake data set.

1 Introduction

The problem of large datasets was once considered a solved issue (Schabenberger and Gotway 2005). By using method-of-moments variograms and moving window kriging, all but the very massive dataset are computationally tractable. In recent years the popularity of and interest in maximum likelihood based algorithms has grown. Problems are encountered computationally with likelihood based methods when more than a few thousand observations are encountered.

Parallel geostatistics is not a new topic and has been considered previously by Pedelty et al. (2003), Gebhardt (2003) and Kerry and Hawick (1998). The basis of these existing techniques is to perform moving window kriging by assigning a prediction area to each processor and predict at the locations for a given area. The authors neglect to discuss parameter estimation in a parallel context. Practically, computing the variogram using a method-of-moments estimator provides few challenges when compared to the computational complexity of prediction since computing the variogram is a $O(n^2)$ process.

The motivation for this study lies in a shift in computer microprocessor design, where uniprocessor microprocessors are being replaced by multi-processor or multi-core architectures. Although this new design does not always result in a speed-up for many geostatistical algorithms. Software typically needs to be written to utilise such architectures. If the software is built upon existing libraries such as BLAS,[1] LAPACK[2] and ATLAS,[3] then these libraries can be replaced with parallel equivalents. The current version of LAPACK comes with configuration options to create multi-threaded versions where the number of threads can be specified. One warning however is that some users have noted decreased computational speeds due to synchronisation and communication between different threads.

In this paper we discuss data parallelism approaches for performing geostatistics. In contrast to task parallelism, data parallelism relies on splitting the data into a number of smaller clusters and performing calculations across a number of processors.

[1] http://www.netlib.org/blas/.

[2] http://www.netlib.org/lapack/.

[3] http://math-atlas.sourceforge.net/.

We utilise the Message Passing Interface (MPI) for intra-node communication since this is the *de facto* standard for parallel programming. Implementations of MPI can be found for most architectures from the largest massively parallel supercomputers to a standard desktop with a dual-core processor. During the development of this software we used the Matlab compatible application Octave[4] and MPITB (Fernández et al., 2004) which is a MPI implementation compatiable with Matlab and Octave. Octave does not have the licensing restrictions that Matlab has so is ideal for using on a multiple processor machine.

2 Methodologies

We discuss and implement two methods in this paper. The first method we consider is that of Vecchia (1988) which can be used to approximate the likelihood function. The second method, the Bayesian Committee Machine (BCM) is used for prediction using all of the data. We compare the results with traditional geostatistics that such as method-of-moments variograms and moving window kriging.

2.1 Vecchia's Approximation

Vecchia (1988) approximation is based on the multiplicative theorem which states for any number of N events: $z_1, \ldots, z_N$ the following relationship holds:

$$p\left(z_1 \cap z_2 \cap \ldots \cap z_N\right) = p\left(z_1\right) \cdot p\left(z_2|z_1\right) \cdot \ldots \cdot p\left(z_N|z_1, z_2, \ldots, z_{N-1}\right) \quad (1)$$

where $p\left(z_a|z_b\right)$ is the conditional probability of z_a given z_b Pardo-Igúzquiza and Dowd (1997).

In the case of a multivariate probability density function, the following relationship is obtained:

$$p\left(Z\left(\mathbf{x}\right)\right) = \prod_{i=1}^{N} p\left(Z\left(\mathbf{x}_i\right)|Z\left(\mathbf{x}_1\right), \ldots, Z\left(\mathbf{x}_{i-1}\right)\right) \quad (2)$$

One then assumes that some of the information in the dataset is redundant and hence instead of conditioning on the whole dataset the observations are conditioned on smaller subsets of size $m < (i-1)$ where i is the current observation of the dataset. This gives the following relationship:

$$p\left(Z\left(\mathbf{x}_i\right)|Z\left(\mathbf{x}_1\right), \ldots, Z\left(\mathbf{x}_{i-1}\right)\right) \cong p\left(Z\left(\mathbf{x}_i\right)|Z\left(\mathbf{x}_1\right), \ldots, Z\left(\mathbf{x}_m\right)\right) \quad (3)$$

[4] http://www.octave.org.

where the approximation becomes almost exact as m approaches the number of observations in the dataset.

Assuming that data is a zero mean multivariate Gaussian, the conditional probability $p\left(Z\left(\mathbf{x}_i\right)|Z\left(\mathbf{x}_j\right)\right)$ where $j = 1,\ldots,m$ is also Gaussian for any observation i and any conditioning subset size m and is given by

$$\mathcal{N}\left(\Sigma_{ij}\Sigma_{jj}^{-1}Z\left(\mathbf{x}_j\right), \Sigma_{ii} - \Sigma_{ij}\Sigma_{jj}^{-1}\Sigma_{ji}\right) \tag{4}$$

where Σ_{ij} refers to the covariance between the observation i and the observations in the conditioning subset j. The following give the mean:

$$\Sigma_{i|j} = \Sigma_{ii} - \Sigma_{ij}\Sigma_{jj}^{-1}\Sigma_{ji} \tag{5}$$

and covariance:

$$\mu_{i|j} = \Sigma_{ij}\Sigma_{jj}^{-1}Z\left(\mathbf{x}_j\right) \tag{6}$$

conditioned on a subset of j observations where Σ_{jj} is a $j \times j$ covariance matrix between the points of vector y_j, Σ_{ij} is a vector of covariances between the ith observation and m points of the vector y_j and y_j are m observations at locations chosen for each subset.

This leads to the following log likelihood approximation:

$$\mathcal{L}(\theta) = -\frac{N}{2}\log(2\pi) - \frac{1}{2}\sum_{i=1}^{n}\log\left|\mathbf{\Sigma}(\boldsymbol{\theta})_{i|j}\right| - \frac{1}{2}\sum_{i=1}^{n}Z(\mathbf{x})^T\,\mathbf{\Sigma}(\boldsymbol{\theta})_{i|j}^{-1}\,Z(\mathbf{x}) \tag{7}$$

which instead of depending the inverse of a covariance matrix $\mathbf{\Sigma}(\boldsymbol{\theta})^{-1}$ of size N, depends on i covariance matrices of maximum size m. Hence the smaller size m the more computationally efficient the algorithm is but at the expensive of yielding a poorer approximation to the *true* probability density function.

Although Vecchia (1988) notes that the orderings of the data makes a difference to the approximation, this is not considered a significant issue and it is not dealt with. A number of years after this approximation method was proposed, Stein et al. (2004) suggested a number of improvements to the algorithm. Firstly it is suggested that the approximation gives better results when the observations are ordered so as to give clustered data. Secondly, by not only conditioning on observations near, but also on some observations far away, the approximation is further improved.

Since the approximate maximum likelihood approach has reduced the calculation to a sum of a number of independent calculations, a parallel implementation follows trivially. A further desirable feature is that all the data need not be sent to each process in the parallel system. How much data sent to each process depends on m, the size of the conditioning data. Particularly accurate approximations to the likelihood can be achieved with large m.

One serial implementation of the approach of Vecchia (1988) was presented by Pardo-Igúzquiza and Dowd (1997). Since the computation of the conditioning subsets is an *embarrassingly parallel* problem, it can be easily parallelised. Figure 1 shows a simple pseudo code parallel algorithm.

```
1. Master to broadcast covariance parameters to each process
   (MPI_Bcast)
2. Master to scatter training data to each process
   (MPI_Scatter)
3. Each node to calculate likelihood of each supplied
   observations conditioned on a subset of the data
4. Master to collect log likelihoods and sum (MPI_Reduce)
```

Fig. 1 Pseudo code for Vecchia approximation

2.2 Bayesian Committee Machine

The Bayesian Committee Machine was proposed by Tresp (2000) as an alternative method for reducing the computational complexity of prediction and has been frequently applied in the context of machine learning. Using this model, the data is split up into submodels or committees and weighted by the inverse variance or precision at the prediction location. The BCM has an equivalence to kriging with a number of additional assumptions.

One important feature to note about the BCM is that it is a *transductive* method rather than an *inductive* method. The term *transductive* means that the method computes a model dependent on a user specified set of prediction locations (Schwaighofer and Tresp, 2003). In this way, knowledge about the covariance between the prediction locations is exploited in the approximation.

It has been shown by Schwaighofer and Tresp (2003) that the BCM method is equivalent to assuming a low-rank covariance matrix where the exact block diagonal structure of the full covariance is retained. As with many low–rank matrix approximations or their equivalents the concept of knots, pseudo inputs or active points are used. For example, assuming a dataset of observations, $(x_i, y_i) : i = 1, \ldots, n$, where a subset of the locations $(x_j) : j = 1, \ldots, m$ are selected and termed the active set. The low-rank covariance, $\hat{\Sigma}$ or approximation to the full covariance matrix Σ is given by:

$$\hat{\Sigma} = c(d)C^{-1}c(d)' \tag{8}$$

where $c(\cdot)$ is the approximate covariance function and C is the covariance matrix between the locations selected for inclusion in the active set.

The BCM assumes that the prediction locations compose the active set which we will denote as Σ_{pred}. The apparent limitation of having to compute the covariance matrix (and the inverse) of the prediction locations is not too restrictive since smaller prediction covariance matrices can be created and the BCM equations can be repeatedly calculated without a growth of the algorithm complexity.

The predictive distribution equations are calculated as:

$$\hat{Z}_{bcm} = \boldsymbol{\Sigma}_{bcm} \sum_{w=1}^{W} \tilde{\Sigma}_w^{-1} \hat{Z}_w \tag{9}$$

where $\mathbf{\Sigma}_{bcm}$ is the predictive covariance is given by:

$$\mathbf{\Sigma}_{bcm}^{-1} = -(W-1)\,\Sigma_{pred}^{-1} \sum_{c=1}^{W} \tilde{\Sigma}_{w}^{-1} \tag{10}$$

where W is the number of committees used and Σ_{pred} is covariance matrix between the prediction locations (Tresp, 2001). An interesting observation is that the BCM predictive mean is constructed from a weighted sum of the individual committee members predictive means:

$$\hat{Z}_w = c(d)_w^T \mathbf{\Sigma}_w^{-1} \mathbf{Z}_w \tag{11}$$

where the matrix $\mathbf{\Sigma}_w$ is the covariance matrix of the observations assigned to committee w. The covariance between the prediction locations and the locations assigned to the committee is denoted by $c(d)_w$. The prediction locations are conditioned on the observed data assigned to a committee w by:

$$\tilde{\Sigma} = \mathbf{\Sigma}_{pred} - c(d)_w^T \mathbf{\Sigma}_w^{-1} c(d)_w. \tag{12}$$

Another observation is that the weights are obtained by the inverse predictive covariance (or predictive precision) at the prediction location. Effectively the BCM scales the contribution of each committee based on how confident it is about the prediction from each committee. Substituting the individual committee members predictive means and variances gives full expressions for the full predictive mean:

$$\hat{Z}_{bcm} = \mathbf{\Sigma}_{bcm} \sum_{w=1}^{W} \left(\mathbf{\Sigma}_{pred} - c(d)_w^T \mathbf{\Sigma}_w^{-1} c(d)_w\right)^{-1} c(d)_w^T \mathbf{\Sigma}_w^{-1} \mathbf{Z}_w \tag{13}$$

and predictive variance:

$$\mathbf{\Sigma}_{bcm} = \left(-(W-1)\,\Sigma_{pred}^{-1} \sum_{w=1}^{W} \left(\mathbf{\Sigma}_{pred} - c(d)_w^T \mathbf{\Sigma}_w^{-1} c(d)_w\right)^{-1}\right)^{-1}. \tag{14}$$

Equations (13) and (14) indicate that there are a number of matrix inversions needed for this calculation. Some of these matrix inversions can be performed independently of other calculations and hence in parallel. The iterations in the sum calculation are completely independent of each other. By assigning these iterations to other processors in a parallel system it is proposed that speed-ups can be achieved since the main bottleneck in this algorithm (and many other geostatistical algorithms) is the matrix inversion.

For the BCM parallel implementation, the individual committee predictive mean and predictive variance will be performed on separate processors. The calculations

```
1. Master to broadcast committee parameters to each process
   (MPI_Bcast)
2. Master to broadcast test data locations to each process
   (MPI_Bcast)
3. Master to scatter training data to each process
   (MPI_Scatter)
4. Each node to calculate the contribution of assigned
   committee
5. Master to collect the mean and variance at the test
   locations from each process and sum results (MPI_Reduce)
```

Fig. 2 Pseudo code for parallel Bayesian Committee Machine

of the predictive mean and predictive variance require the inverse of a matrix of the same size as the number of observed data assigned to each committee. A further inversion is needed to calculate the inverse of the predictive variance which is a matrix of the same size as the number of prediction locations. The basic algorithm for a parallel BCM is given in Fig. 2.

2.3 Moving Window Kriging

One approach to performing kriging with large datasets was introduced by David (1976). A specified search radius from the prediction location is used to select a local neighbourhood of observations to use in the kriging system. This neighbourhood moves according to the location which is being predicted. An alternative approach for selecting the neighbourhood is to select a predetermined number of near observations for each prediction location. As noted by Davis and Culbane (1984), these methods produce spurious behaviour in some of the estimates and hence should be used with caution, this is apparent as observations are added or removed from the moving window. Ad-hoc methods of subsetting the data were formalised by the moving-window approach of Haas (1995), although the local covariance functions fitted within the window may yield incompatible covariances at larger spatial lags. Cressie (1993) states that for datasets that are large, the general feeling is that kriging is impossible and ad-hoc local kriging neighbourhoods are typically used. Isaaks and Srivastava (1989) devote a whole chapter to choosing an effective search strategy. Implementations of kriging tend to use this approach for performing kriging efficiently. Here we use the moving window kriging approach as a means of benchmarking the BCM.

3 Experimental Setup

3.1 Datasets

To test these methods we simulate two large spatial datasets, each with 40,000 observations on a grid of 200 × 200 points. To do this we use the Turning Bands method of simulation (Emery and Lantuéjoul, 2006) since large datasets can be simulated without prohibitive running times. Figure 3 shows the two simulated fields. Both datasets were simulated with an exponential covariance function. The first dataset was simulated with an effective range of 15 m and the second dataset has an effective range of 150 m. We sample 20,000 observations from each dataset using simple random sampling. We use this for learning the model parameters and prediction. We use the remaining 20,000 observations for cross-validation to test our model.

3.2 Software

For the experiments in this paper we used an eight node tightly coupled parallel system where we compared the performance using 1, 2, 4 and 8 processors. We choose LAM/MPI[5] implementation of the MPI standard.[6] ATLAS was compiled with the

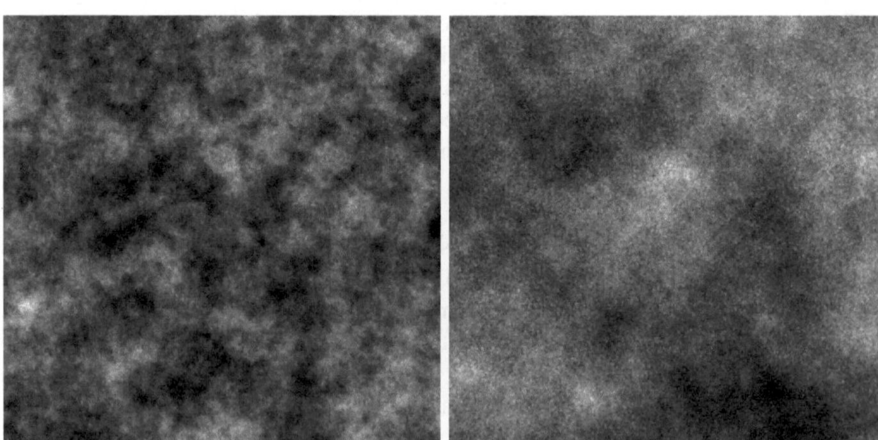

Fig. 3 Plots of simulated datasets with exponential covariance and (*left*) short range parameter (*right*) long range parameter

[5] http://www.lam-mpi.org/.

[6] http://www-unix.mcs.anl.gov/mpi/.

thread support disabled. The software was written using the Matlab language and executed in Octave although we hope to release a standalone implementation soon.

4 Results

The results from the experiments are split into two tables. Table 1 shows the Mean Average Error (MAE) of using the Vecchia method to estimate the covariance parameters and then we predict at the cross-validation locations. The first thing to notice is how the prediction results do not change depending on the number of processors used which is to be expected. As the number of observations used in the conditioning subset, the MAE decreases with both datasets. The effect seems less pronounced with Dataset 2 however. The timings are taken after ten iterations of the conjugate gradient minimisation of the approximate likelihood. For the BCM algorithm we compare this directly to MWK (Moving Window Kriging) in Table 2. By increasing the size of each committee, it can be seen that the computational complexity increases. However, increasing the number of processors reduces the computational burden. As to be expected and as the previous results, the MAE does not change depending on the number of processors used. In this experiment, by increasing the committee size seems to reduce the MAE for Dataset 2 more markedly than with Dataset 1.

The results for MWK show that for Dataset 1, where the range parameter is short with respect to the overall scale of the dataset, that MWK out performs the BCM in terms of prediction accuracy, although the computational speed is significantly slower. This can reduced of course by applying MWK to a parallel processor computer. With Dataset 2 where the range parameter is long with respect to the overall scale of the data, the effect seems less severe. The BCM seems to perform equally

Table 1 Performance of parallel maximum likelihood using Vecchia's method

Processors	Subset Size	Time (s)	MAE Dataset 1	MAE Dataset 2
1	50	50.84	34.44	36.23
2	50	30.75	34.44	36.23
4	50	20.74	34.44	36.23
8	50	16.44	34.44	36.23
1	100	240.12	31.71	34.83
2	100	130.47	31.71	34.83
4	100	70.33	31.71	34.83
8	100	40.34	31.71	34.83
1	200	1,290.38	30.43	34.01
2	200	640.69	30.43	34.01
4	200	340.91	30.43	34.01
8	200	180.34	30.43	34.01

Table 2 Performance of parallel BCM and moving window kriging

Processors	Subset Size	Time (s)	MAE Dataset 1	MAE Dataset 2
1	250	1.84	27.92	22.15
2	250	2.05	27.92	22.15
4	250	1.92	27.92	22.15
8	250	2.14	27.92	22.15
MWK	250	224.23	22.10	22.19
1	500	3.67	27.83	21.62
2	500	3.27	27.83	21.62
4	500	2.84	27.83	21.62
8	500	2.34	27.83	21.62
MWK	500	573.74	21.95	21.22
1	1,000	11.90	27.53	20.76
2	1,000	6.75	27.53	20.76
4	1,000	3.98	27.53	20.76
8	1,000	2.85	27.53	20.76
MWK	1000	1,473.34	22.01	20.90

as well as MWK in terms of prediction accuracy, however in terms of prediction speed, the BCM is many more times more efficient.

5 Conclusions

In this paper we have considered two methods for applying parallel geostatistics. Firstly we looked at approximating the likelihood using a well known technique in geostatistics (Stein et al., 2004). We showed how this was particularly effective when the range parameter was short when compared with the overall scale of the area of interest. When applied to Dataset 2, with a long range parameter, the performance was poorer. Increasing the number of processors reduced prediction time. Using two processors does not exactly half the calculation time due to overheads of distributing the data to the other processor. In terms of the computational complexity of this algorithm, the distribution of data will cause a short delay (depending on the architecture of the system).

The second technique we looked as was the BCM. This was shown to be equivalent to a low-rank covariance matrix approximation with the exact diagonal structure of the true covariance matrix retained. Low-rank methods are particularly useful when the range parameter of the dataset is long when compared with the overall scale of the dataset. Hence it is to be expected that the BCM performs better on Dataset 2. The BCM provides an effective alternative to moving window kriging when large datasets are encountered. For the BCM experiments the covariance function parameters were determined *a-priori*. Another advantage of using the BCM method is that all the data in the dataset is used for prediction rather than a subset. We are aware of an unpublished work that provides an approximation to the BCM likelihood using a Laplace propagation technique. This will be implemented in future versions.

The methods presented here are effective when applied to a specific geostatistical problems. They enable principled geostatistics to be applied to large datasets.

Acknowledgements This work is funded by the European Commission, under the Sixth Framework Programme, by the Contract N. 033811 with the DG INFSO, action Line IST-2005-2.5.12 ICT for Environmental Risk Management. The views expressed herein are those of the authors and are not necessarily those of the European Commission.

References

Cressie NAC (1993) Statistics for spatial data. Wiley, New York

David M (1976) The practice of kriging. Adv Geostatistics in the Min Ind, 31:461

Davis MW, Culbane PG (1984) Contouring very large data sets using kriging. Geostatistics for Nat Resour Characterization 2:599–619

Emery X, Lantuéjoul C (2006) TBSIM: a computer program for conditional simulation of three-dimensional Gaussian random fields via the turning bands method. Comput Geosci, 32(10):1615–1628

Fernández J, Anguita M, Mota S, Cañas A, Ortigosa E, Rojas FJ (2004) MPI toolbox for octave. In: Proceedings of 6th international conference on high performance computing for computational science, Valencia, Spain, 2004. Springer, Berlin Heidelberg

Gebhardt A (2003) PVM kriging with R. In: Proceedings of the 3rd international workshop on distributed statistical computing, Vienna

Haas TC (1995) Local prediction of a spatio-temporal process with an application to wet sulfate deposition. J Am Stat Assoc, 90(432):1189–199

Isaaks EH, Srivastava RM (1989) An introduction to applied geostatistics. Oxford University Press, New York

Kerry KE, Hawick KA (1998) Kriging interpolation on high-performance computers. In: Proceedings of the international conference and exhibition on high-performance computing and networking. Springer Berlin, Heidelberg, pp 429–438

Pardo-Igúzquiza E, Dowd PA (1997) AMLE3D: a computer program for the inference of spatial covariance parameters by approximate maximum likelihood estimation. Comput Geosci, 23(7):793–805(13)

Pedelty JA, Schnase JL, Smith JA (2003) High performance geostatistical modeling of biospheric resources in the Cerro Grande Wildfire Site, Los Alamos, New Mexico and Rocky Mountain National Park, Colorado. NASA Goddard Space Flight Center, Code 930

Schabenberger O, Gotway CA (2005) Statistical methods for spatial data analysis. CRC Press, Boca Raton, FL

Schwaighofer A, Tresp V (2003) Transductive and inductive methods for approximate Gaussian process regression. Adv Neural Inf Process Syst 15:953–960

Stein ML, Chi Z, Welty LJ (2004) Approximating likelihoods for large spatial data sets. J R Stat Soc B 66(2):275–296

Tresp V (2000) A Bayesian committee machine. Neural Comput 12(11):2719–2741

Tresp V (2001) Committee machines, in handbook for neural network signal processing, chapter 5. CRC Press, Boca Raton, FL pp 1–18

Vecchia AV (1988) Estimation and model identification for continuous spatial processes. J R Stat Soc B Met 50(2):297–312

Multiple Point Geostatistical Simulation with Simulated Annealing: Implementation Using Speculative Parallel Computing

Julián M. Ortiz and Oscar Peredo

Abstract Multiple-point geostatistical simulation aims at generating realizations that reproduce pattern statistics inferred from some training source, usually a training image. The most widely used algorithm is based on solving a single normal equation at each location using the conditional probabilities inferred during the training process. Simulated annealing offers an alternative implementation that, in addition, permits to incorporate additional statistics to be matched and imposing constraints based, for example, on secondary information.

This paper focuses on an innovative implementation of simulated annealing to simulate categorical variables, reproducing multiple-point statistics. It is based on a well known paradigm in computer science, namely, speculative computing.

In simulated annealing, categories are initially randomly distributed. Nodes are visited iteratively and a perturbation is proposed to approach the distribution of the categories to some target statistics. A decision is made to accept or conditionally reject the change, depending on an objective function that must approach zero to match the target statistics. Rejection will occur with a probability that changes during the simulation process, as defined in the annealing schedule. Speculative computing consists of using multiple processes in parallel to pre-calculate the next step in the simulation in both situations: accepting or rejecting the change. While the decision is made in the first process, a second level of two processes is used to calculate the two possible cases and subsequent levels can also be initiated. Once the decision is made, processes that do not conform to this decision are dropped and speculations about other possible perturbations at the current simulation stage are initiated.

This implementation of simulated annealing can speed up the process significantly, hence making this algorithm a reasonable alternative to current methods. An example using a geologic data set is provided to demonstrate the improvements achieved and the potential this method has for larger models. Some future work is also proposed.

J.M. Ortiz (✉) and O. Peredo
Department of Mining Engineering, University of Chile, Av. Tupper 2069, 837-0451, Santiago, Chile
e-mail: jortiz@ing.uchile.cl; operedo@dcc.uchile.cl

P.M. Atkinson and C.D. Lloyd (eds.), *geoENV VII – Geostatistics for Environmental Applications*, Quantitative Geology and Geostatistics 16,
DOI 10.1007/978-90-481-2322-3_33,

1 Introduction

Multiple point geostatistical simulation has been a very active research area in recent years. This is caused by the necessity to better reproduce some key features in the construction of numerical models of categorical variables such as facies in the oil industry, rock type distribution in the mining industry or land use allocation in environmental sciences.

Several researchers have proposed algorithms and improvements to impose multiple point statistics in geostatistical simulation. A brief overview of these methods is summarized next.

Object-based modeling: One of the early approaches to capturing structural features beyond the variogram is the object-based modeling. The main goal is to impose structural information into the modeling process, usually by stochastically placing objects into the domain. These objects must conform to the particular features of the geological setting. The main limitations of these methods are that (1) each particular type of object must be parameterized individually, therefore a new implementation is required for each setting, and (2) conditioning to abundant data may be difficult. Nonetheless, these remain as valid and applicable techniques (Deutsch and Wang, 1996; Tjelmeland, 1996).

Conventional simulation with locally varying anisotropies: A more intuitive approach to reproduce some features of the phenomenon is to incorporate locally varying anisotropies in conventional pixel-based simulation (Xu, 1996). The control of structural features can be imposed through angles that change locally and define the curvilinearity of the objects being simulated, as well as locally varying proportions to impose trends in the categories distributions (Zanon, 2004).

Single normal equation simulation: Guardiano and Srivastava's (1993) original proposal was to compute the conditional expectation at an unsampled location using the arrangement of conditioning points as a multiple point event. To determine the probability of the unsampled location to belong to a category, the frequency with which this category was found with a similar configuration of conditioning points was computed from a training image. The method is conceptually quite simple, but its implementation brings several issues. Its implementation in a sequential simulation fashion requires computing the frequency of multiple point events that change at every simulation step, the simulated value is added as a new conditioning value and the training image has to be scanned for constantly changing point configurations. This problem was solved by Strebelle and Journel (2000), making this approach practical for use with larger simulation grids, through the use of a binary tree for storing the multiple point events (see also Strebelle, 2002). Since then, this algorithm has become the most popular methodology for multiple point statistics simulation and many variations and improvements are being developed.

Neural Networks: Caers proposed the use of neural networks to allow inference of conditional distributions given multiple point conditioning information. One of the most interesting features of this approach is the possibility of controlling the degree of overfitting of the realizations by means of a cross validation, where an error function is computed to indicate the quality of the network (Caers et al., 1999; Caers, 1998; Caers and Ma, 2002).

Simulation with patterns: One of the interesting results of the paradigm of the single normal equations is the direct simulation of patterns. Arpat developed an algorithm that allows drawing directly patterns to complete a grid of nodes, using conditioning information and a similarity measure in order to impose the structural information, even when the conditioning pattern is not exactly found in the training information (Arpat and Caers, 2007).

Conventional simulation integrating multiple point statistics: With a slightly different philosophy, Ortiz and Deutsch (2004) proposed an approach to combine multiple point statistics with conventional simulation, by modifying the conditional distributions with multiple point statistics for a set of cutoffs in the case of continuous variables (see also, Ortiz, 2003; Ortiz and Emery, 2005). This approach could be similarly implemented in the context of categorical variables.

Gibbs sampler multiple point simulation: Another approach that suggests a generalization of kriging and of the single normal equation algorithm is multiple point simulation using a Gibbs sampler that can integrate information of any nature to approximate conditional distributions at every location (Boisvert et al., 2007). The algorithm may integrate several multiple point events providing a promising framework.

Simulated annealing: Possibly one of the first "practical" implementations of multiple point geostatistical simulation was proposed by Deutsch in his Ph.D. thesis (Deutsch, 1992) and was already mentioned in the first edition of GSLIB User's Guide (Deutsch and Journel, 1992). This method is discussed in the next section.

In all the cases, the processing time to construct a proper 3D model may be significant, particularly in iterative methods. In this context, the potential of using parallel computing to increase the speed and run more demanding processes is seen as an interesting avenue. In this paper, we present some results related to the implementation of a simulated annealing approach to reproduce multiple point statistics, considering the principles of speculative computing.

2 Simulated Annealing

Simulated annealing is a general optimization algorithm that is capable of incorporating as many statistics and constraints as required to the simulation process (Besag, 1986; Farmer, 1992; Geman and Geman, 1984; Kirkpatrick et al., 1983; Rothman, 1985). The algorithm can honour all of the statistics if they are consistent with each other and the optimization parameters are set correctly. The basic idea is to start with a realization that does not honour the statistics and perturb the nodes until the statistics are close enough to the target. This is done by defining an objective function that corresponds to a weighted sum of component objective functions. Each one of these components corresponds to a measure of mismatch between the target statistics and the current statistics, which are expressed as a mathematical expression.

In general, the objective function is written:

$$O = \sum_{i=1}^{N_C} \omega_i O_i$$

where N_C is the number of components in the objective function, ω_i are the weights assigned to each one of the components, and O_i is the mismatch value for component i. For example, this function could be composed by the mismatch in histogram reproduction, defined as the difference in the cumulative frequencies measured at some quantiles for the model simulated versus the target histogram, a mismatch in variogram reproduction, composed by differences between the target variogram model and the variogram calculated from the realization being perturbed, for a number of lag distances, and a mismatch in the reproduction of multiple point statistics. In general any constraint can be imposed in the objective function, but convergence can be achieved only if these constraints are consistent with each other, providing a *possible* solution.

Additionally, the reproduction of a variogram map, indicator variograms, a histogram of multiple point statistics for some pattern sizes and the requirement of honouring conditional information can be imposed through elements of the objective function.

Starting from a random distribution of the variable over the domain, nodes are visited randomly, a perturbation is proposed, and the change in the objective function is computed. The rule for accepting or rejecting a change is based on the Gibbs or Boltzmann probability distribution, which gives the name to the algorithm, since it was used to model the energy of molecules in the physical process of annealing (Deutsch, 2002). If the change is favourable, the perturbation is accepted, otherwise the perturbation may be conditionally accepted as dictated by a probability distribution defined by an annealing schedule. The fact that some bad changes are conditionally accepted differentiates SA from other optimization algorithms, where all bad changes are rejected. The probability of acceptance, given by the Boltzmann distribution is:

$$P(\text{accept}) = \begin{cases} 1 & \text{if } O_{new} \leq O_{old} \\ e^{\frac{O_{old} - O_{new}}{t}} & \text{otherwise} \end{cases}$$

where t is a parameter equivalent to the product of the Boltzmann constant k_b and the temperature T in the application to the physical process. By analogy, t is called the temperature in SA; O_{old} and O_{new} are the values of the objective function before and after the perturbation, equivalent to the difference in Gibbs free energy ΔE in the physical process of annealing. In SA, the temperature must be lowered as the algorithm runs, emulating the cooling that occurs during the physical process that lets the molecules reorganize to a lower energy state.

The computer implementation requires the specification of some parameters, as described next:

- **Initial Realization** The algorithm perturbs nodes of an initial realization, which is usually random with the target histogram.
- **Objective Function** The components of the objective function dictate the statistics to be reproduced in the simulated models. Convergence can be ensured subject to the consistency of these components. The objective function is a weighted sum of mismatch functions, usually squared differences between current statistics and target statistics. If matched, it should equal zero.
- **Stopping Criteria** Several considerations can be used to stop the algorithm. Firstly, the algorithm should stop if the objective function has reached a value considered low enough, which means that target statistics have been closely matched. A second criterion for stopping is CPU time. Also, the algorithm can be stopped if the objective function does not converge. This may occur because the components of the objective function are not compatible, hence it is not possible to find realizations matching all statistics, or because the algorithm has not attempted a number of perturbations large enough.
- **Perturbation Mechanism** Perturbations of the initial image can be done by changing one node at a time (drawing a value from the global histogram) or swapping nodes randomly selected, which would ensure histogram reproduction. Other alternatives such as drawing from a conditional distribution built by indicator kriging of the surrounding nodes in a given template or calibrating with a secondary variable have also been proposed (Deutsch and Wen, 2001; Deutsch, 2002).
- **Updating of Objective Function** The re-calculation of the objective function can be done by updating the initially calculated value with the changes due to the modification of the node (or nodes) perturbed. This makes the algorithm more efficient in terms of CPU time than re-calculating the entire objective function every time, as illustrated by Deutsch (2002).
- **Annealing Schedule** The annealing schedule refers to the parameters that specify how the temperature is reduced. The temperature parameter t must be lowered to allow convergence. However, convergence is guaranteed only under very restricted conditions. In practice, it depends on how the temperature t is changed during the simulation. As in the physical process, the temperature should be lowered slowly to allow a lower energy state. As the temperature decreases, bad changes will have a lower probability of being accepted, that is, the realization will tend to stay in the same state unless the changes are favourable. In practice the temperature is lowered with some control parameters: (1) An initial temperature t_0 is set to a high value. Some attempts to optimize this temperature have been documented (Norrena and Deutsch, 2000); (2) The temperature is lowered using a reduction factor $\lambda \in (0, 1)$. This rate is a multiplicative factor to reduce the temperature if a maximum number of attempted perturbations K_{max} is reached at the same temperature, or if a maximum number of accepted

perturbations is reached at that temperature; (3) The simulation will stop if K_{max} is reached S times, that is, if the number of perturbations accepted at a given temperature has not been reached in the last S attempted temperatures; (4) The tolerance in the objective function to define convergence ΔO, which should be set to a low value.

To implement simulated annealing using multiple point statistics a histogram of frequencies of multiple point events is constructed from a training image for a given set of pattern configurations. Other statistics such as the connectivity function, reproduction of runs, as well as the implementation considering multiple grids could be applied.

3 Speculative Parallel Computing

Several approaches exist to implement numerical algorithms into a parallel computing framework. However, we have addressed the problem considering a solution called the speculative approach. The basic idea consists of using multiple processes organized in a binary tree to compute in advance the two possible solutions of a decision. This can be further sped up by considering several levels of the tree. In the case of simulated annealing, the decision is to accept or reject the perturbation of a node. Figure 1 shows the decision tree. Process 0 computes the value of the objective function with and without perturbing a node randomly selected. Meanwhile, Process 1 computes the value of the objective function with and without perturbing a subsequent node, subject to having *accepted* the perturbation computed in Process 0. Simultaneously, Process 2 is computing the same solution but subject to having rejected the perturbation on Process 0. A second level of advanced calculations is performed in Processes 3 to 6, each one linked to previous decisions and speculating about the possible decision of the parent node.

Parallel computing can reduce the time of the numerical calculation of SA. The reduction depends on the number of levels of the tree. In general, $2^n - 1$ processes

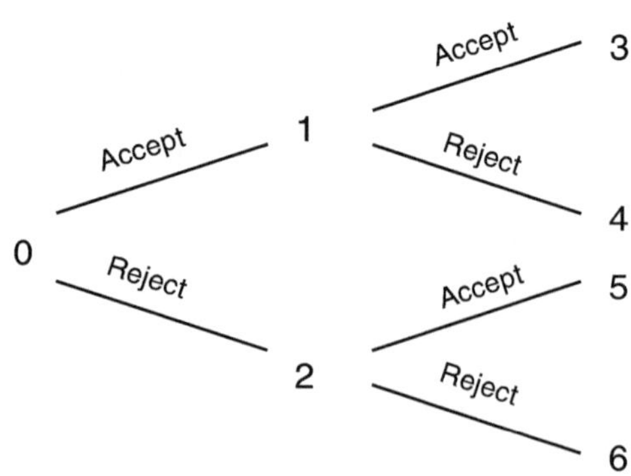

Fig. 1 Tree showing the processes

are used to implement a tree of n levels. Theoretically, this means a speed up in the computations of the order of $\log_2(P + 1)$, where P is the number of processes.

Several authors have studied different issues in the implementation of simulated annealing in parallel computing, addressing issues such as communication overhead between processes (Nabhan and Zomaya, 1995), quantifying and eliminating approximations done in the parallel implementation that generate poorer solutions than the serial implementation (Witte et al., 1991; Chamberlainet al., 1988), and problem independent implementations (Roussel-Ragot and Dreyfus, 1990). Applications are generally restricted to the computer and electrical engineering fields. To the authors' knowledge, the proposed implementation of parallel computation accounting for multiple point statistics is the first documented attempt in a spatial context.

4 Examples

To test the performance of the implementation and the gain obtained by using the parallel approach, several tests were carried out. A training image of a channel setting over a background is considered (Fig. 2). This is a very simple geological setting and it is aimed at evaluating the decrease in processing time that can be achieved, and the capacity of simulated annealing to match multiple point statistics.

For the implementation simple squared patterns are used, and the frequencies of finding each data event (combination of channel and non channel facies) on the

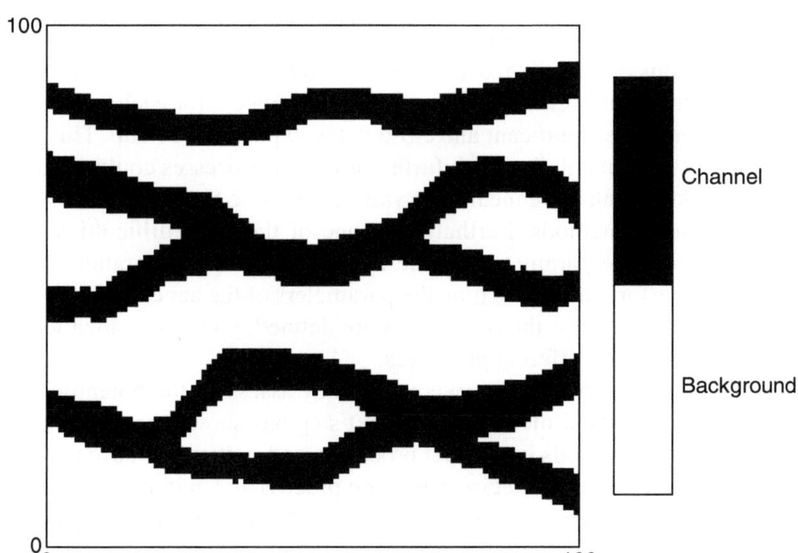

Fig. 2 Training image for test runs

training image are computed. Three pattern sizes are used: 2 by 2, 3 by 3 and 4 by 4 pixels. The implementation does not use a multiple grid approach (Tran, 1994), and does not consider more complex patterns for the objective function. These improvements could be implemented in the future. The first approach considers an objective function that penalizes equally the difference in frequencies of multiple point events, as shown next:

$$O = \sum_{i=1}^{N_{MPE}} (f_i^{\text{target}} - f_i^{\text{model}})^2$$

where f_i^{target} are the frequencies of the multiple point events from the training image, f_i^{model} are the statistics for the same multiple point events in the current state of the simulated model, and N_{MPE} is the total number of possible multiple point events for the pattern size considered. A second implementation is done weighting the frequencies of multiple point events with small frequencies, in order to increase their relative importance in the objective function:

$$O = \sum_{i=1}^{N_{MPE}} \lambda_i (f_i^{\text{target}} - f_i^{\text{model}})^2$$

where λ_i is a standardized weight inversely proportional to the target frequency of the multiple point event.

Figure 3 shows a resulting realization. Although the visual similarity with the training image could be questioned, the reproduction of statistics is deemed satisfactory (Fig. 4). Notice that the scale is logarithmic in both axes to exaggerate the dispersion of low frequency configurations.

Time reductions are illustrated in Table 1, where the speed up is presented in comparison to the implementation through a single process for each pattern size.

Time reductions are significant and close to the expected speed up. This suggests that the parallel implementation with further levels of processes could achieve high speeds and make simulated annealing a valid alternative to current multiple point statistics simulation methods. Furthermore, one of the main difficulties in SA is setting the appropriate parameters in the annealing schedule. Parallel computing could be use to perform test runs, tune the parameters of the annealing schedule and evaluate convergence. Once the parameters are defined, multiple realizations could be run independently in different processes.

Many implementation decisions were driven by assessing the potential of parallelization for geostatistical methods. This first step has shown that it is possible to implement iterative methods in parallel processes and particularly, use the speculative paradigm to structure the decision making process and gain in time reduction.

Many possible avenues of research open with these results and should be explored in the future. Industrial applications of geostatistical simulation could see the benefit of faster construction of numerical models with advanced tools.

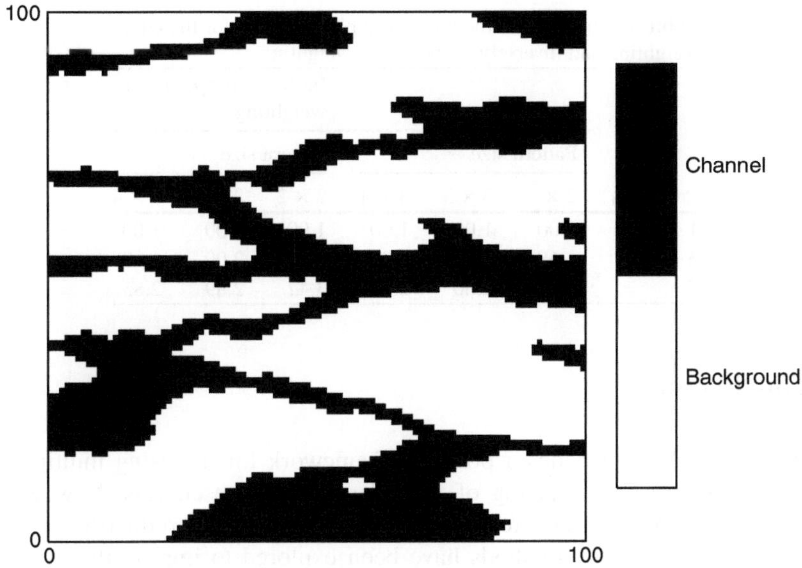

Fig. 3 Realization obtained by simulated annealing with the parallel computing implementation, considering a multiple point statistics from a pattern of 3 by 3 nodes

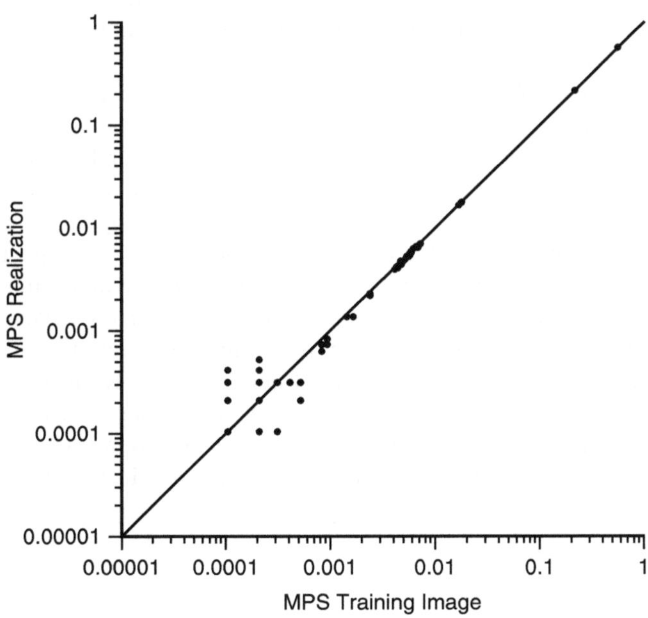

Fig. 4 Comparison of multiple point frequencies of the training image (X axis) versus multiple point frequencies of the realization obtained by simulated annealing, for data events in a 3 by 3 nodes pattern

Table 1 Speed up of the parallel implementation for the equal weighting and inversely proportional weighting

	Equal weighting			Inversely proportional weighting		
	Pattern size			Pattern size		
Processes	2 × 2	3 × 3	4 × 4	2 × 2	3 × 3	4 × 4
1	1.00	1.00	1.00	1.00	1.00	1.00
3	1.68	1.70	1.89	1.87	2.00	1.88
7	2.63	2.60	2.78	3.42	2.89	2.85

5 Conclusions

Simulated annealing provides a powerful framework for imposing multiple constraints into numerical modeling of variables in the Geosciences. New ways to improve the geological characterization are based on the reproduction of pattern statistics. Although many methods have been explored to impose these multiple point statistics, most of them do not offer the flexibility of simulated annealing. Two important practical difficulties cloud its use: (1) the definition of the annealing schedule is generally difficult and some tuning of the parameters is required, particularly regarding the rate to reduce the temperature and its initial value, and (2) the intensive CPU use due to its iterative nature implies long run times that may make the method impractical.

Parallel computation provides the tools to speed up processes by configuring the numerical steps such that some time gain can be achieved. Speculative computing, in particular, uses the fact that a decision that will condition all subsequent calculations must be made, which can lead to two possible answers. Calculations conditioned to the possible answers are done prior to knowing the actual decision, therefore, speeding up the overall construction of the model. These speculations can be done in several levels, organized as a decision tree in the algorithm, increasing even more the time gain for the computation of the final model. We have shown some simple implementations of SA imposing multiple point statistics in regular square patterns to reproduce features of a training image. Results are consistent with the speed up expected in theory and SA could reproduce with a reasonable degree of precision, the required statistics. This opens a new avenue of research where parallel computing is used for the construction of geostatistical models.

Acknowledgments This research was funded by the National Fund for Science and Technology of Chile (FONDECYT) and is part of the project number 1040690. The authors would also like to acknowledge the funding provided by the Codelco Chair on Ore Reserve Estimation at the Mining Engineering Department, University of Chile.

References

Arpat GB, Caers J (2007) Conditional simulation with patterns. Math Geol 39(2):177–203

Besag J (1986) On the statistical analysis of dirty pictures. J Royal Stat Soc B 48(3):259–302

Boisvert JB, Lyster S, Deutsch CV (2007) Constructing training images for veins and using them in multiple-point geostatistical simulation, APCOM 2007, pp 113–120

Caers J (1998) Stochastic simulation with neural networks. In Report 11, Stanford Center for Reservoir Forecasting, Stanford, CA

Caers J, Ma X (2002) Modeling conditional distributions of facies from seismic using neural nets. Math Geol 34(2):143–167

Caers J, Srinivasan S, Journel AG (1999) Stochastic reservoir simulation using neural networks trained on outcrop data: SPE paper no. 49026

Chamberlain RD, Edelman MN, Franklin MA, Witte EE (1988) Simulated annealing on a multiprocessor. Proceedings of the 1988 IEEE International Conference on Computer Design: VLSI in Computers and Processors, 3–5 October 1988, pp 540–544

Deutsch CV (1992) Annealing techniques applied to reservoir modeling and the integration of geological and engineering (well test) data: Unpublished doctoral dissertation, Stanford University, 306 p

Deutsch CV (2002) Geostatistical reservoir modeling. Oxford University Press, New York

Deutsch CV, Journel AG (1992) GSLIB: Geostatistical software library and user's guide. Oxford University Press, New York, 340 p

Deutsch CV, Wang L (1996) Hierarchical object-based stochastic modeling of fluvial reservoirs. Math Geol 28(7):857–880

Deutsch CV, Wen XH (2001) Integrating large-scale soft data by simulated annealing and probability constraints. Math Geol 32(1):49–68

Farmer CL (1992) Numerical rocks. In: King PR (ed) The mathematical generation of reservoir geology. Clarendron press, Oxford (Proceedings of a conference held at Robinson College, Cambridge, 1989)

Geman S, Geman D (1984). Stochastic relaxation, Gibbs distributions, and the Bayesian restoration of images. IEEE Trans Pattern Anal Mach Intelligence PAMI-6(6):721–741

Guardiano F, Srivastava M (1993) Multivariate geostatistics: beyond bivariate moments. In: Soares A (ed) Geostatistics Tróia'92, vol. 1. Kluwer, Dordrecht, pp. 133–144

Kirkpatrick S, Gelatt Jr. CD, Vecchi MP (1983) Optimization by simulated annealing. Science 220(4598):671–680

Nabhan TM, Zomaya AY (1995) A parallel simulated annealing algorithm with low communication overhead. IEEE Trans Parallel Dist Syst 6(12):1226–1233

Norrena K, Deutsch CV (2000) Using the critical temperature to improve the speed of geostatiostical applications of simulated annealing, in Geostatistics 2000. In Kleingold WJ, Krige DG (eds) Proceedings of 6th International Geostatistics Congress. Cape Town, South Africa, vol. 1, pp 254–262

Ortiz JM (2003) Characterization of high order correlation for enhanced indicator simulation. Ph.D. thesis, University of Alberta, 255 p

Ortiz JM, Deutsch CV (2004) Indicator simulation accounting for multiple-point statistics. Math Geol 36(5):545–565

Ortiz JM, Emery X (2005) Integrating multiple point statistics into sequential simulation algorithms, in Geostatistics Banff 2004. In: Leuangthong O, Deutsch CV (eds), Proceedings of the Seventh International Geostatistics Congress. Springer, Banff, Canada, vol. 2, pp 969–978

Rothman DH (1985) Nonlinear inversion, statistical mechanics, and residual statics estimation. Geophysics 50:2784–2796

Roussel-Ragot P, Dreyfus G (1990) A problem independent parallel implementation of simulated annealing: models and experiments. IEEE Trans Computer-Aided Des. 9(8):827–835

Strebelle S (2002) Conditional simulation of complex geological structures using multiple-point statistics. Math Geol 34(1):1–21

Strebelle S, Journel AG (2000). Sequential simulation drawing structures from training images. 6th International Geostatistics Congress, Cape Town, South Africa, April 2000. Geostatistical Association of Southern Africa

Tjelmeland H (1996) Stochastic models in reservoir characterization and Markov random fields for compact objects: Unpublished doctoral dissertation, Norwegian University of Science and Technology

Tran TT (1994) Improving variogram reproduction on dense simulation grids. Comput Geosci 20(7)1161–1168

Witte EE, Chamberlain RD, Franklin MA (1991) Parallel simulated annealing using speculative computation. IEEE Trans Parallel Dist Syst 2(4)483–494

Xu W (1996) Conditional curvilinear stochastic simulation using pixel-based algorithms: Math Geol 28(7):937–949

Zanon S (2004) Advanced aspects of sequential Gaussian simulation. M.Sc. Thesis, University of Alberta, 65 p

Application of Copulas in Geostatistics

Claus P. Haslauer, Jing Li, and András Bárdossy

Abstract This paper demonstrates how empirical copulas can be used to describe and model spatial dependence structures of real-world environmental datasets in the purest form and how such a copula model can be employed as the underlying structure for interpolation and associated uncertainty estimates.

Using copulas, the dependence of multivariate distributions is modelled by the joint cumulative distribution of the variables using uniform marginal distribution functions. The uniform marginal distributions are the effect of transforming the marginal distributions monotonically by using the ranks of the variables. Due to the uniform marginal distributions, copulas express the dependence structure of the variables independent of the variables' marginal distributions which means that copulas display interdependence between variables in its purest form. This property also means that marginal distributions of the original data have no influence on the spatial dependence structure and can not "cover up" parts of the spatial dependence structure. Additionally, differences in the degree of dependence between different quantiles of the variables are readily identified by the shape of the contours of an empirical copula density.

Regarding the quantification of uncertainties, copulas offer a significant advantage: the full distribution function of the interpolated parameter at every interpolation point is available. The magnitude of uncertainty does not depend on the density of the observation network only, but also on the magnitude of the measurements as well as on the gradient of the magnitude of the measurements. That means for the same configuration of the observation network, interpolating two events with very similar marginal distribution, the confidence intervals look significantly different for both events.

C.P. Haslauer (✉)
Department of Hydrology and Geohydrology, Institute of Hydraulic Engineering,
University of Stuttgart, Germany
e-mail: claus.haslauer@iws.uni-stuttgart.de

J. Li and A. Bárdossy
Institut für Wasserbau, Pfaffenwaldring 61, 70569 Stuttgart, Germany
e-mail: Andras.Bardossy@iws.uni-stuttgart.de

P.M. Atkinson and C.D. Lloyd (eds.), *geoENV VII – Geostatistics for Environmental Applications*, Quantitative Geology and Geostatistics 16,
DOI 10.1007/978-90-481-2322-3_34,

1 Introduction

Generally, the workflow of spatial analysis is to first evaluate the spatial dependence structure of measured data. In a second step, a stochastic model is employed to mathematically describe the empirical spatial dependence structure. This theoretical model is subsequently used for interpolation.

Different disadvantages of traditional geostatistical methods have been recognized in the past, most notably the fact that different percentile values can have different degrees of dependence which cannot be expressed with traditional Gaussian two-point geostatistics (Journel and Alabert, 1989). Additionally, several assumptions have to be made when dealing with spatially distributed data. The most basic assumption of any geostatistical analysis is that the set of measured parameter values $z_1, \ldots, z_n$ is a realization of a random function. At every location there are never enough measurements to determine the characteristics of the distribution function of the parameter, and hence the treatment of measurements as realizations of a random function is necessary. This random function is assumed to be identical at every location.

Furthermore, in traditional geostatistics, second order stationarity is assumed, implying that the two-point covariance exists and depends only on the separation vector $\mathbf{h}$ of those two points. The assumptions when using copulas for spatial analysis are more restrictive than when using traditional geostatistics, because when using copulas strong stationarity is assumed, the multivariate distribution function is taken to be translation invariant.

These more restrictive assumptions require more effort but also offer advantages:

1. The marginal distribution, which might distort the dependence structure is filtered out using copulas. Thus the pure dependence structure of spatially distributed data can be obtained, and this structure is identical, no matter what the marginal distributions of the measured data might be. Frequently applied data transformations (e.g. taking the natural logarithm) do not influence a copula.
2. Different percentile values can have different degrees of dependence. For example, high values might exhibit a strong spatial dependence, low values a weak spatial dependence, and values of a different quantile range yet another degree of dependence.
3. At each location where interpolation is carried out, the full conditional distribution function of the interpolated value can be estimated. The shape of this distribution function is not only dependent on the geometry of the measurement network, but also on the values of the measurements. These factors allow for an improved uncertainty quantification of the interpolation.
4. When using copulas as the underlying model for simulation, then values of similar magnitude are simulated to be neighbors.
5. A full stochastic model is the backbone for geostatistical analysis with copulas.

2 Methods

This section explains the steps necessary to analyze spatially distributed data using copulas, for estimating the parameters of a theoretical copula function, and for interpolation.

2.1 Using Copulas for Spatial Analysis

Any multivariate distribution $F(t_1, \ldots, t_n)$ can be represented with a copula (Sklar, 1959):

$$F(t_1, \ldots, t_n) = C\left(F_{t_1}(t_1), \ldots, F_{t_n}(t_n)\right), \tag{1}$$

where $F_{t_i}(t_i)$ represents the i-th one-dimensional marginal distribution of the multivariate distribution.

Assuming that C is continuous, then the copula density $c(u_1, \ldots, u_n)$ can be written as

$$c(u_1, \ldots, u_n) = \frac{\partial^n C(u_1, \ldots u_n)}{\partial u_1, \ldots \partial u_n}. \tag{2}$$

A bivariate copula expresses a symmetrical dependence with respect to the minor axis $u_2 = 1 - u_1$ of the unit square, if

$$c(u_1, u_2) = c(1 - u_1, 1 - u_2). \tag{3}$$

A Gaussian copula is fully symmetrical; a family of non-Gaussian copulas representing non-symmetrical dependence was introduced in Bárdossy (2006) and Bárdossy and Li (2008).

Empirical copulas can be used to describe the spatial variability. For this purpose, several assumptions are required (Bárdossy and Li, 2008):

1. Similar to the variogram- or covariance functions, the bivariate spatial copula of the random variable $Z(\mathbf{x})$ corresponding to two locations separated by the vector $\mathbf{h}$ is assumed to be only dependent on $\mathbf{h}$. The marginal distribution of $Z(\mathbf{x})$ is supposed to be the same everywhere.
2. The parameterization of the copula should enable any n-dimensional copula corresponding to any selected n points to reflect their spatial configurations.
3. The parameterization of the copula should allow arbitrarily strong dependence.

Gaussian copulas and certain non-Gaussian copulas (as shown by Bárdossy and Li (2008)) fulfill these conditions. Further details on the theory of copulas can be found in the books by Joe (1997) and Nelsen (1999). Details on using copulas with spatially distributed data are given by Bárdossy (2006).

2.2 Empirical Bivariate Copulas

Empirical bivariate two-dimensional spatial copulas describe the dependence structure between random variables independent of marginal distributions. Such empirical copulas can be evaluated for different directions and different angles between pairs of points, and they give insights into the form and the quality of the spatial dependence structure of a field of spatially distributed values. Empirical bivariate spatial copulas can be assessed from measured data $z(x_1), \ldots, z(x_n)$ by first calculating the empirical distribution function $F_n(z)$. Using this distribution function for any given vector $\mathbf{h}$, the set of pairs $S(\mathbf{h})$, consisting of distribution function values corresponding to the parameter at locations $\mathbf{X}$ separated by the vector $\mathbf{h}$, can be calculated:

$$S(\mathbf{h}) = \{F_n(z(\mathbf{x}_i)), F_n(z(\mathbf{x}_j)) \mid (\mathbf{x}_i - \mathbf{x}_j \approx \mathbf{h}) \text{ or } (\mathbf{x}_j - \mathbf{x}_i \approx \mathbf{h})\}. \tag{4}$$

$S(\mathbf{h})$ is thus a set of points in the unit square. Note that $S(\mathbf{h})$ is by definition symmetrical regarding the major axis $u_1 = u_2$ of the unit square, namely, if $(u_1, u_2) \in S(\mathbf{h})$, then $(u_2, u_1) \in S(\mathbf{h})$.

Empirical bivariate copula densities for pairs of points separated by $\sim\mathbf{h}$ are no prerequisite to model a theoretical copula based on measurements! They are a possibility to express and visualize spatial dependence structures. On such density plots, locations associated with low measurements are plotted close to the origin, and points where the measured value is high are plotted far from the origin. If the empirical copula density for a certain quantile is high, then there are a lot of pairs of points separated by the given distance which have the corresponding quantile values. On Fig. 1, an example of a copula density plot, high copula densities are indicated by dark shading.

As an alternative to dealing with multiple plots of empirical bivariate copula densities, two scalar measures can be derived from the empirical copula space:

1. The rank correlation function "Rank" representing the degree of the spatial dependence (Equation 5).
2. A measure for the symmetry of the empirical copula density function representing for which range of quantiles the density is strongest ("Sym", Equation 6). High positive symmetry values indicate strong dependence for high quantiles, high negative symmetry values indicate strong dependence for low quantiles. A Gaussian type dependence would have zero symmetry.

Each of these measures is calculated for a given magnitude and/or angle of anisotropy of the separation vector $\mathbf{h}$. The number of pairs of points for each $\mathbf{h}$ is denoted by $n(\mathbf{h})$.

$$\text{Rank}(\mathbf{h}) = \frac{1}{12\, n(\mathbf{h})} \cdot \sum_{\mathbf{x_i}-\mathbf{x_j}\approx\mathbf{h}} \left(F_n(z(\mathbf{x}_i)) - \frac{1}{2}\right) \cdot \left(F_n(z(\mathbf{x}_j)) - \frac{1}{2}\right) \tag{5}$$

$$\mathrm{Sym}(\mathbf{h}) = \frac{1}{n\,(\mathbf{h})} \cdot \sum_{\mathbf{x_i}-\mathbf{x_j}\approx\mathbf{h}} \left(F_n(z(\mathbf{x}_i)) - \frac{1}{2}\right)^2 \left(F_n(z(\mathbf{x}_j)) - \frac{1}{2}\right) + \\ + \left(F_n(z(\mathbf{x}_i)) - \frac{1}{2}\right) \left(F_n(z(\mathbf{x}_j)) - \frac{1}{2}\right)^2 \quad (6)$$

2.3 Parameter Estimation

The parameterization of a copula model for the description of spatial dependence is not a trivial task. As shown in Section 2.2, the calculation of spatial copulas is not based on independent samples (as observations are accounted for in a number of pairs). Hence the parameterization of a copula model on empirical copulas is not appropriate, and instead Bárdossy and Li (2008) proposed a more rigorous approach. In this approach, the observation set is divided into subsets of arbitrary sizes. The likelihood of the parameter vector θ for each subset is estimated by the copula density of the observations in this subset. The result is a set of optimal parameters as given by the maximum likelihood of the product of the individual subsets.

2.4 Interpolation Using Copulas

The typical goal of an interpolation method is to estimate a random variable at unsampled locations $\mathbf{x}'$. In Section 3.2, results are discussed using two precipitation events as examples; this section describes the interpolation algorithm:

1. The observation network consists of n locations $x_1, \ldots, x_n$. At each location there are observations available, $z_1, \ldots, z_n$, which are transformed to $u_1, \ldots, u_n$ by $F(z_i) = u_i$.
2. In the neighborhood of a x', m observation points are selected.
3. The copula density value corresponding to those m locations and their observation values is calculated: $c_m(u_1, \ldots, u_m)$.
4. For the point x', for a quantile v, the $m + 1$ dimensional copula density $c_{m+1}(u_1, \ldots, u_m, v)$ is calculated.
5. The density function corresponding to x' conditioned on the n observations in the vicinity is calculated:

$$c^*(v) = c(v_k | u_1, \ldots, u_n) = \frac{c_{n+1}(u_1, \ldots, u_n, v)}{c_n(u_1, \ldots, u_n)} \quad (7)$$

6. The conditional copula C^* is calculated from its density c^*.
7. The conditional distribution C^* at x' is transformed back into the space of the measurement values, where

$$\begin{aligned} F_x(z') &= P(z(x) \leq z) = \\ &= P(U(x') \leq F_z(z)) = \\ &= C^*(F_z(z)). \end{aligned} \tag{8}$$

copula based interpolation offers choices for the estimated value: the summed observations weighted by the conditional densities, the observed value corresponding to the 50% conditional distribution value (comparable to the median), or the length of the interval between two quantile as a confidence interval of the estimate (the length of $Q80 - Q20$ as a 60% confidence interval).

3 Results

3.1 Analyzing Spatial Dependence Using Copulas

Plots of empirical copula densities are shown for the geological parameter hydraulic conductivity (Fig. 1a), for the geohydrological parameter pH (Fig. 1b), as well as for the meteorological parameter precipitation for two precipitation events in the Neckar catchment in 1982 (Fig. 1c) and in 1992 (Fig. 1d). For the precipitation events, a monitoring network comprising 950 stations in the German part of the Rhine catchment was available for this study. In all three cases, the empirical copula density plots are not symmetric as defined in Equation (3), and hence the spatial dependence structure does not follow a Gaussian distribution function. It is also shown, that for the given process, the dependence structure is different for different quantile values, indicated by the degree of shading in different areas in the unit square.

The shape of the empirical copula density functions is very similar for both events, whereas traditional variograms are quite different, since the marginal distributions for the two events are different. This might be an indication that empirical copulas represent the physical structure of a given process, without the influence of marginal distribution functions.

3.2 Interpolation and Associated Uncertainty Estimates

In this section, the two precipitation events for which an empirical bivariate copula was shown on Fig. 1c and d are used to illustrate results of spatial interpolation and uncertainty estimates using copulas. For each event the same theoretical copula was fitted using the method described in Section 2.3.

This theoretical copula was subsequently used to interpolate mean precipitation intensities, shown on Fig. 2a and b. Estimates were interpolated on a equidistantly spaced grid for a total of 32,000 points. Each interpolation estimate was conditioned on 12 surrounding measurement values. The advantage of being able to calculate

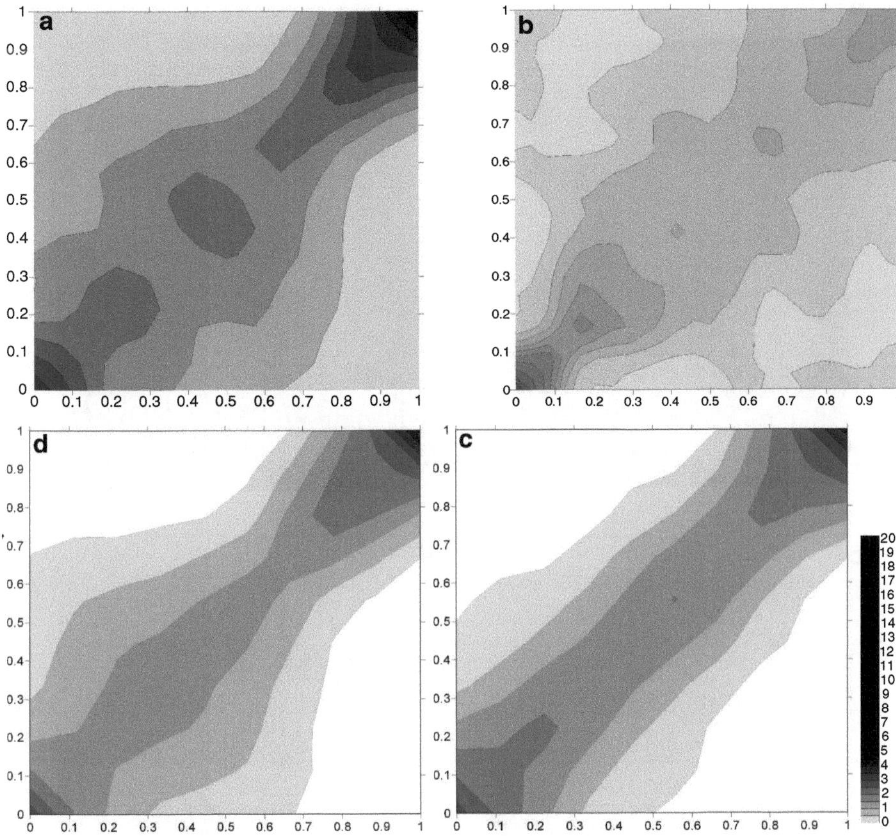

Fig. 1 Empirical Copula density plots for: (**a**) a hydraulic conductivity field along Borden's cross-section AA (Sudicky, 1986) for 0.05 m vertical spacing, (**b**) for the groundwater quality parameter pH based on the monitoring network in the province of Baden-Württemberg, Germany (Bárdossy, 2006), and (**c**), (**d**) two precipitation events in the river Necker catchment, based on 950 stations, for a separation of 5 km

the full conditional distribution function of the estimate at each interpolation point is illustrated on Fig. 2c and d which show the length of the 60% confidence interval, calculated by subtracting the 20% quantile from the 80% quantile.

Generally, in areas where the measured precipitation is high (as indicated by the shading of the dots representing the measurement locations) the uncertainty is low, and vice versa. Additionally, the copula method takes the homogeneity of the interpolated field into account. In areas where the gradient of the measurements is high, also the uncertainty of the interpolation is high. The circle shaped area of high uncertainty to the east of the bow of the river Neckar in 1992 corresponds to a confined area of high precipitation intensities, whereas the other area of high precipitation intensities in 1992, to the west of the river Neckar is a more continuous, larger area, and hence the uncertainties of the interpolation are smaller in that area.

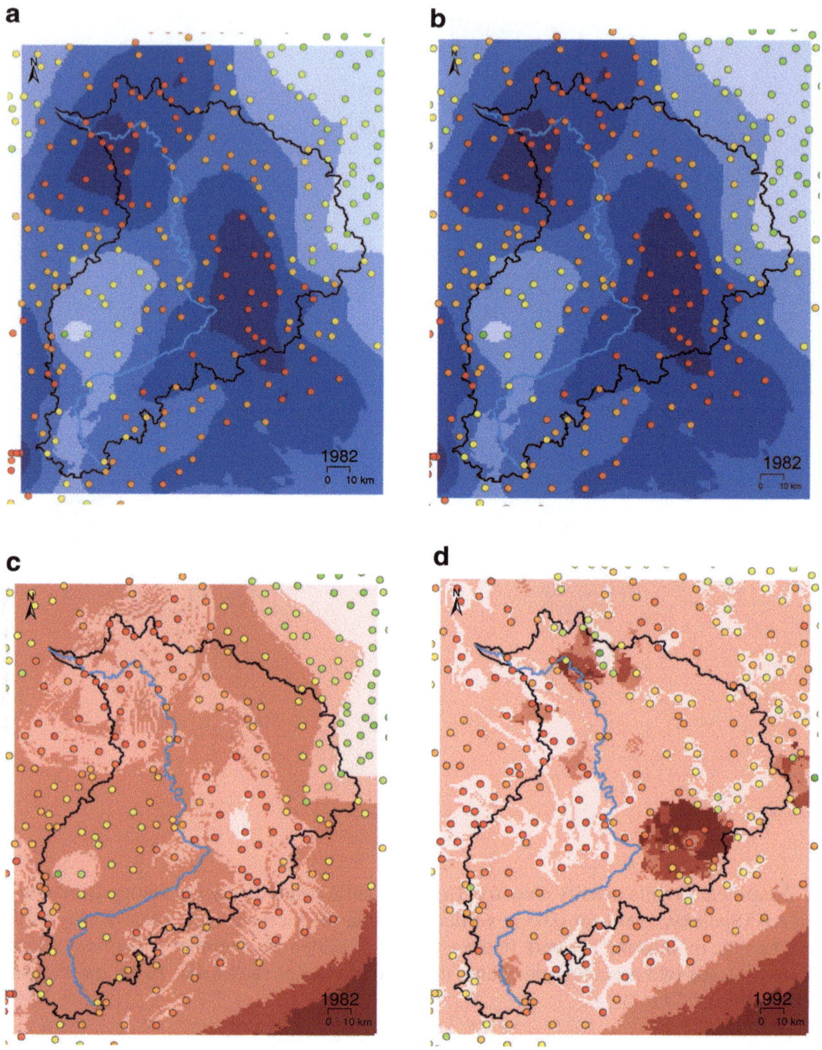

Fig. 2 Maps of precipitation in the Neckar (*blue line*) catchment (*black line*). Measurement locations of precipitation intensity are shown as coloured dots, the colour representing the magnitude of the precipitation intensity. Precipitation intensities are plotted for an event in 1982 on panel (**a**), for an event in 1992 on panel (**b**). The corresponding 60% confidence intervals are shown on panels (**c**) and (**d**). Panels (**e**) and (**f**) show the Ordinary Kriging standard errors ("OK StdEr")

It is important to stress the fact that the shape of the contours, for the same observation network at two different events, is different when using copulas. Figure 3e and 3.1 show Ordinary Kriging prediction standard error maps for the two events. The shadings of both maps have a very similar geometry due to a nearly identical semivariogram. However, the shape of the confidence intervals of interpolation using copulas is significantly different – despite the fact that the same parameters for

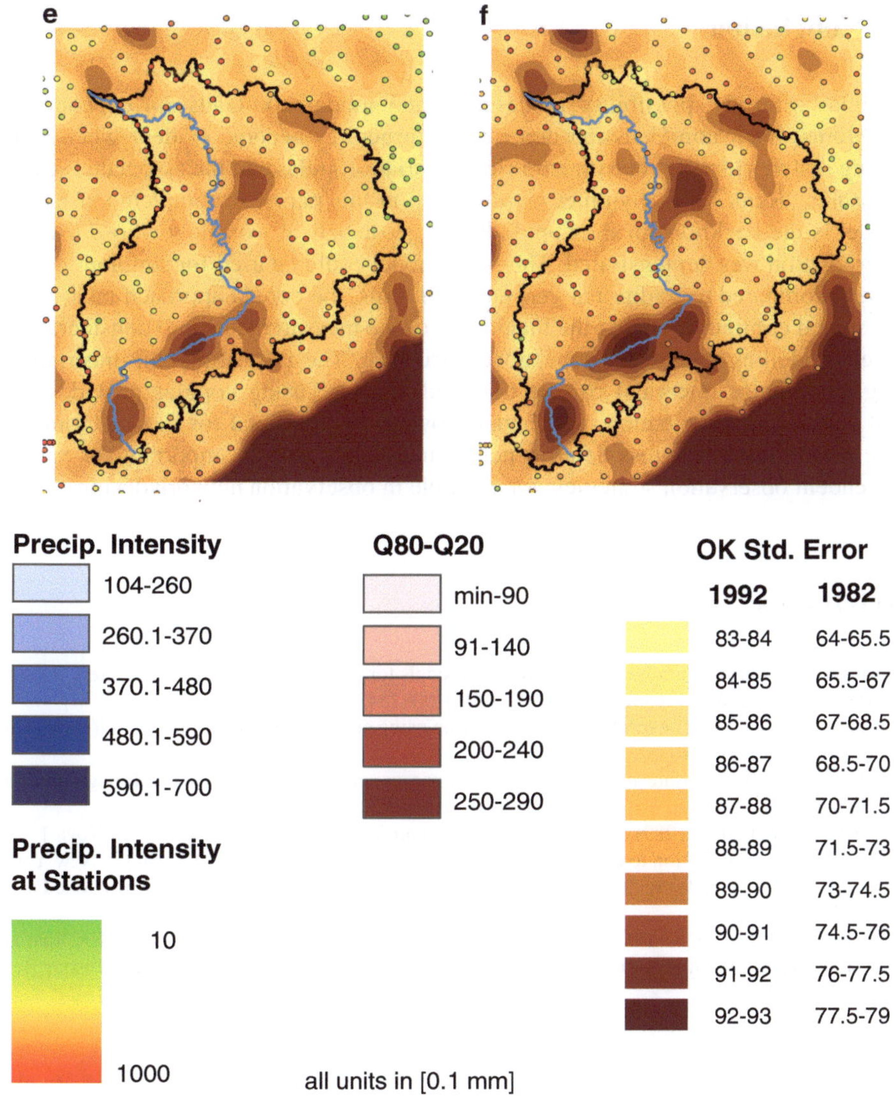

Fig. 2 (continued)

the theoretical copula were used for both events. The "bulls eyes effect" is much less pronounced when using copulas compared to Kriging, but the effect is still recognizable: Near a location where a measurement is available the uncertainty of the interpolated value is small, however it could be that this location happens to be in an area where the gradient of the measurement values is high, causing large uncertainties, and resulting in an overall medium-range uncertainty.

4 Conclusion

Copulas offer the possibility to describe and model non-Gaussian dependence structures. Such non-Gaussian dependence structures become evident when analyzing real world datasets with empirical bivariate copulas and the associated scalar measures "Rank" and "Symmetry" presented in this paper. A complete stochastic model is the backbone of the copula based geostatistical workflow whose use for interpolation was demonstrated. The same model could be used for simulation purposes. Compared to traditional geostatistical tools, the copula approach takes both the spatial configuration *and* the magnitude of the measurements into consideration when modelling the spatial dependence structure. The full estimation uncertainty (e.g. confidence intervals) can be obtained, because using copulas provides the full conditional distribution, which can prove to be beneficial for risk assessment. The possibility to express heterogeneous uncertainty can be important for value-dependent observation strategies, for example in observation network design.

References

Bárdossy A (2006) Copula-based geostatistical models for groundwater quality parameters. Water Resour Res 42(W11416) doi:10.1029/2005WR004754

Bárdossy A, Li J (2008) Geostatistical interpolation using copulas. Water Resour Res 44(W07412) doi:10.1029/2007WR006115

Joe H (1997) Multivariate models and dependence concepts. Number 73 in Monographs on Statistics and Applied Probability. Chapman & Hall/CRC, London

Journel AG, Alabert F (1989) Non-gaussian data expansion in the earth science. Terra Nova 1

Nelsen RB An introduction to copulas. Lecture notes in statistics volume 139 Springer, New York

Sklar M (1959) Fonctions de répartition a n dimensions et leur marges. Publ Inst Stat Paris 8:229–131

Sudicky EA (1986) A natural gradient experiment on solute transport in a sand aquifer: Spatial variability of hydraulic conductivity and its role in the dispersion process. Water Resour Res 22(13):2069–2082

Integrating Prior Knowledge and Locally Varying Parameters with Moving-GeoStatistics: Methodology and Application to Bathymetric Mapping

Cedric Magneron, Nicolas Jeannee, Olivier Le Moine, and Jean-François Bourillet

Abstract The paper aims at presenting an innovative methodology, called M-GS (M-GeoStatistics), which is fully dedicated to the local optimization of parameters involved in variogram-based models. M-GS considers the structural and computational parameters as a set of dependant parameters to be spatially optimized. The optimization process, which may be guided by objective or subjective criteria, is carried out during a M-structural analysis phase that leads to a set of spatially variable structural and computational parameters.

The methodology is applied for bathymetry mapping. The availability of accurate seafloor estimates is essential for numerous oceanographic projects, including hydrographic, oceanographic and biological models, sedimentary processes, etc. Seafloor usually presents strong non stationarity and complex structures, such as small channels with varying orientations, spatially varying measurements errors, local heterogeneities for coastal areas, or deep canyons within general gentle slope for continental margins. The adequacy of the M-GS methodology in this framework is illustrated and compared with classical estimates for the Marenne-Oléron coast (West of France). Moreover such methodology could be used to input different local structures into a general model in the aim of a regional synthesis.

C. Magneron (✉)
ESTIMAGES, 10 Avenue de Québec, 91140 Villebon-sur-Yvette, France
e-mail: cedric.magneron@estimages.com

N. Jeannee
GEOVARIANCES, 49bis Avenue Franklin Roosevelt, BP91, 77212 Avon, France
e-mail: jeannee@geovariances.com

O.L. Moine
IFREMER, Laboratoire Environnement-Ressource des Pertuis Charentais, Avenue de Mus de Loup, 17390 La Tremblade, France
e-mail: olemoine@ifremer.fr

J.-F. Bourillet
IFREMER, Dép. Géosciences Marines, Laboratoire Environnements Sédimentaires, BP70, 29280 Plouzané, France
e-mail: Jean.Francois.Bourillet@ifremer.fr

P.M. Atkinson and C.D. Lloyd (eds.), *geoENV VII – Geostatistics for Environmental Applications*, Quantitative Geology and Geostatistics 16,
DOI 10.1007/978-90-481-2322-3_35,

1 Introduction

Today, most geostatistical methods rely on a global variogram model. The variogram allows to build effective estimation (kriging) and simulation operators by catching the mean spatial correlation inherent to a data set. These methods commonly assume stationarity for the underlying random function. This assumption is too constraining in numerous applications, as soon as the target area becomes large or involves complex structural patterns. Applying stationary approaches in such cases, even locally with a moving neighbourhood, can lead to unsuitable estimates and non stationary approaches are preferable to some extent, provided that one is ready to accept to loose some control on the underlying structural model. Furthermore, even non stationary algorithms hardly handle prior knowledge nor reproduce precisely complex structures, such as local anisotropies, spatially varying small-scale structures or heterogeneity, etc.

The M-GS methodology is suitable for processing data in a wide range of such non stationary contexts.

2 Conventional Variogram-Based Models

2.1 Global Approach

The majority of geostatistical models that are daily implemented in the industry are variogram-based models – see Dubrule (2003), for example. They are used for processing spatially distributed data, especially in natural resources domains such as mines, petroleum and environment. They are mainly devoted to mapping, filtering and uncertainty management applications.

Variogram-based models rely generally on the modelling of a statistical function, the experimental variogram, which depicts the mean spatial correlation between data samples. When data can be considered as the result of a stationary random process, the variogram model is fitted directly to the experimental variogram, which is supposed to be representative of the whole data field or of a well-separated area of the data field. Based on the variogram model, effective estimation (kriging) and simulation operators are built and applied to the data set.

In the second-order stationary case, the variogram-based approach is rather intuitive as some parameters of the model may be related directly to the observation of the data themselves. Non-stationary models, such as IRF-k models (Matheron, 1971; Chilès and Delfiner 1999), are more intricate and lead to less control on the underlying structural model. It justifies the common strategy of transformation for working in a stationary framework as in the universal kriging case, despite the observed bias of the variogram of the residuals (Pardo-Igúzquiza and Dowd, 1998).

2.2 *Variogram-Based Model Parameters*

2.2.1 Structural Parameters

In the stationary case, variogram modelling is driven through a two-steps phase called structural analysis. The first step consists of interpreting the experimental variogram computed from the data. This step is rather likely to involve the user's knowledge about their data set. Based on the first step conclusions, the second step aims at fitting a single or a set of parameterized functions to the experimental variogram, thus defining the variogram model. Broadly speaking, structural parameters are the parameters that are related to the variogram model such as range(s), sill(s), anisotropy coefficient(s), etc.

2.2.2 Computational Parameters

In order to run variogram-based estimation and simulation algorithms, some computational parameters must be tuned. They are mainly tied to the moving neighbourhood used for selecting data points surrounding the target point (the point to be estimated or simulated). In practice, the computational parameters are often utilized for managing processing times, specifically when dealing with large data sets, or for adjusting the neighbourhood according to the samples pattern.

2.3 *Limits*

Variogram-based estimation and simulation results are sensitive to structural and computational parameters. Although sensitivity may be highly variable depending on some data characteristics, such as sampling density or variable continuity for example, it is usually a factor. This point is often not appreciated while running variogram-based models.

More specifically, sensitivity to the parameters can be very problematic when faced with complex structural environment or specific acquisition patterns. In such cases, global stationary models may correspond to local data characteristics and can lead to unexpected poor results.

3 M-GS (Moving-GeoStatistics) Models

3.1 Principle

M-GS methodology is fully dedicated to the local optimization of parameters involved in variogram-based models. M-GS considers the structural and computational parameters as a set of dependant parameters to be spatially optimized. The optimization process, which may be guided by objective or subjective criteria, is carried out during a M-structural analysis phase that leads to a set of spatially variable structural and computational parameters.

3.2 M-Parameters

M-parameters are locally optimized versions of structural and computational parameters of variogram-based models. They vary spatially over the data field. In the past, non-stationarity has been explored for several parameters, such as anisotropy, especially in the environment domain (see Caetano et al., 2004 for example). When dealing with these models the major challenge is to get stable variations of the parameters and as far as possible to automate the parameter determination process.

Several approaches are possible to compute M-parameters. A simple one merely consists in computing local variogram parameters in adjacent areas of the data field and then to smooth the obtained parameters in order to make them available at every target grid node. More sophisticated algorithms currently under development are based on automatic validation techniques. They simplify the determination of the M-parameters and lead to promising results on various real cases that have been tested.

One example of results obtained with an automatic validation approach is presented in Fig. 1, which displays a 2D seismic data set (Fig. 1a) and one associated M-parameter map corresponding to the range variations of an isotropic spherical model (Fig. 1b). An interpolation error criterion has been used for determining the optimal parameters. The north-eastern part of the data field appears to be less structured (range smaller) than the rest of the data field. The M-parameters are used to map the seismic data by ordinary kriging (Fig. 1c).

It should be noted that the M-structural analysis process involves some dependency relationship between several parameters. For example, in the second-order stationary case, the size of the moving neighbourhood in one direction is related linearly to the range of the largest scale structure in that direction. More complex relationships can be introduced into the optimization process.

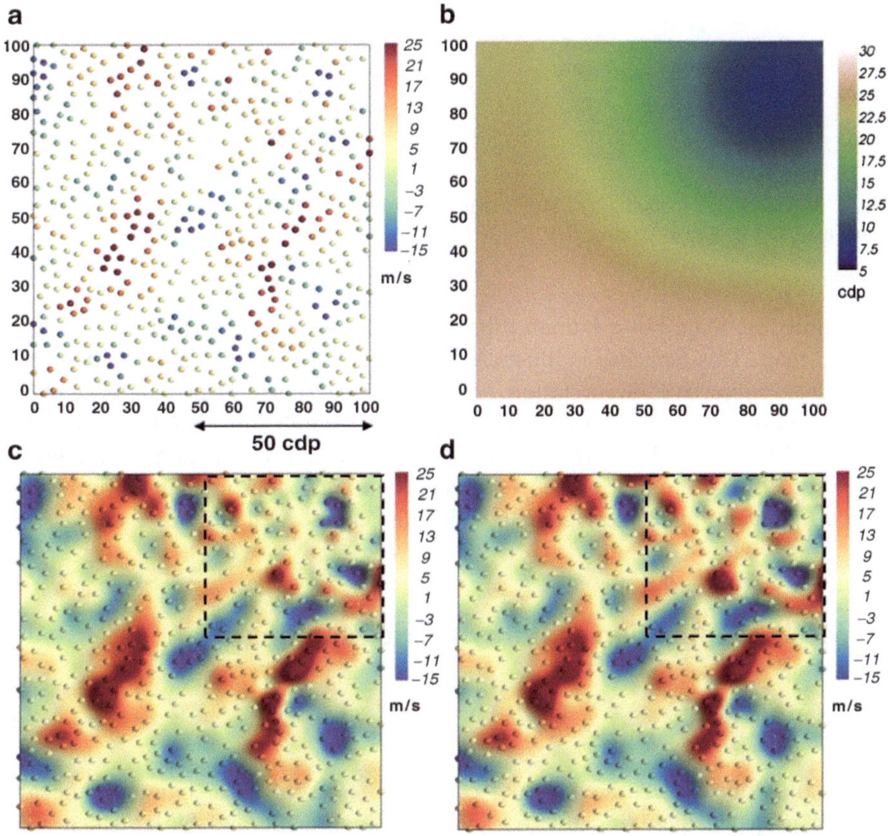

Fig. 1 Seismic data mapping by kriging: (**a**) data set, (**b**) spatially varying range, (**c**) M-GS mapping (**d**) conventional mapping by global kriging

3.3 Advantages

M-GS ensures a better correspondence between the geostatistical model and the data. As a consequence, spatial estimation and simulation results are more precise than those obtained with conventional variogram-based models. Regarding the previous seismic data mapping example, the improvement has been quantified through a cross-validation process. The M-GS map is on average 20% more precise than the conventional kriging map (Fig. 1d) in the north-eastern part of the field. In other words the estimation errors have been reduced by 20%.

Moreover, M-GS opens the way to advanced geostatistical mapping (even simulating) practices by allowing the user to introduce his structural a priori knowledge about the data field directly into the spatial estimation model. In that way geostatistical mapping is no longer a variogram guided process aiming at generating the

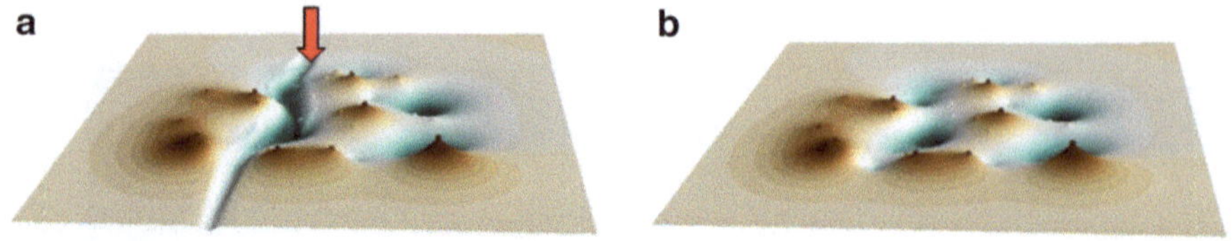

Fig. 2 M-GS guided mapping: (**a**) M-GS mapping guided by channel interpretation, (**b**) conventional mapping by global kriging

most probable map, but a human process aiming at generating the most probable desired map. This last case is illustrated in Fig. 2. Channel information, that could result from subjective interpretation, is translated in terms of M-parameters and then introduced into the kriging model for mapping 25 depth data samples leading to a channel-driven map (Fig. 2a) to be compared with a conventional global approach map (Fig. 2b). The former presents a greater continuity for the channel (red arrow) than the conventionally-derived map which displays several individual depressions.

4 M-GS Application to Bathymetric Mapping

4.1 Context

The availability of accurate seafloor estimates is essential for numerous oceanographic projects, including hydrographic, oceanographic and biological models, sedimentary processes, seismic interpretation of buried channels or canyons, etc. The seafloor usually presents strong non stationarity and complex structures, such as small channels with varying orientations, spatially varying measurement errors, local heterogeneities for coastal areas, or deep canyons within general gentle slope for continental margins.

Conventional variogram-based models often fail to produce consistent maps within such complex structural environment. More advanced models, such as M-GS models, can be applied advantageously.

4.2 Data Set Description

Marenne-Oléron (West of France) is a semi-enclosed Bay, and the first oyster farming zone in Europe. Shellfish culture activity induces silting on large intertidal mud and sandy-mud flats. Several channels incise the inlet between the coast line and Oléron Island. They are mainly controlled by strong tidal currents (up to 1.4 knots during the spring tides) with a residual ebb delta offshore the SW channel. The data set used in this work consists in more than 2,000 sample points, organized along

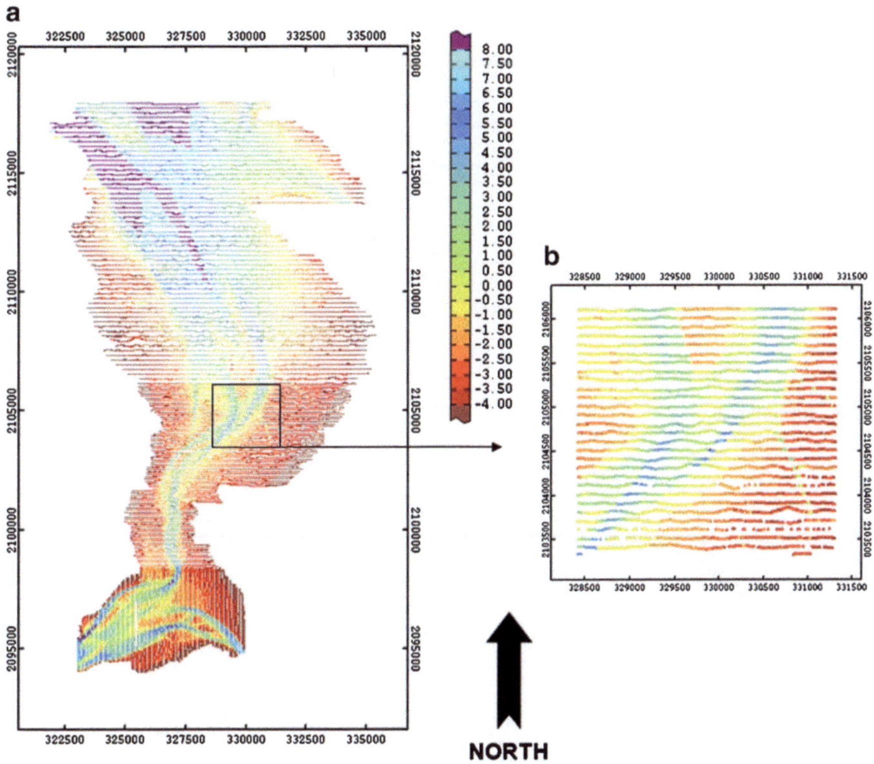

Fig. 3 Marenne-Oléron data set: (**a**) data set, (**b**) target area

lines from West to East (Fig. 3a). Samples are separated by few meters within lines. The (North-South) gap between two lines is about 100 m. Data were acquired with a single beam echoes sounder for the monitoring of the evolution of the muddy layer.

A target area (Figs. 3b and 4) is selected for illustrating conventional and M-GS mapping result differences.

4.3 Conventional Variogram-Based Mapping

For kriging purposes, an experimental variogram is computed within the target area. An anisotropic spherical model (range 800 m along X direction, 1,200 m along Y direction) is fitted to the experimental variogram (Fig. 5) and used to map the depth data.

The resulting bathymetric map is shown in Fig. 6. Major structures have been well imaged. However when looking into detail, the map contains some artefacts on

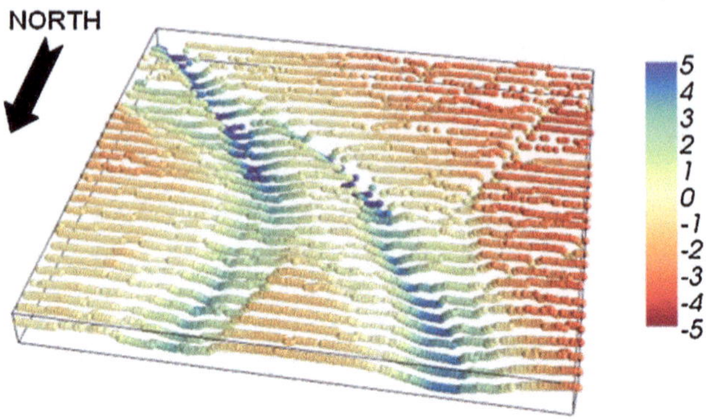

Fig. 4 Target area

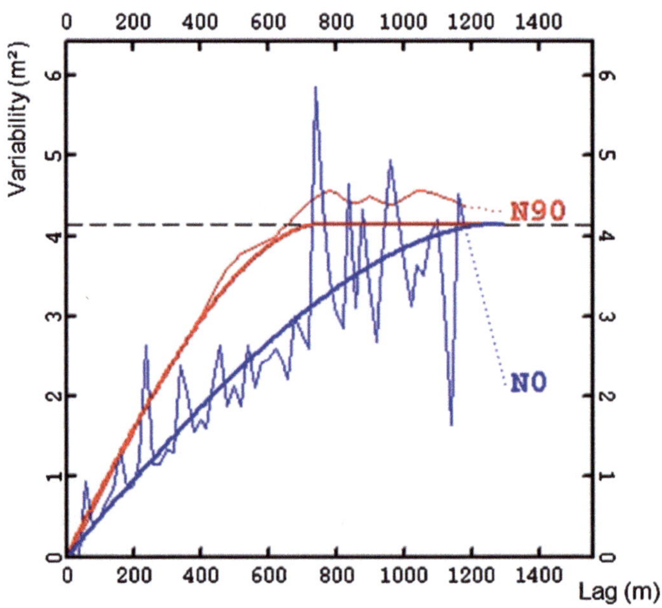

Fig. 5 Global variogram modeling

the walls of the channels which are mainly due to the line-oriented organization of the data within strongly anisotropic areas. Moreover, one micro-channel (red arrow), which is interpretable on the original data set, has not been reproduced at all.

Therefore, a more refined model is needed to reduce the artefacts and to image correctly the interpreted micro-channel.

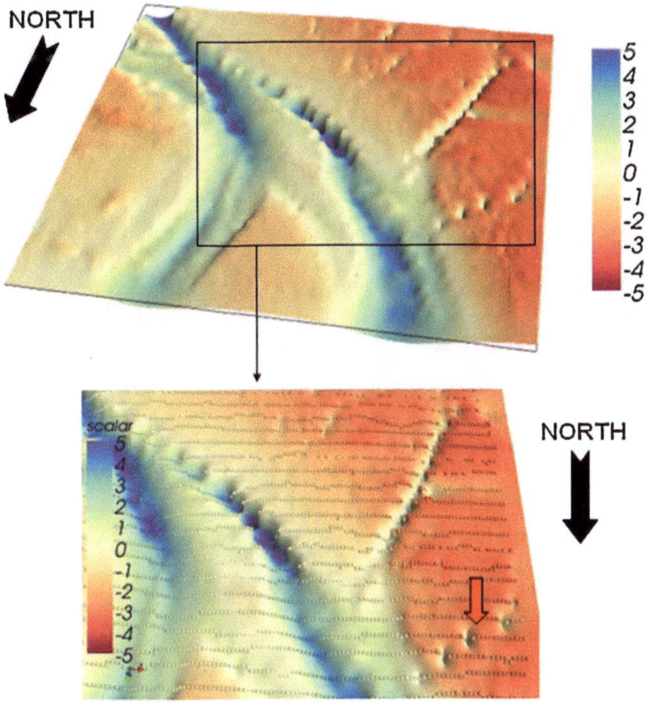

Fig. 6 Conventional mapping results

4.4 M-GS Mapping

The M-GS methodology enables the determination of locally optimized structural and computational parameters. For the current application, a specific emphasis is put on the range, the anisotropy and the related orientation of a generic spherical model. Firstly parameters are optimized during a M-structural analysis step, leading to several M-parameters maps. One resulting M-parameter map is shown in Fig. 7a. This map illustrates the spatial variations of the shortest axes of the anisotropy ellipsoid. Afterwards prior knowledge is integrated into the model: additional information regarding the interpreted micro-channel is introduced into the M-parameter maps. The previous M-range map is transformed as shown in Fig. 7b.

Finally the M-parameters are used to estimate the bathymetry. Mapping results are displayed in Fig. 8. The artefacts identified on the conventional map are no longer visible and the interpreted micro-channel is imaged. In this case it is evident that the M-GS map is of better quality than the conventional map.

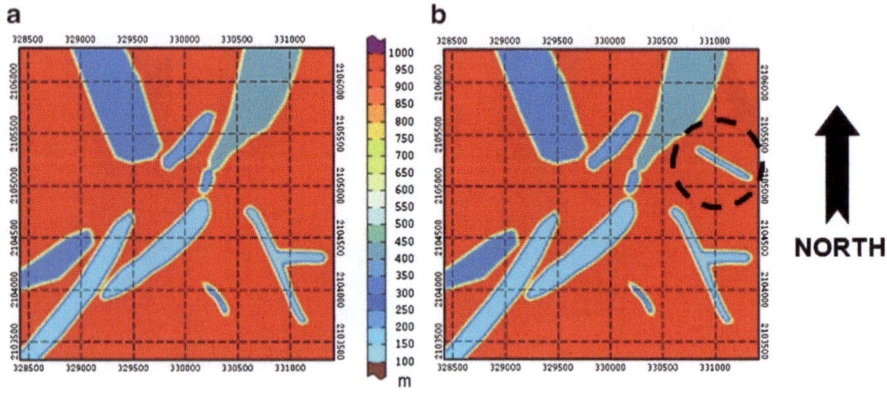

Fig. 7 Short range map: (**a**) short range map without micro-channel interpretation, (**b**) short range map with micro-channel interpretation

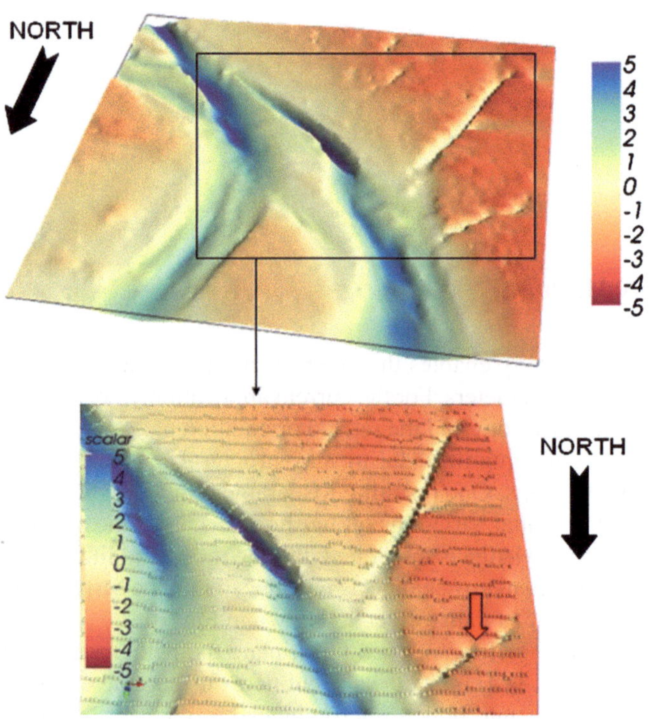

Fig. 8 M-GS mapping results

5 Conclusion

The popularity of stationary variogram-based models is mainly explained by the easy interpretation which is made of the involved parameters. In particular, some structural parameters can be directly linked to the observation of the structural properties of the data. Advanced methodologies, which allow management of spatial variations in these parameters, increase the accuracy of variogram-based model results, especially when processing large data sets and/or areas with complex structural patterns.

In this direction, the M-GS methodology, which is dedicated to the optimization of variogram-based models parameters, has proved to be promising when applied to bathymetric or seismic interpretation data in a complex structural environment. It could be useful too for mapping aquifers' bottom architecture, for example. The adequacy of the M-GS methodology in the framework of bathymetric mapping for Marenne-Oléron coast (West of France) is obvious. Moreover such a methodology could be used to input different local structures into a general model with the aim of a regional synthesis.

Acknowledgments The authors would like to thank Mr. Fazilleau (Port Autonome de La Rochelle) and The "Conseil général de Charente Maritime" for having provided the Marenne-Oléron dataset.

References

Caetano H, Pereira MJ, GuimaràesC (2004) Use of factorial kriging to incorporate meteorological information in estimation of air pollutants. In: Sanchez-Vila X, Carrera J, Gómez-Hernández J (eds) GeoENV IV – geostatistics for environmental applications, Part 2. Kluwer, Dordrecht, pp 55–65

Chilès JP, Delfiner P (1999) Geostatistics: modeling spatial uncertainty. Wiley series in probability and statistics, New York

Dubrule O (2003) Geostatistics for seismic data integration in earth models. Distinguished Instructor Short Course – Distinguished Instructor series, N°6, SEG & EAGE

Matheron G (1971) La théorie des functions aléatoires intrinsèques généralisées. Note Géostatistique N°117. Technical report N-252. Centre de Géostatistique, Fontainebleau, France

Pardo-Igúzquiza E, Dowd PA (1998) The second-order stationary universal kriging model revisited. Math Geol 30(4):347–378 (32)

5 Conclusion

Index

MIX
Papier aus verantwortungsvollen Quellen
Paper from responsible sources
FSC® C105338

If you have any concerns about our products,
you can contact us on
ProductSafety@springernature.com

In case Publisher is established outside the EU,
the EU authorized representative is:
Springer Nature Customer Service Center GmbH
Europaplatz 3, 69115 Heidelberg, Germany

Printed by Libri Plureos GmbH
in Hamburg, Germany